Lecture Notes in Computer Science 1226

Edited by G. Goos, J. Hartmanis and J. van Leeuwen

Advisory Board: W. Brauer   D. Gries   J. Stoer

Springer
Berlin
Heidelberg
New York
Barcelona
Budapest
Hong Kong
London
Milan
Paris
Santa Clara
Singapore
Tokyo

Bernd Reusch   (Ed.)

Computational Intelligence
Theory and Applications

International Conference, 5th Fuzzy Days
Dortmund, Germany, April 28-30, 1997
Proceedings

 Springer

Series Editors

Gerhard Goos, Karlsruhe University, Germany

Juris Hartmanis, Cornell University, NY, USA

Jan van Leeuwen, Utrecht University, The Netherlands

Volume Editor

Bernd Reusch
University of Dortmund, Computer Science I
D-44221 Dortmund, Germany
E-mail: reusch@ls1.informatik.uni-dortmund.de

Cataloging-in-Publication data applied for

Die Deutsche Bibliothek - CIP-Einheitsaufnahme

Computational intelligence : theory and applications ; international
conference ; proceedings / 5. Fuzzy Days, Dortmund, Germany, April
28 - 30, 1997. Bernd Reusch (ed.). - Berlin ; Heidelberg ; New York ;
Barcelona ; Budapest ; Hong Kong ; London ; Milan ; Paris ; Santa
Clara ; Singapore ; Tokyo : Springer, 1997
  (Lecture notes in computer science ; Vol. 1226)
  ISBN 3-540-62868-1 brosch.

CR Subject Classification (1991): I.2.3, F.4.1, F.1.1, I.2, F.2.2, I.4, J.2

ISSN 0302-9743
ISBN 3-540-62868-1 Springer-Verlag Berlin Heidelberg New York

© Springer-Verlag Berlin Heidelberg 1997
Printed in Germany

Typesetting: Camera-ready by author
SPIN 10549674   06/3142 – 5 4 3 2 1 0   Printed on acid-free paper

Preface

The first Fuzzy Days conference in Dortmund was held in 1991. Initially, the conference was intended for scientists and practitioners as a platform for discussion on the theory and application of fuzzy logic. Early on, synergetic links with neural networks were included, and the conference evolved gradually to embrace the full spectrum of what is now called Computational Intelligence (CI). Therefore it seemed logical to launch the 4th Fuzzy Days as a conference for CI – one of the world's first conferences featuring fuzzy logic, neural networks, and evolutionary algorithms together in one event. Following this highly successful tradition the aim of the 5th Fuzzy Days conference was to provide a forum for reporting significant results on the theory and application of CI methods and the potential resulting from their combination.

The papers and poster abstracts contained in this volume and presented at the 5th Fuzzy Days in Dortmund, April 28-30, 1997, were selected from more than 130 submitted papers. The program committee consisted of K. J. Aström, J. F. Baldwin, A. Bastian, H. Berenji, A. N. Borisov, P. Bosc, K. De Jong, A. Di Nola, Ch. Freksa, K. Goser, M. M. Gupta, A. Homaifar, E. Kerre, H. Kiendl, E. P. Klement, L. T. Kóczy, R. Kruse, H. L. Larsen, R. López de Mántaras, E. H. Mamdani, M. Mansour, C. Moraga, K. Morik, R. Palm, Z. Pawlak, B. Reusch (Chairman), E. H. Ruspini, H.-P. Schwefel, P. Sinčák, E. Trillas, N. Tschichold-Gürman, H. Unbehauen, R. Unbehauen, J. L. Verdegay, and T. Yamakawa.

We were pleased that so many outstanding scientists honored the conference with keynote lectures and tutorials. In addition, we have organized tutorials dedicated to significant fields of application. I would like to thank all colleagues and institutions for their contribution to the conference.

For the first time, the following well-known German societies which represent a considerable number of engineers and computer scientists, co-organized and supported the conference:

- Information Technology Society within VDE (ITG)
- VDE/VDI-Society of Microelectronics, Micro- and Precision Engineering (GMM)
- VDI/VDE-Society on Measurement and Control (GMA)
- German Informatics Society (GI).

Further support was provided by a number of important German and international institutions:

- Arbeitsgemeinschaft Fuzzy Logik und Softcomputing Norddeutschland (AFN)
- Berkeley Initiative in Soft Computing (BISC)

- Dachverband Medizinische Technik (DVMT)

- Deutsche Gesellschaft für Logistik e.V. (DGfL)

- Deutscher Verband für Schweisstechnik e.V. (DVS)

- Forschungsgemeinschaft Bekleidungsindustrie e.V.

- Forschungskuratorium Gesamttextil

- Spanish Association for Fuzzy Logic and Technologies (FLAT)

- VDE-Bezirksverein Rhein-Ruhr.

Whereas in the past the conference has aimed at the German-speaking countries, the 5th Fuzzy Days conference was an international forum. Therefore it was merely logical to switch from the Springer series "Informatik aktuell"[1], where the previous proceedings were published, to LNCS.

Finally, we would' like to express our gratitude to the Deutsche Forschungsgemeinschaft and Initiativkreis Ruhrgebiet for providing their financial support to the 5th Fuzzy Days.

Dortmund, February 1997   Bernd Reusch

[1]Reusch, B. (Hrsg.): Fuzzy Logic - Theorie und Praxis, 2. Dortmunder Fuzzy-Tage, Berlin u. a. 1992
Reusch, B. (Hrsg.): Fuzzy Logic - Theorie und Praxis, 3. Dortmunder Fuzzy-Tage, Berlin u. a. 1993
Reusch, B. (Hrsg.): Fuzzy Logic - Theorie und Praxis, 4. Dortmunder Fuzzy-Tage, Berlin u. a. 1994

Table of Contents

Poster Abstracts

A Global Representation Scheme for Genetic Algorithms

J.J. Collins and Malachy Eaton*

Dept. of Computer Science and Information Systems,
University of Limerick, Ireland.

Abstract. Modelling the behaviour of genetic algorithms has concentrated on Markov chain analysis. However, Markov chains yield little insight into the dynamics of the underlying mechanics and processes. Thus, a framework and methodology for global modelling and visualisation of genetic algorithms is described, using tools from the field of Information Theory. Using Principal Components Analysis (PCA) based on the Kullback-Leibic transform, a generation (instance of a population) is transformed into a compact low dimensional state-space representation. A pattern vector (set of weights) is calculated for each population of strings, by projecting it into the eigenspace. A 3D manifold or global signature is derived from the set of connected pattern vectors.

Principal Components Analysis is applied to a GA parameterised by three canonical schema - binary, Kcode and Gray - sided test-platform consisting of twelve functions. The resultant manifolds are described and analysed. The paper is concluded with a discussion of possible interpretation of the derived signal and potential extensions to the proposed methodology.

1. Introduction

High degrees of epistasis crossreferences and to the concept of a GA-easy problem, where cognisance is a measure of the correlation between the GA characteristics ie. genetic operator, probability factors and chromosomal encoding scheme, and the problem domain that constitutes the GA's environment. Poor correlation results in disconnature or a GA-hard problem [7].

Our objective is to derive a methodology and associated tools for modelling and measuring the dynamic behaviour of the underlying GA process. Traditionally, Markov chain analysis techniques [14, 18] have formed the significant tools for analysis into the long term steady state behaviour of large populations of GAs. However, within practical constraints, Markov chain analysis yields little information on the transient behaviour of the GA. Transient behaviour characterises the dynamic interaction of the GA operators and the problem domain. Thus, the authors subscribe to the view that in modelling transience/disconnectanced one needs to elicit and measure the dynamics as opposed to the long term behaviour.

* email: j.j.collins@ul.ie or ...... eaton@ul.ie

A Global Representation Scheme for Genetic Algorithms

J.J. Collins and Malachy Eaton *

Dept. of Computer Science and Information Systems
University of Limerick, Ireland.

Abstract. Modelling the behaviour of genetic algorithms has concentrated on Markov chain analysis. However, Markov chains yield little insight into the dynamics of the underlying mechanics and processes. Thus, a framework and methodology for global modelling and visualisation of genetic algorithms is described, using tools from the field of Information Theory. Using Principal Component Analysis (PCA) based on the Karhunen-Loève transform, a generation (instance of a population) is transformed into a compact low dimensional eigenspace representation. A pattern vector (set of weights) is calculated for each population of strings, by projecting it into the eigenspace. A 3D manifold or global signature is derived from the set of computed pattern vectors.
Principal Components Analysis is applied to a GA parameterised by three encoding schemes - binary, E-code and Gray - and a test platform consisting of twelve functions. The resultant manifolds are described and correlated. The paper is concluded with a discussion of possible interpretations of the derived results, and potential extensions to the proposed methodology.

1 Introduction

High degrees of consonance correspond to the concept of a GA-easy problem, where consonance is a measure of the correlation between the GA characteristics ie. genetic operator, probability factors and chromosomal encoding scheme; and the problem domain that constitutes the GA's environment. Poor correlation results in disconsonance or a GA-hard problem [7].

Our objective is to derive a methodology and associated tools for modelling and measuring the dynamic behaviour of the underlying GA process. Traditionally, Markov chain analysis techniques [14, 18] have formed the significant tools for analysis into the long term, steady state behaviour of large populations of GAs. However, within practical constraints, Markov chain analysis yields little information on the transient behaviour of the GA. Transient behaviour characterises the dynamic interaction of the GA operators and the problem domain. Thus, the authors subscribe to the view that in modelling consonance/disconsonance, one needs to elicit and measure the dynamics as opposed to the long term behaviour.

---

* email: j.j.collins@ul.ie or malachy.eaton@ul.ie

Prior to modelling and analytical evaluation, a set of problems which collectively characterise a significant cross section of operating domains must be elicited. Derivation of such a test platform constitutes an on-going component of current research.

Principal Components Analysis (PCA) is used as the methodology for deriving a model of the transient behaviour of a GA. This approach is based on extracting the significant global features of the data set, and encoding them in a more compact form. In computing the principal components, the eigenvectors for a set of generations or data instances [15, 19, 22] are derived. These eigenvectors constitute the dimensions of the eigenspace for that particular data set. Each generation is then projected into the eigenspace to derive a weight vector. Plotting the first 3 elements of each weight vector yields a manifold, parameterised by the encoding scheme used and the problem domain. This manifold, or parametric eigenspace representation, describes the global dynamic signature of the GA process based on the generated solutions over time [17].

2 Principal Components Analysis

Eigenspaces use global statistical data to encode the *relevant* content of a data instance, by deriving the principal components of the distribution of a set of such instances. The principal components or eigenvectors of the covariance matrix are derived using the using the Karhunen-Loève transform. [15, 19]. The resultant eigenvectors can be thought of as a set of features that together characterise the global statistical variations amongst the data set. The eigenvectors constitute the dimensions of an eigenspace for the specific data set. These eigenvectors form an orthogonal basis set for generating a compact representation of the original data set.

Principal Components Analysis (PCA), from the field of Information Theory, has been applied in the field of computer vision for statistical modelling and classification. In the context of face recognition, the data set consists of a set of face images and the eigenvectors are known as eigenfaces [22]. Each face can be represented exactly in terms of a linear combination of these said eigenfaces. For a set of M face images that have a unified size N, where N = width x height of image, each image is represented as a vector in an N−dimensional vector space (hyperspace). Using PCA, M' eigenvectors are calculated yielding an M' dimensional eigenspace, where $M' \leq M - 1$. The M vectors are then projected into this low dimensional subspace. In the case of recognition, classification is performed by finding a match that minimises the Euclidean distance between the input face image projected into the subspace and a face class represented in this eigenspace.

Given a set of images (snapshot of a population at time t): $\Gamma_1, \ldots, \Gamma_M$; an average image can be computed as:

$$\Psi = \frac{1}{M} \sum_{i=1}^{M} \Gamma_i$$

A new set $\Phi_1, \ldots, \Phi_M$ is given by $\Phi_i = \Gamma_i - \Psi$ which translates the original images by Ψ in the image space. The principal components of the new space given by Φ_i are the eigenvectors of its covariance matrix:

$$C = \frac{1}{M} \sum_{i=1}^{M} \Phi_i \Phi_i^T = AA^T \tag{1}$$

where $A = [\Phi_1 \ldots \Phi_M]$.

The solution to this problem can be computationally intractable due to the size of the covariance matrix. However, one can use singular value decomposition to reduce the matrix size by observing that:

$$(A^T A)V_i = \lambda_i V_i$$
$$A(A^T A)V_i = A(\lambda_i V_i)$$
$$(AA^T)(AV_i) = \lambda_i(AV_i)$$

where (AV_i) are the eigenvectors of AA^T, and λ_i are the corresponding eigenvalues. The eigenvectors of the covariance matrix C are AV_i. Therefore, the *full set* of eigenvectors of C are given by:

$$U_i = AV_i = \sum_{k=1}^{M} v_k^i \Phi_k \qquad \text{where} \qquad i = 1, \ldots, M-1 \tag{2}$$

where v_k^i is the kth element of V_i. Singular value decomposition is implemented by reducing the matrix of images to tridiagonal form using a Householder reduction function. The tridiagonal matrix is then used as input to a function for calculating eigenvalues and eigenvectors using a QL algorithm with implicit shift.

For a given set of eigenvectors $\{U_k\}$, an image Γ is projected onto the eigenvectors by:

$$\omega_k = \frac{U_k^T(\Gamma - \Psi)}{\lambda_k} \qquad k = 1, \ldots, M', \qquad M' \leq M-1 \tag{3}$$

A weight distribution vector or pattern vector is given by $\Omega = [\omega_1 \ldots \omega_{M'}]$. For a new image image, one can calculate the Euclidean distance ε between its pattern vector and the average pattern vector of a known image class (Ω_c) by $\varepsilon = \| \Omega - \Omega_c \|$. This image can then be identified with the known image class if ε falls within a given threshold. Within the context of face recognition, fig.1 depicts a subset of a data set of preprocessed face images, and the average face and eigenfaces derived by applying PCA using the Karhunen-Loève transform. Eigenfaces or eigenimages are calculated by normalising the corresponding eigenvectors.

Fig. 1. Top: A sample set of face images at frontal profile view. Bottom: PCA applied to the set of images in top row yields: an averageface (left) and eigenfaces (first 5 shown) in order of importance from left to right.

3 Encoding Schemes

Encoding schemes for string representation tend to be based on binary, Gray or real alphabet systems. However, with binary or Gray, the effects of 1 bit high order mutation can project a string from one locality in the solution space to a non-adjacent one. The relationship between the performance of a GA and encoding scheme used, is one that requires much work [6]. Some approaches at preserving desirable schemata, while elegant, are computationally difficult because of reliance on the building block hypothesis [3, 23]. Thus, a new code – E-code – was developed, to reduce the obliteration of highly fit low order schemata under the effects of mutation [9, 10]. The average change in value A_n for single bit changes in binary, Gray and E-code is given by:

$$A_n = \frac{\sum_{i=1}^{2^n} \sum_{j=1}^{n} |v_i^\alpha - m_{ij}^\alpha|}{n \times 2^n} \qquad (4)$$

For an n bit string, v_i^a is the assigned value using representation α; and m_{ij}^α is the value assigned to to bit string i with bit j mutated. For binary and Gray coding schemes, while A_n is identical, the standard deviation $\sigma(A_n)$ is greater for Gray.

E-code is derived by assigning to each of the ^nC_r combinations of strings, an increasing number of 1's. Table 1 implies that while the average disruption caused by a single bit change using E-code is higher, the probability of obliteration of fit schemata is very much reduced. Within the field of genetic algorithms, efforts are now being focused on the merits of various encoding schemes. Thus, binary, Gray and E-code encoding schemes are used for string (cchromosome) representation, as one of the variable parameters of the GA implementation. The other being the problem domain.

4 Test Platform

4.1 Considerations

1. Whitley et al. [24] argue that test functions should be non-separable. A function is separable if there are no non-linear interactions between varaibles.

Table 1. The effects of single bit mutation on strings of order 3, 4 and 5, using binary, E-code and Gray genotype encoding schemes.

Bits per String	Code	A_n	σ
	binary	2.33	1.27
3	Gray	2.33	1.93
	E-code	2.50	0.98
	binary	3.75	2.70
4	Gray	3.75	3.76
	E-code	4.36	1.74
	binary	6.20	5.47
5	Gray	6.20	7.25
	E-code	7.88	3.14

If a function is separable [2], an exact search tool such as line search or heuristic based methodologies such as hill climbing or simulated annealing may be computationally more efficient than evolutionary algorithms.

2. Test functions should be scaleable. Nonlinear interactions in non-separable functions should also be sensitive to scaling. However, the landscape characteristics of a scaled test function must be evaluated for an increase in the size of the basin of attraction with increasing dimensionality.

3. Test functions may have symmetric properties, which will contain multiple equivalent solutions. This may induce a failure mode for evolutionary algorithms through the induction of lethals in the propagated offspring. However, a unique global optimum mitigates against the effect of symmetry.

4. The string encoding scheme and granularity (accuracy) of the solution required, constitute another factor when evaluating the efficiency of evolutionary algorithms. It is claimed that Gray codes eliminate the Hamming cliff syndrome and preserve the adjacency found in real representations. However, *Whitley et al.* [24] demonstrate how invariant representation can be achieved under Gray and binary coding schemes using a DeGray matrix.

5. Evaluation metrics should be consistent with evaluating on-line and off-line performance [5].

6. *Goldberg* [11] suggests several other considerations that impact upon problem difficulty. These include isolation, misleadingness, noise, multimodality and crosstalk.

---

[2] If one conceptualises a point in solution space as a pattern or vector, composed of values of function parameters, *Cover's* theorem [4] on the separability of patterns states that a pattern classification problem cast in a high dimensional space is more likely to be linearly separable than in a low dimensional space. Thus, if one projects the function variables into a higher dimensional space to elicit a new problem representation, this transform may yield a linear separable problem. However, now the difficulty is in deriving a suitable basis set to implement the transform.

6

However, the test suite is just one component of the whole test process and as stated by *Hooker* [13], there is a need for more scientific testing as opposed to competitive testing, within the optimization research field. *Whitley et al.* [24] argue that test suites should be hypothesis driven. Underlying this proposal, one must argue that in order to be scientifically driven, a conceptual mathematically sound model of the GA process must be elicited. However, *Goldberg* [11] notes that it has been suggested that one first creates a comprehensive description of problem difficulty with all the notation necessary to be complete, but counter-argues that a rational comprehension of the machinery underlying complex systems is not a realistic objective via this route.

Thus, while such a unified theory will remain elusive for the foreseeable future, the authors view Principal Components Analysis from the field of information theory as a sound and practical mathematical tool in deriving parameterised models of the GA process.

4.2 Test Suite

The literature contains many test suites, the most famous being *De Jong's* [5]. A small selection of the other test suites include *Hinterding et al* [12], *Schaffer et al.* [20], *Mulhlenbein et al.* [16], and *Whitley et al.* [24]. These suites contain many common components and extend *De Jong's* original test platform.

Functions F1 – F5 - taken from *De Jong* [5], the test platform consisted of five problem domains with the following characteristics: 1) continuous v. discontinuous 2) convex v. non-convex 3) unimodal v. multimodal 4) quadratic v. non-quadratic 5) low-dimensionality v. high-dimensionality 6) deterministic v. stochastic. The functions are listed in table 2 and depicted in fig.2.

Table 2. De Jong's five test functions.

Name	Function	Range
F1	$f(x) = \sum_{i=1}^{3} x_i^2$	$-5.12 \leq x_i \leq 5.12$
F2	$f(x) = 100 \cdot \left(x_1^2 - x_2\right)^2 + (1 - x_1)^2$	$-2.048 \leq x_i \leq 2.048$
F3	$f(x) = \sum_{1}^{5} integer(x_i)$	$-5.12 \leq x_i \leq 5.12$
F4	$f(x) = \sum_{i=1}^{30} i x_i^4 + Gauss(0,1)$	$-1.28 \leq x_i \leq 1.28$
F5	$f(x) = 0.002 + \sum_{j=1}^{9} \dfrac{1}{\sum_{i=1}^{2} 1 + (x_i - a_{ij})^6}$	$-65.536 \leq x_i \leq 65.536$

Test function F1 is 3 dimensional, continuous, convex, unimodal, separable and scaleable. F2 is known as Rosenbrock's function and has a deep parabolic valley along the curve $x_2 = x_1^2$. It is non-separable and not scaleable. F3 is a 5D step function, discontinuous, unimodal, separable and scaleable. F4 is a continuous, unimodal, high-dimensional quartic function with Gaussian noise. F5, known as Shekel's function, is a 2D, continuous, multimodal, separable function with local minima (foxholes) at $\{(a_{1,j}, a_{2,j})\}_{j=1}^{9}$.

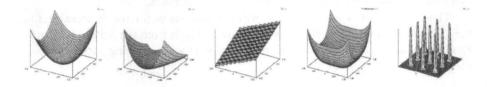

Fig. 2. 3D plots of F1, F2, F3, F4 without noise, and inverted F5.

Functions F6 to F12 - functions F6 (Rastrigin), F7 (Schwefel), F8 (Griewangk) are taken from *Mühlenbein et al.* [16], F9 and F10 from *Schaffer et al.* [20], F11 from *Ackley* [1] and F12 is a modified version of Himmelbalu's function taken from *Deb* [8]. The functions are listed in table 3 and illustrated in fig.3.

Table 3. Test functions F6 to F12.

Name	Function	Range
F6	$f(x) = (N * 10) + \sum_{i=1}^{N} x_i^2 - 10\cos(2\pi x_i)$	$-5.12 \leq x_i \leq 5.12, N = 5$
F7	$f(x) = \sum_{i=1}^{N} -x_i \sin\left(\sqrt{\mid x_i \mid}\right)$	$-512 \leq x_i \leq 512, N = 5$
F8	$f(x) = 1 + \sum_{i=1}^{N} \frac{x_i^2}{4000} - \prod_{i=1}^{N} \cos\left(\frac{x_i}{\sqrt{i}}\right)$	$-512 \leq x_i \leq 512, N = 5$
F9	$f(x) = 0.5 + \frac{\sin^2\left(\sqrt{x_1^2 + x_2^2} - 0.5\right)}{\left(1.0 + 0.001\left[x_1^2 + x_2^2\right]\right)^2}$	$-102.4 \leq x_i \leq 102.4$
F10	$f(x) = \left(x_1^2 + x_2^2\right)^{0.25}\left[\sin^2\left(50\left(x_1^2 + x_2^2\right)^0 .01\right) + 1.0\right]$	$-102.4 \leq x_i \leq 102.4$
F11	$f(x) = \prod_{i=1}^{2} \exp\left(-2\log 2\left(\frac{x_i - 0.08}{0.854}\right)^2\right) \sin(5\pi x_i)^6$	$-5.12 \leq x_i \leq 5.12$
F12	$f(x) = 200 - \left(x_1^2 + x_2 - 20\right)^2 - \left(x_1 + x_2^2 - 15\right)^2$	$-5.12 \leq x_i \leq 5.12$

F6 is known as Rastrigin's function, and is scaleable, separable and of order $O(n)$. F7, known as Schwefel's function, is multimodal, scaleable, separable, and also of order $O(n)$. However, it is characterised by a second best minimum which is misleading in relation to the global optimum. F8, Griewangk's function, is scaleable, nonlinear and non-separable. However, with increasing dimensionality the basin of attraction grows, thus derivation of the solution becomes easier. F9 and F10, known as the sine envelope sine wave function and stretched V sine wave functions, developed by *Schaffer et al.* [20], are non-separable and not scaleable. F11 was derived by Ackley and is a 2D multimodal function. F12, known as Himmelblau's function, is low-dimensional, multimodal and separable. NK landscapes and combinatorial optimisation problems were excluded from this

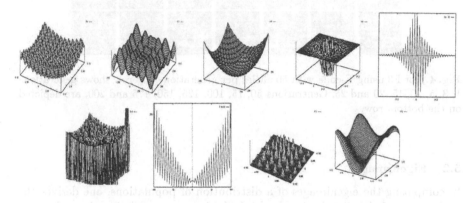

Fig. 3. Top: F6 (Rastrigin), F7 (Schwefel), F8 (Griewank) and F9 (2D and 1D). Bottom: F10 (2D and 1D), F11 (Ackley) and F12 (Himmelblau).

test suite because of the necessity for the incorporation of specialised operators into their computational paradigm.

5   Results

5.1   Populations as Images

For each of the twelve problem domains $\{F1, \ldots, F12\}$, further differentiated by one of the three coding schemes – binary, E-code and Gray – twenty five trial runs were generated. Each run was terminated when 200 generations had been evolved. Each generation or population instance was converted into raster image format and scaled [3] to 256 x 256. Thus, in total, thirty six data sets were generated where each data set contained 5,000 images. For each data set, a mean set of population images was calculated by averaging over the twenty five trials.

---

[3] Scaling was necessary to facilitate correlation of derived eigenspace representation from different domains.

Fig.4 depicts mean population images computed for test function F8 using E-code. The x-axis of an image represents the population size which was set at fifty, and the y axis is a chromosome locus for an allele. A black pixel represents a chromosome locus value of 0, white for 1, and the various shades of grey represents values that are ambiguous ie $\in \{0, 1\}$. Capturing the essence of a population as an image facilitates visual observation of propagation of useful schemata from preceding to proceeding generations (usefulness being dependent on fitness evaluation functions and genetic operators).

Fig. 4. For F3 using E-code with 50 alleles per generation: Top row shows generations 1, 3, 5, 10, 15, 20 and 25. Generations 50, 75, 100, 125, 150, 175 and 200, are depicted on the bottom row.

5.2 Eigenimages

In computing the eigenimages of a distribution of populations, one derives the major axes of the distribution (the principal components), based on the conceptual idea of a population as a point in a hyperspace. For the set of mean populations images derived, applying PCA with $M' = 10$ will yield the most significant ten eigenvectors. In practice, only the first three eigenvectors are needed to generate a manifold. For visualisation, eigenimages were derived by normalising the corresponding eigenvectors.

Fig.5 deppicts the results of applying PCA to the mean population set of problem domain F8 using E-code to encode the genotype. Fig.6 shows the eigenvalues for each of the 10 eigenvectors calculated. Note that the first three are an order of 10 in magnitude greater than the rest, implying that the first three eigenvectors have captured the significant statistical variations amongst the data set.

Fig. 5. For F8 using E-code: the average generation image (right), and the first five most significant eigenimages from left to right.

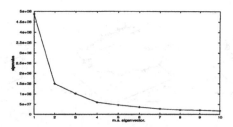

Fig. 6. For F8 using E-code, shown are the eigenvalues of the 10 most significant eigenvectors.

5.3 Manifolds

To generate a genetic algorithm manifold or parametric eigenspace representation, PCA must be first applied to the data set, to yield the eigenvectors of the covariance matrix for the said set. These eigenvectors form an orthogonal basis set with which to represent the data in a more compact form. Pattern or weight vectors are derived for each image by projecting it into its corresponding eigenspace, to yield the expansion coefficients. A 3D manifold is derived by plotting the first three elements of every pattern vector calculated for every image in the data set [17].

Manifolds can be conceptualised as a global signature of the dynamics of the underlying process. Note that even though the manifolds depicted in fig.7 are parameterised by different test functions and encoding schemes, they all display a relatively uniform curve in the x-y axis but vary in the z-axis. Each point p on the manifold, represents the relationship of a population p with all other populations generated. For each manifold shown, generation 1 is represented as the first point on the manifold curve starting at the upper corner of the plot. In general, the E-code manifolds are smoother and have a more regular distribution than either their binary or Gray counterparts.

5.4 Correlation of Manifolds

Each eigenspace is analogous to a specific co-ordinate system based on the data set on which the Karhunen-Loève transform was applied ie. the dimensions of the eigenspace are parameterised by the GA parameters. To evaluate the effects of encoding schemes on the resultant manifolds, for test function F1, all three data sets were projected into the three eigenspaces, as specified by binary, E-code and Gray representations. Fig.8 illustrates the resultant similarity between binary and Gray manifolds, but the E-code manifold does not have a close correlation. Empirical tests show that a binary based alphabet used as the genotype encoding scheme, yields better performance that a real alphabet, for test function F1 [10].

However, projecting a data set into a foreign eigenspace whose parameters differ greatly from the data, will result in a manifold that lacks uniformity. For test function F4, all three data sets were projected into the three eigenspaces,

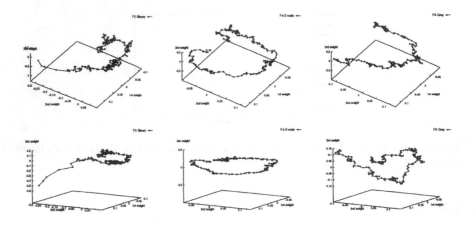

Fig. 7. GA manifolds parametrised by test function and coding schemes. Top: manifold for test function F2 (left) using binary, F4 using E-code (middle), and F6 using Gray (left). The bottom row depicts the same manifolds as the top row, but using a different viewing angle.

as previously specified. Fig.9 illustrates the disimilarity between the manifold. Empirical tests show that a real alphabet will result in a better solution for F4. As the GA parameters are identical except for the genotype encoding scheme, this emphasises the importance of choosing a scheme that is compatable with the function characteristics. However, the necessity for classification of test functions using a rigorous methodology, has just recently emerged in the GA literature [1, 8, 7, 16, 20].

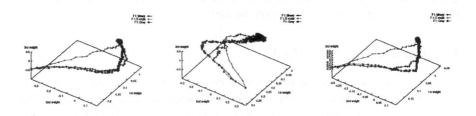

Fig. 8. Data set of images for F1:binary, F1:E-code and F1:Gray, projected into the eigenspace for F1-binary (left), F1:Ecode (middle) and F1:Gray (right)

To evaluate the uniqueness of each manifold, as parameterised by a specific function domain and encoding scheme, each of the thirty six data sets was projected into a specific eigenspace, and manifolds calculated. Each manifold was correlated with the manifold of the data set used to compute the said eigen-

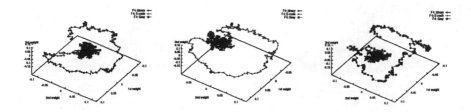

Fig. 9. Data set of images for F4:binary, F4:E-code and F4:Gray, projected into the eigenspace for F4-binary (left), F4:Ecode (middle) and F4:Gray (right)

spaces. This process was repeated for all thirty six eigenspaces to yield a correlation landscape. Two correlation measures were employed. The first, known as the normalised cross correlation coefficient [2], is given as:

$$C(m_1, m_2) = \frac{< m_1 m_2 > - < m_1 >< m_2 >}{\sigma(m_1)\sigma(m_2)} \tag{5}$$

where m_1 and m_2 are the two manifolds to be correlated, and $< m >$ denotes the mean value of m . $m_1 m_2$ represents the point by point product. Standard deviation is used to normalise the energy distribution of the manifolds so that their average and variance match. The second correlation measure used is based on simple linear regression, where the degree of linear relationship among the variables is calculated using:

$$C(m_1, m_2) = \frac{n \sum m_1 m_2 - (\sum m_1)(\sum m_2)}{\sqrt{n \sum m_1^2 - (\sum m_1)^2} \sqrt{n \sum m_2^2 - (\sum m_2)^2}} \tag{6}$$

where n is the number of points on the manifold. The correlation measure lies in the range of $-1 \leq C \leq 1$. A reasonable rule of thumb of thumb is to assume that the correlation is weak if $0 \leq | C | \leq 0.5$, strong ifif $0.8 \leq | C | \leq 1$, and moderate otherwise. The correlation landscapes as illustrated in fig.10, emphasise the uniqueness of manifolds as parameterised by a problem domain. This is evidenced by the diagonal traversing a flat surface. The diagonal represents the correlation of a manifold with itself which is 1, or a weak to moderate correlation between manifolds from the same problem domain but using different genotype encodings.

6 Discussion

Currently, the concept of hypothesis driven versus competitive analysis, problem design, and characterisation are receiving focused attention. In deriving a comprehensive test platform, there is potential for verifying the uniqueness of each formulated problem by projecting data images from other domains into this domains eigenspace and deriving manifolds, as was demonstrated in section 5.4.

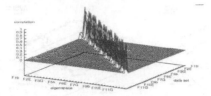 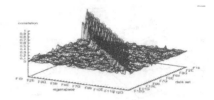

Fig. 10. Left: correlation landscape using cross correlation coefficient as using equation 5, correlation landscape based on correlation measure using equation 6.

The contrasting manifolds in figs. 8 and 9 pose a dilemma. One possible conjecture is that the relationship of a genotype encoding scheme and the interactions of the variables of a test function, are closely linked. Separable functions may be nonlinear in that the function may involve nonlinearities when determining the contribution of a variable to the overall evaluation. Thus, further research is merited on the performance of genotype encoding schemes, using the linearity of the contribution of variables as a classification metric.

As an orthogonal basis set is analogous to a coordinate system, it is pointless for directly comparing manifolds from two different eigenspaces (two different coordinate systems), except for the purpose of system identification. Further investigation is required on the potential for deriving a universal eigenspace or complete representation of the dynamic behaviour of the GA process. It is argued that the expansion of an existing orthogonal basis set would facilitate the use of augmented knowledge about dynamic characteristics of GA process to achieve a more computationally tractable solution in a more efficient manner. This would afford the potential to fine tune the parameters of a GA interactively, such as in the application domain of using GAs to evolve neural architectures and weights. Prior to that, the regularity of the underlying dynamics as represented by the global signature or manifold must be numerically evaluated.

Real value coding schemes were not evaluated as a manifold parameter. Real value schemes perform better than their binary alphabet based counterparts in certain problem domains [10], but guidelines for choosing one over the other remain to be elicited. Currently, the modelling methodology discussed is embedded in a visualisation framework. This imposes constraints on the range of valid values that the encoding scheme can assume. However, PCA itself is not thus constrained.

PCA is based on the discipline of Information Theory and uses global statistical data to encode the significant features of a data set. However, a drawback of the global based approach is the loss of local detail. A possible resolution of the global versus local dilemma is the use of wavelet transforms as a preprocessing tool to capture both local and global characteristics of the data. The wavelet representation could then be used as input into a PCA module to derive the manifolds or parametric eigenspace representation.

7 Conclusion

As a first step in deriving a methodology and associated tool for modelling the GA process, PCA has yielded promising results. The task of converting a set of data instances of generations, into raster format, may yield the evolving schemata upon visual inspection. This is a task which is amenable to traditional image processing techniques such as edge detectors and segmentation using region reconstruction from borders. The manifolds derived are regular with variations along the xy and xz axis, which indicates that the Karhunen-Loève transform (PCA), with further work, may be of value as a tool in modelling the dynamic characteristics of GAs.

The parametric manifolds (parameterised by GA operators, internal representation, etc; and problem domain) are all regular with a parabolic shape. The manifold is a compact representation or a dynamic fingerprint for the GA process. The modelling tool described also has potential in signalling when a GA is trapped at local optimum or reached a global optimum, by identifying when the evolving manifold curve becomes trapped in a local area or is static. Analysis of the efficiency of the GA process could be evaluated by measuring the regularity of the distribution of points on the curve. The E-code manifolds derived are generally smoother and have a more regular distribution than either their binary or Gray counterparts. Manifolds can now be used as a tool for evaluating the effects of various operators (ie. crossover - 1-point, 2-point, uniform) on the overall performance of the GA.

References

1. Ackley, David H. *A Connectionist Machine for Genetic Hillclimbing.* Kluwer Academic Publishers, 1997.
2. Brunelli, Roberto, and Poggio, Tomaso. Face recognition: Features versus templates. *IEEE Transactions on Pattern Analysis and Machine Intelligence,* 15(10):1042–1052, October 1993.
3. Coli, M., and Palazzari, P. Searching for the optimal coding in genetic algorithms. In *Proceedings of the 1995 IEEE Int. Conf. on Evolutionary Computation,* volume 1, pages 92–96, Perth, Western Australia, November 1995. IEEE Computer Society Press.
4. Cover, Thomas M. Geometrical and statistical properties of systems of linear inequalities with applications in pattern recognition. *IEEE Transactions on Electronic Computers,* 14:326–334, 1965.
5. De Jong, Keneth A. *An Analysis of the Behaviour of a Class of Genetic Adaptive Systems.* Phd thesis, Dept. of Computer and Communication Sciences, University of Michigan, Urbana-Champaign, IL, 1975.
6. De Jong, Keneth A., and Spears, William M. On the state of evolutionary computation. In Stephanie Forest, editor, *Proceedings of the Fifth Int. Conf. on Genetic Algorithms,* pages 618–623, Urbana-Champaign, IL, July 1993. Morgan Kaufmann.
7. Spears, William M., De Jong, Keneth A., and Gordon, Diana F. Using markov chains to analyze gafos. In L. Darrel Whitley and Michael D. Vose, editors, *Foundations of Genetic Algorithms:3,* pages 115–137, Estes Park, CO, 1994. Morgan Kaufmann.

8. Deb, Kalanmoy. *Genetic Algorithms in Multimodal Function Optimization.* M.sc. thesis, University of Alabama, Tuscaloosa, Alabama, 1989.

9. Eaton, Malachy. *Genetic Algorithms and Neural Networks for Control Applications.* Phd thesis, University of Limerick, Ireland, 1993.

10. Eaton, Malachy, and Collins, J.J.. A comparison of encoding schemes for genetic algorithms. In *Proceedings of the World Congress on Neural Networks*, pages 1067–1070, San Diego, CA, September 1996. INNS Press.

11. Goldberg, David E. Making genetic algorithms fly: A lesson from the wright brothers. *Advanced Technology for Developers*, 2:1–8, February 1993.

12. Hinterding, R., Gielewski, H., and Peachy, T. C.. The nature of mutation in genetic algorithms. In Larry J. Eschelman, editor, *Proceedings of the Sixth Int. Conf. on Genetic Algorithms*, pages 65–72, University of Pittsburg, July 1995. Morgan Kaufmann.

13. Hooker, J. N. Testing heuristics: We have it all wrong. *Journal of Heuristics*, 1:33–42, 1995.

14. Horn, Jeffrey. Finite markov chain analysis of genetic algorithms with niching. In Stephanie Forest, editor, *Proceedings of the Fifth Int. Conf. on Genetic Algorithms*, pages 110–117, Urbana-Champaign, IL, July 1993. MK.

15. Kirby, M., and Sirovich, L. Applications of the Karhunen–Loéve procedure for the characterization of human faces. *IEEE Transactions on Pattern Analysis and Machine Intelligence*, 12(1):103–108, January 1990.

16. Mühlenbein, H., Schomisch, M., and Born, J. The parallel genetic algorithm as function optimizer. In Richard K. Belew and Lashon B. Booker, editors, *Proceedings of the Fourth Int. Conf. on Genetic Algorithms*, pages 271–278, University of California, San Diego, July 1991. Morgan Kaufmann.

17. Murase, Hiroshi, and Nayar, Shree K. Visual learning and recognition of 3–D objects from appearance. *International Journal of Computer Vision*, 14:5–24, 1995.

18. Nix, Allen E., and Vose, Michael D. Modelling genetic algorithms with markov chains. *Annals of Mathematics and Artificial Intelligence*, 5:79–88, 1992.

19. Oja, E. *Subspace Methods for Pattern Recognition.* Research Studies Press, Hertfordshire, 1983.

20. Schaffer, J. David, Caruna, Richard A., Eschelman, Larry J., and Das, Rajarshi. A study of control parameters affecting online performance of genetic algorithms for function optimization. In J. David Schaffer, editor, *Proceedings of the Third Int. Conf. on Genetic Algorithms*, pages 51–60, Georgee Mason University, 1989. Morgan Kaufmann.

21. Suzuki, Joe. A markov chain analysis on a genetic algorithm. In Stephanie Forest, editor, *Proceedings of the Fifth Int. Conf. on Genetic Algorithms*, pages 146–153, Urbana-Champaign, IL, July 1993. MK.

22. Turk, Matthew, and Pentland, Alex. Eigenfaces for recognition. *Journal of Cognitive Neuroscience*, 3(1):71–86, 1991.

23. Uesaka, Yoshinori . Convergence of algorithm and the schema theorem in genetic algorithms. In D. W. Pearson, R. F. Steele, and R. F. Albrecht, editors, *Proceedings of the Second Int. Conf. on Neural Networks and Genetic Algortihms*, pages 210–213, Ales, france, April 1995. Springer-Verlag.

24. Whitley, D., Mathias, K., Rana, S., and Dzubera, J. Building better test functions. In Larry J. Eschelman, editor, *Proceedings of the Sixth Int. Conf. on Genetic Algorithms*, pages 239–246, University of Pittsburg, July 1995. Morgan Kaufmann.

Genetic Algorithms, Schemata Construction and Statistics

Alexandru Agapie, Doina Caragea

National Institute of Microtechnology, P.O. Box 38-160, 72225,
Bucharest, Romania
agapie@oblio.imt.pub.ro   dcaragea@oblio.imt.pub.ro

Abstract. The paper performs a comparison between two types of binary Genetic Algorithms (GA): Statistical GA vs. Messy GA. They have the same challenge - solving "deceptive" problems - and, up to a point, they are designed on the same paradigm: improving the GA by directing the search using some statistical derived schemata. The Statistical GA that we propose does not use a larger population, but only a real-valued string with the average numbers of "ones" produced on each position of the chromosome during the GA's evolution. Assuming the stagnation of the GA in sub-optimal points (which is usually the case of deceptive problems), we extract - by imposing a threshold on the real-valued string - the schema responsible for stagnation, derive its complementary schema and resume the GA's evolution imposing that schema to all the new chromosomes.

1. Introduction

Genetic Algorithms (GA) are randomised techniques: they operate on populations of strings (especially on binary strings), with the string coded to represent some underlying parameter set. Some operators, called selection, crossover and mutation are applied to successive generations of strings to create new, better valued string populations.

The important role of *schemata* (*building blocks*) in the GA's evolution was pointed out by Holland from the very beginning - [8]: the first result was called *the Schema Theorem* and was a formula for estimating the lower bound of the expected number of offspring generated by an instance of the schema H, under the classical genetic operators: selection, crossover and mutation.

The Schema Theorem models the detrimental effect of crossover and mutation on the propagation of schemata from a generation to another, but does not give a description of the way schema can be constructed. The problem we propose is:

* *How can "good" schemata be determined from a population of chromosomes (not necessarily from the GA working population)?* $\qquad$ *(1)*

The problem of deriving schemata has been tackled in several papers and the proposed solutions rely strongly on statistical computation. The problem of selecting schemata appeared first in papers concerning the Messy Genetic Algorithms (mGA) - [5], and Fast Messy Genetic Algorithms (fmGA) - [9]. In all these, the initial population of the GA was constructed by generating samples for all the schemata of some fixed order (less than the length of the chromosome, of course). In the mGA all these schemata are explicitly enumerated, while in the fmGA this procedure is replaced with a probabilistic technique.

Actually, the problem of finding good schemata appeared as a sub-problem of *sizing the GA populations*, see [6]. The paper considers the effect of stochasticity on the *selection* of schemata and derives a population-sizing equation to ensure a favourable discrimination of the best building blocks. Namely, the authors give an answer to the following question:

- *How can populations be sized to promote the good schemata's selection?* (2)

Comparing questions (1) and (2), at a first glance they seem to refer at different problems. At a closer look, one will find that the solution of (2) is also a good answer for (1): by correct sizing the initial population the "good" schemata will be selected, and they will be explicitly constructed, too. This is the important achievement of all the above mentioned papers. Nevertheless, we must stress a bottleneck of the *sizing of populations* approach:

- In [5] the authors pay little attention to the fitness variance comparison of different schemata, performing the comparison only on the base of mean fitness. We think this permits the possibility of making errors in deciding which schemata is "better", especially in the following situation: schema H_1 is "worse" than schema H_2 (in the *mean fitness* sense provided in [5]), but its best valued instance is better than the best valued instance of H_2. Thus, from the GA point of view, H_1 proves to be better than H_2, so the mean fitness approach leads to a wrong decision.

For surpassing these lacks we propose in the following another statistical approach, but somehow inverse to the one from *mGA*, *fmGA* or the *sizing of populations*. We start with a normal sized initial population, develop a GA and form a statistical population by storing in a separate place all the chromosomes produced during the GA evolution - see also [1]. This additional, statistical population is used for schemata construction, in a very specific way. Although, at this point of our work, we apply the method only for GA stagnation, the procedure for constructing schema proposed in this paper is more general.

The idea of our method is strongly inspired from the way deceptive problems have to be solved: *we propose to direct the GA's search from the neighbourhood of some local optimal point - where the GA usually sticks - to other zones.*

One might ask: *towards which zones of potentially solutions should the search be guided ?* As we generally have no previous information on the function to optimise we are not able to respond at this question. But we may try to answer the "complementary" question: we shall define the zones which are "prohibited" - i.e. the areas which surround some locally optimal points in which the GA had already get stuck - and direct the GA exploration "away" from these zones.

2. Short review of the Messy GA

The messy GA (mGA) is solving problems by combining relatively short, good building blocks to form longer, more complex strings that increasingly cover all the features of a problem to optimise - for details see [4] and [5].

There are some important differences between Simple GA and mGA:
- in mGA it is used a *partially enumerative initialisation* ;
- mGA work with *variable-length* strings that may be over- or under-specified;
- the traditional crossover is replaced by two operators: *cut* and *splice*;
- the evolutionary process is divided into two phases:
  - *primordial phase* (characterised by a variable-size population);
  - *juxtapositional phase* (fixed-size population);
- mGA use *competitive templates* to emphasise the good schemata.

This method of competitive templates is used to solve the problem of under-specification by filling in undefined genes with a locally optimal structure, so that only good schemata have fitness values better than the value of the fixed template.

The general algorithm - mGA - can be described as follows:

STEP 1 - *generate a random template*

STEP 2 -*primordial phase*

  - *the initial population includes all the possible building blocks of a specified length, where the order deceptive nonlinearity suspected in the subject problem.*
  - *the proportion of good schemata is next enriched through a number of generations (by selection only).*
  - *at specified intervals, the population size is reduced by halving the number of individuals.*

STEP 3 - *juxtapositional phase*

  - *cut, splice and other genetic operators are performed.*

Since the chromosome length that must be chosen to encompass the highest order deceptive nonlinearity may not be known before hand, a *level-wise mGA* has been suggested [7]. In a level-wise mGA, a number of *eras* is performed. In the first era, the population is initialised with strings of length one. A random template is used in order to fill up the under-specified strings. The best solution found at the end of the first era is used as a template for the second era, where the population is initialised with all possible strings of length two, and so on.

In order to encompass the difficulties of dealing with a huge initial population, [10] proposed the *Gene Expression mGA* - yielding major improvements in: reducing the population size, reducing the running time by a large factor and eliminating the need for a template solution. The Statistical GA presented below is different from the Gene Expression mGA, but provides in fact the same advantages against the classical mGA.

3. Schemata Construction and the Statistical GA

As an alternative to the family of mGAs we propose another statistical method of schemata extraction. We create an additional population in which we store all the chromosomes created during the GA's evolution. We construct a string S by length ℓ (*the chromosome length*) whose elements are the frequencies of appearance of "one" in the statistical population, on each position respectively. So each component of S is a real number between 0 and 1.

Let $T \in (0.5,1)$ be a real number (a *threshold*). Then we can construct a schema H in the following manner:

```
Procedure Constructing_Schema
for i:=1 to ℓ do
        if S[i]≥T then H[i]:='1'
        else if 1-S[i]≥T then H[i]:='0'
        else H[i]:='x' ;
```

Our fundamental assumption is that *the evolution of the GA is characterised by the string S of appearance frequencies and by the schemata obtained using the above procedure.*

We propose the following algorithm to be applied at the GA's stagnation:

STEP 1 - *appeal the statistical population and derive the string S;*

STEP 2 - *derive the schema H corresponding to the string S, according to the Procedure Constructing_Schema;*

STEP 3 - *make the complementary schema H^C: $H^C(i)=1-H(i)$, for each fixed index i ;*

STEP 4 - *resume the GA's evolution, but applying H^C as a mask (i.e. imposing the fixed values from H^C on every new chromosome);*

The structure of the Statistical-GA is depicted in fig.1.

Notes:

•   In fact, STEP 4 defines a GA with a directed genetic operator: *directed mutation.* The directed mutation resembles with how the evolution strategies work - [2], excepting the aspect that in the evolution strategies the gene values are supposed to be real, while our algorithm works with binary gene values only.

•   In fact, the program works only on the string of appearance frequencies, because there is no reason to store the whole statistical population produced during the GA's evolution.

•   Practically our algorithm takes advantage of the canonical GA when the GA's population is filled up with instances of a single schema, say H. From that iteration the GA's bias is to produce new chromosomes which are instances of H too. One can distinguish two possibilities: *1-* H is leading to the global optimum (in this case the statistical approach is *not* necessary), and *2-* H is leading to a local optimum - as frequently happens in deceptive problems; in this case the application of directed mutation will have a greater disruptive effect on the "bad" schema H than the genetic operators of canonical GA (in fact than the mutation from canonical GA, because the crossover is not really acting in a GA with all the chromosomes instances of a single schema H).

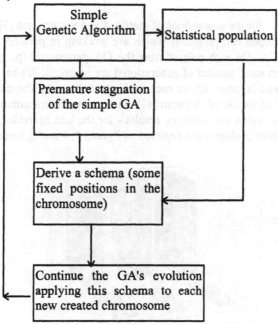

Fig. 1. Logical scheme of the Statistical-GA

4. Numerical Results

We tested the Statistical-GA - vs. mGA and Simple GA - for maximizing three functions: the 30-bit, *order-3 deceptive* problem from [4] and [5] and two hard multi-modale problems from the real-functions test bed of the First International Contest on Evolutionary Optimisation - [3], namely:

1. The problem obtained by linking 10 *order-3 deceptive* problems as the one depicted in fig. 2:

String	000	001	010	011	100	101	110	111
Fitness	28	26	22	0	14	0	0	30

Fig. 2. The *order-3 deceptive* problem

2. $f_S(x) = \sum_{i=1}^{30} \dfrac{1}{\|x - A(i)\|^2 + c_i}$ - the **Shekel's foxholes** problem (max. value ~9.0)

3. $f_L(x) = \sum\limits_{i=1}^{5} c_i \left[e^{\frac{-\|x-A(i)\|^2}{\pi}} \cdot \cos\left(\pi\|x - A(i)\|^2\right) \right]$ - the **Langerman's** function

(max. value ~ 1.4) - for the expressions of matrix A and vector c see [3].

Firstly we developed two programs which are working in parallel: a Simple GA and a Statistical-GA. On each pair of runs the GA parameters (p_c, p_m, population size, initial random seed, number of generations) are identical. We have performed a large number of runs (approx. 50) for each problem. Separately, the same numbers of runs were performed for the mGA (from [4]), on the same optimisation problems.

The comparative *mean* and *variance* results - for the best individual from each of the runs - on the three problems are depicted in figures 3, 4 and 5, respectively.

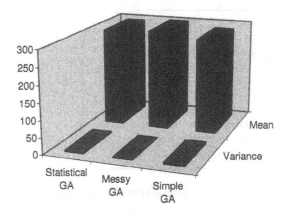

Fig. 3. Results on the o*rder-3 deceptive* problem

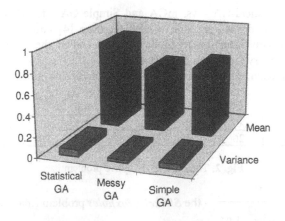

Fig. 4. Results on the *Shekel's foxholes* problem

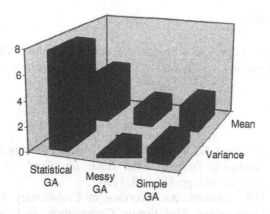

Fig. 5. Results on the *Langerman's* function

One will notice that mGA works better than Statistical GA and Simple GA on the order-3 deceptive problem, but on the two real-valued functions Statistical GA is working better, both in the *mean* and *variance fitness* of the best individual found. We underline that a large variance is preferred in a comparison between two algorithms with close means - as a major scope of optimisation is to obtain better solutions, even in a larger number of runs. In fact, the mGA failed in optimising the last two functions due to its incapacity of finding a "good" template for the primordial phase. So one may conclude that the Statistical GA is more robust than the mGA, providing good results on a wider area of deceptive / multi-modale problems.

5. Conclusions

The Genetic Algorithms (GA) are efficient methods for function optimisation. Since they are not deterministic but probabilistic algorithms, their convergence to the optimal solution is not always achieved. This paper proposes a statistical approach to overcome the premature stagnation of a simple GA. Our method is compared to the messy GA - a classical algorithm for directing the GA's search onto "good" building blocks. We defined a method for deriving a probability string S - which, under our assumption, is responsible for the GA's stagnation - then, on the complementary string S^c we build a schema H. Applying this schema H as a mask (with certain positions fixed in each new created chromosome) we resume the GA's evolution. The experimental results show that this algorithm works better than the messy GA and the simple GA on some hard, real-valued test functions.

Acknowledgements

The first author gratefully acknowledges the support of the staff from Chair Informatics I - University of Dortmund, under the co-operation project RUM X053.4.

References

1. A. Agapie, H. Dediu: GA for Deceptive Problems: Inverting Schemata by a Statistical Approach, Proc. of the IEEE Int. Conf. on Evolutionary Computation (ICEC'96), pp.814-819, 1996.
2. T. Baeck, H.P. Schwefel: An Overview of Evolutionary Algorithms for Parameter Optimisation, Evolutionary Computation, no.1, vol.1, pp.1-23, 1993.
3. H. Bersini et al.: Results of the First International Contest on Evolutionary Optimisation, Proc. of the IEEE International Conference on Evolutionary Computation (ICEC'96) pp.611-615.
4. K. Deb, D.E. Goldberg: mGA in C: A Messy Genetic Algorithm in C, IlliGal Report no. 91008, 1991.
5. D.E. Goldberg, B. Korb, K. Deb: Messy Genetic Algorithms: Motivation, Analysis and First Results, Complex Systems, 3 (1989), pp. 493-530.
6. D.E. Goldberg, K. Deb, J.H. Clark: Genetic Algorithms, Noise and the Sizing of Populations, Complex Systems, 6 (1992), pp. 333-362.
7. D.E. Goldberg, K. Deb, B. Korb: Messy Genetic Algorithms Revisited: Studies in mixed size and scale, Complex Systems, 4 (1990), pp. 415-444.
8. J.H. Holland: Adaptation in Natural and Artificial Systems, University of Michigan Press, Ann Arbor, 1975.
9. H. Kargupta: SEARCH, Polynomial Complexity, and the Fast Messy Genetic Algorithm, PhD thesis, University of Illinois, Urbana, USA, 1995.
10. H. Kargupta: The Gene Expression Messy Genetic Algorithm Proc. of the IEEE Int. Conf. on Evolutionary Computation (ICEC'96) pp.814-819, 1996.

Adaptive Non-uniform Mutation for Genetic Algorithms

André Neubauer

Department of Communication Engineering
Duisburg Gerhard-Mercator-University
Bismarckstraße 81, 47048 Duisburg
Germany

E-mail: neubauer@sent5.uni-duisburg.de

Abstract. A theoretical analysis of MICHALEWICZ' *non-uniform mutation* operator is presented and a novel variant — the *adaptive non-uniform mutation* operator — is proposed. The *non-uniform mutation* operator was developed by MICHALEWICZ' for his modified variant of genetic algorithms *modGA* to tackle numerical parameter optimization problems. As is shown by mathematical analysis, this mutation operator prefers parameter values in the center of the corresponding feasible region. This leads to problems if the optimum is situated near the feasible region's boundaries. In order to avoid this undesirable tendency, the *adaptive non-uniform mutation* operator is proposed, the development of which rests on the mathematical analysis. Experimental results for a standard numerical parameter optimization problem are given that illustrate the superiority and effectiveness of this novel mutation operator for genetic algorithms.

1 Introduction

Genetic operators are important components of *genetic algorithms* [5] that determine to a great deal the algorithm performance in optimization problems. Therefore, a variety of specialized genetic operators have been designed. In order to fully understand a genetic operator's behavior, a theoretical as well as experimental analysis is desirable. Especially, a mathematical analysis can yield insights how to improve a genetic operator.

In this paper, the so-called *non-uniform mutation* operator [6, 8] is analysed. This genetic operator was introduced by MICHALEWICZ for his modified variant of genetic algorithms to tackle numerical parameter optimization problems [6, 8]. It is especially useful in bounded optimization problems [7]. A theoretical analysis of its behavior, however, is missing. Therefore, this paper presents a theoretical analysis of the *non-uniform mutation* operator for numerical parameter optimization problems

$$f(\mathbf{x}) \xrightarrow{\mathbf{x}} \text{minimum} \tag{1}$$

with n-dimensional object parameter vector $\mathbf{x} = (x_1, \ldots, x_n)$ and objective function $f(\mathbf{x})$.

The paper is organized as follows. In Sec. 2, the *modGA* is described and the genetic operators *arithmetical crossover* and *non-uniform mutation* are defined. A theoretical analysis of *non-uniform mutation* is detailed in Sec. 3, leading to the *adaptive non-uniform mutation* operator. The superiority of the adaptive variant is demonstrated in Sec. 4 by experimental results.

2 The Modified Genetic Algorithm

The *modified Genetic Algorithm (modGA)* was proposed by MICHALEWICZ as a variant of the genetic algorithm [5] for numerical parameter optimization problems [8]; special features are a two-step selection procedure and the use of novel genetic operators that are tailored to real parameters. This specific evolutionary algorithm works on a *population*

$$\pi_\lambda(t) := (\, \iota_1(t), \ldots, \iota_\lambda(t)\,) \in I^\lambda \tag{2}$$

consisting of λ *individuals* $\iota_\ell(t)$, $\ell = 1(1)\lambda$, at *generation* t. Each individual is given by

$$\iota_\ell(t) := \mathbf{x}_\ell(t) \in I \tag{3}$$

with the real object parameter vector $\mathbf{x}_\ell(t)$. In general, $I \subset \mathbf{R}^n$ is a bounded subset of $\mathbf{R}^n$, e.g.

$$I := [x_{1,\min}, x_{1,\max}] \times \cdots \times [x_{n,\min}, x_{n,\max}] \ . \tag{4}$$

In Fig. 1, the structure of the *modGA* is given [8].

Notation
$\pi_\lambda(t) := (\, \iota_1(t), \ldots, \iota_\lambda(t)\,) \in I^\lambda$
$\iota_\ell(t) := \mathbf{x}_\ell(t) \in I$
$\mathbf{x}_\ell(t) := (\, x_{\ell,1}(t), \ldots, x_{\ell,n}(t)\,) \in I$

Algorithm
$t := 0;$
initialize $\pi_\lambda(0);$
while termination criterion $\neq$ true **do**
    evaluate $\pi_\lambda(t);$
    select $\pi_\mu(t) := \mathcal{S}_\mu\{\pi_\lambda(t)\};$
    select $\pi_{\lambda-\mu}(t) := \mathcal{S}_{\lambda-\mu}\{\pi_\lambda(t)\};$
    breed $\pi'_\mu(t) := \mathcal{B}\{\pi_\mu(t)\};$
    built $\pi_\lambda(t+1) := \pi'_\mu(t) \cup \pi_{\lambda-\mu}(t);$
    $t := t+1;$
done

Fig. 1. MICHALEWICZ' *modGA*.

In order to create the population $\pi_\lambda(t+1)$, the following steps are carried out. First of all, the *selection* process is subdivided into two consecutive steps. The selection operator $\mathcal{S}_\mu$ selects μ parents which are used to generate μ offspring. These parents are collected in the parent population $\pi_\mu(t)$. In order to select the remaining $\lambda - \mu$ individuals that will be transferred unaltered to the next generation, the selection operator $\mathcal{S}_{\lambda-\mu}$ additionally selects $\lambda - \mu$ individuals. These individuals are stored in the auxiliary population $\pi_{\lambda-\mu}(t)$.

After having selected μ parents with the help of $\mathcal{S}_\mu$, these parents are used to *breed* μ offspring by the operator $\mathcal{B}$. As proposed by MICHALEWICZ, each parent undergoes exactly one genetic operator. The genetic operators considered in this paper are the *arithmetical crossover* operator $\mathcal{R}$ and the *non-uniform mutation* operator $\mathcal{M}$. The numbers $\mu_\mathcal{R}$ and $\mu_\mathcal{M} = \mu - \mu_\mathcal{R}$ of parents that undergo *arithmetical crossover* $\mathcal{R}$ and *non-uniform mutation* $\mathcal{M}$, respectively, are exogenous strategy parameters that have to be specified in advance.

Definition 1 (arithmetical crossover). The parents $\iota_{k_1}(t) = \mathbf{x}_{k_1}(t) \in I$ and $\iota_{k_2}(t) = \mathbf{x}_{k_2}(t) \in I$ are selected for reproduction via *arithmetical crossover*. The offspring $\iota'_{k_1}(t) = \mathbf{x}'_{k_1}(t) \in I$ and $\iota'_{k_2}(t) = \mathbf{x}'_{k_2}(t) \in I$ are created according to the convex combination

$$\mathbf{x}'_{k_1}(t) = (1 - \alpha) \cdot \mathbf{x}_{k_1}(t) + \alpha \cdot \mathbf{x}_{k_2}(t) \ , \tag{5}$$
$$\mathbf{x}'_{k_2}(t) = (1 - \alpha) \cdot \mathbf{x}_{k_2}(t) + \alpha \cdot \mathbf{x}_{k_1}(t) \ . \tag{6}$$

The term α is a real exogenous strategy parameter with $0 \le \alpha \le 1$ [8]. $\diamond$

The *arithmetical crossover* operator $\mathcal{R}$ is applied $\mu_\mathcal{R}/2$ times to the $\mu_\mathcal{R}$ selected parents $\iota_1(t),\dots,\iota_{\mu_\mathcal{R}}(t)$. The remaining $\mu_\mathcal{M} = \mu - \mu_\mathcal{R}$ parents $\iota_{\mu_\mathcal{R}+1}(t),\dots,\iota_\mu(t)$ are mutated with the help of the *non-uniform mutation* operator $\mathcal{M}$.

Definition 2 (non-uniform mutation). Parent $\iota_k(t) = \mathbf{x}_k(t) \in I$ is selected for reproduction via the *non-uniform mutation* operator. The offspring $\iota'_k(t) = \mathbf{x}'_k(t) = (x'_{k,1}(t),\dots,x'_{k,n}(t)) \in I$ is created according to

$$x'_{k,i}(t) = x_{k,i}(t) + \delta_{k,i}(t) \tag{7}$$

using the random variation

$$\delta_{k,i}(t) = \begin{cases} (x_{i,\max} - x_{k,i}(t)) \cdot (1 - [z_{k,i}(t)]^{\gamma(t)}) \\ \qquad \text{with probability } q_{k,i}(t) = \frac{1}{2} \\ \\ (x_{i,\min} - x_{k,i}(t)) \cdot (1 - [z_{k,i}(t)]^{\gamma(t)}) \\ \qquad \text{with probability } 1 - q_{k,i}(t) = \frac{1}{2} \end{cases} \tag{8}$$

$z_{k,i}(t)$ is a random number equally distributed in the interval $[0, 1]$ and

$$\gamma(t) = \left(1 - \frac{t}{t_{\max}}\right)^\beta \tag{9}$$

with exogenous strategy parameter $\beta > 0$ and maximal number of generations $t_{\max}$ [6, 8]. $\diamond$

Originally, MICHALEWICZ proposed $q_{k,i}(t) = \frac{1}{2}$, i.e. both cases are chosen with equal probabilities [6, 8]. As will be shown in the next section, this choice leads to some undesirable effects that can be overcome by an adaptive determination of $q_{k,i}(t)$.

With the help of these genetic operators — the *arithmetical crossover* operator $\mathcal{R}$ and the *non-uniform mutation* operator $\mathcal{M}$ — the offspring population $\pi'_\mu(t) := \mathcal{B}\{\pi_\mu(t)\}$ has been created by the *breeding* operator $\mathcal{B}$.

In the last step, the population $\pi_\lambda(t+1)$ of the next generation $t+1$ is built by collecting all μ offspring from $\pi'_\mu(t)$ and all $\lambda - \mu$ unaltered individuals from $\pi_{\lambda-\mu}(t)$, i.e. $\pi_\lambda(t+1) := \pi'_\mu(t) \cup \pi_{\lambda-\mu}(t)$.

The exogenous parameters that have to be specified for the *modGA* are λ, $\mu_\mathcal{R}$, $\mu_\mathcal{M}$, α, β and $t_{\max}$.

3 Theoretical Analysis of the Non-Uniform Mutation Operator

In this Sec., the *non-uniform mutation* operator is theoretically analysed and an improved *adaptive non-uniform mutation* operator is developed. The generation index t is neglected for convenience. The following theorem characterizes the statistical behavior of the *non-uniform mutation* operator.

Theorem 3. *The probability density function $p_{\delta_{k,i}}(\delta_{k,i})$ of the random variation $\delta_{k,i}$ in the non-uniform mutation operator is given by*

$$
p_{\delta_{k,i}}(\delta_{k,i}) = \begin{cases} \dfrac{1 - q_{k,i}}{\gamma \cdot (x_{k,i} - x_{i,\min})} \cdot \left(1 + \dfrac{\delta_{k,i}}{x_{k,i} - x_{i,\min}}\right)^{\frac{1-\gamma}{\gamma}} \\ \qquad\qquad for \quad x_{i,\min} - x_{k,i} \le \delta_{k,i} < 0 \\[2mm] \dfrac{q_{k,i}}{\gamma \cdot (x_{i,\max} - x_{k,i})} \cdot \left(1 - \dfrac{\delta_{k,i}}{x_{i,\max} - x_{k,i}}\right)^{\frac{1-\gamma}{\gamma}} \\ \qquad\qquad for \quad 0 < \delta_{k,i} \le x_{i,\max} - x_{k,i} \end{cases}
\tag{10}
$$

and $p_{\delta_{k,i}}(\delta_{k,i}) = 0$ for $\delta_{k,i} \notin [x_{i,\min} - x_{k,i}, x_{i,\max} - x_{k,i}]$. $\qquad\square$

Proof:
The condition $0 \le z_{k,i} \le 1$ for the realization $z_{k,i}$ of the random number $z_{k,i}$ leads to $x_{i,\min} - x_{k,i} \le \delta_{k,i} \le x_{i,\max} - x_{k,i}$, i.e. $p_{\delta_{k,i}}(\delta_{k,i}) = 0$ for $\delta_{k,i} \notin [x_{i,\min} - x_{k,i}, x_{i,\max} - x_{k,i}]$. Within this interval the following probability function can be derived.

$$
P_{\delta_{k,i}}(\delta_{k,i}) = \Pr\{\delta_{k,i} \le \delta_{k,i}\}
$$

$$
= \Pr\left\{z_{k,i}^\gamma \ge 1 - \frac{\delta_{k,i}}{x_{i,\max} - x_{k,i}}\right\} \cdot q_{k,i} +
$$

$$\Pr\left\{z_{k,i}^{\gamma} \leq 1 + \frac{\delta_{k,i}}{x_{k,i} - x_{i,\min}}\right\} \cdot (1 - q_{k,i})$$

$$= \left[1 - P_{z_{k,i}}\left(\left(1 - \frac{\delta_{k,i}}{x_{i,\max} - x_{k,i}}\right)^{\frac{1}{\gamma}}\right)\right] \cdot q_{k,i} +$$

$$P_{z_{k,i}}\left(\left(1 + \frac{\delta_{k,i}}{x_{k,i} - x_{i,\min}}\right)^{\frac{1}{\gamma}}\right) \cdot (1 - q_{k,i}) \tag{11}$$

Taking the derivative yields the probability density function

$$p_{\delta_{k,i}}(\delta_{k,i}) = \frac{\partial}{\partial \delta_{k,i}} P_{\delta_{k,i}}(\delta_{k,i}) =$$

$$\frac{q_{k,i}}{\gamma} \cdot p_{z_{k,i}}\left(\left(1 - \frac{\delta_{k,i}}{x_{i,\max} - x_{k,i}}\right)^{\frac{1}{\gamma}}\right) \cdot$$

$$\frac{1}{x_{i,\max} - x_{k,i}} \cdot \left(1 - \frac{\delta_{k,i}}{x_{i,\max} - x_{k,i}}\right)^{\frac{1-\gamma}{\gamma}} +$$

$$\frac{1 - q_{k,i}}{\gamma} \cdot p_{z_{k,i}}\left(\left(1 + \frac{\delta_{k,i}}{x_{k,i} - x_{i,\min}}\right)^{\frac{1}{\gamma}}\right) \cdot$$

$$\frac{1}{x_{k,i} - x_{i,\min}} \cdot \left(1 + \frac{\delta_{k,i}}{x_{k,i} - x_{i,\min}}\right)^{\frac{1-\gamma}{\gamma}} . \tag{12}$$

With the help of the probability density function

$$p_{z_{k,i}}(z_{k,i}) = \begin{cases} 1 & , \quad 0 \leq z_{k,i} \leq 1 \\ 0 & , \quad \text{otherwise} \end{cases} \tag{13}$$

the expression in (10) is obtained. ∎

The corresponding expectation value is given by the next theorem.

Theorem 4. *The expectation value of the random variation $\delta_{k,i}$ in the non-uniform mutation operator is given by*

$$E\{\delta_{k,i}\} = -\frac{\gamma}{1+\gamma} \cdot [(x_{k,i} - x_{i,\min}) - q_{k,i} \cdot (x_{i,\max} - x_{i,\min})] . \tag{14}$$

□

Proof:
The expectation value of $\delta_{k,i}$ can be calculated directly with the help of the

probability density function given in (10). A simpler approach, however, is to calculate

$$
\begin{aligned}
E\{\delta_{k,i}\} &= (x_{i,\max} - x_{k,i}) \cdot (1 - E\{[z_{k,i}]^\gamma\}) \cdot q_{k,i} + \\
&\quad (x_{i,\min} - x_{k,i}) \cdot (1 - E\{[z_{k,i}]^\gamma\}) \cdot (1 - q_{k,i}) \\
&= (x_{i,\max} - x_{k,i}) \cdot \frac{\gamma}{1+\gamma} \cdot q_{k,i} + \\
&\quad (x_{i,\min} - x_{k,i}) \cdot \frac{\gamma}{1+\gamma} \cdot (1 - q_{k,i}) \\
&= -\frac{\gamma}{1+\gamma} \cdot [(x_{k,i} - x_{i,\min}) - q_{k,i} \cdot (x_{i,\max} - x_{i,\min})]
\end{aligned}
\tag{15}
$$

using

$$
E\{[z_{k,i}]^\gamma\} = \int_0^1 z_{k,i}^\gamma \, dz_{k,i} = \frac{1}{1+\gamma} \ .
\tag{16}
$$

∎

Setting $q_{k,i} = \frac{1}{2}$ as proposed by MICHALEWICZ leads to

$$
E\{\delta_{k,i}\} = -\frac{\gamma}{1+\gamma} \cdot \left(x_{k,i} - \frac{x_{i,\max} + x_{i,\min}}{2} \right) \ .
\tag{17}
$$

This leads to

$$
\begin{aligned}
x_{k,i} &< \frac{x_{i,\max} + x_{i,\min}}{2} &:& \quad E\{\delta_{k,i}\} > 0 \\[4pt]
x_{k,i} &= \frac{x_{i,\max} + x_{i,\min}}{2} &:& \quad E\{\delta_{k,i}\} = 0 \\[4pt]
x_{k,i} &> \frac{x_{i,\max} + x_{i,\min}}{2} &:& \quad E\{\delta_{k,i}\} < 0
\end{aligned}
$$

i.e. for $q_{k,i} = \frac{1}{2}$ the *non-uniform mutation* operator $\mathcal{M}$ prefers values of $x'_{k,i} = x_{k,i} + \delta_{k,i}$ in the center of the feasible region $[x_{i,\min}, x_{i,\max}]$.

Equivalently, for a given parameter value $x_{k,i}$ the expectation value of the mutated parameter $x'_{k,i}$ for $q_{k,i} = \frac{1}{2}$ is

$$
\begin{aligned}
E\{x'_{k,i}\} &= x_{k,i} - \frac{\gamma}{1+\gamma} \cdot \left(x_{k,i} - \frac{x_{i,\max} + x_{i,\min}}{2} \right) \\
&= \frac{1}{1+\gamma} \cdot x_{k,i} + \frac{\gamma}{1+\gamma} \cdot \frac{x_{i,\max} + x_{i,\min}}{2} \ .
\end{aligned}
\tag{18}
$$

In the beginning of the search for $t \gtrsim 0$ the term γ is near one: $\gamma \lesssim 1$. Therefore

$$
E\{x'_{k,i}\} \approx \frac{1}{2} \cdot \left\{ x_{k,i} + \frac{x_{i,\max} + x_{i,\min}}{2} \right\} \ ,
\tag{19}
$$

i.e. the search is concentrated in the search space between the center of the feasible region and the parental parameter $x_{k,i}$. Only for $\gamma \gtrsim 0$ at the end of the optimization process ($t \lesssim t_{\max}$) the expected parameter value $\mathrm{E}\{x'_{k,i}\}$ is close to the parental parameter $x_{k,i}$.

For these reasons, an adaptive scheme for the determination of $q_{k,i}$ is presented in this paper. Demanding $\mathrm{E}\{\delta_{k,i}\} \stackrel{!}{=} 0$ — as e.g. in the mutation operator of *Evolution Strategies* [1, 9, 10] — the adaptive scheme becomes

$$q_{k,i} = \frac{x_{k,i} - x_{i,\min}}{x_{i,\max} - x_{i,\min}} . \qquad (20)$$

Equation (20) leads to the following definition of the novel *adaptive non-uniform mutation* operator.

Definition 5 (adaptive non-uniform mutation). Parent $\iota_k(t) = \mathbf{x}_k(t) \in I$ is selected for reproduction via the *adaptive non-uniform mutation* operator. The offspring $\iota'_k(t) = \mathbf{x}'_k(t) = (x'_{k,1}(t), \dots, x'_{k,n}(t)) \in I$ is created according to

$$x'_{k,i}(t) = x_{k,i}(t) + \delta_{k,i}(t) \qquad (21)$$

using the random variation

$$\delta_{k,i}(t) = \begin{cases} (x_{i,\max} - x_{k,i}(t)) \cdot (1 - [z_{k,i}(t)]^{\gamma(t)}) \\ \qquad\qquad \text{with probability } q_{k,i}(t) \\ (x_{i,\min} - x_{k,i}(t)) \cdot (1 - [z_{k,i}(t)]^{\gamma(t)}) \\ \qquad\qquad \text{with probability } 1 - q_{k,i}(t) \end{cases} \qquad (22)$$

and the probability

$$q_{k,i}(t) = \frac{x_{k,i}(t) - x_{i,\min}}{x_{i,\max} - x_{i,\min}} . \qquad (23)$$

$z_{k,i}(t)$ is a random number equally distributed in the interval $[0, 1]$ and

$$\gamma(t) = \left(1 - \frac{t}{t_{\max}}\right)^{\beta} \qquad (24)$$

with exogenous strategy parameter $\beta > 0$ and maximal number of generations $t_{\max}$. ◇

For $q_{k,i}$ as given in (20), the corresponding variance $\mathrm{Var}\{\delta_{k,i}\}$ can be calculated with the help of theorem 3.

$$\mathrm{Var}\{\delta_{k,i}\} = \frac{2\gamma^2}{(\gamma + 1)(2\gamma + 1)} \cdot (x_{i,\max} - x_{k,i}) \cdot (x_{k,i} - x_{i,\min}) \qquad (25)$$

The probability density function $p_{\boldsymbol{x}'_{k,i}}(x'_{k,i})$ of the mutated object parameter $x'_{k,i} = x_{k,i} + \delta_{k,i}$ is $p_{\boldsymbol{x}'_{k,i}}(x'_{k,i}) = 0$ for $x'_{k,i} \notin [x_{i,\min}, x_{i,\max}]$ and

$$
p_{\boldsymbol{x}'_{k,i}}(x'_{k,i}) =
\begin{cases}
\dfrac{1 - q_{k,i}}{\gamma \cdot (x_{k,i} - x_{i,\min})} \cdot \left(\dfrac{x'_{k,i} - x_{i,\min}}{x_{k,i} - x_{i,\min}}\right)^{\frac{1-\gamma}{\gamma}} \\
\qquad \text{for} \quad x_{i,\min} \le x'_{k,i} < x_{k,i} \\[2ex]
\dfrac{q_{k,i}}{\gamma \cdot (x_{i,\max} - x_{k,i})} \cdot \left(\dfrac{x_{i,\max} - x'_{k,i}}{x_{i,\max} - x_{k,i}}\right)^{\frac{1-\gamma}{\gamma}} \\
\qquad \text{for} \quad x_{k,i} < x'_{k,i} \le x_{i,\max}
\end{cases}
\tag{26}
$$

Using $q_{k,i}$ according to (20) the corresponding expectation value is

$$
\mathrm{E}\{x'_{k,i}\} = x_{k,i} \tag{27}
$$

and the variance

$$
\mathrm{Var}\{x'_{k,i}\} = \mathrm{Var}\{\delta_{k,i}\} \le \frac{\gamma^2}{2(\gamma + 1)(2\gamma + 1)} \cdot (x_{i,\max} - x_{i,\min})^2 \ . \tag{28}
$$

4   Experimental Results

In order to verify the theoretical analysis, the *modGA* is applied to the (bounded) *sphere model* [10], i.e. the minimization of the objective function

$$
f(\mathbf{x}) = \sum_{i=1}^{n} (x_i - x_{\mathrm{opt},i})^2 \ . \tag{29}
$$

This objective function is chosen for simplicity. The dimension of the search space is $n = 30$ and each parameter's feasible region is $[x_{i,\min}, x_{i,\max}] = [-10, 10]$. In Tab. 1, the exogenous strategy parameters for the *modGA* are given [8].

Table 1. The exogenous parameters for the *modGA*.

λ	μ_R	μ_M	α	β	$t_{\max}$
50	20	10	0.25	5	10000

For the selection operators $\mathcal{S}_\mu$ and $\mathcal{S}_{\lambda-\mu}$, respectively, BAKER's *linear ranking* methodology with maximal expected number of offspring $\eta_{\max} = 2$ and the *stochastic universal sampling* procedure are used [2, 3, 4]. In each generation t the object parameter vector $\mathbf{x} = \mathbf{x}_{\ell^*}(t)$ of the best individual $\iota_{\ell^*}(t)$ with $f(\mathbf{x}_{\ell^*}(t)) \le f(\mathbf{x}_\ell(t))$ $\forall \ell$ is chosen. 10 simulation runs are carried out for each experiment and the averages $\overline{f(\mathbf{x})}$ are calculated.

Figure 2 shows the probability density function $p_{\boldsymbol{x}'_{k,i}}(x'_{k,i})$ for MICHALE-WICZ' *non-uniform mutation* operator ($--$) and the novel *adaptive non-uniform mutation* operator (——) at generation $t = 2500$ and for $x_{k,i} = 5$. Since $x_{k,i} > (x_{i,\max} + x_{i,\min})/2$ the adaptive scheme leads to probability density function values $p_{\boldsymbol{x}'_{k,i}}(x'_{k,i})$ that are smaller for $x'_{k,i} < x_{k,i}$ and larger for $x'_{k,i} > x_{k,i}$.

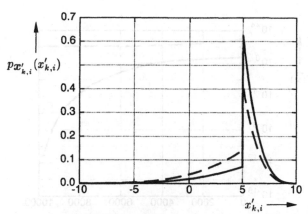

Fig. 2. Probability density function $p_{\boldsymbol{x}'_{k,i}}(x'_{k,i})$.

In the first experiment, the optimal parameters are situated in the center of each parameter's feasible region, i.e. $x_{\text{opt},i} = 0 \quad \forall i$. Figure 3 shows the results for MICHALEWICZ' original formulation of the *non-uniform mutation* operator (– –) and the novel *adaptive non-uniform mutation* operator proposed here (——). With both operators the *modGA* is able to find the optimum $\mathbf{x}_{\text{opt}}$. $\overline{f(\mathbf{x})}$ is nearly identical in both cases.

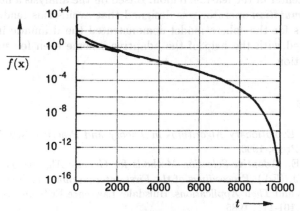

Fig. 3. Averages $\overline{f(\mathbf{x})}$ for $x_{\text{opt},i} = 0 \; \forall i$.

The second experiment considers the case that the optimum $\mathbf{x}_{\text{opt}}$ is near the boundary of the search space: $x_{\text{opt},i} = 8 \quad \forall i$. In Fig. 4, the averages $\overline{f(\mathbf{x})}$ are given for the *modGA* using the two different mutation operators. It is evident that the original *non-uniform mutation* operator (– –) is not able to locate the optimum even for this simple objective function. The adaptive variant of this operator (——), however, performs nearly as good as in the first experiment. It outperforms the original variant by 13 orders of magnitude!

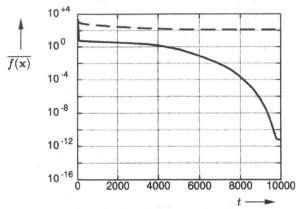

Fig. 4. Averages $\overline{f(\mathbf{x})}$ for $x_{\mathrm{opt},i} = 8\ \forall i$.

5 Conclusions

A theoretical analysis of the *non-uniform mutation* operator proposed by MICHA-LEWICZ for the *modGA* showed its undesirable tendency to prefer parameter values in the center of the feasible region. Based on the analysis a novel *adaptive non-uniform mutation* operator was designed that avoids this tendency. Experimental results for the sphere model demonstrated the dramatic improvement that was gained with the help of the adaptive scheme even for such a simple objective function.

References

1. BÄCK, T.: *Evolutionary Algorithms in Theory and Practice.* New York: Oxford University Press, 1996
2. BAKER, J.E.: *Adaptive Selection Methods for Genetic Algorithms.* In: GREFEN-STETTE, J.J. (Ed.): *Proceedings of the First International Conference on Genetic Algorithms and Their Applications.* Hillsdale: Lawrence Erlbaum Associates Publishers, pp. 101-111, 1985
3. BAKER, J.E.: *Reducing Bias and Inefficiency in the Selection Algorithm.* In: GREFENSTETTE, J.J. (Ed.): *Genetic Algorithms and Their Applications: Proceedings of the Second International Conference on Genetic Algorithms.* Hillsdale: Lawrence Erlbaum Associates Publishers, pp. 14-21, 1987
4. BAKER, J.E.: *An Analysis of the Effects of Selection in Genetic Algorithms.* Ph.D., Vanderbilt University, 1989

34

5. HOLLAND, J.H.: *Adaptation in Natural and Artificial Systems: An Introductory Analysis with Applications to Biology, Control, and Artificial Intelligence.* Cambridge: First MIT Press Edition, 1992
6. JANIKOW, C.Z.; MICHALEWICZ, Z.: *An Experimental Comparison of Binary and Floating Point Representations in Genetic Algorithms.* In: BELEW, R.K.; BOOKER, L.B. (Eds.): *Proceedings of the Fourth International Conference on Genetic Algorithms.* San Mateo: Morgan Kaufmann Publishers, pp. 31-36, 1991
7. MICHALEWICZ, Z.; JANIKOW, C.Z.: *Handling Constraints in Genetic Algorithms.* In: BELEW, R.K.; BOOKER, L.B. (Eds.): *Proceedings of the Fourth International Conference on Genetic Algorithms.* San Mateo: Morgan Kaufmann Publishers, pp. 151-157, 1991
8. MICHALEWICZ, Z.: *Genetic Algorithms + Data Structures = Evolution Programs.* Berlin: Springer-Verlag, 1992
9. RECHENBERG, I.: *Evolutionsstrategie '94.* Werkstatt Bionik und Evolutionstechnik, Band 1, Stuttgart: frommann-holzboog, 1994
10. SCHWEFEL, H.-P.: *Evolution and Optimum Seeking.* New York: John Wiley & Sons, 1995

Structure Identification of Functional-Type Fuzzy Models with Application to Modelling Nonlinear Dynamic Plants

P. Kortmann and H. Unbehauen

Control Engineering Laboratory, Faculty of Electrical Engineering
Ruhr-University Bochum, D-44780 Bochum

Abstract. A new fuzzy model structure identification method, based on orthogonalisation and statistical tests, as well as information criteria to obtain a minimum rule base and a minimum number of membership functions from input-
output data, is proposed. The method is applied to functional-type fuzzy models. The applicability of the proposed method to nonlinear static and dynamic
systems is illustrated by examples.
Keywords. Fuzzy model identification, structure selection, statistical tests,
orthogonalisation

1. Introduction

Nonlinear processes can be described by means of general nonlinear models, as e.g.
Wiener- or *Hammerstein*-models as well as *Kolmogorov-Gabor* polynomial based
models [1, 2, 3]. Another possibility is to derive a set of local linear models scheduled
according to the process operating conditions, described e.g. by the inputs, outputs or
states of the system. The scheduler selection function can be represented both by
classical logical functions and by means of fuzzy rules. The last case is dealt with in
this contribution, where discrete multi-input single-output (MISO) functional-type
fuzzy models according to *Takagi* and *Sugeno* [4-6] are applied to identify nonlinear
dynamic plants.

2. Description of Functional-Type Fuzzy-Models

The reasoning method of such functional-type fuzzy models is associated with a rule
base of a format that is characterised by functional-type conclusions:

$$R_l: \text{IF}\{x_1 \text{ is } \mu_{l1}\} \wedge \ldots \wedge \{x_q \text{ is } \mu_{lq}\} \text{ THEN } y_l = p_{l0} + p_{l1}x_1 + \ldots + p_{lq}x_q, \tag{1}$$

where x_i $i=1,\ldots, q$ are the q inputs, μ_{li} are linguistic values, L is the number of rules, y_l
with $l=1,\ldots, L$ are the conclusions with the parameters p_{li} of the fuzzy system. The
resulting output signal y can be obtained by means of the weighted average algorithm:

$$y = \frac{\sum_{l=1}^{L} a_l y_l}{\sum_{l=1}^{L} a_l} = \sum_{l=1}^{L} \frac{a_l}{\sum_{l=1}^{L} a_l} y_l = \sum_{l=1}^{L} f_l y_l, \qquad (2)$$

where

$$a_l = \mu_{l1}(x_1) \wedge \mu_{l2}(x_2) \wedge ... \wedge \mu_{lq}(x_q) \qquad (3)$$

is called the degree of firing (DOF) of the lth rule and f_l is the fuzzy basis function (FBF) characterising a rule of the rule base. To describe dynamical single-input single-output (SISO) systems the actual input variables of the fuzzy model are the time lagged input and output signals of the process

$$x^T(k) = [u(k), u(k-1), ..., u(k-n), y(k-1), y(k-2), ..., y(k-m)], \qquad (4)$$

m and n describing the maximally time-shifted values of the input and output signal, respectively. Also a pure delay time d can be incorporated. Without loss of generality also a multi-input multi-output (MIMO) process with r outputs can be modelled by r MISO-fuzzy-models.

According to Eq.(2) functional-type fuzzy models can be interpreted as linear models with time varying parameters, where the parameters are determined by the FBFs depending on the rule base and the current inputs of the model [7]. Hence, the parameters are known at each sample and can be calculated as

$$p_i = \sum_{l=1}^{L} p_{li} f_l(x(k)) \qquad (5)$$

which is called dynamic linearisation [8]. Provided that a process is described by a functional-type fuzzy model without mean values, it is possible to design arbitrary linear controllers based on Eq.(5). This is a tremendous advantage of the functional-type fuzzy models over global nonlinear models or artificial neural networks.

3. A New Structure Identification Method for Functional-Type Fuzzy Models

In order to obtain a fuzzy model containing a small number of rules and parameters, respectively, a structure identification has to be applied to select the significant rules. In [5, 6] the structure identification is performed by applying heuristic search methods and a GMDH (general method of data handling) algorithm. Other methods like data clustering methods (e.g. fuzzy-c-means, mountain clustering, subtractive clustering and Gustafson-Kessel-algorithm [9-12]) are also based on heuristic proposals. Another disadvantage is the lack of structure tests, which estimate the complexity of a model and help to obtain a compromise between accuracy and complexity of a fuzzy model.

In this contribution a new identification method based on orthogonalisation and several statistical and information criteria to obtain a minimum rule base and a minimum number of membership functions from measurement data is proposed. Assuming the input variables of the fuzzy model the algorithm determines a statistically significant rule base by choosing the centres of the premise gaussian membership functions (GMF) from the available data, where the standard deviation of the GMF's are constant and predetermined. The parameters of the conclusion parts are estimated by means of the LS-method. The advantage of the proposed method is a significant decreasing number of rules and parameters.

Eq. (2) can be rewritten in vector form as

$$y = m^T p \tag{6}$$

with the signal vector

$$m^T = \begin{bmatrix} v_1 & \cdots & v_z \end{bmatrix} = \begin{bmatrix} f_1 & f_1 x_1 & \cdots & f_L x_q \end{bmatrix}, \tag{7}$$

and the parameter vector

$$p^T = \begin{bmatrix} p_1 & \cdots & p_z \end{bmatrix}. \tag{8}$$

The number of terms is $z = L(q+1)$. The detection of the significant MF centres that leads to the significant rules is based on orthogonalisation. Provided that a mutually orthogonal set of terms $\tilde{v}_i \ i = 0,...,z$ is produced which spans the same space as the original terms v_i, and provided that the LS-method is applied, it can be shown that the sum of squared regressors (SSR_i) of the orthogonal model can be calculated independently.

A maximum reduction of the squared equation error is achieved if the rules corresponding to the largest value of SSR are chosen at each stage.

From eq. (2) it is seen that the FBFs are changing with the number of rules L caused by the sum over all DOFs in the denominator. This means that the same FBF or rule which was significant at an earlier stage and incorporated to the model could be superfluous at a later stage with more rules. Therefore, it is absolutely necessary that every FBF has to be tested for significance at every stage, which is done by means of several statistical tests and information criteria. In addition to the popular F-test, the following information criteria should be used to get more reliable decisions about the selection of significant FBFs for the fuzzy model and to get criteria to quit the selection:

$$FPE = N \ln(V) + N \ln\left(\frac{N+L}{N-L}\right), \qquad \text{(final prediction error)} \tag{9}$$

$$AIC(\phi_1) = N \ln(V) + \phi_1 L, \quad \phi_1 > 0, \qquad \text{(Akaike's information criterion)} \tag{10}$$

$$LILC(\phi_2) = N \ln(V) + 2L\phi_2 \ln[\ln(V)], \phi_2 \geq 1, \qquad \text{(law of iterated log. crit.)} \tag{11}$$

$$BIC = N \ln(V) + L \ln(N), \qquad \text{(Bayesian information criterion)} \tag{12}$$

38

where N is the number of samples, L the number of rules and the loss function V is defined as

$$V = \frac{1}{N}\sum_{k=1}^{N}\varepsilon(k)^2 . \qquad (13)$$

The model error $\varepsilon(k) = y(k) - \hat{y}(k)$ is defined as the difference between the measured output $y(k)$ and the estimated output $\hat{y}(k)$. Besides the performance of the fuzzy model, these criteria estimate the complexity of the rule base also. While the first term of eqs. (9)-(12) decreases with increasing the model complexity, the second term increases with enlarging the number of rules. So the best model is characterised by the smallest values of eqs. (9)-(12). Apart from the above criteria the following multiple correlation coefficients (MCC) are implemented, which also decide on aborting the selection:

$$MCC^2 = \sum_{k=1}^{N}(\hat{y}(k)-\bar{y})^2 \Big/ \sum_{k=1}^{N}(y(k)-\bar{y})^2 , \qquad (14)$$

$$MCC_T = 1-(1-MCC^2)\frac{N-1}{N-L}, \qquad (15)$$

$$MCC_A = 1-(1-MCC^2)\frac{N+L}{N-L}, \qquad (16)$$

where $\bar{y}$ designates the mean value. Equations (15) and (16) show the corrected MCCs of Theil and Amemiya which punish the complexity of a model. The algorithm for selecting the significant terms of the model can then be formulated as follows:

Step 1: Specify the membership function type and the corresponding form parameter which is the standard deviation (σ) for gaussian or the base (b) if triangular membership functions are used. Determine the significance level α of the F-test. In the case of a small number of data the results are doubtful and α should be high. In the case of a large number of data the results are made more reliable and a small α can be applied. Choose one input data vector, whose elements become the first membership function centres. Specify the t-norm (min or product operator). Set $L=1$.

Step 2: Choose one after another the remaining $N-L$ data vectors to be the next membership function centres, where N is the number of samples, and calculate the SSR_i by means of orthogonalisation, using the QR-deviation.

Step 3: Select the membership function centre that yields the highest quotient,

$$f_{opt_L}: \quad SSR_{opt} = \max_i[SSR_i]$$

and set $L = L+1$.

Step 4: Calculate the information criteria and MCCs according to eqs. (9)-(16) for the obtained fuzzy model.

39

Step 5:   Compare the values of the information criteria and MCCs with their values gained for L-1 rules of the previous model. The procedure quits if the MCCs give smaller or the information criteria give greater values than those for the previous model or if the F-test rejects the recent model.

Step 6:   Use the partial structure tests. Compute for every rule in the model the partial F-test value

$$PF_i = \frac{\sum_{k=1}^{N}(\hat{y}(k)-\bar{y})^2\Big|_L - (\hat{y}(k)-\bar{y})^2\Big|_{L-1}}{\left(\sum_{k=1}^{N}\varepsilon(k)^2\Big|_L\right)\Big/(N-L)}, \quad i=1,...,L-1, \quad L>2, \tag{17}$$

$\varepsilon(k)$ being the model error, and the partial information criteria

$$FPE_i, AIC_i, LILC_i, \text{and } BIC_i, \quad i=1,...,L-1, \quad L>2, \tag{18}$$

according to eqs. (9)-(12) where the number of rules in the model is decreased to L-1. That means reject the first rule of the fuzzy model and calculate the necessary quantities. Next introduce again the first rule, reject the second one and compute the above quantities and so on. Continue until rule number L-1. These variables can be applied to test the alternative hypotheses

$$\begin{aligned} H_0: &\ R_i \text{ excluded,} \\ H_1: &\ R_i \text{ included,} \end{aligned} \quad \text{for } i=1,...,L-1. \tag{19}$$

Since PF_i is F-distributed with $(1, N\text{-}L)$ degrees of freedom, it is possible to formulate the decision procedure as

$$\begin{aligned} &\text{accept } H_0 \text{ if } \min_i[PF_i] \le F(1,N-L,1-\alpha), \\ &\text{accept } H_1 \text{ if } \min_i[PF_i] > F(1,N-L,1-\alpha), \end{aligned} \tag{20}$$

where $\min_i[PF_i]$ is the smallest test value calculated by eq. (17). Concerning the partial information criteria and MCCs the decision procedure can be formulated as:

$$\text{accept } H_0 \text{ if } \{\min_i[FPE_i] \le FPE_L\} \vee \{\min_i[AIC_i] \le AIC_L\} \vee \cdots$$

$$\cdots \{\min_i[LILC_i] \le LILC_L\} \vee \{\min_i[BIC_i] \le BIC_L\}, \tag{21}$$

$$\text{else accept } H_1,$$

where $\min_i[...]$ are the smallest calculated quantities according to eqs. (9)-(12) and FPE_L, AIC_L, $LILC_L$, BIC_L are the criteria evaluated in Step 4.

Notice that the test quantities PF_i, and the partial information criteria represent a measure of the influence of a rule in the fuzzy model.

Step 7: Examine the quantities of step 6 according to eqs. (20)-(21). If one of the conditions for acceptance of the hypothesis H_0 is fulfilled, the examined rule is statistically not significant and rejected. L is set to L-1. Continue with step 2.

4. Results

Three examples illustrate that the proposed modelling method is appropriate for nonlinear dynamic systems.

4.1. Example 1: Nonlinear time discrete system

It is assumed that the SISO-system to be modelled is of the form

$$y(k) = \frac{y(k-1)y(k-2)y(k-3)u(k-2)\left[y(k-3)-1\right]+u(k-1)}{1+y^2(k-2)+y^3(k-3)}. \tag{22}$$

The input variables of the fuzzy model are $x1=y(k-1)$, $x2=y(k-2)$, $x3=y(k-3)$, $x4=u(k-1)$, and $x5=u(k-2)$. By means of the proposed selection algorithm the following rule base was identified:

1. If $(x1$ is $Z) \wedge (x2$ is $Z) \wedge (x3$ is $Z) \wedge (x4$ is $Z) \wedge (x5$ is $Z)$ then $(y$ is $y1)$,
2. If $(x1$ is $NB) \wedge (x2$ is $NB) \wedge (x3$ is $NB) \wedge (x4$ is $N) \wedge (x5$ is $N)$ then $(y$ is $y2)$,
3. If $(x1$ is $P) \wedge (x2$ is $P) \wedge (x3$ is $P) \wedge (x4$ is $P) \wedge (x5$ is $P)$ then $(y$ is $y3)$,
4. If $(x1$ is $NS) \wedge (x2$ is $NS) \wedge (x3$ is $NS) \wedge (x4$ is $Z) \wedge (x5$ is $Z)$ then $(y$ is $y4)$,

with

$$y1 = 0.017+0.670x1-0.075x2-0.227x3+0.148x4-0.063x5,$$
$$y2 = 0.296+0.389x1+0.314x2+0.489x3-0.150x4+0.183x5,$$
$$y3 = 0.334+0.723x1-0.176x2+0.111x3-0.477x4-0.177x5,$$
$$y4 = -0.278+0.958x1-0.622x2+0.301x3-0.364x4+0.360x5.$$

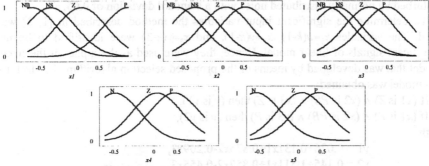

Fig. 1. Membership functions of the fuzzy model

The membership functions of the premise part are shown in Fig. 1. The simulation results presented in Fig. 2 show the approximation of the nonlinear process according to Eq. (22) by a third order linear and a fuzzy model. The excellent approximation accuracy of the fuzzy model is quite obvious.

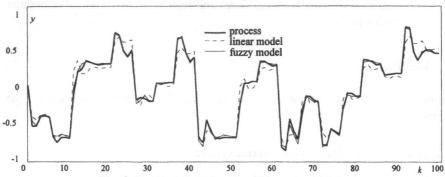

Fig. 2. Input/ output behaviour of process and models

4.2. Example 2: *Wiener*-system

The second system is characterised by a *Wiener*-structure, where

$$G(z) = \frac{X(z)}{U(z)} = \frac{0.1333z^{-1} + 0.0667z^{-2}}{1 - 1.5z^{-1} + 0.7z^{-2}} \tag{23}$$

describes the dynamic and

$$y(k) = x(k) + 3x(k)^2 + 1.5x(k)^3 \tag{24}$$

the static behaviour of the plant. Two data series of 200 measuring values were generated. One series was applied for estimation and one for prediction. The input signals of the estimation and prediction sets consist of white uniformly distributed noise with the amplitudes $u = \pm 0.5$. The output signal of the estimation series was disturbed by normally distributed noise with a standard deviation of 0.05.

To determine the significant input variables the method described in [1, 2] was used. The variables $x_1 = u(k-1)$, $x_2 = y(k-1)$, and $x_3 = y(k-3)$ were identified to be the significant signals of linear model. Hence, they were used as the inputs of the fuzzy model that was developed by means of the proposed selection algorithm. The following model was obtained:
1. If $(x1$ is $Z) \wedge (x2$ is $PS) \wedge (x3$ is $Z)$ then $(y$ is $y1)$,
2. If $(x1$ is $P) \wedge (x2$ is $PB) \wedge (x3$ is $P)$ then $(y$ is $y2)$,
with

$$y1 = 0.026 + 0.131x1 + 0.875x2 - 0.209x3,$$
$$y2 = 0.145 + 1.511x1 + 0.857x2 - 0.455x3.$$

The membership functions of the premise part are shown in Fig. 3.

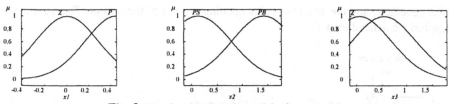

Fig. 3. Membership functions of the fuzzy model

Moreover a pure linear and a *Kolmogorov-Gabor*-polynomial based nonlinear model were estimated by the method of [1, 2], where the linear model consists of four terms including a bias and the nonlinear model of seven. The results of all models obtained by parallel simulation considering the prediction data series are shown in Fig. 4. It can be concluded that the fuzzy model as well as the nonlinear model is appropriate to describe the nonlinear behaviour of the plant, where the linear model is quite bad, as expected.

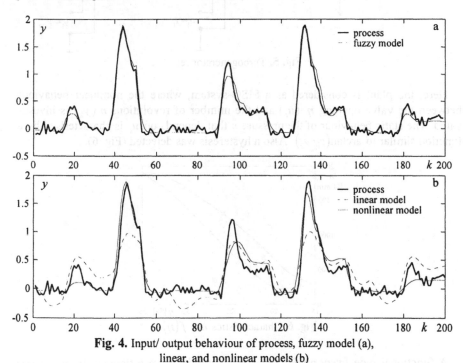

Fig. 4. Input/ output behaviour of process, fuzzy model (a),
linear, and nonlinear models (b)

4.3. Example 3: Turbogenerator set

The turbogenerator set (Fig. 5) consists of an air pressure turbine driving a small synchronous generator. Two output signals, the speed n (y_2) and the phase voltage V_{GR} (y_1), are excited by two input signals, the field current I_f (u_1) of the synchronous generator, and the degree of valve opening η (u_2) of the turbine. The generator supplies a balanced load R_l.

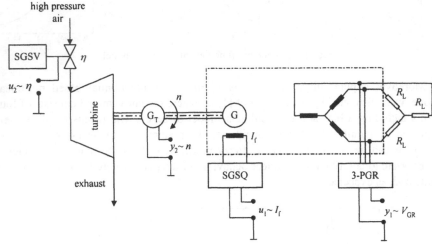

Fig. 5. Turbogenerator set

Here, the plant is considered as a SISO-system, where the nonlinear behaviour between the valve opening η (u_2) and the number of revolutions n (y_2) is investigated. The static behaviour of the pressure p that depends on u_2 is characterised by a function similar to $\arctan(c_1 \cdot u_2)$. Also a hysteresis was detected (Fig. 6).

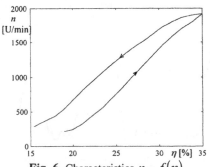

Fig. 6. Characteristics $n = f(\eta)$

A functional type fuzzy model and a linear model were estimated. In both cases two data sets consisting of 250 samples were generated with a sampling rate of 0.4 sec. in the operating point $I_f = 0.6$ A, $\eta = 30\%$, and $n = 1350$ r.p.m. One series was applied for estimation and one for prediction. The input signal of the estimation and prediction sets consist of white uniform distributed noise with the amplitude ±0.75, corresponding to a valve opening change of ±15%, but different bandwidth.

First of all a linear model was estimated by the method described in [1, 2] including the variables $u(k)$, $u(k-1)$, $u(k-2)$, $u(k-3)$, $y(k-1)$, $y(k-4)$ and an additional bias. Then a fuzzy model was developed with the same variables consisting of three rules that yielded 21 adjustable parameters, estimated by the least squares algorithm. The rule selection was stopped because of deterioration of the information criteria according to eqs. (9)-(12).

Table 1. Verification with N=250 samples

	fuzzy model	*linear model*
V	0.008	0.034
parameters	21	7

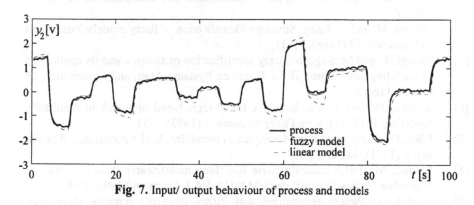

Fig. 7. Input/ output behaviour of process and models

The predicted output signal of the fuzzy model (Fig. 7, Table 1) indicates that the selected rule structure and the estimated conclusion parameters approximate very well the static and dynamic behaviour of the considered system while a pure linear model is not sufficient to describe the static characteristics.

5. Conclusions

A self-selection method for structure detection in the identification of nonlinear time-discrete static and dynamic systems has been described, where the systems are represented by functional-type fuzzy models. The method is based on orthogonalisation and a combination of several statistical tests, as well as information criteria, which are essential. Identification of nonlinear static and dynamic simulated systems and a turbogenerator set as case studies have shown the significant improvement in prediction accuracy compared to linear modelling.

The results are quite similar to those obtained by a general polynomial based approach as in [1, 2], but the advantage of first order functional type fuzzy models over other nonlinear models is that because of the local linear functions, scheduled by the rule base, local *linear* controllers can be designed ([13, 14]) and applied to the process.

6. References

[1]   Kortmann, M.: Die Identifikation nichtlinearer Ein- und Mehrgrößensysteme auf der Basis nichtlinearer Modellansätze. VDI-Verlag, Düsseldorf 1989.

[2]   Kortmann, M. and H. Unbehauen: Structure detection in the identification of nonlinear systems. APII Automatique productique informatique industrielle, 22 (1988), 5-25.

[3]   Unbehauen, H.: Some new trends in identification and modeling of nonlinear dynamical systems. J. of Applied Mathematics and Computation, 78 (1996), 279-297.

[4]   Sugeno, M. and G. Kang: Structure identification of fuzzy models. Fuzzy Sets and Systems, 28 (1988), 15-33.

[5]   Takagi, T. and M. Sugeno: Fuzzy identification of systems and its application to modeling and control. IEEE Trans. on Systems, Man, and Cybernetics, 15 (1985), 116-132.

[6]   Sugeno, M. and T. Yasukawa: A fuzzy-logic-based approach to qualitative modeling. IEEE Trans. on Fuzzy Systems, 1 (1993), 7-31.

[7]   Filev, D.: Fuzzy modeling of complex systems. Int. J. of Approximate Reasoning, 5 (1991), 281-290.

[8]   Fischer, M.: Fuzzy-modellbasierte Regelung nichtlinearer Prozesse. Proc. 6. Workshop "Fuzzy Control" des GMA-UA 1.4.2, Dortmund 1996, 29-42.

[9]   Bezdek, J.: Pattern recognition with fuzzy objective function algorithms. Plenum Press, New York 1981.

[10]   Yager, R. and D. Filev: Generation of fuzzy rules by mountain clustering. J. Intelligent and Fuzzy Systems, 2 (1994), 209-219.

[11]   Chiu, S.: Fuzzy model identification based on cluster estimation. J. Intelligent and Fuzzy Systems, 2 (1994), 267-278.

[12]   Gustafson, D. and W. Kessel: Fuzzy clustering with a fuzzy covariance matrix. Proc. IEEE-CDC (Conference on Decision and Control) 1978, 761-766.

[13]   Wang, H., K. Tanaka and M. Griffin: An analytical framework of fuzzy modelling and control of nonlinear systems: Stability and design issues. Proc. of American Control Conference (ACC), Seattle, Washington 1995, 2272-2276.

[14]   Babuška, R. and H. Verbruggen: Model-based methods for design of fuzzy control systems, Journal A, 36 (1995), 56-61.

Identification of Systems Using Radial Basis Networks Feedbacked with FIR Filters

Luciano Boquete, Rafael Barea, Ricardo García, Manuel Mazo, J. A. Bernad

Departamento de Electrónica. Universidad de Alcalá. 28801. Spain
E-mail: boquete@depeca.alcala.es

Abstract. A new model of a radial basis neural network is presented in this article which is fedbacked with a FIR filter. Using various neurons of this type, it is possible to construct a recurrent neural network, where the coefficients of each filter and the synaptic connections are adjusted to minimize an error function. The simulations carried out show the validity of this method for identifying systems with memory.

1 Introduction

One of the possibilities for identifying the behaviour of a system consists in using neural networks [1,2]. If the physical system to be identified has a memory (the output at a given moment depends on previous inputs and outputs), it is necessary to use a recurrent network to minimize the identification error. There are various models of recurrent networks which are generally multi-layer: Jordan, Ellman, taped delay, with a memory associated with the neuron, etc. One of the problems in using recurrent networks is that the training time can be very long. Because of this, one possibility consists in using radial basis networks (RBF), which only have one hidden layer. Their capacity as universal approximators of functions has been demonstrated. Various modifications can be made to their architecture so that an RBF can take previous states into account: tape delay, Chng model [3], with synaptic connections with FIR/IIR filters [4], memories associated with each neuron [5] - in this case a multi-layer network was applied - etc.

A new definition of a recurrent neuron is proposed in this article where the output of the neuron (normally a Gaussian function) with feedback towards its own input, is added at the output of the neural network. This feedback is carried out by means of a FIR filter, so that it is possible to memorize previous states. It is possible to construct a neural network capable of identifying dynamic systems with delays by using a sufficient number of units of these characteristics.

This article has been divided into the following sections: Part 2 describes the new model of a radial basis function, Part 3 a neural network formed with neurons of this type, Part 4 shows some simulated results and lastly Part 5 put forward the main conclusions of this work.

2 A radial basis function with feedback using a FIR filter

The neuron model in Figure 1 will be considered. The outputs of the neuron and the

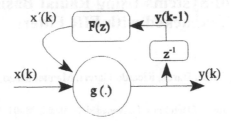

Fig. 1. Model of a neuron fedbacked by a FIR filter

filter are respectively:

$$y(k) = e^{-\frac{[x(k) + x'(k) - C]^2}{\sigma^2}} \quad ; \qquad x'(k) = \sum_{i=1}^{N} a_i.y(k-i) \qquad (1.a, 1.b)$$

The variation in the output of the neuron with respect to each of the filter coefficients is:

$$\frac{\partial y(k)}{\partial a_i} = \frac{\partial y(k)}{\partial x'(k)} \cdot \frac{\partial x'(k)}{\partial a_i} + \frac{\partial y(k)}{\partial x(k)} \cdot \frac{\partial x(k)}{\partial a_i} + \frac{\partial y(k)}{\partial C} \cdot \frac{\partial C}{\partial a_i} \qquad i = 1,...N; \qquad (2)$$

In the previous expression, the second and third sums are null, as the inputs $x(k)$ and C do not depend on any of the "a_i" coefficients. The variation in the output of the neuron with respect to the filter output is:

$$\frac{\partial y(k)}{\partial x'(k)} = -2.[\frac{x(k)+x'(k)-C}{\sigma^2}].e^{-\frac{[x(k) + x'(k) - C]^2}{\sigma^2}} \equiv g'(k) \qquad (3)$$

And the variation from the filter output with respect to the coefficients is:

$$\frac{\partial x'(k)}{\partial a_i} = y(k-i) + \sum_{j=1;}^{N} a_j.\frac{\partial y(k-j)}{\partial a_j}; \qquad i=1,...N; \qquad (4)$$

So a recurrent relathionship is established between the derivative in the current instant and in previous ones:

$$\frac{\partial y(k-j)}{\partial a_j} = \frac{\partial y(k-j)}{\partial x'(k-j)} \cdot \frac{\partial x'(k-j)}{\partial a_j} = g'(k-j).\frac{\partial x'(k-j)}{\partial a_j} \qquad (5)$$

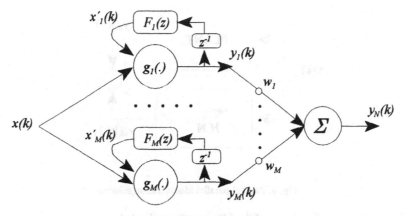

Fig. 2. Neural network model

3 Neural network

The architecture of a neural network formed by basis functions such as those defined is shown in Figure 2. The same is formed by basis M functions with feedback, where the sum of their respective weighted outputs produces a global network output. The network output is:

$$y_N(k) = \sum_{t=1}^{M} w_t.y_t(k) \tag{6}$$

If a supervised learning method is used, inputs are applied to the network and the corresponding output (y_N) is calculated, which is compared with the ideal output (y_d) producing the error: y_d-y_N. If the error to be minimized is defined as:

$$E = \frac{1}{2}.(y_d - y_N)^2 \quad \rightarrow \quad \frac{\partial E}{\partial y_N} = -(y_d - y_N); \tag{7}$$

For each operational cycle, the same signal is applied to the input of the system to be identified and of the neural network; with the outputs of the system and of the network the error (equation 7) is obtained and the coefficients "w_t" are adjusted according to the following expression:

$$\Delta w_t = -\alpha_1.\frac{\partial E}{\partial w_t} = \alpha_1.(y_d\text{-}y_N).g_t(k) \qquad \alpha_1 > 0 \tag{8}$$

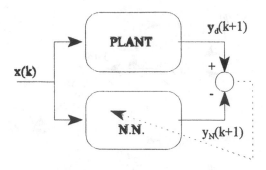

Fig. 3. Parallel-parallel identification scheme

The coefficients of each one of the filter are then adjusted:

$$\Delta a_i^{\ s} = -\alpha_2 \cdot \frac{\partial E(k)}{\partial a_i^{\ s}} = -\alpha_2 \cdot \frac{\partial E(k)}{\partial y_N} \cdot \frac{\partial y_N}{\partial y^s} \cdot \frac{\partial y^s}{\partial a_i^{\ s}} = \alpha_2 \cdot (y_d - y_N) \cdot w_i \cdot \frac{\partial y^s}{\partial a_i^{\ s}}; \qquad \alpha_2 > 0; \qquad (9)$$

where s=1, ...M; and i = 1,...N;

Concerning with the centres of the functions, the distribution of these is uniform and the standard deviation can be calculated by (10):

$$\sigma = \frac{M_{x(k)} - m_{x(k)}}{2.M} \qquad (10)$$

4 Simulations

In order to check the validity of the method described, 3 models of non linear physical systems are used (equations 11,12 and 13) to which the inputs of equation 14 are applied respectively. In all the cases, a "parallel-parallel" identification scheme is used (Fig. 3).

$$y_d(k+1) = \frac{y_d(k) \cdot y_d(k-1) \cdot [y_d(k) + 2,5] \cdot [y_d(k) - 1]}{1 + y_d^2(k) + y_d^2(k-1)} + x(k) \qquad (11)$$

$$y_d(k+1) = \frac{0.85 y_d(k) \cdot y_d(k-1) + 0.16 x(k) + 0.25 \cdot x(k-1)}{1 + y_d^2(k)} \qquad (12)$$

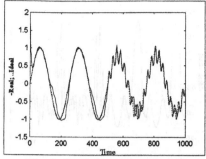

Fig. 4. Simulations of equation 11

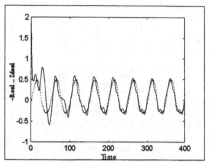

Fig. 5. Simulations of equation 12

$$y_d(k+1) = \frac{y_d(k).[y_d(k)+0,3]}{1+y_d^2(k)} + x(k).[x(k)+0,8].[x(k)-0,5] \tag{13}$$

In all cases, the input is:

$$u(k) = \sin(2.\pi.k/250); \qquad 0 \le k \le 500;$$
$$u(k) = 0.8.\sin(2.\pi.k/250) + 0.2.\sin(2.\pi.k/25); \qquad k > 500; \tag{14}$$

As shown in Figures 4, 5 and 6, after a reduced number of cycles, the network is capable of identifying the systems proposed with great accuracy. In the cases shown, the number of neurons in the network is 5, and in each one of them N=2 is complied with in the associated filters.

The figure 7 shows the output of the model (13) and the identification model after 5 iterations of the equation 14.

5 Conclusions

A new model of a neural network has been implemented which enables dynamic systems with a memory to be identified, in a relatively small number of cycles, which permits them to be used in systems which do not have very rapid changes. The final aim of this line of research is to develop the identification and the adaptive control of an industrial lift truck (GU-100-B stacker), so designed that full control of its position is retained no matter how much its working conditions change (increase or reduction of load, run-down batteries, variations in ground conditions, etc).

The main advantage of the proposed method is the use of the parallel-parallel identification model in any case, implying a simplification of the identification process and, additionally, eliminating the need od delayed inputs or outputs.

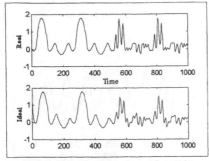

Fig. 6. Identification of equation (13)

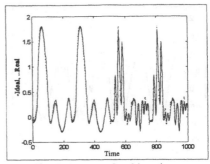

Fig. 7. Outputs after 5 iterations.

6 Acknowledgments

The authors would like to express their gratitude to the Comisión Interministerial de Ciencia y Tecnología (CICYT) - Spanish Interministerial Science and Technology Commission, for the aid given through the project TAP94-0656-C03-01, thanks to which this research is being carried out.

References

1. A.G. Barto: Connectionist Learning for Control: An Overview. COINS Technical Report 89-89 (1989)
2. K.S. Narendra, K. Parthasarathy: Identification and control of dynamic systems using neural networks. IEEE Transactions on Neural Networks, Vol. 1, N° 1, 4-27 (1991)
3. E. S. Chng, S. Chen and B. Mulgrew: Improving the Radial Basis Function Networks for Homogeneous Nonstationary Time Series Prediction. European Signal Processing Conference VII 94 Edinburgh, 1819-1822 (1994)
4. I. B. Ciocoiu : Radial Basis Function Networks with FIR/IIR Synapses. Neural processing Letters 3, 17-22 (1996)
5. P.S. Sastry, G. Santharam, K.P. Unnikrishnan: Memory Neuron Networks for Identification and Control of Dynamical systems. IEEE Transactions on Neural Networks, Vol. 5, N° 2 (1994)

A Novel Neurocomputing Approach to Nonlinear Stochastic State Estimation

M.B.MENHAJ   and   F.RAJAII SALMASI
Electical Engineering Departmant
Amirkabir University of Technology
Tehran 15914, Iran
Phone: (+09821) 6466009
Fax: (+09821) 6406469
E-Mail: menhaj@CIC.AKU.AC.IR

Abstract. This paper presents a novel neuro-computing approach to the problem of state estimation by means of an hybrid combination of Hopfield neural network whose capability of solving certain optimization problems is well-known and feedforward multilayer neural net which is very popular because of its universal approximation property. This neuro-estimator is very appropriate for the real-time implementation of linear or/and especially nonlinear state estimators. Simulation results shows the effectiveness of the proposed method.

1. Introduction

State variable models are widely used in control and communication theory and also in signal processing. Sometimes, we need the entire state vector of a dynamical system in order to implement an optimal control law for it, or to implement a digital signal processor. Usually, we cannot measure the entire state vector and our measurements are corrupted by noise. In state estimation our objective is to estimate the entire state vector from a limited collection of noisy measurements. The most common approach to the state estimation problem in linear systems is the Kalman Filtering. In this approach, if the process and measurement noises are mutually uncorrelated jointly Gaussian white noises, the Kalman Filter will be an optimal estimator[1]. For nonlinear systems, the state estimation problem is still a challenging problem. In this case, the most common approach is the Extended Kalman Filtering (EKF). This method is effective only in a class of nonlinear systems whose behavior is close to the corresponding linearized systems[1,2].

In this paper, we introduce an hybrid neuro-structured estimator based on the Hopfield and multilayer feedforward neural networks . The Hopfield network has the capability of solving some classes of optimization problems based on the steepest descent algorithm[3,4] . Also, it's well known that any nonlinear continuous function can be approximated by a multilayer feedforward neural net with a desired of accuracy[5]. Our proposed approach is very appropriate for the real-time implementation of state estimators especially it becomes more highlighted while a non-linear estimation is concerned. Simulation results help to judge the merit of the proposed method. The paper is organized as follows: In section 2, we formulate the problem. The Hopfield neural model is introduced in section 3. We will see that the outputs of the network do follow gradient descent paths on the network's energy landscape. In section 4, we design the Hopfield estimator consisting of a Hopfield network , a single stage predictor and a unit for minimization of the estimation error covariance. After determining the proper values of the Hopfield net parameters, weights and biases, and programming the network with these parameters, we observe that the outputs of the network will converge to real state vector. The single stage predictor is a nonlinear mapping that can be implemented by a multilayer feedforward neural network. In section 5, we perform some simulations to show the effectiveness of the proposed method. Section 6 concludes the paper.

2. The Hopfield Neural Model

The hopfield network consisits of a number of mutually interconnected processing units called neurons whose outputs are monotonically increasing functions of their states. In the discrete form, neurons change their states according to the network dynamic equation

$$n_{i+1} = n_i + \sum_{j=1}^{N} w_{ij} a_j - \frac{n_i}{R_i} + I_i \; , \tag{2-1}$$

where n_i is the state variable of the i th neuron, a_j is the output of the j th neuron, I_i is the bias input of the i th neuron and w_{ij} is the synaptic weight connecting the i th to the j th neuron. The output equation of the j th neuron is defined by $a_j = g(\lambda n_j)$. The nonlinear function $g(\cdot)$ is a monotonically increasing function like $\tanh(\cdot)$. Now, Let the network energy function E be defined as follows:

$$E = -\frac{1}{2} \sum_{i=1}^{N} \sum_{j=1}^{N} w_{ij} a_i a_j - \sum_{i=1}^{N} I_i a_j + (\frac{1}{\lambda}) \sum_{i=1}^{N} \frac{1}{R_i} \int_0^{a_j} g_i^{-1}(a) da \; . \tag{2-2}$$

It can be shown that [6] if $g^{-1}(.)$ is positive and linear in a region, then we may have

$$a_i(t+1) = a_i(t) - \gamma \frac{dE}{da_i} \; . \tag{2-3}$$

Because the network energy function has a negative time gradient, the outputs of the network do follow gradient descent paths on the energy landscape.

3. Statement of the Problem

Consider the nonlinear discrete-time stochastic dynamic system described by the following state and output equations

$$x_{k+1} = \phi(x_k, u_k) + \Phi x_k + \Gamma w_k \tag{3-1}$$

$$y_k = h(x_k) + H x_k + v_k \; , \tag{3-2}$$

where x_k and y_k represent the n and m -dimensional state and output vectors, $\mathbf{w}_k$ and $\mathbf{v}_k$ are uncorrelated zero-mean Gaussian white noise sequences with the covariance matrices $S_k = \text{var}(\mathbf{w}_k)$ and $T_k = \text{var}(\mathbf{v}_k)$, respectively. Φ and H are $n \times n$ and $m \times n$ matrices. ϕ and h are two nonlinear mappings which satisfy Lipschits condition on a bounded region, i.e.

$$\|\phi(\mathbf{x}) - \phi(\mathbf{y})\| \le k_T \|\mathbf{x} - \mathbf{y}\| \tag{3-3}$$

$$\|h(x) - h(y)\| \le l_T \|x - y\|, \tag{3-4}$$

where k_T and l_T are two Lipschits constants[7].

We wish to design a neuro-structured state estimator for the nonlinear dynamic system described by (3-1). The problem is to obtain an estimate of the real state vector $\mathbf{x}_k$ at the k th step, $\hat{\mathbf{x}}_{k/k}$, such that the following cost function

$$J_k = \frac{1}{2}\left\{\left\|\mathbf{y}_k - h(\hat{\mathbf{x}}_{k/k}) - H\hat{\mathbf{x}}_{k/k}\right\|_{Q_k}^2 + \left\|\hat{\mathbf{x}}_{k/k} - \hat{\mathbf{x}}_{k/k-1}\right\|_{R_k}^2\right\} \tag{3-5}$$

is minimized. $\hat{\mathbf{x}}_{k/k-1}$ is the single stage predictor, i.e.

$$\hat{\mathbf{x}}_{k/k-1} = \phi(\hat{\mathbf{x}}_{k-1/k-1}, \mathbf{u}_{k-1}) + \Phi\hat{\mathbf{x}}_{k-1/k-1}. \tag{3-6}$$

The nonlinear state estimation is indeed a nonlinear programming problem which may be solved by any gradient based methods.

4. The Neuro-Structured State Estimator

As mentioned before, the Hopfield network representing an steepest descent algorithm has the capability of solving some classes of optimization problems. Therefore, we utilize it to design the state estimator as stated in section 3. The structure of our proposed estimator is shown in Fig.1. The estimator is composed of a single stage predictor, a Hopfield network and a unit for minimization of the estimation error covariance. The output vector of the Hopfield network is the state estimation vector . The network might be trapped in a local minima of it's energy surface. The minimizing unit computes R_k given in (3-5) such that an upper bound for the estimation error covariance is minimized. The output of this unit adjusts the Hopfield networks' parameters. In the next subsections , we show the design procedures for each component of the proposed estimator for the system described by (3-1) and (3-2) .

4.1. The Single Stage Predictor

Single stage predictor is a nonlinear mapping as in (3-6) which represents the transition function of the state equation. In this problem, the predictor is implemented by a multilayer feedforward neural network. As it is well known, this network can approximate any continuous function with a desired accuracy.

4.2. Designing the Hopfield Network

In this part , we compute the proper values of the Hopfield net's parameters to minimize the estimation cost function in (3-6). Substituting J_k into (2-3) as the network energy function E and denoting $\hat{x}_{k/k}$ as the output of the Hopfield network, we find that

$$\hat{x}_{k/k} = \hat{x}_{k/k-1} - \gamma_k^* \nabla_{x_k} J_k \Big|_{x_k = \hat{x}_{k/k-1}}$$

$$= \hat{x}_{k/k-1} + \underbrace{[-\gamma_k^* \Xi_k^T Q_k H] \hat{x}_{k/k-1}}_{W_k} + \underbrace{[\gamma_k^* \Xi_k^T Q_k y_k - \gamma_k^* \Xi_k^T Q_k h(\hat{x}_{k/k-1})]}_{I_k}, \qquad (4\text{-}1)$$

where $\Xi_k = (H + \dfrac{\partial h}{\partial x_k}\Big|_{x_k = \hat{x}_{k/k-1}})$. Consequently, comparing (2-3) and (4-1), we may

have the synaptic weight matrix and the bias input of the network as

$$W_k = -\gamma_k^* \Xi_k^T Q_k H \qquad (4\text{-}2)$$

$$I_k = [\gamma_k^* \Xi_k^T Q_k y_k - \gamma_k^* \Xi_k^T Q_k h(\hat{x}_{k/k-1})], \qquad (4\text{-}3)$$

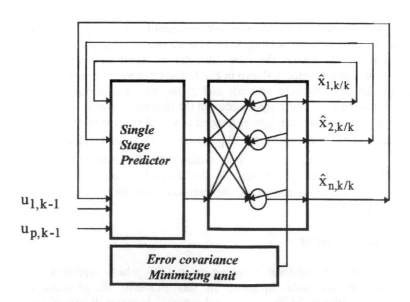

Fig.1. Structure of the Proposed Neuro-Estimator

where $\gamma_k^* = [R_k + \Xi_k^T Q_k \Xi_k]^{-1}$ is the convergence rate parameter which minimizes

$$J_k(\hat{x}_{k/k-1} - \gamma_k \nabla_{\hat{x}_{k/k}} J_k \Big|_{\hat{x}_{k/k-1}}).$$

Programming the hopfield network with the determined associated parameters, the estimation vector converges to an stationary point of the cost function.

4.3. Developement of a Recursive Upper Bound for the Estimation Error Covariance

Let us define the estimation error covariance P_k as

$$P_k = E\{\tilde{x}_k \tilde{x}_k'\} \qquad (4\text{-}4)$$

where $\tilde{x}_k = x_k - \hat{x}_{k/k}$ is the estimation error and $E(\cdot)$ is the expected value operator. Because of the nonlinear nature of the system equations, it is not possible to determine a precise recursive equation for the estimation error covariance. Therefore, we use some matrix inequality lemma's to compute a recursive upper bound equation for P_k. Now, substituting (4-1) into (4-4) and based on the derivation given in the appendix, we find that[8]

$$P_k \leq (I - \Lambda_k H) A_{k-1} (I - \Lambda_k H)' + \Lambda_k B_{k-1} \Lambda_k' + k_T^2 \, \mathrm{tr}(P_{k-1}) I + \Phi P_{k-1} \Phi', \quad (4\text{-}5)$$

where

$$A_{k-1} \overset{\Delta}{=} \left\{ 2 \left[k_T^2 \mathrm{tr}(P_{k-1}) I + \Phi P_{k-1} \Phi' \right] + \Gamma S_k \Gamma' \right\}, \qquad (4\text{-}6)$$

$$B_{k-1} \overset{\Delta}{=} 5 l_T^2 \mathrm{tr}(A_{k-1}) + H \left[k_T^2 \mathrm{tr}(P_{k-1}) I + \Phi P_{k-1} \Phi' \right] H' + T_k \qquad (4\text{-}7)$$

and $\Lambda_k = \gamma_k^* \Xi^T Q_k$.

So, the recursive upper bound equation can be written as

$$P_k^{upper} = (I - \Lambda_k H) A_{k-1}^{upper} (I - \Lambda_k H)' + \Lambda_k B_{k-1}^{upper} \Lambda_k'$$
$$+ k_T^2 \mathrm{tr}(P_{k-1}^{upper}) I + \Phi P_{k-1}^{upper} \Phi' \qquad (4\text{-}8)$$

with $P_0^{upper} = P_0$. It can be easily shown that $P_k^{upper} \geq P_k$, therefore P_k^{upper} is an upper bound for the estimation error covariance.

To obtain the optimal Λ_k, we set the first derivation of P_k^{upper} (with respect to Λ_k) to zero. This yields the optimal Λ_k :

$$\Lambda_k^0 = A_{k-1}^{\text{upper}} H' \left(H A_{k-1}^{\text{upper}} H' + B_{k-1}^{\text{upper}} \right)^{-1} \tag{4-9}$$

Finally, the optimal R_k, R_k^0, can be computed as a function of Λ_k^0, i.e.

$$R_k^0 = \Xi_k^T Q_k \Xi_k (\Lambda_k^0 \Xi_k)^{-1} (I - \Lambda_k^0 \Xi_k) \tag{4-10}$$

5. Simulation Results

In this section, we apply the proposed method to a practical example. Consider a missle which only moves vertically as shown in Fig.2. The state equations describing the model can be written as follows

$$r_{k+1} = r_k + T(v\sin(\theta_k))$$

$$\theta_{k+1} = \theta_k + T(\frac{v}{r}\cos(\theta_k))$$

$$v_{r_{k+1}} = v_{r_k} + T(\alpha\sin(\theta_k) + \frac{v^2}{r}\cos(\theta_k))$$

$$v_{\theta_{k+1}} = v_{\theta_k} + T(\frac{\alpha r - v^2 \sin(\theta_k)}{r^2}\cos(\theta_k) - \frac{v^2}{2r^2}\sin(2\theta_k))$$

$$\tag{5-1}$$

where r, θ, v_r and v_θ are the radial and angular distances and velocities, respectively. T is the sampling period and v and α are velocity and acceleration constants. In this model, only radial distance which is corrupted with a sequence of white noise can be measured . So, we ought to estimate the other state variables. For this purpose, we utilize our proposed neuro-estimator where sampling period is 0.1, k_T equals to 1.1 and the variance of measurement noise is 0.1. Simulation results are shown in Fig.3. These figures demonstrate the fast convergance and efficiency of the neuro-estimator.

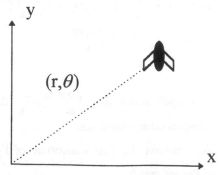

Fig. 2. A Missle Moving Vertically

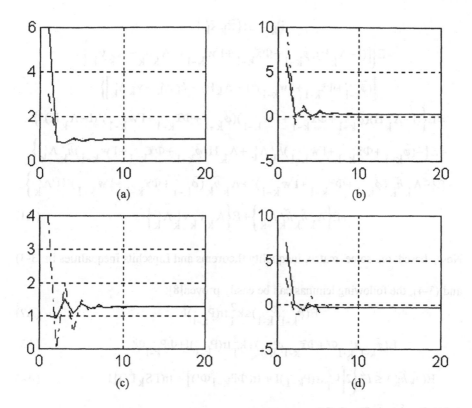

Fig.3. Simulation Results: (---) Real State Variable, (-.-.) Estimated State Variable

(a) r (b) θ (c) v_r (d) v_θ

6. Conclusion

A novel neuro-computing approach to estimation of state variables of a nonlinear dynamic system has been presented. This method is based on the Hopfiled neural network. Our proposed approach is suitable for real-time implementation. Simulation results show the effectiveness of the proposed method.

Appendix

In this appendix we intend to determine the recursive upper bound for the estimation error covariance as given in (4-8). Defining $\tilde{\phi}_{k-1} \overset{\Delta}{=} \phi(x_{k-1}, u_{k-1}) - \phi(\hat{x}_{k-1/k-1}, u_{k-1})$

and $\tilde{h}_k \overset{\Delta}{=} h(x_k) - h(\hat{x}_{k/k-1})$, P_k is computed as follows

$$P_k = E\{\widetilde{x}_k \widetilde{x}_k'\}$$

$$=E\left\{\left[(I-\Lambda_k H)(\widetilde{\phi}_{k-1}+\Phi\widetilde{x}_{k-1}+\Gamma w_{k-1})-\Lambda_k \widetilde{h}_k -\Lambda_k v_k\right]\right.$$

$$\left.\left[(\widetilde{\phi}_{k-1}+\Phi\widetilde{x}_{k-1}+\Gamma w_{k-1})'(I-\Lambda_k H)'-\widetilde{h}_k'\Lambda_k' - v_k'\Lambda_k'\right]\right\}$$

$$=E\left\{(I-\Lambda_k H)(\widetilde{\phi}_{k-1}+\Phi\widetilde{x}_{k-1}+\Gamma w_{k-1})(\widetilde{\phi}_{k-1}+\Phi\widetilde{x}_{k-1}+\Gamma w_{k-1})'(I-\Lambda_k H)'\right\}$$

$$+E\left\{-(\widetilde{\phi}_{k-1}+\Phi\widetilde{x}_{k-1}+\Gamma w_{k-1})\widetilde{h}_k'\Lambda_k' +\Lambda_k H(\widetilde{\phi}_{k-1}+\Phi\widetilde{x}_{k-1}+\Gamma w_{k-1})\widetilde{h}_k'\Lambda_k'\right\}$$

$$+E\left\{-\Lambda_k \widetilde{h}_k (\widetilde{\phi}_{k-1}+\Phi\widetilde{x}_{k-1}+\Gamma w_{k-1})' +\Lambda_k \widetilde{h}_k (\widetilde{\phi}_{k-1}+\Phi\widetilde{x}_{k-1}+\Gamma w_{k-1})'H'\Lambda_k'\right\}$$

$$+E\left\{\Lambda_k \widetilde{h}_k \widetilde{h}_k'\Lambda_k'\right\}+E\left\{\Lambda_k v_k v_k'\Lambda_k'\right\} \tag{A1}$$

Now, based on some matrix inequality theorems and Lipschits inequalities in (3-3) and (3-4), the following lemmas can be easily proven[8]:

$$E(\widetilde{\phi}_{k-1}\widetilde{\phi}_{k-1}')\leq k_T^2 \mathrm{tr}(P_{k-1})I \tag{A2}$$

$$E(\widetilde{\phi}_{k-1}\widetilde{x}_{k-1}'\Phi'+\Phi\widetilde{x}_{k-1}\widetilde{\phi}_{k-1}')\leq k_T^2 \mathrm{tr}(P_{k-1})I+\Phi P_{k-1}\Phi' \tag{A3}$$

$$E(\widetilde{h}_k \widetilde{h}_k') \leq l_T^2 \left\{2\left[k_T^2 \mathrm{tr}(P_{k-1})I + \mathrm{tr}(\Phi P_{k-1}\Phi')\right] + \mathrm{tr}(\Gamma S_k \Gamma')\right\}I \tag{A4}$$

$$E(H\Phi\widetilde{x}_{k-1}\widetilde{h}_k'+\widetilde{h}_k \widetilde{x}_{k-1}'\Phi'H')$$
$$\leq l_T^2 \left\{2\left[k_T^2 \mathrm{tr}(P_{k-1})I + \mathrm{tr}(\Phi P_{k-1}\Phi')\right] + \mathrm{tr}(\Gamma S_k \Gamma')\right\}I + H\Phi P_{k-1}\Phi'H' \tag{A5}$$

$$E(H\widetilde{\phi}_{k-1}\widetilde{h}_k'+\widetilde{h}_k \widetilde{\phi}_{k-1}'H')$$
$$\leq l_T^2 \left\{2\left[k_T^2 \mathrm{tr}(P_{k-1})I + \mathrm{tr}(\Phi P_{k-1}\Phi')\right] + \mathrm{tr}(\Gamma S_k \Gamma')\right\}I + HH'k_T^2 \mathrm{tr}(P_{k-1})I \tag{A6}$$

$$-E(\widetilde{\phi}_{k-1}\widetilde{h}_k'\Lambda_k' +\Lambda_k \widetilde{h}_k \widetilde{\phi}_{k-1}')$$
$$\leq \Lambda_k \Lambda_k' l_T^2 \left\{2\left[k_T^2 \mathrm{tr}(P_{k-1})I + \mathrm{tr}(\Phi P_{k-1}\Phi')\right] + \mathrm{tr}(\Gamma S_k \Gamma')\right\}I + k_T^2 \mathrm{tr}(P_{k-1})I \tag{A7}$$

$$-E(\Phi\widetilde{x}_{k-1}\widetilde{h}_k'\Lambda_k' +\Lambda_k \widetilde{h}_k \widetilde{x}_{k-1}'\Phi')$$
$$\leq \Lambda_k \Lambda_k' l_T^2 \left\{2\left[k_T^2 \mathrm{tr}(P_{k-1})I + \mathrm{tr}(\Phi P_{k-1}\Phi')\right] + \mathrm{tr}(\Gamma S_k \Gamma')\right\}I + \Phi P_{k-1}\Phi' \tag{A8}$$

Therefore, applying the above inequalities to (A1), the equation (4-5) can be obtained.

References

1. J.M. Mendel, Lessons In Digital Estimation Theory, Prentice Hall, 1987.

2. A.H. Jazwinski, Stochastic Processes and Filtering Theory, Academic Press, 1970.

3. J.J. Hopfield, "Neural Networks and Physical systems with Emergent Collective Computational Abilities", Proc. Natl. Acad. Sci, vol. 79, pp. 2554-2558, 1982.

4. J.J. Hopfield, "Neurons with Graded Response Have Collective Computational Properties Like Those of Two-State Neurons", Proc.Natl.Aad.Sci., vol.81 , pp. 3088-3092, 1984.

5. K.S. Narendra and K. Parthasarathy, "Identification of Systems and it's Application to Modelling and Control", IEEE Trans on Neural Networks, vol. 1, No. 1, 1990.

6. F.Rajaii Salmasi and M.B.Menhaj, " State Estimation Using Hopfield Neural Network", IEE & IPM Proc. of International Conference on Intelligent and Cognitive Systems, Tehran, Iran, September 23-26,1996.

7. M. Vidyasagar, Nonlinear Systems Analysis, Prentice Hall, 1978.

8. F.Rajaii Salmasi, M.Sc. thesis, State Estimation Using Neural Networks, Amir-Kabir University of Technology, Tehran, Iran, 1996.

A Toolset for the Design of Analogous Fuzzy–Hardware

S. Triebel, J. Kelber, G. Scarbata

Technical University Ilmenau
Faculty of Electrical Engineering and Information Technology
Department of Microelectronic Circuits and Systems

Abstract. In this paper, we present a module generator which is able to create fuzzy hardware for on–chip implementation. Based on the description of a desired behaviour, which may be generated by a commercial development tool and available in a standard format (FPL), our tool will be capable to generate all necessary data to describe a hardware cell, providing the specified behaviour. These data, of course, must include the netlist and the layout of the cell. In this presentation we consider analogous circuits only.

1 Fuzzy Hardware

1.1 Implementation Alternatives for Fuzzy Systems

Information processing based on several methods is becoming more and more important. One of the key problems is the specification and implementation of non–linear functionality. Here the fuzzy technology is a good alternative. It allows the designer to describe problems in terms of human–like language constructs and is able to approximate non–linear functions with only little restrictions. Furthermore, it supplies a methodology to transform such constructs into formal expressions to be implemented by well–known systems, e.g. standard software.

Today such fuzzy systems are mainly designed using one of the fuzzy development tools available on the market. Usually these tools are easy to handle and offer different specification and simulation capabilities. Using such tools the specified behaviour of a developed system is expressed in terms of well–defined language constructs.

Starting from a behavioural description different ways are possible to implement a fuzzy controller:

(a) software on standard computers,
(b) programmable fuzzy chips and
(c) dedicated fuzzy hardware.

The common way is to implement the fuzzy system on a standard computer using programming languages like "C". Furthermore, there are some special fuzzy chips available which can be adapted to the given problem by programming internal memory [Eich95] [Wat90]. The disadvantage of solution (a) is that it runs slowly in comparison to (b) and (c). Furthermore, dedicated timing constraints are very difficult to be taken into account. Solution (b) normally comes with a relatively large overhead because such chips are designed to meet most of the possible requirements.

1.2 Applications for Fuzzy Hardware

In some cases it can be very attractive to have a dedicated hardware for a fuzzy system designed for a special application. A solution like this becomes interesting if the designer is faced with:

1. hard timing restrictions: optimization possibilities for systems with fixed hardware are very limited
2. mobile applications: decrease of power consumption due to the absence of programming overhead
3. high production volume: decrease of chip area

A very important field is the use of special hardware in embedded systems. The electronic parts of such systems normally consist of both analog and digital components including processor cores and interfaces to sensors and actors. Here the fuzzy approach can be used to model non–linearities required for the overall function. Beside the implementation in the processor core the non–linearities can also reside inside the interfaces of analog or mixed analog–digital nature. Caring about special environmental properties including timing restrictions leads to special hardware solutions for these interfaces in many cases. So the inclusion of fuzzy components in the interfaces is much more attractive if dedicated hardware solutions are available. This hardware should be based on analog techniques since the interfaces themselves are analog. The design process of an embedded system including such fuzzy hardware can be described in a very simple manner as shown in Fig. 1.

The disadvantage, of course, is the dedicated design and production of the hardware which leads to relatively high design costs. To minimize them, the generation of the hardware–specific design data should be performed automatically. Therefore the concept of a module generator, known from other applications (e.g. RAM, PLA), is an interesting solution. Based on a specified behaviour it can generate all necessary design data.

2 A Design Environment for Analog Fuzzy Hardware

2.1 General Design Flow

The principles of a module generator based design process adapted to fuzzy hardware are illustrated in Fig. 2.

The generation result is a cell in a given chip technology. Among other things, it must contain all necessary information about the physical layout, the internal

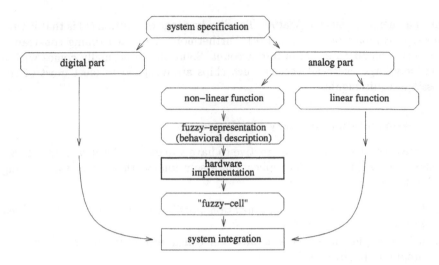

Fig. 1. inclusion of fuzzy hardware in the design of embedded systems

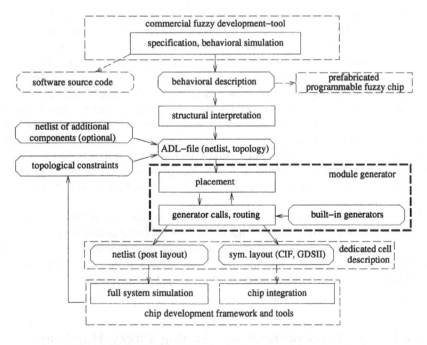

Fig. 2. design of fuzzy hardware by a module generator

structure, and the simulation models of that cell. This set of data can be used as an input to common chip CAD systems for further processing (e.g. design rules check, post layout simulation) or for incorporation into more complex chips. In this paper we want to focus on a module generator for analogous fuzzy hardware.

2.2 The Link Between Fuzzy Specifications and Chip Layout Generation

The first step in the design flow is a structural interpretation of the specified behaviour. This is done by a simple assignment of a component to a particular operation. For example a dedicated multi-input MAX-operator is provided for each required MAX-operation. Instead of this procedure a real synthesis approach will be discussed in the future but is not subject of this paper. The result of the structural interpretation is a netlist of fuzzy components denoted in an ADL-file[1] (see Fig. 3). Each of the established components must have a

```
include "defuzzy.adl"
include "rulebase002.adl"
include "lingvar.adl"

controller()
  {
  terminal EK     : RIGHT, INPUT;
  terminal DEK    : RIGHT, INPUT;
  terminal UK     : RIGHT, OUTPUT;
  terminal U1     : RIGHT, INPUT;
  terminal U5     : RIGHT, INPUT;

  abutment over;
    call COG        UK  (0 0.17 0.5 0.83 1);
    call RuleBase002 RB ();
    call LingVar    DEK (0 0.375 0.375 0.44
                     0.56 0.625 0.625 1 prec=1);
    call LingVar    EK  (0 0.25 0.25 0.5 0.5
                     0.75 0.75 1 prec=1);

  orientation RB MY;

  i = 1;
  while (i<=5) {
    connect RB.uk[i] UK.i[i];
    connect RB.ek[i] EK.out[i];
    connect RB.dek[i] DEK.out[i];
    i = i + 1;
  }

  connect EK.in EK;
  connect DEK.in DEK;
  connect UK.v1 U1;
  connect UK.v5 U5;
  connect UK.out UK;

  connect *.vdd vdd LEFT;
  connect *.gnd gnd LEFT;
  netattrib width gnd 3*size("Metal2");
  netattrib width vdd 3*size("Metal2");
  }
```

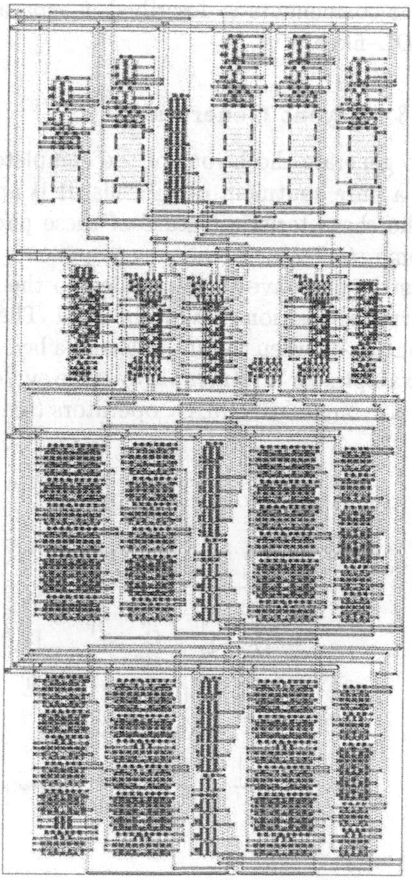

Fig. 3. example of an ADL-file and the resulting layout

---

[1] *Abutment-oriented Layout Description Language*

representation in the generator library of the development system.

The ADL-format is a special language to describe the internal structure of functional units. It has been developed at our institute and is capable to include component instantiation, connectivity description and geometrical (topological) information, for example. More complex functional units including simple fuzzy systems with membership functions, rule bases and defuzzyfication may be specified. The language is similar to "C" and contains only a few dedicated constructs. So it is very easy to learn and to understand.

In Fig. 3 an example of an ADL–file together with the resulting layout is shown. The whole cell consists of four parts which are stacked vertically. This is described by the "abutment over"–expression. At the bottom there are two linguistic variables with five trapezium fuzzy sets each. In the upper part, the rule base and the defuzzification circuit is located. The internal structure of the four components is described in separate files which are included in the overall ADL–file of Fig. 3.

2.3 Layout Generation

To generate the layout of the complete cell three strategies are used. The first is a pure abutment of subcells. It is applied to the regular parts of the circuit. The characteristic feature of these parts is the possibility to divide them into some similar or identical components realized by subcells. In this case, some restrictions have to be applied to the layout of the subcells. For example, the terminal positions have to be fixed. The advantages are the absence of an explicit routing between the subcells and a layout with nearly optimal area consumption. Examples for abutment layout are current mirrors for membership functions and multi–input MIN/MAX operators (see Fig. 4).

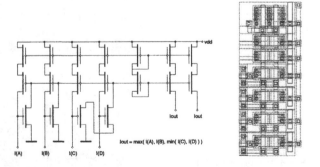

Fig. 4. schematic and layout of a multi–input MIN/MAX-operator

The second layout strategy is a slicing layout, known from previous work at different institutions. This strategy is very suitable for analog layouts. It allows several components with different properties, shapes and sizes to be considered. Nevertheless matching and other constraints can be taken into account. In this strategy routing channels are introduced between the subcells. Third total irregular layouts can be included but must be designed by hand using a symbolic layout description.

The placement optimization is executed by simulated annealing. The designer can define topological constraints explicitly. This is strongly recommended because the cost function of the annealing algorithm does actually not take all necessary information into account.

The focus of our work was on the development of the generator library and the analog routing algorithms. The generator library consists of two parts: built-in generators and ADL-macros. The built-in generators are written in C++ using an own library called "WOOD". The following are actually available: transistors, differential transistor pairs, cascaded current mirror input and output, resistors, capacitors, MAX-, MIN- and MIN/MAX-gates (Fig. 4). The generators offer a couple of useful properties. So a fast algorithm for the calculation of the sizing parameters was implemented. Furthermore, the generators distinguish virtual from real interface terminals. For floorplanning purposes all possible terminal positions are available. The "higher–level" tools decide, which terminals are really to be generated. So the optimization space is not restricted additionally and the existence of unused wires will be avoided.

The implemented channel router is based on the results in [Gyu89]. This approach provides the possibility of different wire width and distances. The channel boundaries can be irregular. The algorithms were improved to handle terminals at different layers and to allow the description of "wire–clusters" for example pairs of wires with symmetrical geometry (see Fig. 5).

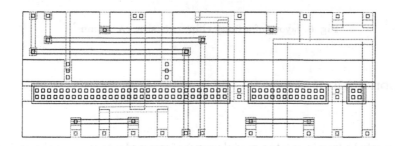

Fig. 5. example of a channel

For example, the generated layout of a fuzzy controller is shown in Fig. 6. It was implemented in a 2.4μm technology.

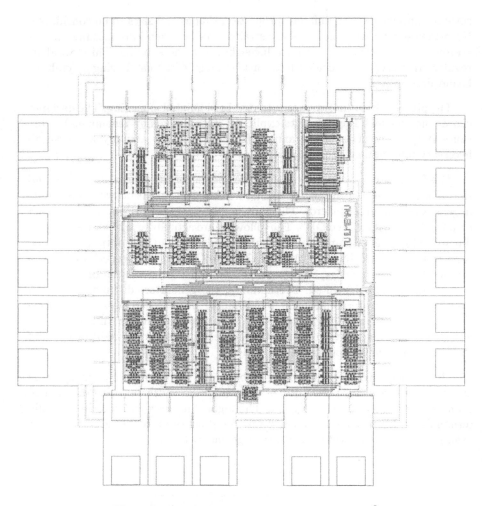

Fig. 6. layout of a fuzzy chip, core size: ca. 1.5mm^2

3 Conclusions

We presented a way to create a module generator in order to support the design of analogous hardware for fuzzy controllers. Such a generator seems to make a good link between the world of fuzzy development tools on one side and chip design systems on the other side. Of course there are many problems left to be solved. Examples are the implementation of real synthesis algorithms to replace the simple structural interpretation and the improvement of the included placement solutions.

Acknowledgement

The authors would like to thank the "Deutsche Forschungsgemeinschaft" supporting this work in the "Sonderforschungsbereich 358".

References

[Gyu89] R. S. Gyurcsik, J.-C. Jeen
A generalized approach to routing mixed analog and digital signal nets in a channel
IEEE J. Solid-State Circuits 24 (2): 436–442, April 1989

[Eich95] H. Eichfeld, T. Künemund, M. Menke
Architecture of a General–Purpose 12 Bit Fuzzy Coprocessor
Proc. 3. EUFIT, Aachen, Sept. 1995, pp. 1815–1819

[Kel94] J. Kelber, S. Triebel, G. Scarbata
Automatic Generation of Analogous Fuzzy Controller Hardware Using a Module Generator Concept
Proc. 2. EUFIT, Aachen, Sept. 1994, pp. 1562–1569

[Wat90] H. Watanabe, W. Dettloff, K. Yount
A VLSI fuzzy logic controller with reconfigurable, cascadable architecture
IEEE journal of solid-state circuits, Vol. 25, April 1990, 2, pp. 376

Genetic Algorithms with Optimal Structure Applied to Evolvable HardWare (EHW)

Dan Mihaila

Lab.: Microsystems Architectures based on Computational Intelligence
National Research & Development Institute in Microtechnology
PO Box 38-160, 72225, Bucharest, ROMANIA
E-Mail: danm@oblio.imt.pub.ro, danm@org.meganet.ro

Abstract.
Evolvable HardWare (EHW) is a concept that describes those hardware circuits that adapt themselves as a response to environment changes or hardware faults. Such kind of circuit is a FPGA (Field Programmable Gate Array) or a PLD (Programmable Logic Device) whose program is being evolved by a GA (Genetic Algorithm)
Actually there are two types of EHW's: with *on-line evolution* and with *off-line evolution*. The on-line evolution type can be used to evolve hardware circuits that act as a fault tolerant backup circuits [1][4] or as digital controllers for autonomous robots [5]. The off-line evolution type can act as an evolutionary approach for VLSI (Very Large Scale Integration) circuits design or non-standard FPGA's or PLD's programming technique [3]. In this case the evolved circuit can not react to the environmental changes or hardware faults that occur after the design (programming) phase. In this paper there are described the best operators in terms of GA convergence for the off-line PLD programming.

1. Introduction.

In this paper I studied the optimal structure of a possible hardware implementation of a GA that can evolve the configuration for a PLD circuit. For the PLD I choose a 16V8 GAL (Generic Array Logic) that has eight macro-cells. Although the macro-cells can be interconnected one to another, I ignored this capability of the circuit and presumed that the cells are not. Each macro-cell can implement logic functions that have maximum eight product terms of sixteen variables so the number of functions that can be implemented using only one cell has this limitation. The most unfortunate case is implementation of arithmetic functions that requires a big number of product terms. The GA's, that are studied, are applied to the evolution of a XOR circuit with four inputs. This circuit can be used to compute the parity of any word of four bits. To implement this digital function 8 product terms are required.
In first chapter will be presented the genotype and the basic structure of all algorithms that were studied. The next steps will attempt to study the influence of various operators, like selection, crossover and mutation, on the convergence of hardware GA's devoted to evolve digital circuits. An adaptive mutation operator is proposed in order to speed-up the convergence of the GA.
Conclusions deal with optimal structure of a possible hardware GA implementation.

2. The genotype and the basic structure of the GA's

The simplified structure of 16V8 GAL macro-cell is presented in figure 1.

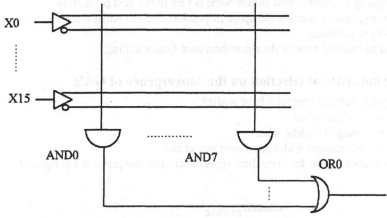

Fig. 1. The simplified structure of a 16V8 GAL macro-cell

The 16V8 GAL circuit has 8 such macro-cells that can be interconnected, the possibility of programming the input and the output pins, and the possibility of realizing sequential circuits using registers. In this paper I ignored the full facilities of the circuit and I simulate only one circuit macro-cell on witch I tried to evolve a simple digital function of four inputs.

For this macro-cell there are 8*32=256 programmable connections. The state of each connection was represented by a bit with the following signification:

  0- unconnected
  1- connected

The gene represents the state of a connection.
Each chromosome is represented by a bit string of length 256.
The GA's that were tested have the following basic structure:

generation = 0
Initialize P (generation)
Evaluate P (generation)
While (not (stop condition))
        begin
        selection of the parents
        apply crossover to the parents in order to generate offspring's
        apply mutation to the offspring's
        evaluate offspring
        Insert offspring's in P (generation)
        generation=generation+1
        end

In order to maintain the GA complexity at a reasonable level the replacement strategy was kept as simple as possible. The population is completely and unconditional replace by its offspring's. Only the best chromosome is kept in the next population.

This two requirement restricts the space of possible GA's to elitist algorithms with non-overlapping populations.

For the same reasons' none of the algorithms used fitness scaling.

3. The influence of selection on the convergence of GA's

Three types of selection operator have studied:
- roulette wheel
- integral roulette wheel
- tournament with tournament size of two

The convergence graphs for these three types of selection are presented in figure 2.

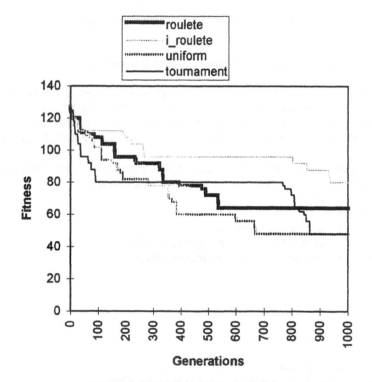

Fig. 2. The influence of selection operator on the convergence

In case of GA's that do not use any form of fitness scaling the best selection method was tournament selection.

4. The influence of crossover on the convergence of GA's

Three types of crossover operators were studied:
- one point crossover
- two point crossover
- uniform crossover

All crossover operators are applied with probability of crossover of one.

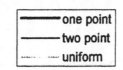

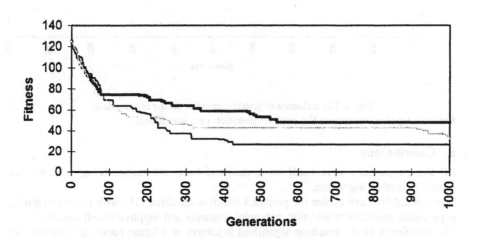

Fig. 3. The influence of crossover operator on the convergence

The best results were obtained for two point crossover.

5. The influence of mutation on the convergence of GA's

Two types of mutation were tested:
- bit invert mutation
- adaptive re-init mutation (a new mutation operator was proposed)

For the adaptive re-init mutation a modified structure of the algorithm is needed. First of all a new function must be appealed at the end of each generation. This function will maintain a statistic table of the chromosomes that were generated. The mutation operator that I proposed uses this statistic table for generating new chromosomes mainly in zones that are poorly explored. This technique is highly effective in cases of GA's that works with small populations.

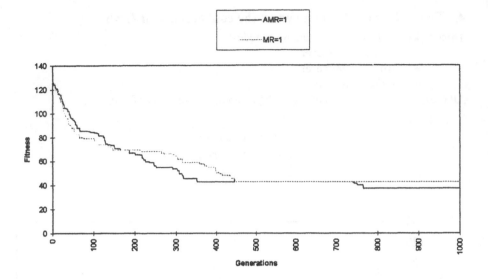

Fig. 4. The influence of mutation operator on the convergence
For both types of mutations the optimal mutation rate was found.

6. Conclusions

All the comparisons were made on a statistical base, the results being averaged on
several runs of an algorithm.
The best GA that can evolve the proposed circuit is an elitist GA, with non-overlapping
populations, tournament selection, two point crossover and adaptive re-init mutation.
The complexity of the resulting algorithms is subject of a future paper, so I cannot say
which of the operators studied are optimal in terms of both the performance and the
complexity.

7. References

1. Higuchi, M. Iwata, I. Kajitani "Evolvable Hardware and Its Application to Pattern
   Recognition and Fault-Tolerant Systems", "Towards Evolvable Hardware", (Sanchez
   E., Tomassini M. eds.), Springer, 1995, pp.119-135
2. Kitano "Morphogenesis for Evolvable Hardware", "Towards Evolvable Hardware",
   (Sanchez E., Tomassini M. eds.), Springer, 1995, pp.99-117
3. Sanchez "Field Programmable Gate Array Circuits", "Towards Evolvable Hardware",
   (Sanchez E., Tomassini M. eds.), Springer, 1995, pp.1-18
4. Thompson, "Evolving Fault Tolerant Systems", Proc. 1st IEE/IEEE IC on GA's in
   Eng. Sys. (GALESIA'95), IEE Conf. Pub 414, 1995, pp. 524-529
5. Thompson, "Evolving Electronic Robot Controllers that Exploit Hardware
   Resources", Advances in Artificial Life: Proc 3rd ECAL (ECAL95), ed. Moran F., et
   al, Springer Verlag Lecture Notes in AI (929), 1995, pp. 640-656

Position Control System with Fuzzy Microprocessor AL220

Bohdan Butkiewicz

Warsaw University of Technology, 00-665 Warsaw, Poland

Abstract. Processor of new generation, so called fuzzy microprocessor, is used to control the position of header with infrarerd LED's. In its operations the processor use fuzzy logic rules and approximate reasoning. Hole system can be consider as intelligent infrared sensor, or system which track the infrared source and turn the header in the source direction. If it is necessary, intensity of infrared lights can be measured. In the paper design method of the system is described.

1 Introduction

New generation of microprocessors has been developed in last few years. Its use fuzzy logic principles in its operations [1] . The first one FP1000 was introduced by NEC OMRON. In this time there are some others OMRON microprocessors as FP6000 and FP9000 (first fuzzy analog processor). Also other manufactures entered in fuzzy logic market. We have microprocessor FC110 produced by Togai Infralogic Inc., NSM91U112 by Oki Semiconductor, NLX230 and AL220 by Adaptive Logic (formerly American NeuraLogic Inc.), 81C99 by Siemens Corp., Fuzzy-166 by Inform GmbH, WARP by SGS Thomson. Some of them can work as coprocessors interfaced to most microcontrollers. Stand alone configuration is sometimes also available (example for AL220). All fuzzy processor can store membership functions and rules in its or external memory, provide operations of fuzzyfication, rules evaluation, approximate reasoning using minimum and maximum (sometimes more complicated as algebraic product), and defuzzification procedure. Some manufactures, for example MOTOROLA, developed software tools, which can be used for classic microprocessors to provide fuzzy logic operations. This solution is however 10 to 100 times slower in operation. Japan first introduced fuzzy logic to industry. It simplify user interface and can reduce the time required to produce a product.

2 Positioning System or Intelligent Sensor

In this paper we describe a position control system with fuzzy microcontroller AL220. It is simple one and very cheap. It has four 8-bit digital or analog inputs four 8-bit digital or analog outputs, 10MHz clock, support 40 control rules, during approximate reasoning it use minimum and maximum. Each output depends on one winning rule with maximal weight. Block diagram of the controller is presented in the Fig. 1 . The core of processor consists of four units: fuzzifier, rule and parameter memory, fuzzy reasoning block, and defuzzifier. The fuzzifier compares input data (samples for analog inputs) with membership functions, built before during design process and saved in memory unit, to calculate the degree of membership for each input in appropriate fuzzy sets. Fuzzy reasoning unit calculates weights of control rules and finds winning rule with maximum membership value. Conclusion set of any rule can be only singleton. The defuzzifier sends the singleton value of winning rule to output. More detailed description of microcontroller features can be find in [2] , [3] . The positioning system con-

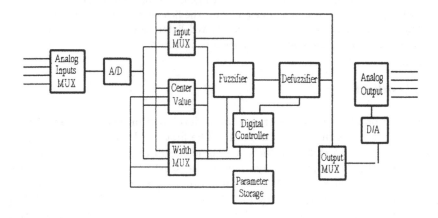

Fig. 1. AL220 block diagram.

sists of a header and fuzzy logic controller (FLC) Fig. 2 . Main purpose of the header is tracking a source of infrared radiation. Thus, the header has possibility to rotate in both direction around vertical axis. The header is build in conventional way. Two photodiodes are used as sensors of infrared radiation. Both signals are applied to differential amplifier. Other operational amplifiers supply driving current for direct current motor. If tracking in horizontal direction is also necessary, two similar systems can be applied. Thus, differential signal is used to drive the header in the source direction. Common signal, i.e. sum of diode signals, can be used to measure intensity of source radiation, so system can be considered as intelligent sensor and intensity meter of infrared radiation. Good property of common signal is its relative independence on angular error of positioning. Second part of the system is fuzzy logic control device with microcontroller AL220. Diagram of closed loop system is presented in the Fig. 2 . Programming the microcontroller is not so easy problem. Moreover, good tuning of membership functions shapes and parameters, and also control rules, is necessary. Thus, we first use simulation program, and after emulation program, before introducing the data in microcontroller memory.

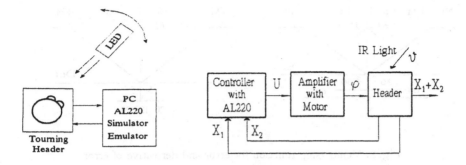

Fig. 2. Header positioning and appropriate closed loop system.

3 Membership Function and Control Rules

The most popular shape of membership functions in first applications had triangular shapes. This shape is not useful in our case. The AL Inc. introduced so called floating membership functions in his microcontroller. It is a very good conception. The center value for the membership function of a fuzzy set can be not fixed. The centers can move along the value of other input or output. We can realize a difference between inputs in easy hardware way. We have in our system two inputs X_1, X_2 - signals from diodes. Angular error is proportional to the difference

$$Err = X_1 - X_2 \qquad (1)$$

We can consider X_1 as input of closed loop system and X_2 as loopback signal *Out* from output. If $X_1 = X_2$ then we have steady-state, so the system has classic closed loop form. We use for linguistic variable *Err* floating membership functions to compare X_1 with X_2. Membership functions are presented in the Fig. 3 . For example, fuzzy set PM (Positive Medium) is declared as: center - *Floating X_2*, type - *Left Exclusive*, width - *Fixed* 30. Fuzzy set PS (Positive Small) is declared as: center - *Floating X_2*, type - *Left Exclusive*, width - *Fixed* 0. Thus, if $Err = X_1 - X_2 = 40$ then membership values $\mu_{PM} = \mu_{PS} = 40$ and zero for other linguistic values. Third, internal input variable is derivative *Der*

$$Der = \frac{dErr}{dt} \qquad (2)$$

It represent derivative of position error of the control system. Here we use classic membership functions with fixed center. Each digital input value of the AL220 must be positive in the range 0-255, so we chose 120 as center value which represent zero of *Der*. For membership values we have only 5 bits, so the range 0-31.

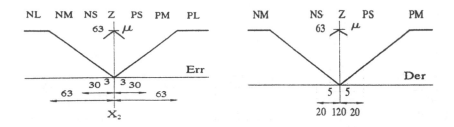

Fig. 3. Membership function for error and derivative of error.

4   Simulation and Emulation

Next step of design is a choice of control rules. We start with rules of Mamdani type [4]. Conclusion of each rule concerns the change of control action. For example we have rule

if Err is PL and Der is NS then ΔU is -15

where ΔU is a change of U. It is so called FLC of PI type. It has property that position error will tend to zero in steady-state. But we have motor and angular position as output signal, so we have integration in the system. Thus, the controller can be PD type. Hence, we chose *Immediate* type of action for first rule and the form

if Err is Anything then U is U_{ref}

where U_{ref} is initial reference (offset) value. It must be equal to the reference voltage in the system. Typical value is 2.5V, ie. 128, but we have chose 120 for Der. The set *Anything* has membership always equal to 63, ie. maximal membership. The value of Der can be calculate internally in AL220. Additional (internal) output $LastErr$ can be introduced to store old Err value, and $Der = Err - LastErr$, so Der can have membership function floating with $LastErr$. In this case additional rule, with *Immediate* action, must be introduced

if Err is Anything then $LastErr = Err$

In general fuzzy theory order of rules has no influence on result of reasoning and control action. It is not the case for our microcontroller. We can have unexpected effects during simulation. The control action is produced for winning rule, but we have situation where two or more rules have same membership value. In this case win first rule in the order of rules referencing to one output. Thus, in our case rules with terms PL, NL must precede rules with PM, NM, and rules with PM, NM must precede rules with PS, NS. Moreover, the first rule must be separated from other rules referencing to U. Thus, no action rule referencing to other output $LastErr$ we introduce as second

if Err is Anything then $LastErr = LastErr + 0$

Third and next rules are presented in the Fig. 4 . All the rules are *Accumulate* type, ie. output U is modified by control action ΔU of winning rule, $U = U_{ref} + \Delta U$. Only one of this rules wins, because it is an ordering set of rules referencing to same output U. As last rule we introduce the rule concerning *LastErr* value, mentioned before, to calculate the derivative. Now we can

	Err						
	NL	NM	NS	ZE	PS	PM	PL
NM	40	30	10	3	0	-5	-15
NS	30	20	5	1	-1	-10	-15
Der ZE	20	15	3	0	-3	-15	-20
PS	15	10	1	-1	-5	-20	-30
PM	15	5	0	-3	-10	-30	-40

	Err						
	NL	NM	NS	ZE	PS	PM	PL
NM	40	30	10	5	3	0	-5
NS	30	20	5	1	-1	-8	-15
Der ZE	20	15	3	0	-3	-15	-20
PS	15	8	1	-1	-5	-20	-30
PM	5	0	-3	-5	-10	-30	-40

Fig. 4. The control rules: before (left) simulation, and after simulation (right).

begin tuning all system. A good practice is first simulate the system. For system simulation and emulation of the AL220 we use program INSIGHT IIe [2], prepared for this purpose by manufacturer. The program allows introducing simple equations describing system behavior, and input signals. Using equation editor we can simulate step function, sinus and some other functions. Our plant can be considered approximately as first order system with integrator. Time response $\vartheta(t)$ of the plant we calculate as

$$\frac{d\vartheta}{dt} = (U - U_{ref})(1 - e^{-\frac{t}{\tau}}) \tag{3}$$

where τ is time constant. Time constant of the header is about 0.3s. With 1MHz clock of AL220 we have time step 1.024ms. We chose time scale 1:20, so for simulation we have $\tau = 15$. System simulation is provided with set of equations

$$X_1 = X_{10} + theta; \tag{4}$$
$$X_2 = X_{20} - theta; \tag{5}$$
$$Err = X_1 + U_{ref}; \tag{6}$$
$$Der = Err; \tag{7}$$
$$theta = theta + (U - U_{ref})/tau; \tag{8}$$
$$frame = frame + 1; \tag{9}$$

with constants: initial values for inputs $X_{10} = 30$, $X_{20} = 120$, reference $U_{ref} = 120$ and $tau = 15$. Variable $frame$ represent step time t.

After simulation we have change a little the rules. Modified rules are shown in the Fig. 4 . Tuning the system we find optimal values of parameters. We not use any sophisticated method of optimization, because the system is nonlinear and equations for simulation give only approximation. The rise time and overshoot are observed, and correctness of rules.

Next step of design is connection of real header to the Development System INSIGHT IIe for emulation of the AL220. Configuration of the system is shown in the Fig. 5 . It is very important step, because of system nonlinearity. The motor, we use direct current motor, has friction and zone of insensitivity. We can tune once more membership functions of fuzzy sets and control rules using our plant. Very small changes of output U near U_{ref} (current is too small to start the motor) have no influence on ϑ. Thus, the values of ΔU for sets PS, NS are increased (see Fig. 5). Digital or analog inputs and outputs can be used during

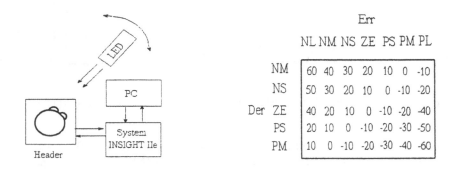

		NL	NM	NS	ZE	PS	PM	PL
	NM	60	40	30	20	10	0	-10
	NS	50	30	20	10	0	-10	-20
Der	ZE	40	20	10	0	-10	-20	-40
	PS	20	10	0	-10	-20	-30	-50
	PM	10	0	-10	-20	-30	-40	-60

Fig. 5. System configuration for emulation (left), and rules after emulation (right).

emulation. We use analog, with sampling period $T = 0.01\ s$ and 1.024 MHz clock. The microprocessor emulation program has special setup for emulation, and the parameters can be changed in easy way. The system step angular response, and output control signals are presented in the Fig. 6 . Also, the controller of PID type can be implemented on the AL220. The design steps are similar, but the rules more complicated. The microcontroller allows only 40 rules. Sometimes it is not sufficient, specially for PID. In this case, output is modified sequentially. For example, first rules calculate output U depending on integral Int, next depending on error Err, and finally depending on derivative Der. It is of course not exactly as for fuzzy PID, but it is possible to do something like classic PID. If any linguistic value is described by 5 fuzzy sets, we have 5x5x5=125 rules for full fuzzy PID, and only 5+5+5=15 (plus some separation and additional) rules in classic case. We have tested this possibility during simulation. There are some troubles at the beginning to attain stability of the system.

Finally, after emulation we use the INSIGHT IIe compiler and programmer to program the fuzzy logic controller AL220. We not present the results obtained in this case, because we have some problems with exact measuring of angular movement of the header, and transferring the results to computer. It was not possible, so we present in the Fig. 7 curves obtained after final tuning with

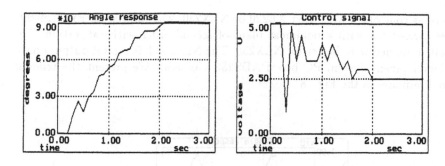

Fig. 6. Angular step response (left), and control signal (right) with emulated AL220.

emulated AL220. For classic PD similar angular movement curve is smoother, but response is slower. Exact comparison was difficult because of measuring problem.

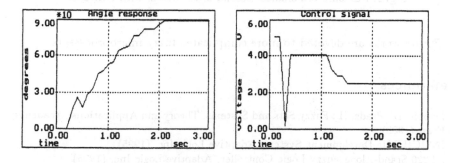

Fig. 7. Final step response (left), and control signal (right) after tuning.

5   Conclusion Remarks

Using the AL220 we had difficulties with lack of information about details of microcontroller operations. Moreover, in emulation program there are some mistakes. After second call of the program neither positions of curves presented in the *LineGraph*, nor colors of the curves were not true, and changes seemed to be random. We have some experience with other fuzzy microcontroller NLX230. In this case we had not such problems, and we arrived to good solution easier. The NLX230 allows 64 rules, and control can be more precise. The results of control, where as controllers were used AL220 or NLX230, was compared. Experiments were carried out with a nonlinear plant of second order with saturation. The results obtained were better for NLX230. The NLX230 has digital outputs only, so Development System PC board ADS230 was used. We present the triangle signal response in the Fig. 8 .

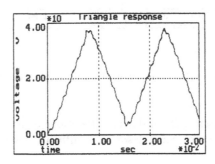

Fig. 8. Response of nonlinear second order system with NLX230.

Future works are devoted to more complicated fuzzy microprocessors.

References

1. Dubois, D., Prade, H.: Fuzzy Sets and Systems, Theory and Applications. Academic Press (1980)
2. INSIGHT IIe Development System. Adaptive Logic Inc. (1995)
3. AL220 Stand-Alone Fuzzy Logic Controller. Adaptive Logic Inc. (1995)
4. Yager, R.R., Filev, P.: Essentials at Modeling and Control. John Wiley & Sons Inc. (1994)

Fuzzy-Neural Computing Systems:
Recent Developments and Future Directions

Madan M. Gupta
Intelligent Systems Research Laboratory
College of Engineering
University of Saskatchewan
Saskatoon, Sask.
Canada, S7N 5A9
Phone: (306) 966-5451, FAX: (306) 966-5427
Email: GUPTAM@SASK.USASK.CA.

ABSTRACT

Recently, several significant advances have been made in two distinct theoretical areas. These theoretical advances have created an innovative field of theoretical and applied interest: fuzzy neural systems. Researchers have provided a theoretical basis in the field while industry has used this theoretical basis to create a new class of machines using the innovative technology of fuzzy neural networks. The theory of fuzzy logic provides a mathematical framework for capturing the uncertainties associated with human cognitive processes, such as thinking and reasoning. It also provides a mathematical morphology for emulating certain perceptual and linguistic attributes associated with human cognition. On the other hand, computational neural network paradigms have evolved in the process of understanding the incredible learning and adaptive features of neuronal mechanisms inherent in certain biological species. The integration of these two fields, fuzzy logic and neural networks, has the potential for combining the benefits of these two fascinating fields into a single capsule. The intent of this paper is to describe the basic notions of biological and computational neuronal morphologies, and to describe the principles and architectures of fuzzy neural networks.

KEYWORDS:

Neural Systems, Fuzzy Systems, Fuzzy Logic, Neural Fuzzy Computing.

1. INTRODUCTION

1.1 Motivation

The incredible flexibility and adaptability of biological neuronal control mechanisms may be used as a plausible source of motivation and a framework for the design of intelligent and autonomous robots. Unlike most conventional control techniques, biological control mechanisms are non-model based, and such non-model based mechanisms are quite successful in dealing with uncertainty, complexity, imprecision and approximate data. For example, we can reach a destination with vague and approximate information: from here how to *get to my office:*

"Go straight for about 30 meters, and then take a left turn and go straight for about 25 meters, turn right, and then in about 30 seconds you will find my office ."

With this fuzzy information, our carbon-based computer, the brain, will generate motor commands and smoothly coordinate many degrees of freedom during the execution of manipulative tasks in an unstructured environment. Biological control mechanisms are usually very complex and do not depend upon exact mathematical formulation for their operations. They carry out complex tasks without having to develop their mathematical models or that of the environment and without solving in an explicit form any integral, differential or complex mathematical equations.

Conversely, to make a mobile robot perform the same task, using vague and imprecise information, is an extremely complex task for it involves the fusion of most of the existing control methodologies such as adaptive control, knowledge engineering, fuzzy logic and computational neural networks. The computations required to coordinate different robot joints to produce a desired trajectory may be obtained by solving complex trigonometric relationships between different structural members of the robot. The control methodology developed in this traditional way may completely fail should the desired task or the environment change.

It is our hypothesis that if the fundamental principles of neural computations used by biological control systems are understood, it seems most likely that an entirely new generation of control methodologies can be developed which are more robust and intelligent, and far beyond the capabilities of the present techniques based upon explicit mathematical modeling.

In this process of understanding biological computational power, and the desire of system scientists to capture this power, the two most powerful fields in modern technology, *fuzzy logic* and *neural networks*, along with *genetic algorithms*, have emerged [1-6]. During the past decade, these two fields have grown, however, independently to form distinct branches. More recently, the integration of these two fields has presented system designers with another powerful computational tool called fuzzy neural networks.

1.2 Integration of Fuzzy Logic and Neural Networks

Fuzzy logic provides an inference morphology that enables approximate human reasoning capabilities to be applied to knowledge-based systems [2, 3]. The theory of fuzzy logic provides a mathematical strength to capture the uncertainties associated with human cognitive processes, such as thinking and reasoning. It also, provides a mathematical morphology to emulate certain perceptual and linguistic attributes associated with human cognition.

While fuzzy theory provides an inference mechanism under cognitive uncertainty, computational neural networks offer advantages such as learning, adaptation, fault tolerance, parallelism and generalization. Computational neural networks, comprised of processing elements called neurons, are capable of coping with computational complexity, nonlinearity and uncertainty. In view of this versatility of neural networks, it is believed that they hold great potential as building blocks for a variety of behaviors associated with human cognition.

To enable a system to deal with cognitive uncertainties in a manner more like humans, one may incorporate the concept of fuzzy logic into neural networks. Although fuzzy logic is a natural mechanism for modeling cognitive uncertainty, it may involve an increase in the amount of computation required (compared with a system using classical binary logic). This can be readily offset by using fuzzy neural network approaches having the potential for parallel computations with high flexibility.

A fuzzy neuron is designed to function in much the same way as a non-fuzzy neuron, except that it reflects the fuzzy nature of a neuron and has the ability to cope with fuzzy information. Inputs to the fuzzy neuron are fuzzy sets $(x_1, x_2, ..., x_n)$ in the universe of discourse $(X_1,$

X_2, ..., x_n) respectively. These fuzzy sets may be labeled by such linguistic terms as *high, large, warm, medium*, etc. The fuzzy inputs are then 'weighted' in synapses in a much different way from that used in a non-fuzzy case. The weighted fuzzy inputs are then aggregated not by the summation, but by the fuzzy aggregation operations (fuzzy union, weighted mean, or intersection).

An important difference between the computational aspects of a non-fuzzy neuron and those of a fuzzy neuron is in the definition of the mathematical operations. The mathematical operations in a non-fuzzy neuron may be defined in terms of confluence operation (usually, an inner product) between adjustable synaptic weights and neural inputs. Any learning and adaptation occurring within the neuron involves modifying these synaptic weights. In a fuzzy neuron, the synaptic connections are represented by a two-dimensional fuzzy relation between the synaptic weights and neural inputs.

The term 'fuzzy neural network (FNN)' has existed for more than a decade. However, the recent resurgence of interest in this area is motivated by the increasing recognition of the potential of fuzzy logic and neural networks as two of the most promising approaches for exploring the functioning of a human brain. Many researchers are currently investigating ways and means of building fuzzy neural networks by incorporating the notion of fuzziness into a neural network framework [1, 4-10].

To provide motivation and some basic notions of 'neurons', we present a basic description of biological neuronal morphology This is followed by descriptions of some basic mathematical operations in terms of synaptic and somatic operations or equivalently, in terms of the confluence and activation operations of a single computational neuron. This mathematical neuronal morphology of a single neuron is extended to fuzzy neurons.

2. BIOLOGICAL NEURONAL MORPHOLOGY

In this section, we describe briefly the biological neuronal morphology (structure) which forms the basis for the study of computational neural networks.

The basic building block of the central nervous system (CNS) is the neuron, the cell that processes and communicates information to and from various parts of the body. From an information processing point of view an individual neuron consists of the following three parts each associated with a particular mathematical function.

(i) the synapses are a storage area of the past experience (knowledge base) and receive information from other neurons;

(ii) the cell body, called the soma, receives synaptic information and provides further processing of information; and

(iii) the neuron transmits information to other neurons through a single fiber called the axon.

The junction point of an axon with a dendrite is called a synapse. Synapses provide long term memory (LTM) of the past accumulated experience and are a storage area for the knowledge base. A single biological neuron may have, on the average, 10,000 synaptic connections.

From a systems theoretic point of view, the neuron can be considered as a multiple-input-single-output (MISO) system. From a mathematical point of view, it may be concluded that the processing of information within a neuron involves two distinct operations:

(i) *the synoptic operation:* The strength (weight) of the synapse is a representation of the stored knowledge. The synaptic operation provides a weight to the neural inputs. Thus, the synaptic operation assigns a relative weight (significance) to each incoming signal according to the past experience (knowledge or memory) stored in the synapse;

(ii) *the somatic operation:* This provides aggregation, thresholding and nonlinear activation to the synaptic inputs. If the weighted aggregation of the neural inputs exceeds a certain threshold, the soma produces an output signal.

3. COMPUTATIONAL NEURONAL MORPHOLOGY

A biological neuron consists of synapses (junction points) and a soma - the main body of the neuron. The numerous synapses which adjoin a neuron receive neural inputs from other neurons and transmit modified (weighted) versions of these signals to the soma via the dendrites. Each soma receives, on the average, 10^4 dendritic inputs. The role of the soma is to perform a spatio-temporal weighted aggregation (often a summation) of all these inputs. If this weighted aggregation is greater than an intrinsic threshold, then the weighted aggregation is converted into an action potential yielding a neural output. These action potentials are transmitted along the axon to the other neurons for further processing.

From a signal processing point-of-view, the biological neuron has two key elements, the synapse and soma, which are responsible for performing computational tasks such as learning, acquiring knowledge (LTM of the past experience), and recognizing patterns. Each synapse is a storage element that contains some attribute of the past experience. The synapse learns by continuously adapting its strength (weight) to the new neuronal inputs. The soma combines the weighted inputs such that if its weighted aggregation exceeds a certain threshold, then the neuron will fire. This axonal (output) signal undergoes a nonlinear transformation prior to leaving the axonic hillock in the soma. Mathematically, the synapses and the early stage of the soma provide a confluence operation between the fresh neuronal inputs and the stored knowledge (past experience). The latter part of the soma provides a nonlinear bounded activation operation to the aggregated signals.

4. FUZZY NEURAL NETWORK ARCHITECTURES

The concept of graded membership in fuzzy sets was introduced by Zadeh in 1965. This notion of graded membership was introduced in order to provide mathematical precision to information arising from our cognitive process. The theory of fuzzy sets provides a mechanism for representing linguistic constructs such as 'many', 'low', 'medium', 'often', 'few'. In general, the fuzzy logic provides an inference structure that enables approximate human reasoning capabilities [2]. On the contrary, the traditional binary set theory describes crisp events that either do or do not occur. It uses probability theory to explain if an event will occur, measuring the chance with which a given event is expected to occur. The theory of fuzzy logic is based upon the notion of relative graded membership and so are the functions of mentation and cognition. Thus, the utility of fuzzy sets lies in their ability to model uncertain or ambiguous data so often encountered in real life.

Neural network structures can deal with imprecise data and ill-defined activities but subjective phenomena such as reasoning and perceptions are often regarded to be beyond the domain of conventional neural network theory. Fuzzy logic, however, is another powerful tool for modeling uncertainties associated with human cognition, thinking and perception. In fact, the neural network approach fuses well with fuzzy logic [1, 4, 5] and some research endeavors have given birth to the field of 'fuzzy neural networks' or 'fuzzy neural systems'.

Paradigms based upon this integration are believed to have considerable potential in the areas of expert systems, medical diagnosis, control systems, pattern recognition and system modeling.

The computational process envisioned for fuzzy-neural systems is as follows. It starts with the development of a 'fuzzy neuron' based on the understanding of biological neuronal morphologies, followed by learning mechanisms. This leads to the following three steps in a fuzzy-neural computational process:

(i) development of fuzzy neural models motivated by biological neurons;
(ii) models of synaptic connections which incorporate 'fuzziness' into neural networks; and
(iii) development of learning algorithms (that is, the method of adjusting the synaptic weights).

5. LEARNING SCHEME: ADAPTING THE KNOWLEDGE BASE

The weighting and spatio-temporal aggregation operations performed by the synapses and soma, respectively, provide a similarity measure between the input vector $\mathbf{X}_a(t)$ (new neural information) and the synaptic weight vector $\mathbf{W}_a(t)$ (accumulated knowledge base). When a new input pattern that is significantly different from the previously learned patterns is presented to the network, the similarity between this input and the existing knowledge base is small. As the neural network learns this new pattern, by changing the strength of the synaptic weights, the distance between the new information and accumulated knowledge decreases. In other words, the purpose of learning is to make $\mathbf{W}_a(t)$ very similar to a given pattern $\mathbf{X}_a(t)$. Most of the neural network structures undergo a 'learning procedure during which the synaptic weights (connection strengths) are adapted.

6. CONCLUSIONS

This paper is a tutorial presentation on the principles of biological and conventional neuronal morphologies. Biology does provide a motivation and framework for the development of computational neural structures. Biological neuronal principles can be extended to generate several neural topologies and algorithms for both non-fuzzy and fuzzy situations. In this paper, we have emphasized the basic principles rather than giving some advanced structures of neural networks which are available in the literature.

Our emphasis in this paper, both from the mathematical structure and information processing point of view, has been on operations such as confluence and nonlinear activation. The confluence operation provides a measure of similarity between the neural inputs and accumulated stored experience in synaptic weights, and the activation function provides a graded output to the similarity measure.

It should be noted that more research endeavors are necessary to develop a general topology of fuzzy neural models, learning algorithms, and approximation theory so that these models are applicable to system modeling and control of complex robotics and other systems.
The area of fuzzy neural networks is still in its infancy, and is a very fertile area of
theoretical and applied research.

REFERENCES

1. Gupta, M.M., (1992). Fuzzy Logic and Neural Networks, *Tenth Int. Conf. on Multicriterion Decision Making*, Taipei, July 19-24, 281-294.

2. Zadeh, L.A., (1973). Outline of a New Approach to the Analysis of Complex Systems and Decision Process, *IEEE Trans. Systems, Man and Cybernetics*, 3(1), 28-44.

3. Kaufmann, A. and Gupta, M.M., (1991). *Introduction to Fuzzy Arithmetic: Theory and Applications*, 2nd Edition (Van Nostrand Reinhold, New York).

4. Cohen, M.E. and Hudson, D.L., (1990). An Expert System on Neural Network Techniques, in I.B. Turksen, Ed., *The Proceedings of NAFIP*, Toronto, June, 112-117.

5. Yamakawa T. and Tomoda S., (1989). A Fuzzy Neuron and its Application to Pattern Recognition, *Proc. of the Third IFSA Congress*, Seattle, Aug., 30-38.

6. Hayashi, I., Nomura, H. and Wakami, N., (1989). Artificial Neural Network Driven Fuzzy Control and its Application to Learning of Inverted Pendulum System, in: J.C. Bezdek, Ed., *Proc. of the Third IFSA Congress*, Seattle, 610-613.

7. Bezdek, J.C., (1991). *Pattern Recognition with Fuzzy Objective Function Algorithms* (Plenum Press, NY).

8. Simpson, P.K., (1992). Fuzzy Min-Max Neural Networks - Part I: Classification, *IEEE Trans. on Neural Networks*, **3**(5), Sept., 776-786.

9. Gupta, M.M. and Qi, J., (1991). Design of Fuzzy Logic Controller Based on Generalized T-Operators, *Fuzzy Sets and Systems*, **40**(3), 473-489.

10. Yamakawa, T., (1993). A Fuzzy Inference Engine in Nonlinear Analog Mode and Its Application to a Fuzzy Logic Control, *IEEE Trans. on Neural Networks*, **4**(4), 496-522.

11. Gupta, M.M., and Sinha, N.K. [Editors], (1995). Intelligent Control Systems: Theory and Applications, *A Volume of Invited Chapters, IEE Neural, Networks Council, IEEE-Press*, New York, 820 pages.

12. Gupta, M.M. and Rao, D.H., (1994). On the Principles of Fuzzy Neural Networks, *Fuzzy Sets and Systems*, Vol. 61 (1), Jan., p 1-18.

13. Gupta, M.M. and Qi, J., (1991). On Fuzzy Neuron Models, *Int. Joint Conf. on Neural Networks*, (IJCNN), Seattle, 431-456.

14. Carpenter, G.A., Grossberg, S., Markuzon, N., Reynolds, J.H. and Rosen, D.B., (1992). Fuzzy ARTMAP: A Neural Network Architecture for Incremental Supervised Learning of Analog Multidimensional Maps, *IEEE Trans. on Neural Networks*, **3**(5), Sept., 698-713.

15 Zadeh, L.A., (1965). Fuzzy Sets, *Information and Control*, **8**, 338-353.

16. Gupta, M.M., (1991). Uncertainty and Information: The Emerging Paradigms, *Int. J. of Neuro and Mass-Parallel Computing and Information Systems*, **2**, 65-70.

17. Pal, S.K. and Mitra, S , (1992). Multilayer Perceptron, Fuzzy Sets and Classification, *IEEE Trans. on Neural Networks*, **3**(5), Sept. 683-697.

18. Gupta, M.M., (1992). Fuzzy Neural Computing Systems, *2nd Int. Conf. on Fuzzy Logic and Neural Networks*, July 17-22, Japan.

19. Zimmermann, H.-J., (1991). *Fuzzy Set Theory and Its Applications* ,(Dordrecth, Kluwer Academic Press).

20. Gupta, M.M. and Knopf, G.K., (1993). Dynamic Neural Network for Fuzzy Inference, *SPIE Conf. on Applications of Fuzzy Logic Technology*, Boston, Sept. 7-10.

21. Gupta, M.M. and Qi, J., (1991). Connections (AND, OR, NOT) and T-Operators in Fuzzy Reasoning, in I.R. Goodman, M.M. Gupta, H.T. Nguyen and G.S. Rodgers (Eds.), *Conditional Logic in Expert Systems*, (North-Holland, Amsterdam), 211-233.

22. Hebb, D.O., (1949). *The Organization of Behavior,* (John Wiley and Sons, New York).

23. Hammerstrom, D., (1993). Working with Neural Networks, *IEEE Spectrum*, July, 46-53..

24. Li, Hua and Gupta, M.M. [Editors], (1995). Fuzzy Logic and Intelligent Control Systems, *(Kluwer Academic Publisher,)* Nokwell, U.S.A., 400 pages.

25. Fuzzy Logic and Neural Systems: Theory and Applications, *Journal of Intelligent and Fuzzy Systems*, Special Issue, Vol. 3, No. 1, 1995, pp. 1-103.

26. Fuzzy Neural Control, *Fuzzy Sets and Systems, Special Issue,* Vol. 71, No. 3, May 1995, pp. 255-369.

Optimizing the Self-Organizing-Process of Topology Maps

Karin Haese

Universität der Bundeswehr, Hamburg
Deutsche Forschungsanstalt für Luft- und Raumfahrt, Braunschweig*

Abstract. This contribution proposes, how the self-organizing process of feature maps can be improved.

The self-organizing process converges to a map, which preserves the neighbourhood relations of the input data, if the learning parameters, learning coefficient and width of the neighbourhood function, are chosen correctly. In general, the parameters are chosen empirically, dependent on the distribution of the training data and the network architecture [3]. Consequently, some experience with the algorithm and the training data is needed to choose proper courses of learning parameters. To avoid time consuming parameter studies a system model of the self-organizing process is developed and a linear Kalman filter used to estimate the learning coefficient. To estimate the width of the neighbourhood function the process of neighbourhood preservation during the training is modelled for the first time successfully. This process is then followed by an extended Kalman filter algorithm, which estimates the width of the neighbourhood function.

In case of fast self-organizing algorithms, as published in [1], the proposed parameter estimation method is essential for the training of data with unknown density distribution.

1 Introduction

The *self-organizing topological map* (SOM) of Kohonen [7], [8] is an important tool to map high dimensional data sets $\mathcal{M}$ of unknown density distributions onto a low $n_{\mathcal{A}}$-dimensional discrete lattice of units (*neurons*). In Kohonen's algorithm the nearest neighbour to the input vector m has to be found at each time step j among all $N^{n_{\mathcal{A}}}$ weights w_r with locations $r \in \mathcal{A}$:

$$\|w_{r'}(j) - m(j)\| = \min_{r \in \mathcal{A}} \|w_r(j) - m(j)\|. \qquad (1)$$

Then, the weights w_r are updated according to the well known learning rule

$$w_r(j) = w_r(j-1) + \Delta w_r(j) \qquad (2)$$

$$\Delta w_r(j) = \epsilon(j) \cdot h_{rr'}(j) \cdot [w_r(j-1) - m(j)], \qquad (3)$$

---

* new address

with the neighbourhood function

$$h_{rr'}(j) = exp\left(\frac{\|r - r'\|^2}{2 \cdot \sigma^2(j)}\right), \tag{4}$$

where $\rho^{\mathcal{A}}(r, r') = \sqrt{(r_1 - r_1')^2 + \cdots + (r_{n_{\mathcal{A}}} - r_{n_{\mathcal{A}}}')^2}$. In general, the maps dimension $n_{\mathcal{A}}$ is greater than one. For this case the learning process cannot be proven to converge to a map, which preserves the neighbourhood relations of the input data. The conditions $\lim_{j \to \infty} \sum_{k=0}^{j} \epsilon(k) = \infty$ and $\lim_{j \to \infty} \epsilon(j) = 0$ ensure the convergence only, if the process is already near the equilibrium state [10]. Conditions for $\sigma(j)$ are not known. *Bouton* and *Pages* [3,4] explained that the convergence depends on the distribution of the training data and the network architecture. Therefore, the width $\sigma(j)$ of the neighbourhood function $h_{rr'}(j)$ as well as the learning coefficient $\epsilon(j)$ must be chosen in accordance with the distribution of the training data and the network architecture.

Often, time consuming empirical studies have to be performed for each training data set and network architecture.

Furthermore, using fast self-organizing algorithms [1,6] the parameter studies become more difficult, because these algorithms are more sensitive to the choice of the learning parameters. Therefore, a method which determines the learning parameters automatically is desired.

In the special case of one-dimensional maps $(n_{\mathcal{A}} = 1)$ the here proposed estimation method lowers the numbers of learning steps needed to get to a well organized state of the map [11].

2 Optimal learning parameter estimation

Estimating the learning coefficient $\epsilon(j)$, a system model of the learning process is developed following the models known from the process in error backpropagation learning [9,5]. This leads in case of the SOM to a linear system model, which can be followed by a linear Kalman filter algorithm. The Kalman gain matrix is then interpreted as the learning coefficients. In order to estimate the width $\sigma(j)$ of the neighbourhood function the enhancement of the neighbourhood preservation of the input data on the feature map is modelled successfully on the basis of the 'wavering product'. Then an extended Kalman filter is able to estimate the width of the neighbourhood function.

2.1 Estimating the learning coefficient

The Kalman filter approach to train a feature map leads to the calculation of the learning coefficients. Therefore, the weights w_r of the network are considered as the states of the system to be estimated. Since they do not have any dynamics the state and measurement equations can be written as

$$w(j) = w(j-1) + q_w(j-1) \tag{5}$$
$$o(j) = H(j)w(j) + q_o(j), \tag{6}$$

where $w(j)$ is the vector of all weights in the map, $q_w(j-1)$ and $q_o(j)$ are vectors of white Gaussian noise sequences, $H(j)$ is a matrix with the values of the neighbourhood function on its diagonal and the outputs of the map are gathered in the vector $o(j)$. Using the standard form of the linear Kalman filter, the Kalman gain matrix is easily identified with the learning coefficients for every weight of the map.

2.2 Estimating the width of the neighbourhood function

The idea of this first approach to estimate the width of the neighbourhood function is based on the 'wavering product' [2], a measure for the neighbourhood preservation of the input data on the map. Therefore first the 'wavering product' is introduced.

Denoting the nearest neighbours to the neuron at location r (see also equation (1)) as follows by

1. $\kappa_k^{\mathcal{A}}(r)$, if the kth nearest neighbour to the neuron at location r is measured by the distance in the output space $\mathcal{A}$, i.e.:

$$\kappa_1^{\mathcal{A}}(r): \quad \rho^{\mathcal{A}}(r, \kappa_1^{\mathcal{A}}(r)) = \min_{\hat{r} \in \mathcal{A} \backslash \{r\}} \rho^{\mathcal{A}}(r, \hat{r}), \tag{7}$$

$$\kappa_2^{\mathcal{A}}(r): \quad \rho^{\mathcal{A}}(r, \kappa_2^{\mathcal{A}}(r)) = \min_{\hat{r} \in \mathcal{A} \backslash \{r, \kappa_1^{\mathcal{A}}(r)\}} \rho^{\mathcal{A}}(r, \hat{r}), \tag{8}$$

etc.

and

2. $\kappa_k^{\mathcal{M}}(r)$, if the kth nearest neighbour to the neuron at location r is measured by the distance in the input space $\mathcal{M}$, i.e.:

$$\kappa_1^{\mathcal{M}}(r): \quad \rho^{\mathcal{M}}(w_r, w_{\kappa_1^{\mathcal{M}}(r)}) = \min_{\hat{r} \in \mathcal{A} \backslash \{r\}} \rho^{\mathcal{M}}(w_r, w_{\hat{r}}), \tag{9}$$

$$\kappa_2^{\mathcal{M}}(r): \quad \rho^{\mathcal{M}}(w_r, w_{\kappa_2^{\mathcal{M}}(r)}) = \min_{\hat{r} \in \mathcal{A} \backslash \{r, \kappa_1^{\mathcal{M}}(r)\}} \rho^{\mathcal{M}}(w_r, w_{\hat{r}}), \tag{10}$$

etc.

the ratio

$$Q_{\mathcal{M}}(r, k) = \frac{\rho^{\mathcal{M}}(w_r, w_{\kappa_k^{\mathcal{A}}(r)})}{\rho^{\mathcal{M}}(w_r, w_{\kappa_k^{\mathcal{M}}(r)})} \tag{11}$$

of distances in input space $\mathcal{M}$ and the ratio

$$Q_{\mathcal{A}}(r, k) = \frac{\rho^{\mathcal{A}}(r, \kappa_k^{\mathcal{A}}(r))}{\rho^{\mathcal{A}}(r, \kappa_k^{\mathcal{M}}(r))} \tag{12}$$

of distances in output space $\mathcal{A}$ are combined by their geometric mean to a product $P_3(r, k)$:

$$P_3(r, k) = \left(\prod_{l=1}^{k} Q_{\mathcal{M}}(r, l) \, Q_{\mathcal{A}}(r, l) \right)^{1/(2k)}. \tag{13}$$

Then the mean of the logarithm of $P_3(r, k)$ over all k nearest neighbours to a neuron at location r describes the neighbourhood preservation with respect to one neuron at location r:

$$P_2(r) = \frac{1}{(N^{n_A} - 1)} \sum_{k=1}^{N^{n_A}-1} \log P_3(r, k) \tag{14}$$

Finally, the measure of the neighbourhood preservation on the whole map is found by taking the mean of $P_2(r)$ over all locations r. This leads to the 'wavering product' $P = \frac{1}{N^{n_A}} \sum_{r \in A} P_2(r)$.

In order to develop a simple model of self-organizing process equation (14) is evaluated for the winner neuron at location r' (see also equation (1)). $P_2(r')$ is split into four components

$$P_2(r') = -D^{\mathcal{M}}(r') + \hat{D}^{\mathcal{M}}(r') - D^{\mathcal{A}}(r') + \hat{D}^{\mathcal{A}}(r') \,, \tag{15}$$

with

$$D^{\mathcal{M}}(r') = \frac{1}{N^{n_A} - 1} \sum_{k=1}^{N^{n_A}-1} \frac{1}{k} \sum_{l=1}^{k} \log \rho^{\mathcal{M}}(w_{r'}, w_{\kappa_l^{\mathcal{M}}(r')}), \tag{16}$$

$$\hat{D}^{\mathcal{M}}(r') = \frac{1}{N^{n_A} - 1} \sum_{k=1}^{N^{n_A}-1} \frac{1}{k} \sum_{l=1}^{k} \log \rho^{\mathcal{M}}(w_{r'}, w_{\kappa_l^{\mathcal{A}}(r')}), \tag{17}$$

$$\hat{D}^{\mathcal{A}}(r') = \frac{1}{N^{n_A} - 1} \sum_{k=1}^{N^{n_A}-1} \frac{1}{k} \sum_{l=1}^{k} \log \rho^{\mathcal{A}}(r', \kappa_l^{\mathcal{M}}(r')) \text{ and} \tag{18}$$

$$D^{\mathcal{A}}(r') = \frac{1}{N^{n_A} - 1} \sum_{k=1}^{N^{n_A}-1} \frac{1}{k} \sum_{l=1}^{k} \log \rho^{\mathcal{A}}(r', \kappa_l^{\mathcal{A}}(r')). \tag{19}$$

The first component $D^{\mathcal{M}}$ is a measure of the neighbourhood relations between the input data in the input space. This ordering is desired to be obtained in the output space as well. Therefore, it is considered as a measurement of the actual neighbourhood relations. The actually reached ordering is described by the second term $\hat{D}^{\mathcal{M}}$, i. e. $\hat{D}^{\mathcal{M}}$ quantifies the order of the neurons in the input space. If this ordering is nearly perfect - that means, the neurons, mapped into the input space, show the ordering of the lattice of the feature map - then $D^{\mathcal{M}}$ can be expressed in terms of the neighbourhood function. Assuming that the learning process is near its equilibrium, it is expected that

$$w_{\kappa_l^{\mathcal{M}}(r')}(j) = \epsilon(j) h_{\kappa_l^{\mathcal{M}}(r'), r'}(j) \; w_{r'}(j). \tag{20}$$

Using equation (20) the enhancement in neighbourhood preservation $\Delta D^{\mathcal{M}}(r', j)$ at learning step j is expected to be

$$\Delta D^{\mathcal{M}}(r', j)$$

$$= \frac{1}{N^{n_{\mathcal{A}}} - 1} \sum_{k=1}^{N^{n_{\mathcal{A}}}-1} \frac{1}{k} \sum_{l=1}^{k} \log \rho^{\mathcal{M}}(\boldsymbol{w_{r'}}, \epsilon(j) \, h_{\kappa_i^{\mathcal{M}}(r'), r'}(j) \, \boldsymbol{w_{r'}})$$

$$= \frac{1}{N^{n_{\mathcal{A}}} - 1} \sum_{k=1}^{N^{n_{\mathcal{A}}}-1} \frac{1}{k} \sum_{l=1}^{k} \log \left(1 - \epsilon(j) \, h_{\kappa_i^{\mathcal{M}}(r'), r'}(j)\right) \|\boldsymbol{w_{r'}}\|. \qquad (21)$$

Now, the weights $\boldsymbol{w_{r'}}$ are assumed to be normalized to one, so that the measure of neighbourhood preservation enhancement on the map can be modelled by:

$$\Delta D^{\mathcal{M}}(r', j) = \frac{1}{N^{n_{\mathcal{A}}} - 1} \sum_{k=1}^{N^{n_{\mathcal{A}}}-1} \frac{1}{k} \sum_{l=1}^{k} \log \left(1 - \epsilon(j) \, h_{\kappa_i^{\mathcal{M}}(r'), r'}(j)\right). \qquad (22)$$

Further on, we have to interpret the third term $D^{\mathcal{A}}$ in equation (15). $D^{\mathcal{A}}$ (see equation (19)) is a measure for the ordering of the neurons on the map. Its value depends only on the underlying lattice of the map and is therefore constant during the learning process. The corresponding actually reached ordering of the input space mapped onto the lattice of neurons is described by $\hat{D}^{\mathcal{A}}$. $\hat{D}^{\mathcal{A}}$ will be the measure of the observed enhancement in neighbourhood preservation. These interpretations directly lead to the following state space model of the process in neighbourhood preservation enhancement.

The model is linear in its state equation (23) without any outer excitation and it is nonlinear in its measurement equation (24):

$$\boldsymbol{x}(j) = \boldsymbol{x}(j - 1) \qquad (23)$$

$$\boldsymbol{y}(j) = \boldsymbol{C}(\boldsymbol{x}, j), \qquad (24)$$

choosing the state vector

$$\boldsymbol{x}(j) = \begin{bmatrix} \sigma(j) \\ D^{\mathcal{A}}(r', j) \end{bmatrix} \qquad (25)$$

and the measurement vector

$$\boldsymbol{y}(j) = \begin{bmatrix} \Delta D^{\mathcal{M}}(r', j) \\ D^{\mathcal{A}}(r', j) \end{bmatrix}. \qquad (26)$$

Then $\boldsymbol{C}(\boldsymbol{x}, j)$ follows from equations (26), (22) and (19):

$$\boldsymbol{C}(\boldsymbol{x}, j) = \begin{bmatrix} \dfrac{1}{N^{n_{\mathcal{A}}} - 1} \displaystyle\sum_{k=1}^{N^{n_{\mathcal{A}}}-1} \frac{1}{k} \sum_{l=1}^{k} \log \left(1 - \epsilon(j) \, h_{\kappa_i^{\mathcal{M}}(r')r'}(j)\right) \\[2em] \dfrac{1}{N^{n_{\mathcal{A}}} - 1} \displaystyle\sum_{k=1}^{N^{n_{\mathcal{A}}}-1} \frac{1}{k} \sum_{l=1}^{k} \log \rho^{\mathcal{A}}(r', \kappa_i^{\mathcal{A}}(r')) \end{bmatrix}. \qquad (27)$$

At last, to estimate parameters in nonlinear systems the system model has to be linearized and the extended Kalman filter algorithm has to be applied to estimate the state space vector x. The extended Kalman filter equations need the differential measurement matrix, which is deduced from equation (27):

$$\frac{\partial}{\partial x} C(x, j) =$$

$$\left[\begin{array}{ccc} \frac{1}{(N^{n_A} - 1)} \sum_{k=1}^{N^{n_A}-1} \frac{1}{k} \sum_{l=1}^{k} \frac{-\epsilon(j)\, h_{\kappa_l^A(r)r'}(j)}{(1 - \epsilon(j) h_{\kappa_l^A(r)r'}(j))} \frac{\rho^A(r', \kappa_1^A(r'))}{\sigma^3(j)} & & 0 \\ & 0 & & 1 \end{array} \right].$$

The estimation results using the linear and extended Kalman filter are demonstrated in the following.

3 Estimation Results

At first, the estimation results of the learning coefficient and the width of the neighbourhood function are shown for a two-dimensional map trained with Gaussian distributed input data around three centers. The weights of the map are trained with the original learning algorithm [8] (see equation 1 to 4). Figure 1 shows the training result using the estimation method for the parameter courses. Figure 2 shows the training result using the standard parameter courses. The

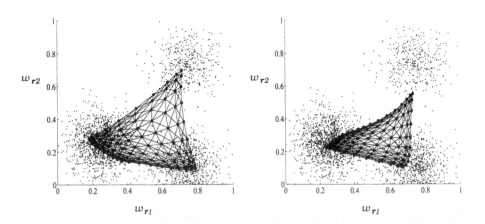

Fig. 1. Trained weights w_r plotted in input space using the original learning algorithm with optimized estimated parameters

Fig. 2. Trained weights w_r plotted in input space using the original learning algorithm with standard parameter choice

corresponding estimated and standard parameter courses are shown in figure 4

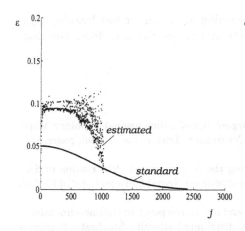

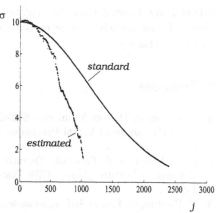

Fig. 3. Comparison of standard and estimated learning coefficients $\epsilon(j)$ during the training with the original learning algorithm

Fig. 4. Comparison of standard and estimated width $\sigma(j)$ of the neighbourhood function during the training with the original learning algorithm

and 3. A comparison of figure 1 and 2 shows, that the weights trained with the estimated parameters quantize the input data better than the weights trained with standard parameter choice. These better training results are gained automatically, so that exhaustive parameter studies are not necessary anymore. Furthermore, it can be deduced from figure 4 and 3, that the proposed estimation method also optimizes the number of learning steps needed for the self-organizing process.

In a last example the proposed estimation method is applied in the fast self-organizing algorithm "Growing Quick SOM", proposed in [1]. This fast algorithms has found to be very sensitive to the right choice of the learning parameters. But using the learning parameter estimation method, "Growing Quick SOM" is easily applied to all input data and network architecture. It converges to very good mappings, which is demonstrated in figure 5.

The courses of the estimated learning parameters are shown in figure 6 and 7. The learning coefficient is estimated nearly constant during the training. The estimated width of the neighbourhood decays nearly exponentially.

4 Conclusions

The parameter estimation method for the learning algorithms of self-organizing feature maps works on the basis of two process models. The first process, which is modelled, is the learning process. The second process is the neighbourhood preservation enhancement on the feature map. Both processes are followed by Kalman filters. They estimate the learning coefficient and the width of the neighbourhood function during the learning. This leads to good learning results in an optimized number of learning steps. The method can be applied to n-dimensional feature maps trained with the original learning algorithm or fast learning algorithms. Consequently, the proposed method releases the user from laborious parameter studies.

References

1. K. Haese, H.-D. vom Stein: Fast Self-Organizing of n-dimensional Topology Maps. In: VIII European Signal Processing Conference. Trieste, Italy: 1996, pages 835–838

2. H.-U. Bauer, K.R. Pawelzik: Quantifying the Neighborhood Preservation of Self-Organizing Feature Maps. IEEE Transactions on Neural Networks, 3(4):570–579, 1992

3. C. Bouton, G. Pagès: Self-organization and a.s. convergence of the one-dimensional Kohonen algorithm with non-uniformly distributed stimuli. Stochastic Processes and their Applications, 47:249–274, 1993

4. C. Bouton, G. Pagès: Convergence in distribution of the one-dimensional Kohonen algorithm when the stimuli are non uniform. Advances in Applied Probability, 26:80–103, 1994

5. P. S. Chandran: Comments on the "Comparative Analysis of Backpropagation and the Extended Kalman Filter for training Multilayer Perceptrons". IEEE Transaction on Pattern Analysis and Machine Intelligence, 16(8):862–863, 1994

6. Y. P. Jun, H. Yoon , J. W. Cho: L*-Learning: A Fast Self-Organizing Feature Map Learning Algorithm Based on Incremental Ordering. IEICE Transaction on Information & Systems, E76-D(6):698–706, June 1993

7. T. Kohonen: Self-organized Formation of Topologically Correct Feature Maps. Biological Cybernetics, 43:59–69, 1982

8. T. Kohonen: Self-Organization and Associative Memory. Springer Series in Information Sciences 8, Heidelberg, 1984

9. D. W. Ruck, S. K. Rogers, P. S. Kabrisky, M. E. Oxley: Comparative Analysis of Backpropagation and the Extended Kalman Filter for training Multilayer Perceptrons. IEEE Transaction on Pattern Analysis and Machine Intelligence, 14(6):686–691, 1992

10. H. Ritter, T. Martinetz, K. Schulten: Neuronale Netze. Addison-Wesley (Deutschland) GmbH, Bonn, 1991, 2. erweiterte Auflage

11. E. Erwin, K. Obermayer, K. Schulten: Self-organzing maps: ordering, convergence properties and energy functions. Biological Cybernetics, 67:47–55, 1992

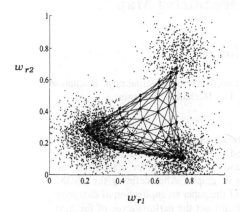

Fig. 5. Weights w_r plotted in input space after the training with "Growing Quick SOM" and the proposed parameter estimation method

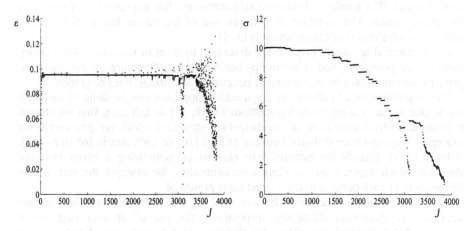

Fig. 6. Estimated learning coefficient $\epsilon(j)$ during the training with "Growing Quick SOM"

Fig. 7. Estimated width $\sigma(j)$ of the neighbourhood function during the training with "Growing Quick SOM"

Syntactical Self-Organizing Map

Ovidiu Grigore

Faculty of Electronic and Telecommunication, University "Politehnica" Bucharest
B-dul Armata Poporului nr. 1-3, Bucharest, Romania

Abstract: In this paper a new neural network structure, called *Syntactical Self-Organizing Map* (SSOM), is introduced. SSOM is obtained from classical (numerical) Kohonen neural network and is specifically for classifying the syntactical structures, like: strings, trees or graphs. After defining the SSOM structure and algorithm, in the third part of the paper an application of character recognition is solved using SSOM. To point out the performances of the new neural network, a comparison of results obtained using the SSOM and the Fu and Lu's clustering algorithm [10] for the same application is done. Moreover, we show that the syntactical Kohonen map have also the topological feature like the numerical one.

1 Introduction

In statistical pattern recognition [1], a pattern is represented by a vector called feature vector. The similarity between two patterns is often expressed by a metric in the feature space. The selection of features and of the metric has a fairly strong influence on the results of cluster analysis [3, 4].

In a syntactical approach, patterns are described by a set of constituent elementary parts (called primitives) and relationships between these. Therefore, in this case, the patterns are represented by mathematical structures like: strings, trees or graphs.

The discrimination of patterns in a syntactical approach can be done in different ways. One of this is using clustering methods [5, 10, 17]. In this case, first we choose a distance (metric) suitable to the mathematical structures which are processed, for example: Wagner-Fisher distance between strings [16] or Lu's metric for trees [11]. After this, we classify the patterns into clusters by optimizing a given criterion function, which depends on the classes concentration, for example the sum of the distances from each patterns to the nearest class prototype.

This paper contains five parts. In section 2 of this paper we introduce a new approach of syntactical clustering, respectively the use of unsupervised neural networks for syntactical structures classification. In the next part of the paper a handwriting character recognition application is implemented using the new method. Also, in this section is given a comparison between the obtained performances of SSOM algorithm and the results of the Fu and Lu's algorithm [10]. The last part of the paper is dedicated for conclusions and remarks.

2 Syntactical Self-Organizing Map

Like classical (numerical) Kohonen neural network [8, 9], our SSOM is composed from:
- the input layer,
- the output layer, which can be a 1-dimensional, 2-dimensional (figure 1) or even 3-dimensional structure of neurons

Dimension of the input layer is equal to the maximum number of components which composed the structural representation of the processed patterns. The output layer dimension depends on the desired number of classes.

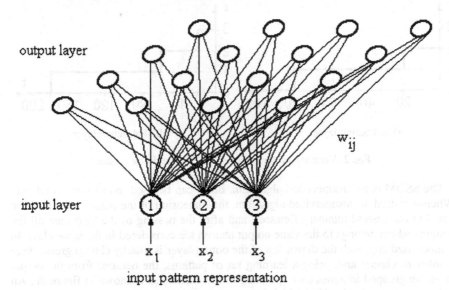

Fig. 1. The structure of SSOM network

We assume that all input patterns can be decomposed using a finite set of primitives $V=\{v_1,v_2,..,v_n\}$.

After a random initialization of the weights w_{ik} with possible primitives v_i, including the null symbol λ, we determine the optimum weights w_{ik} using a training process, which consists of two main steps:

1. the determination of the winner neuron, of which weights of connection with the input layer forms the nearest syntactical structure of the input structures;
2. adaptation of the weights both for the winner neuron and for all the other neurons existing in a certain neighborhood of it.

For the first step we use a distance suitable to the input patterns representation, for example the Wagner-Fisher distance [16] for strings or the Lu's metric [11] for trees.

In the second step we modify the weights w_{ik} of the processed neuron k, thus the structure associated with the k neuron to be closer to the input structure. The closer degree depends on the learning rate $\eta(t)$, which decrease while increasing step's number t :

$$\eta(t) = 1 / \sqrt{t}$$

Laterally interaction between the neurons of the output layer is simulated by the adaptation of the weights of the neurons in the neighborhood of the winner neuron.

The neighborhood is defined as the totality of all neurons which stay at a distance smaller than the radius r in respect to the winner neuron. The dimension of the radius r decreases with the number of steps t. From all the variety of radius r's variation that was experimented, it had been chosen that from in figure 2.

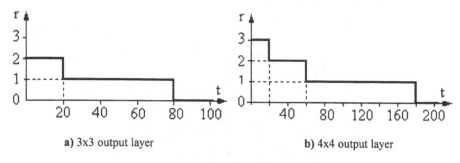

a) 3x3 output layer   b) 4x4 output layer

Fig. 2. Variation of the radius r for SSOM training process

The SSOM is an unsupervised algorithm, but it can be used too as supervised one. When it is used as unsupervised algorithm, the dimension of the output layer must be equal to the desired number of classes, and after the passing of the algorithm all the patterns which belongs to the same output neuron are considered in the same class. In a supervised approach the dimension of the output layer is usually chosen greater than number of classes and, using a learning set of patterns, the neurons from the output layer are grouped in zones corresponding to the classes, as is shown in figure 8b. An unknown pattern in classified taking into account this zones.

The discussed neural network has topological feature, because as you'll see in the next part, the setting up of the winner neurons at the output is done keeping the topological relations existing between the syntactical structures introduced at input.

3. Application

To point out the performances of our method, we implemented the same application from Fu and Lu's paper [10]. We use a set of 51 characters, which are from nine different classes: D, F, H, K, P, U, V, X and Y. Each character is a line pattern on a

Fig. 3. The four primitives used in pattern representation

104

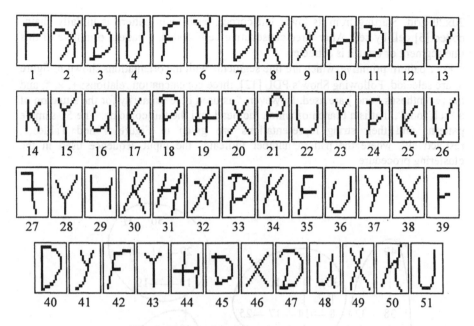

Fig. 4. The set of the input character patterns

Pattern No.	String Representation	Pattern No.	String Representation	Pattern No.	String Representation
1	bbb+(b+dddbdd)*	18	bb+(bbb+dddbaa)*	35	bb+(bb+d)xdd
2	aaa+cddxaaxcbb	19	b+bbbxd+bxdxb	36	cbbxdaabb
3	(bbb+ddcbaa)*	20	aa+cbxaaxcc	37	ba+bcxaa
4	cbbbxdabbbb	21	bbb+(bbadcbad)*	38	aa+ccxaaxcc
5	bb+(b+dd)xd	22	bbbxddabb	39	bb+(bb+dd)xdd
6	bbbb+ccxbb	23	bbb+bcxaa	40	(bbbbb+dddcbaad)*
7	(bbb+dxddcbbaa)*	24	b+(bb+ddcbad)*	41	dab+cbxba
8	ba+bbxaaxcc	25	b+bbbxaaxbc	42	bb+(b+ddd)xd
9	aaa+cxaaxb	26	cbbbxabbba	43	bbb+cxa
10	b+bbxdddd+bbxb	27	bbb+(dx(b+dd))xdd	44	b+dxabxdd+bxbb
11	(bbb+dxddbbad)*	28	bb+cbxaaa	45	(bb+ddcbdd)*
12	bb+(bb+dd)xd	29	bb+bbxddd+axbb	46	aa+cxaaxcc
13	bbbbbxabaa	30	baa+bbxaaxcc	47	bbb+ddaabcddd
14	bb+bbxaaxcc	31	ba+abxdd+axbbb	48	bbbxda+bbxb
15	bbb+cxa	32	aa+ccxaxc	49	baa+ccxbbxccc
16	cbbxda+bbxb	33	bb+(bb+dxddcaad)*	50	ba+bxa+abxbbb
17	bb+bbbxaaxbcc	34	bb+bbxa+axcb	52	cbbxdabbbb

Table 1. PDL representation of the 51 character patterns

20x20 grid. Starting from its lower left corner, each input pattern is initially chain-coded [6] cell by cell. After three consecutive cells coded, a pattern primitive of this line segment or branch is extracted.

Four pattern primitive, which are line segments with the orientations shown in figure 3, are selected. Following Show's PDL [12], three concatenation relations: +, ×, *, and the parentheses (and) are used. However, * is used here primarily for the situation of a "self loop", that is a branch of which the head and the tail coincide. The 51 sample patterns and their string representations are given in Figure 4 and Table 1, respectively, where the pattern number indicates the input sequence used in the clustering procedure.

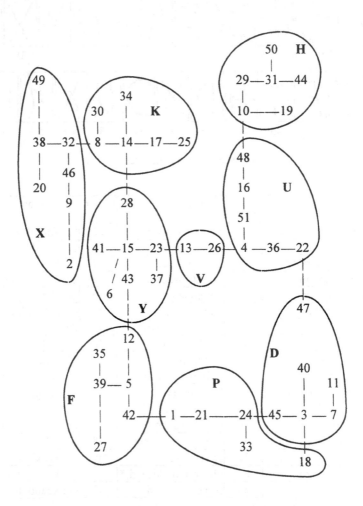

Fig. 5. Minimum spanning tree and the true
classification of the 51 character pattern

C(i,j)	Λ	a	b	c	d	x	+	*	(	)
Λ	0	0.6	0.7	1.3	1.2	1	1.2	0.9	0.9	0.9
a	0.6	0	1.3	2	1	-	-	-	-	-
b	0.7	1.3	0	1.2	1	-	-	-	-	-
c	1.3	2	1.2	0	1	-	-	-	-	-
d	1.2	1	1	1	0	-	-	-	-	-
x	1	--	-	-	-	0	3	3	3	3
+	1.2	-	-	-	-	3	0	3	3	3
*	0.9	-	--	-	-	3	3	0	3	3
(	0.9	-	-	-	-	3	3	3	0	3
)	0.9	-	-	-	-	3	3	3	3	0

Table 2. Costs associated with symbols transformations

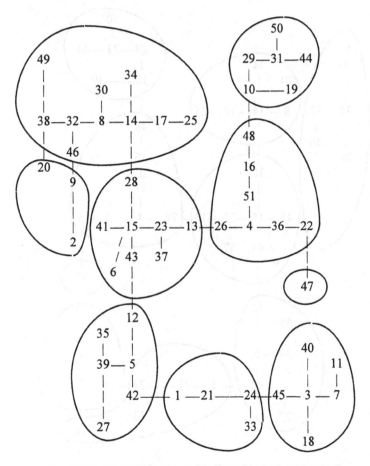

Fig. 6. Patterns classification using Fu and Lu's algorithm (t=4)

In figure 5 is shown a 2-dimensional representation (minimum spanning tree) of the entire set of patterns and the true clusters are circled on it.

Fu and Lu's algorithm with a threshold $t = 4$ lead to the classification shown in figure 6.

In this application we have patterns from 9 classes, and from this reason we choose a 3x3 output layer structure for the SSOM. In figure 8a is presented the configuration of the output layer for the unsupervised 3x3 SSOM and in figure 7 are circled the clusters corresponding to each output neuron.

For both algorithms was used the Wagner-Fisher string distance and the cost between primitives are given in Table 2. When two symbols are not compatible (we can't change one with the other) in the Table 2 it is used the symbol "-". Also, in the table is used the symbol "Λ", null character, for inserting and eliminating operations.

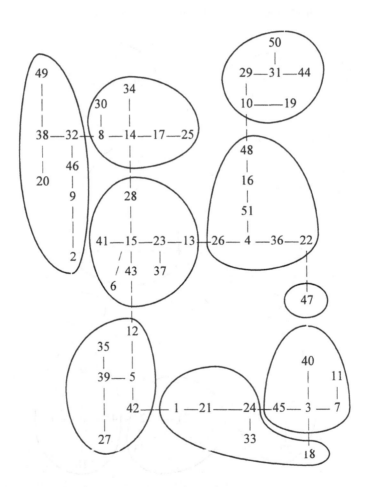

Fig. 7. Clusters obtained applying SSOM

To proof the topological character of the SSOM, we present in figure 8 the setting up of the neurons corresponding to the patterns classes in the network with 3x3 and 4x4 output layer.

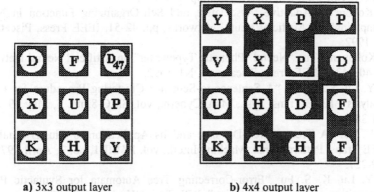

a) 3x3 output layer   b) 4x4 output layer

Fig. 8. Output layer neurons clustering

4 Conclusions and Remarks

It was studied a new method for syntactical structures clustering, respectively the use of unsupervised neural networks. As is shown in the diagrams, the results for SSOM are better than Fu and Lu's algorithm. Moreover, though the topological character of the SSOM, the distribution of the output neurons corresponding to the input structures gives us more information about relationships between patterns. Also, from the same reason, this network can be used to represent in a 2-dimensional space a set of syntactical structures.

The proposed clustering procedure, though only applied to an example of pattern represented by strings, can be easily extended to that of patterns represented by trees using the tree distance define in [11] or [12].

References

1. H. C. Andrews: Introduction to Mathematical Techniques in Pattern Recognition, Wiley, New York, 1972.
2. N. M. Cheung, A. B. Horner: "Group Synthesis with Genetic Algorithm", Journal of the Audio Engineering Society, vol. 44, no. 3, March 1996, pp. 130 - 147.
3. E. Diday, J. C. Simon: "Clustering Analysis", Digital Pattern Recognition, K. S. Fu Ed. Springer-Verlag, New York, 1976.
4. R. O. Duda, P. E. Hart: Pattern Classification and Scene Analysis, Wiley, New York, 1972.
5. J. Fehlauer, B. A. Eisenstein: "Structural Editing by a Point Density Function", IEEE Trans. Syst., Man, Cybern., vol. SMC-8, no. 5, May 1978, pp. 362-370.

6. H. Freeman: "On the Encoding of Arbitrary Geometric Configuration", IEEE Trans. Electon. Comput., vol. EC-10, 1961, pp. 260-268.
7. G. P. Geaves: "Design and Validation of a System For Selecting Optimized Midrange Loudspeaker Diaphragm Profiles", Journal of the Audio Engineering Society, vol. 44, no. 3, March 1996, pp. 107-119.
8. T. Kohonen: "Adaptive, Associative and Self-Organizing Function in Neural Computing", in Artificial Neural Networks, pp. 42-51, IEEE Press, Piscataway, NJ, 1992;
9. T. Kohonen: "The "Neural" Phonetic Typewriter", in Artificial Neural Networks, pp. 409-421, IEEE Press, Piscataway, NJ, 1992;
10. S. Y. Lu, K. S. Fu: "A Sentence-to-Sentence Clustering Procedure for Pattern Analysis", IEEE Trans. Syst., Man, Cybern., vol. SMC-8, no. 5, May 1978, pp. 381-389.
11. S. Y. Lu: "A Tree-to-Tree Distance and Its Application to Cluster Analysis", IEEE Trans. Patterns Anal. Machine Intell., vol. PAMI-1, no. 2, April 1979, pp. 219-227.
12. S. Y. Lu, K. S. Fu: "Error-Correcting Tree Automata for Syntactic Pattern Recognition and Image Processing", RPI, Troi, NY., June 6-8 1977.
13. V. Neagoe, O. Stanasila: Teoria Recunoasterii Formelor, Ed. Academiei Romane, Bucuresti, 1992
14. P. K. Simpson: "Foundation of Neural Networks", in Artificial Neural Networks, pp. 3-24, IEEE Press, Piscataway, NJ, 1992;
15. A. C. Show: "A Formal Picture Description Scheme as a Basis for Picture Processing Systems", Inform. Contr., vol. 14, 1969.
16. R. A. Wagner, M. J. Fisher: "The String to String Correction Problem", J. Ass. Comput. Mach., vol. 21, Jan. 1974.
17. Z. Wu, R. Leahly: "An Optimal Graph Theoretic Approach to Data Clustering: Theory and Its Application to Image Segmentation", vol. 15, no. 11, Nov. 1993, pp. 1101-1113.

On-Line Learning Based on Adaptive Similarity and Fixed Size Rule Base

Joana Matos Dias[1], António Dourado Correia[2]
[1]Research Student, [2]Professor
CISUC-Centro de Informática e Sistemas da
Universidade de Coimbra
Departamento de Engenharia Informática
PÓLO II da Universidade de Coimbra -Pinhal de Marrocos
P 3030 COIMBRA PORTUGAL
Phone: +351 39 7 000 000 Fax: +351 39 701 266
e-mail: joana@student.dei.uc.pt , dourado@dei.uc.pt

Abstract: In this paper a methodology is developed to control linear and non-linear processes using a fuzzy approach with the main assumption that the output of the process is monotone with respect to the input. Beginning with an empty rule base, a fuzzy model is on-line built. The rule base has a fixed number of rules determined *à priori* and not depending on the complexity of the process. The controller experiences a learning phase during which it learns how to control the process, that is repeated whenever there is some change in the process behaviour. The inference and defuzzification mechanisms have their background on the Fuzzy Equality Relations Theory, using an adaptive degree of similarity. The proposed controller was successfully applied in simulation for linear and non-linear systems and practical essays were made on a real non-linear thermal process, for both the regulation and the tracking problem.

1 Introduction

Modern industrial processes are becoming each time more and more complex. Most of them have non-linear features partially unknown that make the use of mathematical models difficult and sometimes inefficient. Neural network controllers try to overcome this problem, functioning as a black box that is trained to control the process apparently without knowing much about its dynamics. But if the process modifies its behaviour in time the results will not be satisfactory and the network will have to be trained again.

In fuzzy controllers, one has the opportunity to include the operator's knowledge in the controller algorithm in a non-mathematical way. But the time spent trying to tune the parameters of the controller correctly can be unbearable.

The proposed controller is a fuzzy controller that does not need the expert knowledge about the process. Its rule base is empty at first, and is constructed based on the values the controller gets from and sends to the process. The construction of the rulebase is on-line and is a never ending process, so that the controller can react whenever there is a change in the process behaviour. When the rule base is empty, or if there is a change in the process parameters or dynamics, the controller experiences a learning phase during which it learns how to control the process. One important characteristic of the controller is that its rule base has a fixed maximum number of rules (FMNR), determined by the operator and that gives him a more accurate control over the sampling time.

In section 2, assumptions will be made that have to be fulfilled in order to the controller to behave properly. In section 3 the way rules are constructed and introduced in the rule base is explained. In section 4, 5 and 6 the fuzzification, inference and defuzzification mechanisms are described. In section 7 other aspects that can influence the algorithm behaviour are introduced, in section 8 the Fuzzy Equality Relations is referred, in section 9 some results are shown and finally in section 10 some conclusions are drawn.

2 Assumptions

It is desired that one has the minimum possible knowledge of the process as possible. In this work the minimum knowledge needed is given by the following assumptions:

i) The process can be modelled by an NARX model:

$$y_k = f(y_{k-1}, y_{k-2}, \dots, u_{k-T}, u_{k-T-1}, \dots, e_k) \qquad (1)$$

with f a linear or non-linear function, time variant or not, y_k, y_{k-1},... the outputs of the system and u_k, u_{k-1},...,the inputs to the system at the k, $k-1$, ... instants. T represents the discrete pure time delay. e_k is a stochastic disturbance.

ii) The pure time delay T is known or, at least, its upper bound.

iii) The reference is known at least $T+1$ sampling instants in advance.

iv) The memory of the system its known (at least its upper bound).

v) The relation, one out of two possible, between the change in the input signal and the change in the output signal is known:

      1 - The output increases/decreases with the increase/decrease of the input;
      2 - The output increases/decreases with the decrease/increase of the input;

If assumption iii) is dropped, we can still control the system, but a longer learning phase and a not so good performance is expected. The main objective of the assumption is to give the controller the characteristic of antecipate so that it can deal with the pure time delay of the process. In regulation problems this assumption has no importance.

3 The Rule Base

3.1 The Rules

Considering (1), solving for u_{k-T}, (2) is obtained:

$$u_{k-T} = g (y_k, y_{k-1}, \ldots, u_{k-T-1}, \ldots, e_k) \qquad (2)$$

Then the rules format is derived from the linguistic transformation of (2) as shown by (3) in the form of a FARX - Fuzzy Auto-Regressive with Exogenous Input system

If y_k is A_1 and y_{k-1} is A_2 and ... and u_{k-T-1} is B_1 and ... then u_{k-T} is C.   (3)

$A_1, A_2, \ldots, B_1, \ldots, C$ are linguistic values and, in the ith iteration, are substituted by *"approximately y_i", "approximately y_{i-1}", ..., "approximately u_{i-T-1}"* , and so on, with y_i, y_{i-1}, u_{i-T-1}, the inputs sent to and the outputs read from the process. Each situation experienced by the process is considered a valid rule. If we add $T+1$ to each and every index, each rule gives us a value for u_{k+1}, if the antecedent is true (in a fuzzy way). We consider the future values of the process output the same as the future (known) values of the reference; implicit in this consideration is the assumption that the controller will be capable of making the output follow the reference and so, by this simple way, some feedforward effect is introduced in the control signal.

If assumption iii) is not fulfilled, then we can consider the future values of the reference equal to its past values. Despite the fact that this is not true for the tracking problem, it is the best information we have. This leads to an increase in the learning period and the controller will no longer be capable of dealing properly with the pure time delay of the process. However for the regulation problem the controller still behaves well. This behaviour is an unsolved (and probably unsolvable) problem. When future values of the reference are not known there is no way to introduce any feedforward effect (only feedback is possible).

3.2 The construction of the rule base

At every sampling time an input signal is sent to the process and an output signal is received from it. Based on these new values and on the values memorised, the rules are construeted accordingly to the experiments. The rule base is nothing more than a matrix where each row is a rule and each column is a linguistic value. Each rule is continuously added to the rule base, until the FMNR is reached. At this point the new rules are written over the older ones, beginning with the rule that occupies the first place in the matrix, so that the FMNR is never exceeded. This gives the algorithm the characteristic of easily forgetting one system condition and adapting to a new one.

3.3 The choice of the FMNR

The choice of the FMNR must be considered with some care. The number of rules in the rule base is directly related with the sampling time needed to compute the algorithm. In the proposed method, it is a function of the desired sampling time and of the reference. The following principles give a guide line to calculate this value:

i) - If the reference is a periodic signal with period P and the sampling time is Ts, then

$$FMNR = \frac{P}{Ts}.$$

ii) - If the FMNR that results from applying 1 is too high for the desired Ts (the computational time needed is directly related to the number of rules in the rule base), then the FMNR calculated is divided by n, $n \in N$, and the situations experienced by the system are included as new rules just when the number of the iteration is divisible by n.

iii) - If we want the system output to follow two or more periodic references with different mean and amplitude values, a FMNR is calculated for each different reference, and the rule base is structured so that each reference has its own space. When the system output is following one reference, the rules originated will be placed only on that reference space, so that no information about the other references is lost.

iv) - If the reference is not periodic, the rule base is constructed, when this is feasible in a way similar to 3, using all the information available about the reference. When this is not feasible, the maximum number of rules is fixed and the rule base is permanently refreshed.

4 The fuzzification mechanism

The fuzzification follows the method described in [2]. The membership functions used are the triangular ones defined by (4):

$$\mu_{ui} = \begin{cases} 1 + \dfrac{u_k - u_i}{b - a} & a \leq u_k \leq u_i \\ 1 - \dfrac{u_k - u_i}{b - a} & u_i \leq u_k \leq b \\ 0 & otherwise \end{cases} \tag{4}$$

with $u_k \in [a,b]$. This interval includes all possible values for the variable. All the fuzzy sets overlap on the entire interval [a,b]. Every crisp u_k value defines the unity vertex of a membership function and all the membership functions have the same slope. These facts reduce to a minimum the information required to define these functions : this minimum is simply the crisp value. On the other hand with only one rule the rule base is complete.

5 The Inference mechanism

Considering, for instance, a 1st order process each rule has the format: *If y_k is approximately y_{1i} and y_{k-1} is approximately y_{2i} then u_{k-T} is approximately u_i".* If the future values of the reference are R1 and R2 (respectively in $k+T+1$ and $k+T$ sampling times), the degree of equality between them and each rule antecedent is calculated as [2]:

i) $D_i = \sqrt{(y_{1i} - R1)^2 + (y_{2i} - R2)^2}$ (5)

ii) $\omega_i = -\dfrac{1}{D \max} * D_i + 1$ (6)

This is called the degree of similarity. *Dmax* represents the maximum distance considered. If D_i is greater than *Dmax* it is considered that there is no similarity.

iii) The consequent fuzzy set for each rule is an interval of values *[p1,p2]*, where

$\mu_{ui}(u(k)) \geq \omega_i, \forall uk \in [p1, p2]$. (7)

The total output is taken as the intersection of all the intervals (one for each rule). So, at the end of the inference mechanism, one has an interval of possible values for u_{k+1} that we could name *learning interval*.

$\Delta u = [\min, \max]$ (8)

The justification for the choice of the intersection operator can be found in the Theory of Fuzzy Equality Relations [1].

5.1 The adaptive parameter Dmax

If *Dmax* has a value that is too low the system output will oscillate. If its value is too high the system output will have steady state error. Considering a time window during which the behaviour of the output is observed, these two situations can be identified. Then it is easy to adapt the *Dmax* parameter. Its value is increased in the presence of the first situation and decreased in the presence of the second, as shown by the equations (9) and (10):

$$Dmax = Dmax * (1 + pecentage\ of\ oscillation) \qquad (9)$$

if the *Dmax* has to be increased;

$$Dmax = Dmax * (1 - pecentage\ of\ error) \qquad (10)$$

if the *Dmax* has to be decreased.

5.2 Changes in the process dynamics

In the presence of changes, particularly heavy changes, the algorithm as described would not react immediately. This problem can be overcame by replacing (5) by a new equation that contains information from both the reference values and the real output values as shown in (11) . This is a simple way of eliminating the non-reaction period of the controller.

$$D_i = \frac{D_i + \sqrt{(y_{1i} - y(k))^2 + (y_{2i} - y(k-1))^2 + (u_i - u(k-T)\ )^2}}{2} \qquad (11)$$

5.3 The intersection results in an empty set

In the inference mechanism, one has to take the intersection of all the consequent intervals, one for each rule. It is possible that this intersection results in an empty set. It takes only one rule with one empty interval for this to happen. In this case, the best thing to do is just ignore the rule. The conclusion reached by the experiences made was that this rule is insignificant, it illustrates a transitory state of the system and should not be taken into account.

6 Defuzzification mechanism

The output of the inference mechanism is an interval *[min,max]* (8). Now a crisp value for u_{k+1} must be computed. Based on [2], a simple but efficient method is proposed, using the concept of *error in advance* that is the difference between the actual output and the desired reference after the pure time delay of the process:

i) *error_in_advance = reference(k+T+1) - yk.*

ii)
- Considering the relation 1) of the assumption v):

    if *error in advance* > 0 then

$$u(k + 1) = \frac{\text{max} + u(k - T)}{2} \qquad (12)$$

$$\text{else} \quad u(k + 1) = \frac{\text{min} + u(k - T)}{2}. \qquad (13)$$

- Considering the relation 2 of the assumption v):

if *error_in_advance* > 0 then

$$u(k + 1) = \frac{\text{min} + u(k - T)}{2} \qquad (14)$$

$$\text{else} \quad u(k + 1) = \frac{\text{max} + u(k - T)}{2}. \qquad (15)$$

If the assumption iii) (section 2) is not fulfilled, one can use the error instead of the *error_in_advance* and the value for u_k instead of the value for u_{k-T}.

In this way the pure time delay of the process tends to be compensated since the control signal acquires the predictive characteristics intrinsic in the *error in advance*. This fact was experimentally proved.

7 Other aspects

The performance of the algorithm, at least in its early iterations, can be influenced by other factors, like the interval considered as being the domain of the input variable. The smaller the interval, the faster is the convergence of the algorithm.

If, in the learning phase, the process output oscillates in an unreasonable way, the defuzzification mechanism can be changed and the calculation of the variable value can be made with a weighted average with past control values, where the heaviest weight is put on the preceding values of the input variable, instead of (12)-(13) or (14)-(15). This can lead to an increase in the learning phase time but will result in a much smoother control signal.

8 Fuzzy Equality Relations Theory

This theory is implicit in everything that has been said. For instance, when in the kth iteration a value $y_k = y_i$ is read from the process, one can not take this value as an exact one. Introducing this value in the rule base is the same as saying that the value measured is *approximately equal to* y_i. In [1] the following is defined: A mapping $E:XxX => [0,1]$ is an equality relation if E satisfies the following axioms:

i) $E(x,x) = 1$;

ii) $E(x,x') = E(x',x)$;

iii) $E(x,x') + E(x',x'') - 1 \leq E(x,x'')$.

The greater the value of $E(x,x')$ the more x and x' are similar.
If one considers now a mapping φ from X to Y, the following definition arises: Let E and F be equality relations on X and Y respectively. A mapping $\varphi: X => Y$ is called extensional if $E(x,x') \leq F(\varphi(x), \varphi(x'))$, for all $x, x' \in X$. This definition says that if x and x' are equal to a certain degree in X, then the images $\varphi(x)$ and $\varphi(x')$ should also be equal to some degree in Y.

Each rule has an antecedent and a consequent. If we consider the antecedent as being x and the consequent as being $\varphi(x)$, the algorithm proposed calculates the degree of equality between x and the desired situation (x'). What is known is $\varphi(x)$ but what one would like to know is $\varphi(x')$ (the correct input to the desired output). To find out a possible range of values for $\varphi(x')$, the algorithm performs the following:

1) calculates $E(x,x')$ for each and every rule;
2) As one knows that
$$E(x,x') \leq F(\varphi(x), \varphi(x')),$$ (16)
knowing $\varphi(x)$ (rule's consequent) one has to consider each and every value for $\varphi(x')$ that fulfils the condition;
3) The condition (16) has to be fulfilled for each and every rule so one has to take the intersection of all the intervals calculated using the inference mechanism (one for each rule).
The extension property referred in [1] is, as stated in the article, very difficult to be achieved. This corresponds to the cases of empty intersection.

9 Results and Discussion

9.1-Simulations

Case 1: Tracking of a non-linear non disturbed system
Considering the non-linear system from [4]

$$y(k) = \frac{y(k-1) * y(k-2) * (y(k-1) + 2.5)}{1 + y(k-1)^2 + y(k-2)^2} + u(k-1)$$ (17)

to be controlled with a sinusoidal reference with period *25s*, and a sampling time of *100ms*, with a FMNR equal to *100* (by the principle ii) of 3.3) the result is as shown in Figure 1. The algorithm makes the output of the system rapidly follow the reference. In the long run, the two signals became practically indistinguishable.

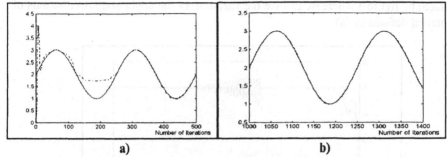

a)   b)

Figure 1 - The algorithm is learning how to control the system of Case 1.
---- System Output ; —— Reference

Figure 1 shows the on-line learning phase of the controller. Between iteration *1* and *250* the rules are built (starting from an empty rule base). After this instant they are replaced on-line in such a way that as time goes by the performance increases, as can be seen in figure 1b).

Case 2: Regulation of a time variant non-linear system

Considering now two non-linear systems from [3]

$$\begin{cases} \dot{x}_1 = \dot{x}_2 \\ \dot{x}_2 = -9.25x_1 - x_2 - 0.1x_1^3 + 9.25u \\ y(t) = x_1(t - 0.1) \end{cases} \quad \text{until } t = 40 \text{ seconds} \quad (18)$$

and

$$\begin{cases} \dot{x}_1 = 1.5\,\dot{x}_2 \\ \dot{x}_2 = -11.25x_1 - x_2 - 0.1x_1^3 + 9.25u \\ y(t) = x_1(t - 0.1) \end{cases} \quad \text{after 40 seconds} \quad (19)$$

representing a time varying system (the coefficient of the first equation and one coefficient of the second).

A sampling time of *100ms* and a FMNR of *40* are considered. If we want the controller to control firstly the system (18) and latter the system (19), with a constant reference with amplitude *5* the results are shown in Figure 2.

The change in the parameters happens in the 400th iteration. The controller responds immediately to this situation, and a new learning phase begins. The learning ability is illustrated by figure 3 where we can see the variation of the upper and lower bounds of the interval defined by (8).

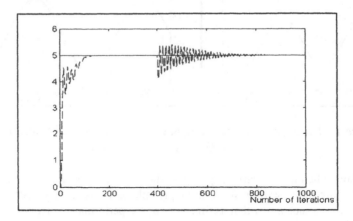

Fig.2 - (Case 2).Changing the systems parameters
.---- System Output ; ——⁻ Reference

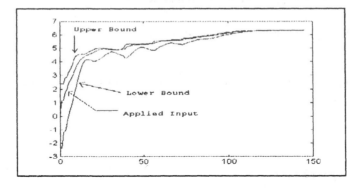

Fig. 3 - (Case 2).The interval (8) of the Learning Ability.

9.2 Application to a real process

The process is a small scale laboratory one intended to produce a certain quantity of heated air (fig. 4). The air enters the process due to a blower. Inside the tube there is an energized electrical resistance that heats the air (the actuator). The goal is to control the air temperature along the tube, which can be read by a sensor placed in one of the three possible positions. The angle of the air overture can also be changed (introducing load disturbances).

The sampling time used in all these experiments was *150ms*.

In figures 5 are the results obtained with a sinusoidal reference of period *30s*,

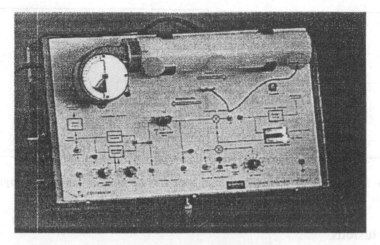

Figure 4 -The real process PT326.

and *FMNR=100* and a triangular reference with the same *FMNR*. Despite the existence of disturbances, the algorithm is capable of controlling the process with a smooth input signal. After the learning phase shown in fig.5, the performance increases with time in a way similar to fig.1.

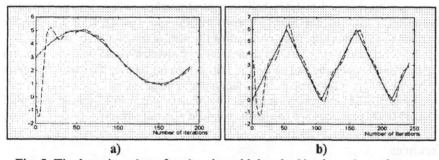

a)   b)

Fig. 5. The learning phase for a) a sinusoidal and a b) triangular reference.

If we change the angle of the air overture (a load disturbance), the result is illustrated in figure 6a), where the steady state *y=2* is regulated. Changing the sensor position from the third to the first position counting from left to right, a change in the process pure time delay is introduced. As shown in figure 6b) the controller performs well and is able to recover from the situation were the sensor was being changed (in some sense this is a situation of a sensor failure).

Vertical scales in figures 5 and 6 are in volts representing input to the electronic actuactor (for the control variable) and electrical signal from sensor/transducer for the process output.

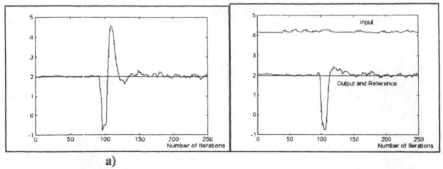

a)

Figure 6- Disturbances in the process: a) load disturbance (changing the angle of the air overture); b) changing the dynamics (position of the sensor).

10 Conclusions

The most important and original characteristic of this controller is that the FMNR is a parameter to the controller given by the programmer so he can have a more accurate control over the sampling time. The assumptions made can be fulfilled by the majority of systems and do not require much knowledge of the process. The on-line construction of the rule base as well as the adaptive similarity degree makes it easy for the controller to forget one process and to learn how to control another when process dynamics change or other disturbances appear. It also tries to reduce to the least possible the intervention of the human operator, eliminating the need of trial and error procedures.

Problems were found trying to control unstable systems and those of non-minimum phase. Unstable systems do not allow the initial learning phase. The non-minimum phase systems do not fulfil completely the assumptions made (v).

Research was carried out to study the possibility of using this method for the control of MISO and MIMO processes and the results obtained were satisfactory.

References

[1] Klawonn, Frank; Gebhardt, Jörg and Kruse, Rudolf;(1995) **Fuzzy Control on the Basis of Equality Relations with an Example from Idle Speed Control**, IEEE Transactions on Fuzzy Systems, vol 3,n°3, pags.336-349

[2] Park, Young-Moon; Moon, Un-Chul and Lee, K.Y.;(1995) **A Self-Organizing Fuzzy Logic Controller for Dynamic Systems Using a Fuzzy Auto-Regressive Moving Average Model**, IEEE Transactions on Fuzzy systems, vol 3, n°1, pags. 75-82

[3] Patyra, M.J. and Mlynek,D.M.(1996) **Fuzzy Logic Implementation and Applications** , Wiley Teubner

[4] Wang,Li Xin; (1994) **Adaptive Fuzzy Systems & Control, Design and Stability Analysis**, Prentice Hall

Ability of Fuzzy Constraint Networks to Reduce the Dynamic of Changes in Product Development

Walter Eversheim[1], Axel Roggatz[1]
Robert E. Young[2]

[1] Laboratory for Machine Tools and Production Engineering
Chair of Production Engineering
Aachen University of Technology
52056 Aachen, Germany

[2] Workgroup for Intelligent Systems in Design and Manufacturing
Department of Industrial Engineering
North Carolina State University
Raleigh, NC 27695-7906, USA

Abstract. In the product development process, customer requirements as well as constraints concerning the entire product life cycle have to be satisfied. Introducing the concept of Concurrent Engineering (CE), interdependencies of constraint sets increase due to the parallelization of activities. Constraint systems are used to attack this problem. Fuzzy Constraint Networks (FCN) have been introduced to expand the employment of constraint systems to the early phases of the product development process.

Aside of the ability to model and solve constraints of the product development process in a straight forward approach, the capability to handle disturbances (e.g. changes) is needed. In underdamped systems, changes are propagated through the entire system and therefore effect nearly all the decisions made. In this article, a possibility to reduce unnecessary propagation of changes will be discussed. Therefore, a two step application of Fuzzy Constraint Networks is introduced. The two steps consist of a straight forward approach, which takes advantage of the structure of product development processes using phases and milestones, and a stepwise backward propagation used to find valid solutions in case a change occurs.

1. FUZZY CONSTRAINT NETWORK

During the design process, vague and imprecise specifications defined by customers will be transformed to a complete and consistent product description with exact values. In order to do so, a great number of constraints of many different areas has to be satisfied. Using the methodology of Concurrent Engineering (CE), the requirements of the design process are intensified. The basic idea of CE is to parallelize activities in order to reduce the development lead time and thus the time to market. This goal can best be achieved, if the parallelization is not only based on exact information. Many activities can be started even if their predecessor have not been completed. This requires the early pass on of information through the process chain including the management of uncertain and incomplete information [1]. In addition to that, the goal of reducing the product lead time creates the need to consider the whole product life cycle, which starts with the idea of the product and ends with its disposal or recycling [2]. Therefore, the number of constraints which have to be satisfied in the design process increases.

To solve the constraints in the design process, several constraint systems have been developed. To expand their applicability in CE problems, they most recently have been expanded and now offer the use of fuzzy technology [3,4]. Fuzzy technology has proven to be a useful mean to present and to process quantitative and qualitative information as well as inconsistency and imprecision. Thus, fuzzy technology is an effective way to describe engineering knowledge [5]. Giachetti et al. [6] show, that fuzzy constraint networks are able to make the transition from imprecise goals and requirements to the precise specifications needed to manufacture the product.

Fuzzy constraint networks are an extension of constraint networks. They allow the additional use of linguistic terms and fuzzy sets within the descriptions of the constraints. On top of that, fuzzy relationships can be modeled with these networks. A fuzzy constraint network can be seen as a fuzzy constraint satisfaction problem. This is defined as follows (adapted from [7]):

A fuzzy constraint network problem consists of a set of n variables, $\tilde{X} = \{\tilde{X}_1, \tilde{X}_2, ..., \tilde{X}_n\}$, and a set of m constraints, C = {C_1, C_2, ... , C_m}. A fuzzy variable $\tilde{X}_i$ has its domain, Ω_i, which defines the set of values that the variable can have. A constraint C_i is a *k-ary* relation on $\tilde{X}'_i = \{\tilde{X}_{i1}, \tilde{X}_{i2}, ..., \tilde{X}_{ik}\} \subseteq \tilde{X}$, i.e., $C_i(\tilde{X}_{i1}, \tilde{X}_{i2}, ..., \tilde{X}_{ik})$, and is a subset of the Cartesian product $\Omega_{i1} \times \Omega_{i2} \times ... \times \Omega_{ik}$.

In this formulation each constraint is satisfied to a degree, $\mu_{Ci} \in [0,1]$. In the following it is referred to this value as the *degree of satisfaction* (DoS). A network is considered to be solved, if values are assigned to all its variables and if the constraints are satisfied. The constraints are satisfied when $\mu_{Ci} \geq \alpha_S$ where α_S is the system truth threshold. It is the lowest level of satisfaction a constraint must have to be considered as valid by the designer. This value is set apriori by the user. It is an important tool for analysis of the problem. Its use will not be discussed here.

The constraint processing system used is called FuzCon [8]. The constraint processing system's objective is to make all the constraints "satisfied". It performs this by interacting with the user. A user supplies values for a variable and the system propagates those values through the system inferring unknown values and calculating the state of the constraints wherever possible. Constraint networks are linked by variables they share. Consequently, this system supports a user in finding a feasible solution. It does not perform automatic constraint satisfaction. In the context of engineering design, constraints represent the requirements an artifact must satisfy. The designer then interacts with the system, testing different design alternatives in a solution space bounded by the constraints.

Fuzzy sets are used to represent ambiguous and imprecise information. They offer to contain elements to a certain degree. The membership of element x in a fuzzy set is represented by $\mu(x)$ and is normalized so that the membership degree of element x is in $[0,1]$. Fuzzy sets can represent linguistic terms and imprecise quantities. Fuzzy quantities are described as LR fuzzy numbers [9] and represented as a triple. The fuzzy number <2, 3, 4> contains all numbers between 2 and 4. The membership to which a number is related to this set is $\mu(x) = \begin{cases} x-2, & x \in [2,3] \\ -x+4, & x \in (3,4] \\ 0, & else \end{cases}$. Linguistic terms are used to model the imprecision of natural language statements like 'short' and 'inexpensive'. This can best be demonstrated at an example: In order to define the linguistic variable 'price' the fuzzy number <2, 3, 4> can be applied to the basic variable '$' and be labeled with the term 'low'. In addition to this, other fuzzy numbers can used to describe the terms 'medium' and 'high'. This way, the set of 'amount of $' can be mapped to the linguistic descriptions of the variable 'price'.

When only imprecise information is available, the assumption of crisp values for parameters in order to solve a problem can produce undesirable results and may impose unnecessary restrictions. The use of linguistic descriptions and fuzzy sets allows to represent the level of knowledge that is actually available when the problem has to be solved. Hence, a more realistic model can be build.

2. IMPACT OF CHANGES

Changes within the design process have tremendous effects due to the high number of interdependencies within this process. External and therefore inaffectable changes are propagated through the entire system. The wide range of effects of these changes can be shown, using Quality Function Deployment (QFD). QFD is a set of matrixes, which map customer requirements to product features, and those to product component specifications and process definitions. Changes in one of these matrixes have to be propagated omni-directional and affect nearly all aspects of the QFD. This change explosion is due to reflection and refraction of changes in the matrixes [10]. A small external change will initiate several internal changes, implementing a self introduced dynamic into the product development process.

3. REDUCTION OF UNNECESSARY PROPAGATION OF CHANGES

Even if it was possible to avoid internal changes completely, external influences during the product development process would cause revisions and adjustments of fixed decisions. In underdamped systems, changes are propagated through the entire system and effect all decisions that have been made. To reduce the impact of necessary adjustments, the capability to handle disturbances is needed. Therefore it is necessary to model dependencies. This can be done using constraint networks [6]. In this section, a possibility to reduce unnecessary propagations of changes in the product development process will be discussed.

The method introduced in the following, structures the cause-and-effect chain in a way that allows to control the consequences of changes. It applies Fuzzy Constraint Networks in a straight forward approach, which takes advantage of the structure of product development processes using phases and milestones. Besides it uses stepwise backward propagation in order to find valid solutions in case a change occurs. The benefits of using this structure for the propagation of changes are examined at an example.

3.1 APPLICATION OF FUZZY CONSTRAINT NETWORKS IN CONCURRENT ENGINEERING

In general, the development of new products is an ill-structured problem [11]. Neither the path to a valid solution nor possible problems on this path can be predicted. Thus, Concurrent Engineering projects are structured in several phases or stages, which are connected by milestones. At a milestone, the results of the previous phase will be reviewed and released, if they are consistent with the goal-set of this phase. Furthermore, the procedure for the next phase is outlined. The review is mostly done by a cross functional team. Even if the number of stages varies depending on company and branch, this structure is generally well suited for the employment in product development [12]. Applying this procedure, parallel activities can be coordinated more easily because their results have to be integrated at the milestones. Besides, reviewed milestones are fixed points in the development process. If significant changes occur, it is possible to step back to the milestone reviewed last and to restart the process.

The use of phases and milestones is possible because two different levels of the product development process are modeled. The first level reflects the project as a whole. The second contains the single activities which have to be solved within one stage. Fuzzy Constraint Networks can be structured analogously.

At the overall level of the project, the development process is reflected holistically. Networks used on this level are oriented along the process chain. They will be called *horizontal application*. Horizontal applications are used to ensure the consistency of the solutions chosen. Networks on the second level are used to support certain activities within a stage. They will be called *vertical application*. Both applications have a hierarchical structure and are closely linked to each other. In the following these applications and their effects on the propagation of changes will be discussed.

The vertical application. Let us assume a milestone has been reviewed in a current product development project. The information and the requirements specified define the starting point of all activities of the next project phase. Besides, the specifications narrow the solution space for the activities of the next project phase. Solving a certain activity using a Constraint Network is called a vertical application. A typical example of a vertical application is a design problem. Three steps can be identified to solve such an activity:

1. Design of the Fuzzy Constraint Network for the specific activity by using the information given as fixed preconditions.
2. Solving of the new Fuzzy Constraint Network. If necessary, the network can be driven to more detail or to a more specific state (concreteness) as stated below.
3. Evaluation of the degree of satisfaction the results provide. This has to be done considering the predefined requirements.

The horizontal application. A constraint network which describes the interaction of vertical networks in a specific project stage is called horizontal application. In general, a horizontal application can be defined for three different purposes. First, it can be used to support the recognition of dependencies and conflict sets of constraints within different activities. Thus it helps to fulfill specific tasks. Second, a horizontal application can be used at the end of a project stage to check the consistency of the acquired results. Thus, it supports the milestone review. Third, these applications can be employed after a project stage is finished. In this case they allow to propagate changes along the whole development process. Therefore it becomes possible to identify the effects of the results gained on vertical applications.

In order to use Fuzzy Constraint Networks on the project level, five steps have to be undertaken:
0. Review of the previous milestone. The information and requirements set in this review describe a solution space which covers all solutions of activities of this phase.
1. Continuous solving of the vertical applications as described above.
2. Aggregation of the highest levels of the vertical applications into one constraint network. This can be done by identifying shared variables and by defining their correlation. Thus, new horizontal applications are build up.
3. Verification of the consistency of results acquired from vertical applications. This is achieved by assigning the results in the horizontal applications.
4. Evaluation of the solutions found within the activities. This can be done using the degree of satisfaction of the solutions in the horizontal application. The integration of the results guides to the next milestone.
5. Review of the milestone.

Degree of detail and concreteness. The procedure described above should be applied at all milestones. This way, the network of any phase is the result of the previous phase. The network of the first phase sets the boundaries of the solution space for the whole project. This is generally done using linguistic variables. The resulting network defines the frame in which the activities related to the project have to be solved. Following the process chain, two directions of development can be distinguished, namely the *degree of detail* and the *concreteness (Fig. 1)*. The degree of detail is closely related to the number of modules or components the product consists of. It describes whether an activity is applied to product modules, segments, or single parts and thus defines the level of specification. This number increases during the lead time of the project. At the example of the milestone "Product Definition", the milestone can be subdivided into "Definition of Mechanical Parts" and "Definition of Electrical Parts" at a higher degree of detail. Concreteness describes how refined the activities are that are applied to the components of the product. This can be related to switching from linguistic variables to fuzzy sets or from fuzzy sets to the definition of tolerances. In general, more refined activities are applied to a certain part of a product at a higher level of concreteness, whereas at a higher degree of detail a certain activity is applied to a more specified part of the product.

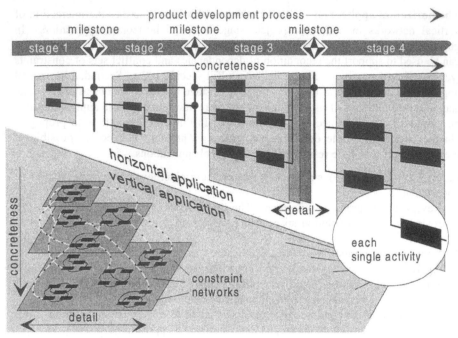

Fig. 1. Horizontal and Vertical Application

Effects on the propagation of changes. The use of horizontal and vertical applications guarantees a consistent solution. Starting from a FCN that defines the general solution space ensures, that the solutions of all subnetworks are valid. Thus, internal disturbances can be avoided along the straight forward process. However, the applications shown should also be able to reduce the effects of changes made to the model due to external influences. The need to deal with external influences arises because of the dynamic character of the product development process.

The hierarchical organization of networks and their solutions allows to damp the effects of changes. A subnetwork covers only a certain part of the solution space defined in a network higher up in the hierarchy. Thus, a change in a network does not necessarily require changes in subnetworks and vice versa. Under the assumption that the allowed range of a certain variable has been changed and that this requires changes in other networks, the effects entailed by the change can be propagated in two different ways. First, they can be propagated in the direction of the continuous process. This is called forward propagation. Second, they can be propagated in the opposite direction, which is called backward propagation. A change at a certain point of the product development process narrows, widens or simply moves the solution space of an affected network. If such a change occurs it has to be verified, that the change itself is within a valid solution space. This is done using backward propagation. The solution spaces of the networks higher up the hierarchy have to be checked, until one arrives at a network that includes the changes made in its solution space. This network defines the upper border of all changes. No changes have to be made beyond this network. The networks in between the one that was changed and the one that verifies the change have to be reviewed. In the best case, the solution spaces of those networks can be adjusted to the changes made. This can be done by either widening or moving their solution spaces. If both proves to be impossible, the solution space offered by the network originally changed has to be narrowed. This in a way, that makes the change in the network a valid solution of the networks higher up the hierarchy. In case this is impossible, too, the changes lead to an invalid solution and require to restart the design process from the upper border.

The impact of the changes lower down the hierarchy have to be considered next. In order to do so, the solutions of the subnetworks have to be reevaluated and the solution spaces found within these subnetworks have to be adjusted. This has to be done until a solution space is reached, that is not affected by the change. This solution space is the lower border of propagation. The definition of networks using linguistic variables or fuzzy sets supports the procedures of forward and backward propagation. Linguistic networks allow to cover a wide range of values and can therefore effectively damp the propagation of changes. A requirement for damping the propagation of changes is the storage of both, the solutions of the networks and their structures. This way, the effects of changes can be managed and a trace of history can be maintained.

3.2 EXAMPLE

The propagation of changes can best be demonstrated at an example. The constraint network used in this example is structured in the way described above. It models parts of the product development process of a cellular phone and is an extension of the example shown in [6, 13]. Three different stages of the product development process are covered: the clarification_of_task, the conceptual stage and the specifying stage. To explain the concepts above, the stages 'clarification_of_task' and 'specifying stage' are simplified and represented as one network respectively. During the conceptual stage the activities 'conceptual design', 'selection of process-technology', 'battery selection', and 'cost check' are distinguished (*Fig. 2*). The horizontal application in this example covers all three stages. Vertical applications are only shown in the conceptual stage. The vertical application 'selection of process technology' is described in more detail in the subnetwork 'supplier selection' which covers a "Make or Buy" decision. Conflicting constraints are expected within the different networks. In this example, constraints with a degree of satisfaction lower than 0.5 are regarded as violated.

CLARIFICATION OF TASK: During the clarification_of_task stage, marketing, production, technology, and cost requirements are taken into consideration. These requirements are interconnected through shared variables which are defined using linguistic terms. This is done according to the level of knowledge available at this early stage in the design process. Linguistic terms and their interconnections can be represented in a fuzzy constraint network. The related variables are 'Phone weight', 'Battery life', 'Battery Technology', 'Time-to-Market', and 'Selling Price'. A solution of the clarification of task with a degree of satisfaction of 0.5 is as follows [13]:

Phone weight:	below average	(0.28 to 0.6 kg)
Battery life:	above average	(6 to 9.2 hours)
Manufacturing cost:	lower than average	($27 to $55)
Battery technology:	nickel-cadmium type 2	
Time-to-market:	average	(13.5 to 19.5 months)
Selling price:	lower than average	($65 to $125)

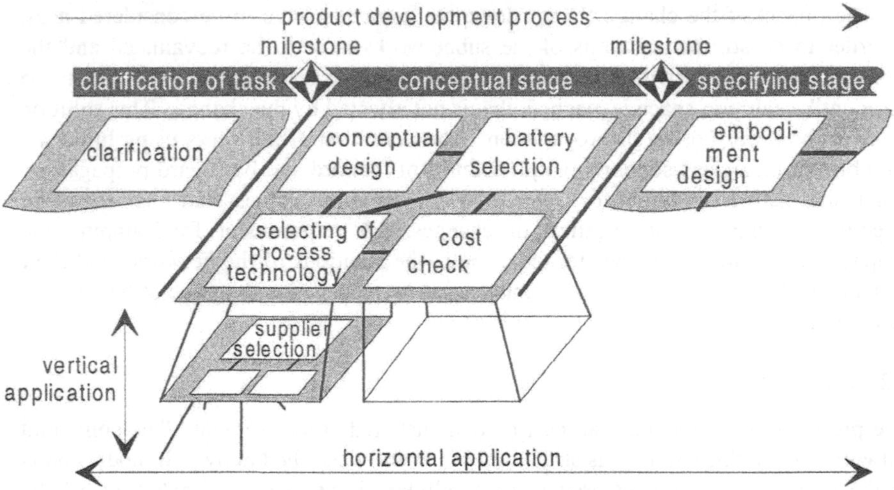

Fig. 2. Structure of the Example

CONCEPTUAL STAGE: The goal of the conceptual stage is to accomplish a stable product and process description on a rough level. Starting point of this stage are the results of the clarification_of_task stage after they have been confirmed in a milestone review. In our simplified example, four activities have to be fulfilled during the conceptual stage. Each of them can be described as a constraint network and solved separately (vertical application). Neglecting the interrelationship between these activities, inconsistencies can appear due to the uncoordinated use of shared variables. To avoid these inconsistencies, a coordinating network (horizontal application) has to be installed. The different networks are described subsequently with a special emphasis to the calculated values of shared variables.

CONCEPTUAL DESIGN: The objective of the conceptual design in this example is to calculate the dimensions of the body of the cellular phone. The related variables are 'weight', 'length', 'width', 'thickness'. The Information taken into consideration is the density of a typical cellular phone and the dimensions of a shirt pocket, the location such a device is usually kept in. The battery weight, calculated in the clarification stage, is used as a starting point to solve the fuzzy constraint network. The result of the conceptual design activity is as follows [13]:

length:	<16.9, 21, 25.5> cm
width:	<5, 7, 9> cm
thickness:	<1.5, 2, 2.5> cm

The degree of satisfaction for this solution is 0.57 [13].

BATTERY SELECTION: The battery selection in this example is reduced to a search in a configuration table (*Table 1*). This table represents the inventory data used to select a battery. Information needed from the clarification_of_task stage is 'battery technology', 'phone weight', and 'battery life'. Thus imprecise information modeled as fuzzy sets are used to select a battery described by crisp values. The result of this activity is a set of batteries with different dimensions and prices. In this example, the battery 'B_03' leads to a solution with a degree of satisfaction of '0.82'. This solution satisfies all the constraints. An alternative is the choice of battery 'B_04', which solves the network with a degree of satisfaction of '0.75'. As an independent decision, battery 'B_03' would be the adequate choice.

Part number [PN]	Battery Technology	Battery Life [h]	Weight [kg]	Length [cm]	Width [cm]	Thick-ness [cm]	Price [$]
B_01	NiCd_1	5	0.3	14	9	2.5	6
B_02	NiCd_2	6	0.1	7	5	1.5	11
B_03	NiCd_2	8.5	0.2	11	5	1.8	15
B_04	NiCd_2	7.2	0.15	11	7	2	8
B_05	NiCd_2	5	0.3	7	5	1.5	6

Table 1: Configuration Table for the Battery Selection

SELECTION OF PROCESS TECHNOLOGY: In general, two different technologies can be chosen to produce the body of the cellular phone. One is called thermoforming, the other one is injection molding. The choice of the technology is affected by the requirements 'set up time', 'investment', 'unit price' and by the number of units that are to be produced. Another important influence is the experience in using a certain technology that is available in a company. The solution spaces of the 'manufacturing cost' and of the predicted 'number of units' are provided as results of the clarification_of_task stage. Due to the fact that the knowledge available in the conceptual stage still is very imprecise, the requirements for the selection of the technology are described using linguistic terms (*Table 2*).

Technology	set up time [month]	investment [100,000 $]	unit price [$]	number of units [1,000 #]	experience
thermo-forming	middle [3.5, 4, 5]	middle [1.8, 2, 2.1]	middle [1.5, 2, 2.5]	high [18, 30, 42]	more or less
injection molding	long [5, 6, 7]	high [2.2, 2.5, 2.7]	high [2.5, 3, 4.5]	very high [40, 49, 62]	more

Table 2: Descriptions of technologies

The most important constraints are the ones concerning the costs:

$$(unit\_price + investment / number\_of\_units) * manufacturing\ supplement$$
$$\sim < 30\% * manufacturing\_cost$$

and the time-to-market:

$$set\_up\_time \sim < 30\% * time\_to\_market,$$

where 'number_of_units', 'manufacturing_cost' and 'time_to_market' are given by the clarification_of_task stage. The choice of the technology thermoforming satisfies the constraints with a degree of 0.778, injection molding leads to a degree of satisfaction of 0.523. Evaluating these results, thermoforming is regarded as the appropriate technology.

SUPPLIER SELECTION: On the second level of this vertical application, it is evaluated, whether the body of the phone is manufactured in the company or bought from a supplier. In case the decision is made to buy the part, a supplier has to be chosen. The factors taken into account for these decisions are the price per part, the time needed to start manufacturing or to receive the first delivery, the maximum number of units that can be produced, the flexibility to react to changes in the demand and the quality according to process audits. In this example, two suppliers are described in more detail.

The network 'supplier selection' is linked to the one for the selection of the technology. Shared variables are 'unit-cost' (calculated using the variables 'unit price', 'investment', and 'number of parts') and 'time-to-market'. Both are defining upper limits for the selection of a supplier.

The constraints of this application mainly match the information about the suppliers to the necessities defined in the process-selection network. The information about the suppliers and the degree to which they satisfy the constraints are shown in *Table 3*. Evaluating the results of this application, supplier 'B' turns out to be the best choice.

COST CHECK: The constraint network 'cost check' helps to avoid overpriced products. To keep this example as simple as possible, the check consists of only one formula:

$$((unit\_price + investment / number\_of\_units) * manufacturing\_supplement +$$
$$battery\_price + price\_for\_electronics) * labor\_supplement \sim < manufacturing\_cost.$$

Supplier name	cost [$]	time [month]	flexibility [% of units per month]	quality	number of units [1,000 #]	overall DoS
self	[6, 7.5, 10] DoS: 0.875	[3, 4.2, 5.5] DoS: 0.909	[8, 10, 20] DoS: 1	middle DoS: 0.692	[55, 65, 75] DoS: 0.618	0.618
A	[5, 6, 7] DoS: 1	[3.6, 4, 5] DoS: 1	[3, 6, 9] DoS: 0.5	high DoS: 1	[65, 72, 78] DoS: 0.75	0.5
B	[6.5, 8, 10] DoS: 0.75	[3, 4, 5] DoS: 1	[8, 10, 20] DoS: 1	high DoS: 1	[70, 74, 80] DoS: 0.813	0.75

DoS: Degree of Satisfaction

Table 3: Supplier selection

This network is linked to the network of the clarification_of_task stage and uses the descriptions of the available batteries 'B_03' and 'B_04' as well as the description of the different technologies. Besides, different price levels for electronic parts are taken into consideration. These price levels are described using linguistic terms. In this survey, not only the technology thermoforming is checked, injection molding is revised, too. This is done because both technologies proved to be valid solutions in the network 'technology selection'. *Table 4* describes the solutions space of the cost check network.

Technology		Battery	Prices for	Degree of
investment [$]	unit price [$]	battery price [$]	electronics	satisfaction
middle [180000, 200000, 210000]	middle [1.5, 2, 2.5].	very high [12.5, 15, 20]	low	0.521
			middle	0.298
			high	0.125
		middle [7, 8, 10]	low	0.959
			middle	0.689
			high	0.49
high [220000, 250000, 270000]	high [2.5, 3, 4.5]	very high [12.5, 15, 20]	low	0.377
			middle	0.172
			high	0.009
		middle [7, 8, 10]	low	0.771
			middle	0.525
			high	0.34

Table 4: Solution space for cost check

Five valid solutions are found that solve the cost check network. Considering only the solutions of this network, the battery 'B_04', 'low price' electronic and the technology 'thermoforming' are the foremost choice.

RESULT: The results of the presented vertical applications 'conceptual design', 'battery selection', technology selection', and 'cost check' are listed below (*Table 5*). They were solved independent of each other but based on the solutions of the clarification_of_task stage. All results obtained are valid solutions. However, the consistency of these vertical applications is less than clear. A comparison between the solution preferred when selecting the battery and the one regarded as the best one in the cost check makes the inconsistencies become obvious. These inconsistencies are due to the neglecting of constraints in-between these applications. Therefore the need for a horizontal application arises. Horizontal applications model the constraints in-between vertical applications and thus guarantee a consistent solution.

Conceptual Design (DoS: 0.57)		Battery Selection (DoS: 1)	
Phone-Length	[16.9/ 21/ 25.5] cm	Battery-Length	11 cm
Phone-Width	[5/ 7/ 9] cm	Battery-Width	5 cm
Phone-Thickness	[1.5/ 2/ 2.5] cm	Battery-Thickness	1.8 cm
Phone-Weight	[253.5/ 882/ 2295] g	Battery-Weight	200 g
		Battery-Price	15 $
Technology Selection (DoS: 0.53)		Cost Check (DoS: 0.926)	
Technology	thermoforming	Battery-Price	[7/8/10] $
Investment	[180000/ 200000/ 210000] $	Investment	[180000/ 200000/ 210000] $
Unit Price	[1.5/ 2/ 2.5] $	Unit Price	[1.5/ 2/ 2.5] $

DoS: Degree of Satisfaction

Table 5: Independently selected results of the vertical applications

THE HORIZONTAL APPLICATION: The horizontal network allows to verify the consistency of all vertical applications in one stage of the product development process. In order to achieve this, all variables that are shared by vertical applications have to be controlled at a certain level. The constraints between the vertical applications in this example are illustrated in *Fig. 3.*

To prove the consistency of the results of the vertical applications, their solutions have to be assigned in the horizontal application and their degree of satisfaction in this application has to be evaluated. In this example, the constraint C4 is violated and thus the degree of satisfaction of the horizontal application is zero. This means that the solutions obtained in the conceptual stage are invalid even though they are based on the results of the clarification_of_task stage.

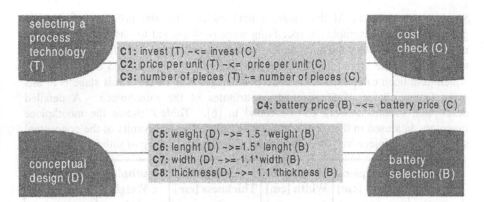

Fig. 3: Constraints in-between the vertical applications

To achieve consistency, the battery prices in the constraint network 'battery selection' and 'cost check' have to be adjusted. One possibility is to choose the battery price that was regarded as the best solution in the application 'battery selection' and to check, whether the 'cost check' network still leads to a valid solution. This is done by assigning the linguistic variable "very high" as the battery price in the cost check network. According to the results presented in table 3 this leads to a degree of satisfaction of 0.505. Thus, a valid solution can be found without having to reconsider the choice of the technology or the conceptual design (*Table 6*).

Horizontal application (DoS: 1)			
Conceptual Design (DoS: 0.57)		**Battery Selection (DoS: 1)**	
Phone-Length	[16.9/ 21/ 25.5] cm	Battery-Length	11 cm
Phone-Width	[5/ 7/ 9] cm	Battery-Width	5 cm
Phone-Thickness	[1.5/ 2/ 2.5] cm	Battery-Thickness	1.8 cm
Phone-Weight	[253.5/ 882/ 2295] g	Battery-Weight	200 g
		Battery-Price	15 $
Technology Selection (DoS: 0.53)		**Cost Check (DoS: 0.505)**	
Technology	thermoforming	Battery-Price	[12.5, 15, 20] $
Investment	[180000/ 200000/ 210000] $	Investment	[180000/ 200000/ 210000] $
Unit Price	[1.5/ 2/ 2.5] $	Unit Price	[1.5/ 2/ 2.5] $

DoS: Degree of Satisfaction

Table 6: Consistent Results of the Conceptual Stage

SPECIFYING STAGE: At this stage, a detailed product and process definition is generated. In this example the specifying stage is simplified to one single constraint network. It deals with the embodiment design, in particular with the mouthpiece of the cellular phone. Input variables are the values for the 'width' and the 'thickness', which have been calculated at the conceptual stage. The result of this stage is in this example the choice of the geometric attributes of the mouthpiece. A detailed description of this network can be found in [6]. *Table 7* shows the mouthpiece inventory data used in the constraint network. Based on the results of the conceptual stage, the mouthpiece MP-01 is chosen due to its highest degree of satisfaction.

Part Number	Mouthpiece Length [cm]	Mouthpiece Width [cm]	Mouthpiece Thickness [cm]	Mouthpiece Weight [g]	Degree of Satisfaction
MP-01	4	4	1.25	100	0.7
MP-02	5	5	1.25	125	0.7
MP-03	6	6	1.5	150	0.2
MP-04	6.5	6.5	1.75	150	0
MP-05	7	7	1.75	200	0

Table 7: Mouthpiece inventory data

CHANGES IN A VERTICAL APPLICATION: A possible change in a vertical application is a change of the conditions under which company 'B' supplies the bodies of the cellular phone. Company 'B' chooses to retract its initial offer. Instead it suggests to either raise the prize per unit to <7.5, 8.5, 10.5> $ (option B-I) or to delay the first delivery to <4, 5, 6> months (option B-II).

When dealing with a change like this, the first step is to find out which networks are directly affected by the change. In this example, only the constraint network 'supplier-selection' has to be changed. The reason for this is, that none of the variables that are subject of the change are directly related to any other vertical or horizontal application. Therefore the network 'supplier-selection' can be identified as the only starting point for the propagation of the change. However, other networks can still be affected indirectly.

In step two, the identified network has to be solved using the changed information. The results are shown in *Table 8*. According to the new results, the choice of company 'B' is no longer opportune. Now, the best solution is to manufacture the body instead of buying it.

The third step deals with the indirect affection of other applications. In general, all constraint networks which are directly connected to the one changed have to be checked. In vertical applications, shared variables occur in the networks above and below the one changed due to the structure of these applications. In this example the networks below 'supplier-selection' are not described. Hence, only the constraint network 'technology-selection' has to be checked. The variables this network shares with the one for the selection of the supplier have not been affected by the changes made. Therefore, the choice of the technology does not have to be reconsidered.

supplier name	cost [$]	time [month]	flexibility [% of units per month]	quality	number of units [1,000 #]	DoS
self	[6, 7.5, 10] DoS: 0.875	[3, 4.2, 5.5] DoS: 0.909	[8, 10, 20] DoS: 1	middle DoS: 0.692	[55, 65, 75] DoS: 0.618	0.618
A	[5, 6, 7] DoS: 1	[3.6, 4, 5] DoS: 1	[3, 6, 9] DoS: 0.5	high DoS: 1	[65, 72, 78] DoS: 0.75	0.5
B-I	[7.5, 8.5, 10.5] DoS: 0.571	[3, 4, 5] DoS: 1	[8, 10, 20] DoS: 1	high DoS: 1	[70, 74, 80] DoS: 0.813	0.571
B-II	[6.5, 8, 10] DoS: 0.75	[4, 5, 6] DoS: 0.5	[8, 10, 20] DoS: 1	high DoS: 1	[70, 74, 80] DoS: 0.813	0.5

DoS: Degree of Satisfaction

Table 8: Supplier selection

The introduction of vertically structured applications allows to identify the direct effects of a change and helps to control these. Besides, the consistency of the solution chosen can be proofed. As a consequence, the propagation of changes is significantly damped and thus the effects of changes are reduced.

CHANGES IN A HORIZONTAL APPLICATION: Let us assume that market developments force us to lower the estimated price of the product by 25% and that this increases the marketing potentiality by 10%. In this case it has to be examined, whether or not the constraint networks are affected by these changes. In our example, the variable 'price' is used in the clarification_of_task network. Thus, the change-propagation has to start there. After the review of the clarification_of_task stage, starting points for the review of the networks in the conceptual stage have to be defined. It has to be checked, whether the top levels of the vertical applications in this stage are directly influenced by the changes of the prior stage. Then, the networks of the affected applications, including their subnetworks at lower levels have to be solved individually. The next step at the conceptual stage is to prove the consistency of the changes made. To do so, the horizontal application has to be reviewed. Finally, the necessary changes in the network of the specifying stage have to be identified and made. In the following, the propagation of changes in this example will be discussed in more detail.

In the clarification_of_task network, the variable 'price' has to be changed. The use of linguistic terms to describe the price makes it possible to simply map the terms used to different values, e.g. the term 'medium' is changed from <95, 125, 155>$ to <72.5, 95, 117.5>$. The constraints using these linguistic terms are still valid. This avoids major changes in the network and simplifies its reuse.

The constraint that is mainly affected by the change is: $\dfrac{price}{manufacturing\_cost} \geq \tilde{2}$

Due to this, the manufacturing costs have also to be reduced by 25%. Having made these changes, the network can be solved with a degree of satisfaction of 0.5. The results are:

Phone weight:	below average	(0.28 to 0.6 kg)
Battery life:	above average	(6 to 9.2 hours)
Manufacturing cost:	lower than average	($20 to $41)
Battery technology:	nickel-cadmium type 2	
Time-to-market:	average	(13.5 to 19.5 months)
Selling price:	lower than average	($50 to $95)

Comparing these results to the solution obtained before, one realizes that the variables 'number of units' <45.000, 55.000, 65.000> units, 'price' <50, 72.5, 95>$, and manufacturing cost <20, 30.5, 41>$ have to be propagated in the conceptual stage. Thus, the constraint networks 'battery selection', 'technology selection', and 'cost check' are affected directly. The application 'conceptual design' does not contain any of the variables that had to be changed and therefore does not have to be reviewed.

BATTERY SELECTION: The selection of the battery still has to be made according to the configuration table (*Table 1*). Because of this, the designed network does not need to be changed. Applying the changes in manufacturing costs to this constraint network, only battery 'B_04' is a valid solution with a degree of satisfaction of 0.75.

SELECTION OF PROCESS TECHNOLOGY: Information relevant for the selection of the appropriate technology is the maximum manufacturing costs and the number of units that are to be produced. Again, no changes have to be made to the design of the network. Applying the new information, technology thermoforming obtains a degree of satisfaction of 0.688.

Technology injection molding can no longer be regarded as a valid solution since it only leads to a degree of satisfaction of 0.394. This is mainly due to a low satisfaction of the cost constraint. Since information about costs is a vivid input in the next level of this vertical application, the subnetwork 'supplier selection' has to be analyzed.

SUPPLIER SELECTION: The changes made in the cost constraint result in a new solution at the second level of the vertical application, too. Supplier 'B', who had been selected in the network before does no longer satisfy the constraint dealing with the price per unit. The option to manufacture the parts within the company does not lead to a valid solution, either. The choice of supplier 'A' is the only possible solution in this network.

COST CHECK: Solving the constraint network 'cost check' applying the new variables 'number of units' and 'manufacturing cost', a single solution can be found. The combination of middle unit price, middle investment, middle battery price, and low price for the electronics leads to a degree of satisfaction of 0.511.

HORIZONTAL APPLICATION: To verify the new solutions of the vertical applications in the conceptual stage the horizontal application has to be solved. The horizontal application mainly checks the shared variables. In this example, the calculated results are consistent and satisfy this application to a degree of 0.6.

A valid solution in the horizontal application indicates, that the review of the conceptual stage is completed. *Table 9* summarizes the result of the changes:

Horizontal application (DoS: 0.6)			
Conceptual Design (DoS: 0.57)		Battery Selection (DoS: 0.75)	
Phone-Length	[16.9/ 21/ 25.5] cm	Battery-Length	11 cm
Phone-Width	[5/ 7/ 9] cm	Battery-Width	7 cm
Phone-Thickness	[1.5/ 2/ 2.5] cm	Battery-Thickness	2 cm
Phone-Weight	[253.5/ 882/ 2295] g	Battery-Weight	150 g
		Battery-Price	8 $
Technology Selection (DoS: 0.688)		Cost Check (DoS: 0.511)	
Technology	thermoforming	Battery-Price	[7, 8, 10] $
Investment	[180000/ 200000/ 210000] $	Investment	[180000/ 200000/ 210000] $
Unit Price	[1.5/ 2/ 2.5] $	Unit Price	[1.5/ 2/ 2.5] $

DoS: Degree of Satisfaction

Table 9: Consistent Results of the Conceptual Stage

The constraint network in the specifying stage does not share any variables with the ones changed in the conceptual stage. Therefore, the changes at the conceptual stage do not have to be propagated in this network.

This example shows the benefits of structuring and modeling the product development process using fuzzy constraint networks. The propagation of changes can be controlled and the limits of the propagation can easily be found by identifying the variables networks share. On top of that, the consistency of the solutions chosen can be proved.

4. CONCLUSION

In this article, a possibility to reduce the dynamics that are brought into the product development process through changes is described. Using the structure of the development process and fuzzy constraint networks as introduced here, the propagation of changes can be controlled. Furthermore, an upper and a lower border of the impact of a change can be identified. This is demonstrated at an example. It is shown, that changes and their effects can be made transparent even in a complex environment. To do so, vertical and horizontal applications are distinguished in the product development process. The former allow to damp the propagation of changes whereas the latter can be used to proof the consistency of the solutions chosen. This way, the dynamics of a product development process can be reduced and its efficiency is improved.

5. ACKNOWLEDGMENT

The results presented in this article were obtained in the common research project 361 "Models and methods for an integrated product and process development" funded by the German Research Community (DFG).

The collaboration between the North Carolina State University (USA) and the RWTH Aachen (FRG) was supported by a DAAD-fellowship HSP II financed by the German Federal Ministry of Education, Science, Research and Technology.

6. LITERATURE

1   Eversheim, W., Roggatz, A., Zimmermann, H.-J. and Derichs, Th., (1995). "Kurze Produktentwicklungszeiten durch Nutzung unsicherer Informationen", it+ti, vol 37., pp. 47-53.

2   Laufenberg, L. (1995). "Methodik zur integrierten Projektgestaltung für die situative Umsetzung des Simultaneous Engineering", Dissertation, Aachen, 1995.

3   Jackson, S., D., Sutton, J., C., Zorowski, C., F. (1993). "Design for Assembly Using Fuzzy Sets", DE-vol. 52, Design for Manufacturability, ASME, pp. 117-122.

4   Otto, K., N., Antonsson, E. K., (1994). "Modeling Imprecision in Product Design", 1994 IEEE International Conference on Fuzzy Systems, vol. 1, pp. 346-351.

5   Pearce, R., Cowley, P., H., (1996). "Use of Fuzzy Logic to Describe Constraints Derived from Engineering Judgment in Genetic Algorithms", IEEE Transactions on Industrial Electronics, vol. 43, n. 5, pp. 535-540.

6   Giachetti, R., E., Young, R., E., Roggatz, A., Eversheim, W., Perrone, G. (1996). "A Methodology for the Reduction of Imprecision in the Engineering Design Process", to appear in European Journal of Operations Research.

140

7   Dechter, R. and Pearl, J., (1988). "Network-Based Heuristics for Constraint-Satisfaction Problems", Artificial Intelligence, vol. 34, pp. 1-38.

8   Young, R., E., Giachetti, R., E., Ress, D., A. (1996). "Fuzzy Constraint Satisfaction in design and manufacturing", IEEE Conference on Fuzzy Systems (Fuzz-IEEE'96), New Orleans, pp. 8-10.

9   Dubois, D., and Prade, H. (1988). "Possibility Theory", Plenum Press, New York.

10   Young, R., E., Eversheim, W., Roggatz, A., Nöller, C. (1995). "Dynamic Change Propagation within Concurrent Engineering System", Proceedings of the 5th International FAIM Conference, Schaft, Ahmad, Sulivan, Jacobi (editors), Begell House, New York, N.Y., pp. 724-733.

11   Wildemann, H. (1993). "Optimierung von Entwicklungszeiten: Just-in-Time in Forschung & Entwicklung und Konstruktion", Transfer-Centrum-Verlag GmbH.

12   Kleinschmidt, E.; Geschka, H.; Cooper, R. (1996). "Erfolgsfaktor Markt - Kundenorientierte Produktinnovation", Springer Verlag.

13   Young, R., E.; Perrone, G.; Eversheim, W.; Roggatz, A.:
     Fuzzy Constraint Satisfaction for Simultaneous Engineering;
     Production Engineering Vol. II/2 (1995), S.: 181-184.

A Novel Neural Network Technique for Modelling Data Containing Multiple Functions

Owen M. Lewis & J.Andrew Ware

Division of Mathematics and Computing, University of Glamorgan
Pontypridd, Mid Glamorgan, UK.

Abstract.

Increasingly neural network techniques are being applied to a wide range of pattern recognition and classification problems. However, there is often insufficient information available to facilitate optimal operation. This problem can lead to a situation where the data exhibits signs of containing multiple underlying functions. For example, if location is not included as a feature when modelling residential property appraisal, the data will appear to map across more than one underlying function. The methodology proposed in this paper uses a form of data stratification to overcome this problem. The premise followed is that it is better to produce multiple models that are specific to - and accurate within - certain scenarios, rather than a single model that is too general and therefore inaccurate.

1. Introduction

One of the major unresolved issues affecting the use of neural networks is the selection of a suitable input vector that will facilitate the correct determination of the output vector. Moreover, users of neural networks often have available an insufficient subset of the information that would enable effective modelling of the mapping function between input and output space.

When there is sufficient *a priori* knowledge, selecting input features is a relatively simple task. For example, when attempting to model house prices, common sense dictates that there is no need to include 'door colour' in the input vector. Similarly, it is intuitive that if the 'number of bedrooms' or 'location' is missing from the input vector then the model will only have limited use in prediction.

When input space does not include the full complement of features that would enable effective mapping to take place, then it resembles a situation in which the input space contains more than one underlying function. For example, two identical houses may differ in value by £50,000 because of their location - if locational information is not available then a single function cannot model the scenario successfully - whereas two separate functions might. Therefore, this paper considers an input space containing missing features to be equivalent to an input space generated by multiple functions.

The problem of modelling data containing multiple underlying functions can be ameliorated by stratifying the data set into a number of subsets that each contain a single underlying function. Consider the simple following example: a data set containing records produced by the following functions:

$$x = 6a + 5b - c$$
$$y = 4a - b + 3c$$
$$z = 2a + 3b + 7c$$

Table 1 shows an example of such a data set.

Table 1 - An extract from the generated data set.

a	b	c	result
1	2	3	13
1	2	3	11
1	2	3	29
:	:	:	:
:	:	:	:
:	:	:	:
2	2	2	20

If a complete data set of this type was analysed using either multiple regression analysis or neural networks, the ability to predict the output given the values of a, b and c would be poor. However, if the records could be split up into three separate data sets (a data set for those records produced by equ.1, a data set for those records produced by equ.2 and finally a data set for those records produced by equ.3) then the problem could be solved.

In order to understand how to achieve this data stratification a number of scenarios related to clustering in input and output spaces are now considered. Figure 1 shows two of these scenarios.

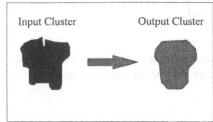

Figure 1a - Useful Input Clusters

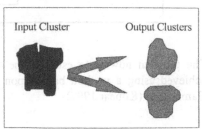

Figure 1b - Useless Input Clusters

Figure 1a shows a cluster of similar input vectors. When the corresponding data in output space is examined all the examples describe similar output values. For example, if the input cluster describes houses that have three bedrooms; semi-detached; under 2 years old - then the cluster in output space says they all have similar property values. Conversely, Figure1b shows a situation where the data can only be modelled using two functions.

In order to determine whether the whole of the input and output space is suitable for being modelled with a single MLP - or whether data stratification needs to be performed - the following algorithm can be applied:

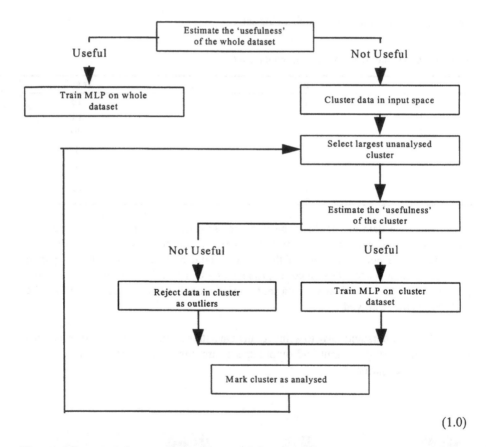

(1.0)

The problem now is to estimate the 'usefulness' of a given cluster. This can be achieved using a method based upon a variance estimation routine known as the Gamma test (Koncar 1997).

2. The Gamma Test

The Gamma test attempts to estimate the best mean square error that can be achieved by any smooth modelling technique using the data. If y is the output of a function then the Gamma test estimates the variance of the part of y that cannot be accounted for by a smooth (differentiable) functional transformation. Thus if $y = f(x) + r$, where the function f is unknown and r is statistical noise, the Gamma test estimates Var(r).

Var(r) provides a lower bound for the mean squared error of the output y, beyond which additional training is of no significant use. Therefore, knowing Var(r) for a data set allows prediction beforehand of what the MSE of the best possible neural network trained on that data would be.

Applying this test to the data set shown in table 1 would obviously result in a high variance. However, the Gamma test provides a method of determining the quality of the data stratification - a good stratification technique will result in a low value of Var(r) for each subset.

3. Methodology

In order to implement algorithm (1.0) a Kohonen Self Organising map can be used to cluster the data in the input space. The framework shown in Figure 2 illustrates the methodology.

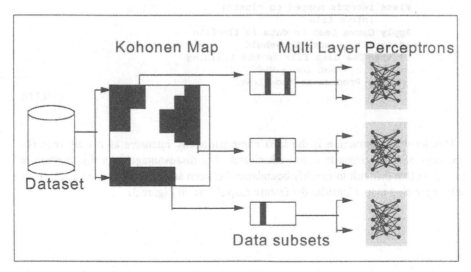

Figure 2 - Methodology used for stratification of data.

Once the Kohonen network has converged, the data from each significant grouping becomes the training set for a MLP. Each MLP is trained using a save best approach. This involves periodical testing of the network with a validation set and, if the current state produces better results than the previous best, the current state becomes the new best state. At the end of training the current best state is adopted as the network to be used during day-to-day operations.

The methodology divides into three abstraction levels:

- Cluster level
- Node Level
- Record Level

Data stratification is achieved at cluster level or at node level, depending on the ease at which cluster boundaries can be determined. The record level gives an indication of outliers.

3.1 Cluster Level Analysis

This can be achieved thus:

```
Identify  Cluster boundaries in Kohonen map
For every cluster
  Place records mapped to cluster
        into a file
  Apply Gamma test to data in the file
  If Var(r) <= some Threshold
      then Use data file as the training
                set for a MLP
      else Process at Node Level
```

$$(2.0)$$

This level of abstraction is the least computationally intensive as it only requires one pass of the Gamma test for each cluster. The disadvantage with this method is that it is often difficult to identify boundaries between adjacent clusters on a Kohonen self organising map. Consider the feature map shown in Figure 3.

146

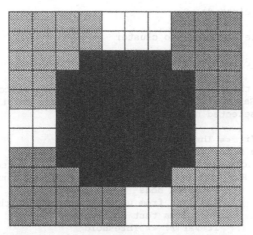

Figure 3 - An Example Feature Map for a Trained Kohonen Network.

There appears to be five classes within the data set, but there are regions of uncertainty relating to the boundaries of each cluster. The boundaries may be estimated visually but this may result in a loss of accuracy.

If binary input variables are used, class boundaries can be found by inspecting the records mapped to each node and grouping together nodes that contain the same values. Unfortunately, using binary inputs with a Kohonen network, often gives groupings that are no better than those achieved using a simple sort procedure. However, if continuous input variables are used to represent the data, this inspection method will no longer work as the similarities between records will not always be apparent. A solution to this problem is to examine the Euclidean distance from the subject node to the nearest identified centroids, and attribute the subject node to the cluster that it is closest to (Zurada 1992).

3.2 Node Level Analysis

At this abstraction level, the methodology attempts to identify useful clusters by selecting a centroid and adding neighbouring nodes - where the addition of a node increases the variance significantly it is subsequently removed. This process iterates until the cluster size is maximised within a specified variance threshold. This is achieved thus:

```
number_of_clusters:=0

While there are nodes to cluster

    number_of_clusters:= number_of_clusters + 1

    Select the unclusterd node with the largest record count
    Apply Gamma test to estimate the variance for the data in the
          selected node

    If Var(r) <= Threshold
       then nodes_of_interest:=None
            cluster includes only the data from selected node
            For each unclusterd node immediately surrounding
                selected node
                Add data from unclustered node to the cluster
                Run gamma test on cluster
                If Var(r) <= Threshold
                   then Add unclusterd node number to
                             nodes_of_interest
                   else Remove data from the unclustered node
                             from the cluster

            While nodes_of_interest <> None
                Select c_node from nodes_of_interest
                Remove c_node from nodes_of_interest
                For each unclusterd node immediately
                    surrounding c_node
                    Add data from unclustered node to the
                        cluster
                    Run gamma test on cluster
                    If Var(r) <= Threshold
                       then Add unclusterd node number to
                                 nodes_of_interest
                       else Remove data from the unclustered
                                 node from the cluster

            Record the boundaries of this cluster

       else Process at Record Level
```

$$(3.0)$$

This algorithm identifies useful clusters on a 2D Kohonen map. The boundary detection algorithm for a 1D Kohonen map is very similar except neighbouring nodes are selected progressively further away from the left and the right of the centroid node.

This level of analysis is more computationally intensive than algorithm (2.0) as it require $m*\sum n_i$ passes of the Gamma test, where 'i' is the number of nodes investigated for cluster 'n' for a Kohonen map containing 'm' clusters.

If using either the cluster level analysis (2.0) or the node level analysis (3.0), useful clusters have been identified, it is then possible to train an independent MLP on each subset. The Kohonen map is then used to select the appropriate MLP on which to predict the value of a previously unseen example. The resulting system is closely related to a panel judgement system.

However, if the methods described in (2.0) and (3.0) have still resulted in poor training sets (useless clusters) then the analysis is taken to the most detailed abstraction level, that is the record level.

3.3 Record Level Analysis

The record level analysis is the most computationally intensive. The purpose of this level of the methodology is to identify data subsets from examples that have mapped to the same node on the Kohonen map. This facilitates extraction of outliers from a data set as well as giving some indication as to the examples that require additional features.

The algorithm developed for this level of analysis is very similar to that shown in (3.0). However, this time it is sets of records that are iteratively analysed using the Gamma test. This is achieved thus:

```
For each node in the Kohonen
  Apply Gamma test to estimate the variance for the data in node
  If Var(r) > Threshold
    then For each record at node
          Remove record from data set
          Apply Gamma test to estimate the variance for the
                data in node
          If New Var(r) < Previous Var(r)
            then Mark record as outlier
            else Add record back into data set
    else Proceed at Node Level
```

$$(4.0)$$

This level of analysis will identify the need for additional features and highlight records that may be classed as outliers.

4. Conclusions

It is envisaged that the methodology outlined will be useful in many application areas. For example, it has been tested using property data from the UK with the objective of predicting a property price given a set of attributes. The methodology compared very favourably with neural network models trained on the whole data set. The results show an average increase in prediction accuracy of 10% (Lewis et al. 1997). Similar results were achieved when using the methodology as a predictor in a digital elevation model data compressor (Ware et al. 1997).

5. References

Koncar N: Optimisation Methodologies for Direct Inverse Neurocontrol, Ph.D. Thesis, Department of Computing, 180 Queen's Gate London, SW7 2XZ, U.K, 1997.

Lewis OM, Ware JA, Jenkins DH: A Novel Neural Network Technique for the Valuation of Residential Property, Journal of Neural Computing and Applications, Springer Verlag, 1997.

Zurada, JM: Introduction to Artificial Neural Systems, West Publishing Company (ISBN 0-314-93391-3) p58, 1992.

Ware JA, Lewis OM, Kidner DB: A Neural Network Approach to the Compression of Digital Elevation Models, 5th GISRUK Research Conference - Leeds, 1997.

Using Genetic Engineering to Find Modular Structures and Activation Functions for Architectures of Artificial Neural Networks

Christoph M. Friedrich[1] and Claudio Moraga[2]

[1] University of Witten/Herdecke; Institute for Technology Development and Systems Analysis; Alfred-Herrhausen Str. 50; D-58448 Witten, Germany; E-mail: chris@uni-wh.de; URL: http://www.tussy.uni-wh.de/~chris
[2] University of Dortmund; Department of Computer Science; D-44221 Dortmund; Germany; E-mail: moraga@ls1.informatik.uni-dortmund.de

Abstract. An Evolutionary Algorithm is used to optimize the architecture and activation functions of an Artificial Neural Networks (ANN). It will be shown that it is possible, with the help of a graph-database and Genetic Engineering, to find modular structures for these networks. Some new graph-rewritings are used to construct families of architectures from these modular structures. Simulation results for two problems are given. An analysis of the data in the database suggest the usage of symmetric activation functions.

1 Introduction

One of the major problems using ANN's is the design of their architecture. The architecture of an ANN greatly influences its learning capability and generalization ability. Beside of constructive and pruning techniques, Evolutionary Algorithms have been suggested by many scientists to find good performing architectures for Artificial Neural Networks. For a summary of recent results reading of [2] is suggested. Most of these methods have the problem of scalability. In [4] a method was proposed to find building-blocks for architectures of Artificial Neural Networks and to create families of modular architectures. In this paper an extension of this method is presented that additionally to the architecture optimizes the kind of activation functions of the neurons. Furthermore new graph-rewriting methods are suggested and tested.

2 Cellular Encoding

A method to optimize the architecture and weights of boolean neural networks, where the weights are restricted to the values -1 and 1, was suggested by Gruau [5] who named it Cellular Encoding. In this encoding technique, the information to develop an architecture is obtained by interpreting the information from a grammar-tree. The nodes of this tree encode information

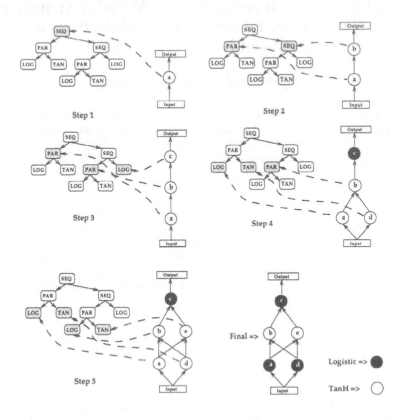

Fig. 1. Six steps to develop an 2-2-1 network with mixed activation functions

about graph rewriting operations. The development of a neural network starts with a graph consisting of one node (cell), having ingoing connections from the input area and outgoing connections to the output area. Every cell has a reading head pointing to one node of the grammar-tree (cellular code). The nodes of the cellular code contain symbols defining the graph rewriting operation. This technique is comparable to a Turing machine. Instead of writing to a tape in Cellular Encoding the cells are changed. The operation #par: for example symbolizes a parallel cell division, both cells inherit the input and output connections from the mother cell. The reading heads of the new created cells are moved to the left and right subtree of the grammar-tree. For biological plausibility, the development of the cells should be parallel. This can be achieved by using a FIFO Queue. The development of a network ends, if all cells evaluate the operations, located as terminals at the leaves of the grammar-tree. The evaluation of the terminal #log: for example determines the final activation function of the node. Figure 1 shows six steps to develop the architecture of an XOR network with mixed activation functions.

As mentioned above, the cellular code is given as a grammar-tree. In this case it is possible to use Genetic Programming [8] to optimize the cellular code. Genetic Programming is a Genetic Algorithm, that uses trees instead of Bitstrings as representation. The recombination operator crossover is realized by exchanging the subtrees of two parent individuals. The mutation operator brings some variations into the genome and is realized by inclusion of randomly initialized trees.

Gruau has proved some properties of his encoding technique, among them are completeness, compactness, closure, modularity and scalability. He got these properties by using a recursion operator that makes it possible to repeat parts of his cellular Code. In [6] he used Automatic Defined Functions (ADF) to get modular architectures.

3   Modified Cellular Encoding

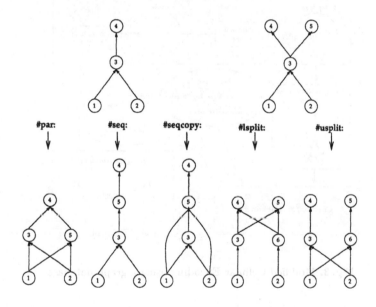

Fig. 2. The used rewriting operations for the modified Cellular Encoding

For the optimization of the architecture of ANN's with real-valued weights, the Cellular Encoding method was modified. Some new Development operators were developed and tested. The set of operators {#par:, #seq:, #seqcopy:, #usplit:, #lsplit:} was used to develop the architectures. Figure 2 shows the effects of rewriting the hidden node of a network with the development operators. In contrast to some work of Gruau in this work the input and output nodes are not coded in the cellular code and are given by the problem. As an extension to [4], the set of terminals {#tanh:, #tanhDiv2: #log: #gaussian:} was used to develop architectures with mixed activation functions. Following activation functions were used:

$$f_{\#tanh:}(x) = tanh(x) \qquad (1)$$

$$f_{\#tanhDiv2:}(x) = tanh(\frac{x}{2}) \qquad (2)$$

$$f_{\#log:}(x) = \frac{1}{1 + e^{-x}} \qquad (3)$$

$$f_{\#gaussian:}(x) = e^{-x^2} \qquad (4)$$

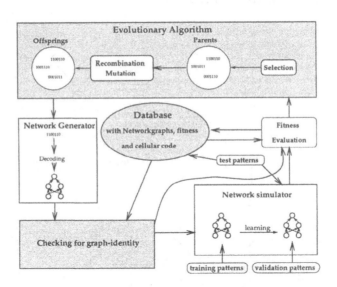

Fig. 3. Modified Cellular Encoding using a graph-database

One property of the created networks should be their correctness. A created graph is a correct architecture for a feedforward neural network, if it contains only feedforward connections and if all hidden nodes are on a path from an input node to an output node. All operators were designed to hold this property, so only correct network graphs could be created. Different to Gruau's work, it is possible with the operator #seqcopy:, to create networks with shortcut connections.

The second modification to the evolutionary development cycle was the use of a graph-database. This database contains the cellular code, network graphs and fitness parameters like the number of necessary epochs for learning, classification error on the learning, validation and test set and other parameters of the networks. Figure 3 shows the working structure of the evolutionary cycle.

Before the fitness calculation of a developed network, it is checked, if the fitness of this network was calculated five* times before. If not, the fitness is calculated again and the results are saved in the database. The resulting fitness of a network is the mean of all fitness evaluations of this network. This method allows it to minimize the fitness distortion resulting from the random initialization of the weights. This increases the validity and robustness of the obtained results. Several authors document only solutions which are results of a single fitness evaluation. In this work, only results are given for architectures tested five times.

3.1 Genetic Engineering

Optimizing the architecture of an ANN using Evolutionary Algorithms leads to proper performing architectures for problems. But it gives no insight into the problem, how good architectures are built or what activation functions should be used in general. It would be more interesting to find building principles for good architectures to create modular architectures.

In his work on Genetic Programming, Koza [8] suggested Automatic Defined Functions as a possible solution to this problem. Unfortunately for the task of architecture Optimization the fitness evaluation is very computing intensive, so it is not possible to use big population sizes and many epochs. Especially for Genetic Programming, where the representation of a problem is a grammar-tree, Altenberg [1] suggested a method to find good modular structures that are subtrees of the representation. He called this technique *Genetic Engineering*. It depends on the assumption, that some parts of the genetic program have a higher impact on the complete fitness of the genetic program than others. It is a problem to attach a fitness value to sub-programs, because it is only possible to obtain the fitness of the complete phenotype.

In this work the fitness of modular structures of the genetic programs were found by statistically analyzing the cellular code in the graph-database, which was build during the formerly described optimization process. This makes it possible to obtain information about several optimization runs that start with different populations. The fitness of a subtree is equivalent to the frequency of this subtree in the database. The found modular structures are called modules and noted as a preorder traversal of the subtree.

---

* The number five was used as a compromise between statistical evaluation possibility and necessary computing time. It would be better to use the fitness distribution as information.

3.2 Graph-Rewriting

Having found good modular structures for neural network architectures, it would be interesting to see how architectures built of this modules perform. One possibility is to include these subtrees as encapsulated functions in the next evolutionary optimization process.

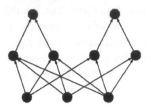

Fig. 4. Resulting network of $(1, 1)_{\#par:}$-rewriting of a 3-1-2 network with the module #usplit:#par:#lsplit:

In contrast to this, in this work some special graph-rewriting methods were used. The $(i, k)_{op}$-rewriting of a network graph with a module starts with a fully connected $n - 1 - m$ network, where n, m are determined by the problem. Then the hidden node will be rewritten i-times with the development-operator op. The resulting hidden nodes are rewritten k times with the rewriting operations defined by the module. For $k = 0$ and $op = \#par$: this results in standard architectures with one hidden layer and i hidden nodes. For $op=\#$seqcopy: this results in deep networks like that of the Cascade Correlation algorithm [3]. With this method it is possible to create families of architectures consisting of good performing modules. Figure 4 shows an example of a network created by $(1, 1)_{\#par:}$-rewriting of a 3-1-2 network with the module #usplit:#par:#lsplit:. The hidden node will be first rewritten with a #usplit: operation and the resulting nodes will be rewritten with the operation #par: or #lsplit:.

4 Experimental Results

The data for the first experiment are taken from the benchmark suite from Prechelt [11]. The task for the diabetes problem is to classify from some diagnostic data (e.g. blood pressure, result of glucose tolerance test etc.), whether a female Pima indian is diabetes positive or not. This problem with 8 input and 2 output parameters is difficult, because some values for the input patterns are not available and set to zero. This decepts the classification algorithm, but is realistic for real world tasks. In contrast to [4] in this work cross validation with 3 datasets (576/96/96) and weight-decay was used to find good performing architectures. This improvement makes it possible to obtain better results than in that paper. Additionally there is no bias towards the test set by the evolutionary algorithm, because the fitness function does not use the test set. Prechelt and Michie et al. [10] found networks with an classification error on the test set of 24.8 %. The learning set was learned with RPROP up to 500 epochs, but usually cross validation stopped learning before reaching this maximum. As classification method WTA (winner takes all) was used. In the first tests only $tanh(x)$ was used as the activation function. In a second test, mixed activation functions were allowed. The population size for the Genetic Algorithm was 100. The fitness function was a linear combination of the classification error on the validation set (factor 1) and the learning set (factor 0.3).

Table 1. Classification error for the diabetes problem with the module #par:#usplit:#usplit:

Architecture	Size	classification error			Epochs
	Nodes/Weights	learn set	vali. set	test set	
Best Evo	33/148	15.0 %	22.4 %	23.1 %	180
Best Evo Mixed	21/95	17.0 %	21.3 %	23.8 %	120
$(13,0)_{\#par:}$	23/130	23.8 %	24.5 %	25.0 %	145
$(5,1)_{\#par:}$	26/160	20.1 %	23.1 %	23.1 %	212
$(2,1)_{\#seqcopy:}$	18/100	20.3 %	21.8 %	25.6 %	135

Several optimization runs with and without mixed activation functions were made. Altogether about 120,000 architectures were tested, this needed about 20 days of an UltraSparc computing time. This is much time, but it is possible to parallelize the evolutionary algorithm. The application of Genetic Engineering on the data in the database found the module #par:#usplit:-#usplit: with the highest frequency. Table 1 summarizes the results. All results are the mean of five different runs from the best members of any graph-rewriting family. Figure 5 shows the best architecture with mixed activation functions found by the evolutionary algorithm. It can be seen, that the evolutionary algorithm found a network with a strict hierarchical structure.

It can be concluded that for this problem, the evolutionary algorithm found networks which are better than standard architectures and networks found in literature. The performance of networks with mixed activation functions is comparable to that of uniform $tanh(x)$ activation functions but their architecture is smaller. The architectures built by $(i,1)_{\#par:}$-rewriting with the best found module show better performance than standard architectures. The architecture built by $(2,1)_{\#seqcopy:}$-rewriting (compare Figure 6) gave good results on the learning and validation set, but the bias towards the patterns in the validation set leads to bad results for the test set. All other rewriting classes except the #seq:-rewritings gave results comparable to standard architectures.

The second tested problem was the well known two-spirals problem [9]. The task is to learn to discriminate between two sets of training points which lie on two distinct spirals in the x-y plane. These spirals coil three times around the origin and around one another. This problem is very difficult for standard backpropagation networks and only Cascade Correlation [3] gave

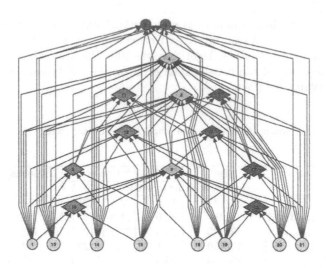

Fig. 5. Best evolved network for the diabetes problem with mixed activation functions

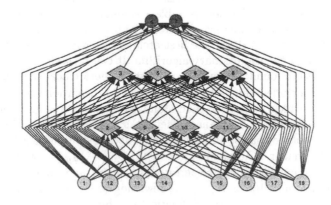

Fig. 6. network created by $(2,1)_{\#seqcopy:}$-rewriting with the module #par:#usplit:#usplit: for the diabetes problem

good results, their network needed about 1700 epochs to solve this problem. In this work the output activation function was the $tanh(x)$ function and as classification method the correctness of the sign of the outputs was used. This is a less strict classificaton method than the 402040 technique, but good for boolean classification tasks. For all evolutionary created networks the number of hidden nodes was bounded at 40 and RPROP was used to present the learning set up to 3000 epochs.

The first spirals test was the usage as an approximation problem with the classic 194 points and no cross validation. The evolutionary algorithm with the $tanh(x)$ activation function found a network with 25 nodes and 116 weights that approximates, in the mean of five runs, the spirals in 1788 Epochs. It should be noted, that not all five runs were successful in learning all points. The application of Genetic Engineering found the module #seqcopy:#usplit:#usplit:. The family of architectures build by $(i, 1)_{\#par:}$- rewriting gave only results comparable to that of standard architectures with comparable size. Whether this modular architectures nor standard architectures with up to 40 nodes could learn this function in 3000 epochs.

Table 2. Classification error for the second spirals test with the module #par:#lsplit:#par:

Architecture	Size	classification error			Epochs
	Nodes/Weights	learn set	vali. set	test set	
Best Evo	39/123	3.3 %	31.4 %	40.2 %	2157
Best Evo Mixed	18/55	5.2 %	38.8 %	42.7 %	1728
$(40, 0)_{\#par:}$	43/122	5.1 %	47.0 %	52.0 %	2736
$(8, 1)_{\#par:}$	35/82	7.2 %	41.3 %	47.8 %	2423
$(7, 1)_{\#seqcopy:}$	31/300	2.5 %	45.2 %	49.3 %	1008

In the second spirals test, the network was learned with 194 points and tested on 584 unseen points of the spirals to measure generalization ability. The validation and test set consists of 292 points each. This time experiments with and without mixed activation functions were made. The fitness function for the evolutionary algorithm was a linear combination of classification error on the validation set (factor 1) and the classification error on the learning set (factor 0.3). The learning set was learned with RPROP up to 3000 epochs, but usually cross validation stopped learning before.

Again Several runs with and without mixed activation functions were made. Altogether about 60,000 architectures were tested. The application of Genetic Engineering on the data in the database found the module #par:- #lsplit:#par: with the highest frequency. Table 2 summarizes the results for the second spirals test. The presented results are from the best networks in

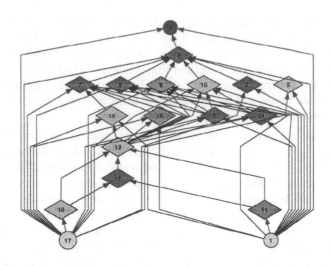

Fig. 7. best evolutionary found network with mixed activation functions for the second spirals test

every graph-rewriting class. Figure 7 shows the network graph of the best evolutionary found network with mixed activation functions.

It could be shown, that networks created by the modified Cellular Encoding technique or by graph-rewriting have better generalization abilities than standard architectures of comparable size. Evolutionary created networks with mixed activation functions show comparable performance than uniform $tanh(x)$ activation functions, but again architectures are smaller. It could be seen that the two spirals problem is a very hard problem for standard feedforward networks. The architectures tend to memorize the learning set, so big network graphs with many weights (degrees of freedom) were mainly created. No network architecture shows satisfying generalization abilities for this problem.

The database was statistically analyzed to get information about the used activation functions. Table 3 gives an overview about the used activation functions. It can be concluded, that for the tested problems the usage of symmetric activation functions (not gaussian) gives better results for evolutionary created networks.

Table 3. Percentages of the used activation functions in all evolved network graphs with mixed activation function

Problem	#gaussian:	#tanH:	#tanHdiv2:	#logistic:
Diabetes	7.9 %	42.8 %	27.9 %	21.4 %
Spirals	1.4 %	46.6 %	11.8 %	40.2 %

An analysis of the last population of every evolutionary optimization process shows a bias towards architectures that have good classification performance on the validation set, but worse classification on the unseen test set. Therefore it is necessary to be very careful to choose pattern sets that are representative and big enough for the problem domain and to adjust the parameters for cross validation and the evolutionary process.

5 Conclusion and Outlook

It was shown that it is possible to find good performing architectures for nonboolean ANN's, with a modified Cellular Encoding method. All evolved architectures show better performance for the optimized criteria than comparable standard architectures or architectures found in literature. With the help of a graph-database it was furthermore possible to find modular structures for the architecture of ANN's. Architectures created with a graph-rewriting method from this modular structures show better performance than standard architectures. The usage of mixed activation functions in an architecture show similar performance than uniform activation functions. From data in the database it can be concluded, that symmetric activation functions result in better performance. This technique can be used as an alternative to Automatic Defined Functions for problems that restrict the number of possible fitness evaluations. Using this method, it is possible to create a library of good modular structures for problem-specific modular network architectures.

In later works the possibility should be tested, if the modules found by Genetic Engineering could be used for the mixture of expert networks architecture proposed in [7]. Another possibility will be the usage in Cascade Correlation networks, where not nodes but modules will be cascaded.

References

1. L. Altenberg. The evolution of evolvability in genetic programming. In K. E. Kinnear, editor, *Advances in Genetic Programming*. MIT Press, 1994. http://pueo.mhpcc.edu/ altenber/PAPERS/Papers2.html.
2. J. Branke. Evolutionary algorithms for neural network design and training. In *Proceedings of the 1st Nordic Workshop on Genetic Algorithms and its Applications*, 1995. ftp://ftp.aifb.uni-karlsruhe.de/pub/jbr/Vaasa.ps.
3. S. Fahlmann and C. Lebiere. The cascade-correlation learning architecture. In *Advances in Neural Information Systems -2*, pages 525–532. Morgan Kaufmann, 1990.
4. C. M. Friedrich and C. Moraga. An evolutionary method to find good building-blocks for architectures of artificial neural networks. In *Proceedings of the Sixth International Conference on Information Processing and Management of Uncertainty in Knowledge Based Systems (IPMU '96)*, pages 951–956, Granada, Spain, 1996. ftp://archive.cis.ohio-state.edu/pub/neuroprose/friedrich.ipmu96.ps.Z.

5. F. Gruau. Cellular encoding of genetic neural networks. Technical Report 92-21, Ecole Normale Superieure de Lyon, Institut IMAG, 1992. ftp://lip.ens-lyon.fr/pub/Rapports/RR/RR92/RR92-21.ps.Z.
6. F. Gruau. Genetic synthesis of modular neural networks. In Stephanie Forrest, editor, *Proceedings of the Fifth International Conference on Genetic Algorithms*, pages 318–325. Morgan Kaufmann, 1993.
7. R. A. Jacobs, M. I. Jordan, and A. G. Barto. Task decomposition through competition in a modular connectionist architecture: The what and where vision tasks. Technical Report COINS Technical Report 90-27, University of Massachusetts, Amherst, MA, 1990. ftp://archive.cis.ohio-state.edu/pub/neuroprose/jacobs.modular.ps.Z.
8. J. R. Koza. *Genetic programming, on the programming of Computers by means of natural selection*. MIT Press, 1992.
9. K. J. Lang and M. J. Witbrock. Learning to tell two spirals apart. In *Proceedings of the 1988 Connectionist Models Summer School*, 1988.
10. D. Michie, D. J. Spiegelhalter, and C. C. Taylor. *Machine Learning, neural and statistical classification*. Ellis Horwood Ltd., 1994.
11. Lutz Prechelt. PROBEN1 — A set of benchmarks and benchmarking rules for neural network training algorithms. Technical Report 21/94, Fakultät für Informatik, Universität Karlsruhe, D-76128 Karlsruhe, Germany, September 1994. ftp://ftp.ira.uka.de/pub/papers/techreports/1994/1994-21.ps.Z.

Genetic Algorithms for Solving Systems of Fuzzy Relational Equations

Marius Giuclea, Alexandru Agapie

National Institute of Microtechnology, P.O. Box 38-160, 72225,
Bucharest, Romania
marius@oblio.imt.pub.ro, agapie@oblio.imt.pub.ro

Abstract: We propose in this paper a unified method for approximating the solution of a System of Fuzzy Relational Equations (SFRE). The method is essentially based on the use of Genetic Algorithms (GA) and on a probabilistic algorithm for solving a SFRE - presented elsewhere. This approach is useful both in classical SFRE problems and in dynamic system identification. Some numerical results regarding both aspects show that our method can be successfully applied.

1 Introduction

The problem of solving fuzzy relational equations is not new. It was already tackled by classical methods of numerical analysis (in [2] it was proposed a Newton method) or by clustering techniques (Isodata and Fuzzy-C-means in [3]). Probabilistic algorithms for approximating the solution of a SFRE were developed with good results in [5]. Besides the theoretical problem, the major interest in solving SFRE comes from the area of identification and control of industrial processes. In [10] a recursive identification algorithm based on the prediction error method is derived to optimal solving the fuzzy relational equations. In [11] a synthesis of fuzzy closed-loop control system based on relational equation is performed in order to assure stability properties.

In this paper we present a method for approximating the solutions of a SFRE based on GA. Starting from the probabilistic algorithm developed in [5] we took the idea a step further by applying the GA on the space generated by the solutions of the individual fuzzy relational equations.

The next section of the paper presents the GA-based algorithm for solving a SFRE: first the general case is considered and second our method is applied in modeling of dynamic systems. The numerical results from section 3 illustrate the advantages of the GA search in both situations.

2 GA for solving SFRE

2.1 Solving the classical SFRE problem

In this section we propose a GA method for approximate solving a SFRE of the following type: $\{X_k oR = Y_k, k=1,m\}$ (1)
where: " o" is max-min composition; X_k, Y_k are fuzzy sets defined on finite universes of discourse, namely: $X=\{x_1,..., x_n\}$ and $Y=\{y_1,..., y_n\}$; while R stands for a fuzzy relation defined in the Cartesian product XxY -i.e. R is a "nxn" matrix (corresponding to the number of input-output pairs (x_i,y_i), $i=1,n$). The system of equations has to be solved with respect to R, treating all the pairs (X_k,Y_k) as provided. Under the assumption that the system has a non empty set of solutions, the SFRE theory proves, [1], that a solution of the system, say R, is computed as the intersection of α-compositions between the respective fuzzy sets (X_k, Y_k), i.e.:

$R=\bigcap_{k=1,n} R_k$ where $R_k = X_k \alpha Y_k$ is the greatest solution of the "k" equation $(X_k$ o R $= Y_k)$ and "α" stands for the α-composition. The above stated assumption is poorly satisfied in most cases, so we considered the following situation: each individual equation X_k o R $= Y_k$ of the system has got at least a solution (say R_k defined above), but the entire system has no solution.

We start our search from a set of solutions of the individual equations (as in [5]), and using their frequencies of appearance we try to find a better approximate solution for the entire system. For this goal we apply a searching method based on GA. The GA are probabilistic algorithms which start with an initial population of likely problem solutions, and then evolve towards better solutions. The main elements of a GA are (as in [4], [6]): the chromosomes codification, the genetic operators, the option for the initial population, the fitness function and the STOP conditions. We applied the GA described bellow to obtain one column of the matrix representing the approximate solution R. To find the whole matrix we repeat the GA for each column.

We coded the chromosomes as "nxm" binary matrix (as in [7], [8]) for the following reasons: let $R_k=X_k \alpha Y_k$ be the solution of the "k" equation, as considered; for each pair (x_i, y_j) from $X \times Y$ we obtain a string $\{\alpha^1_{ij},...,\alpha^m_{ij}\}$, where $\alpha^k_{ij}=X_k(x_i)\alpha Y_k(y_j)$, k=1,m; some of $\{\alpha k_{ij}, k=1,m\}$ may be not distinct, so we arrange the string such that the first m'_{ij} elements shall not repeat, and form a shorter string $A_{ij}=\{\alpha^1_{ij},...,\alpha^{m'ij}_{ij}\}$ with their associate appearance frequencies $P_{ij}=\{p^1_{ij},...,p^{m'ij}_{ij}\}$, $m'_{ij} < m$. For fixed "j" we determine the "j" column of the matrix R:

• the "i" line of the chromosome's matrix will correspond to the A_{ij} vector in computing the fitness function;

- the genetic operators used in the algorithm are: selection, crossover, mutation and simplification;
- the initial population is randomly generated and the size of population remains fixed during the algorithm.

Let $C=\{c_{ik}, i=1,n, k=1,m\}$, $c_{ik} \in \{0,1\}$ be a matrix representing a chromosome in the population, at a certain iteration; we calculate the fitness function as follows:

- $R'(x_i,y_j)$ is computed as the average between $\{\alpha^1_{ij},...,\alpha^{m'ij}_{ij}\}$, with the weights $\{p^k_{ij}*c_{ik}, k=1,m'_{ij}\}$, for $i=1,n$, namely:

$$R'(x_i,y_j)=\sum_{k=1,m'ij}(p^k_{ij}*c_{ik}*\alpha^k_{ij})/\sum_{k=1,m'ij}(p^k_{ij}*c_{ik}).$$

- For this chromosome the fitness function is computed by the following formula:

$$Q_j(C)= (1/m)\sum_{k=1,m}|Y_k(y_j)-\max_{i=1,n}(\min\{X_k(x_i),R'(x_i,y_j)\})|$$

The STOP conditions are: the diminution of the performance index Q under a certain threshold, or running through a given maximal number of iterations

2.2 A SFRE model for dynamic system identification

Let us consider a discrete time dynamic process characterized by: $X_koR=Y_k$, where $k=1, 2,...$; X_k is the fuzzy input, Y_k the fuzzy output, R a fuzzy relation and "o" is a fuzzy composition operator. To obtain a fuzzy relational model for the process we start from a set of measured (crisp) input-output data: $\{(x_i, y_i), i=1,n\}$. We define $X=\{x_1,..., x_n\}$, $Y=\{y_1,..., y_n\}$ the universes of discourse associated to the input and output respectively, and fix the fuzzy sets $(X_k)_{k=1,m}$, $(Y_j)_{j=1,p}$ on X, res. Y. The number of fuzzy labels and the shapes of the membership functions have to be set empirically or by a clustering technique. In addition, in this approach the membership functions defined on Y are supposed to be symmetric.

We propose to use for the dynamic process a SFRE model of the following type (see [10]):

$$\{ X_koR =Y_j, k=1,m, j=1,p \} \qquad (2)$$

where: "o" is max-min composition, and R stands for a fuzzy relation on $X \times Y$. It is known (for instance, see [10]) that the SFRE (2) is equivalent to a rule base of the following form: IF input is X_k THEN output is Y_j. The existence of this rule is characterized by the (i,j) component of matrix R. We must underline that in this case R is an "mxp" matrix - i.e. corresponding to the number of fuzzy sets , while in (1) R was defined as a "nxn" matrix - i.e. corresponding to the number of crisp input-output examples. One can see that we have considered the maximal number of fuzzy relational equations - mxp - as there is no previous information on the process. The problem we tackle in this section consists in solving the system of equations (2) with respect to R^p, treating all the pairs (X_k, Y_j) as provided. When solving a SFRE in the finite case one is usually looking for a "nxn" matrix R - as in *section 2.1* - to optimize a performance index Q (e.g. the averaged Hamming-fuzzy distance between the desired and the obtained output fuzzy sets: $Q = (1/m^2) *\sum_{k,j=1,m}|Y_j - R'oX_k|$, R' is the approximate solution).

But in the identification problem tackled in this paper the evaluation has to be reported to the set of crisp examples $\{(x_i, y_i), i=1,n\}$. Thus R is assumed to be an "m×p" matrix and the objective function is computed via a fuzzy inference procedure, as follows:

1. for each $i=1,m$ and $k=1,n$ compute $a^i_k=X_i(x_k)$ *(fuzzification)*;
2. for a fixed fuzzy relation matrix $R=(r^{ij})_{i=1,m, j=1,p}$ each $k=1,n$ and each $j=1,p$, compute $b^j_k=\max_{i=1,m}(\min(a^i_k,r^{ij}))$ *(inference mechanism)*;
3. $y'_k=(\sum_{j=1,p} b^j_k*y^c_j)/(\sum_{j=1,p} b^j_k)$, where y^c_j is the centre of the fuzzy set Y_j, respectively *(defuzzification)*;
4. $Q=\sum_{k=1,n} (y'_k-y_k)^2 /\sum_{k=1,n} y_k^2$ is the objective function.

From the procedure depicted above yields that we need for evaluation the whole matrix R - so a separate column calculation (as in *section 2.1*) is no longer possible. Inspired from the previous algorithm in approximating the solution of a SFRE using GA ([5], [8]), we propose to compute the m×p elements of the matrix R using the results of the α–composition, in the following manner:

- let us consider (i,j) fixed; for every $k=1,n$ compute the crisp values $Xi(x_k)\alpha Yj(y_k)$; it yields a n-length string $\{\alpha^1_{ij}, \ldots , \alpha^n_{ij}\}$; let us consider only the (shorter) string of non-zero values, arranged in increasing order: $\{\alpha^1_{ij},..., \alpha^{q(i,j)}_{ij}\}$, with their associate appearance frequencies $P_{ij}=\{p^1_{ij},...,p^{q(i,j)}_{ij}\}$, where $q(i,j) \leq n$;
- for each pair (i,j), for $i=1,m$, $j=1,p$ we associate a binary, $\{0,1\}$-valued string of length $q(i,j)$, which will be the GA codification for the r^{ij} element of the matrix R.
- the decodification : $r^{ij} = \sum_{k=1,q(i,j)} (p^k_{ij} * c^{ij}_k * \alpha^k_{ij}) /\sum_{k=1,q(i,j)}(p^k_{ij} * c^{ij}_k)$, where $C=(c^{ij}_k)$ is the chromosome string of length $I=\sum_{i=1,m;j=1,p} q(i,j)$. Applying this GA model for the SFRE we approximate the solution matrix R to optimally fit the set of input-output crisp examples.

3 Numerical results

3.1 Results of solving classical SFRE

Let us consider a set of pairs of fuzzy sets: (X_k, Y_k), for $k=1,6$ (m=6)
$X_k : X=\{x_1, x_2, x_3, x_4\}\rightarrow[0,1]$ (n=4), $Y_k : Y=\{y_1, y_2, y_3, y_4\} \rightarrow[0,1]$
X=[[1 0.4 0.3 0.1], [0.4 1 0.6 0.4], [0.5 0.5 0.8 0.9], [0.6 0.4 0.7 1], [0.9 0.6 0.4 0.3],[0.5 0.6 0.8 1]]
Y=[[0.5 0.5 0.7 1], [0.3 0.3 1 0.9], [0.4 0.2 0.2 0.6], [0.5 0.6 1 0.3], [0.4 0.3 0.5 0.7],[1 0.6 0.3 0.2]]

We examine the system of induced equations $\{X_koR=Y_k , k=1,6\}$. Each equation treated separately has a solution, but the whole system has not. Applying a classical algorithm for approximating the solution (presented in [5]) it was obtained a fuzzy relation:

R= [[0.4 0.3 0.7 0.7], [0.4 0.3 1.0 0.9], [0.4 0.3 1.0 0.6], [0.4 0.6 1.0 0.6]] with the performance index calculated like an averaged Hamming distance between fuzzy sets: $Q=\sum_{j=1,4} (Q_j)$ = 0.15+0.12+0.26+0.17=0.70. Applying the GA-method from section 2.1 (for each column of the matrix R) we obtained the following matrix: R= [[0.4 0.37 0.7 1], [0.37 0.3 1 0.9], [0.35 0 0.25 0.37], [0.5 0.6 0.25 0.6]] with the corresponding $Q=\sum_{j=1,4} (Q_j)$ = 0.13 + 0.12 + 0.20 + 0.15 = 0.60

3.2 Results in system identification

Next we have tested the proposed method on a dc (direct current) motor identification. The input of this system is the motor current and the output is the motor speed, and both are considered fuzzy variables. Our goal is to construct a fuzzy model for this process using a set of crisp examples (see [12]) as follows:
- the set of pairs of fuzzy sets: (A_k, Y_j), for k=1,6, j=1,5 (m=6, p=5),

$A_k : X=\{x_1, x_2,.., x_9, x_{10}\}\rightarrow[0,1]$, $Y_j : Y=\{y_1, y_2,..., y_9, y_{10}\}\rightarrow[0,1]$;

A_k are defined like in [9] and the output space is divided into five uniform fuzzy labels - like in [12] ;
- from the set of crisp examples, applying the fuzzification procedure one will obtain the matrixes X and Y:
X=[[1 0 0 0 0 0], [0.25 0.76 0 0 0 0], [0 0.76 0.58 0 0 0], [0 0 0.88 0 0 0], [0 0 0.76 0.33 0 0], [0 0 0.17 1 0 0], [0 0 0 0.33 0 0], [0 0 0 0 0.66 0], [0 0 0 0 0.93 0], [0 0 0 0 0.26 1]];
Y=[[0 0 0 0 1], [0 0 0 0.63 0.37], [0 0 0 0.1 0.9], [0 0 0.56 0.44 0], [0 0.09 0.91 0 0], [0 0.49 0.51 0 0], [0 1 0 0 0], [0.41 0.59 0 0 0], [0.55 0.45 0 0 0], [0.68 0.32 0 0 0].
Applying the GA-method from *section 2.2* we obtained the following matrix :
R =[[0 0 0 0 0.9], [0 0 0.1 0.173 0.9], [0 0.2 0.2 0.444 0.09], [0.2 0.5 0 0.1 0], [0.886 0.545 0.1 0 0], [0.968 0.232 0.1 0 0]], with $Q(10)=\sum_{k=1,n} (y'_k-y_k)^2$ $/\sum_{k=1,n} y_k^2 = 1.78*10^{-5}$. This result is better than the one obtained in [12] for relation optimization ($Q(10)= 1.72*10^{-4}$), and comparable to the one obtained in the same paper, by simultaneously optimization of the relation and the membership functions ($Q(10)= 5.68*10^{-6}$). In this way one can obtain a fuzzy relational model for an arbitrary process if the measured input-output data are available.

4 Conclusions

Genetic algorithms are probabilistic algorithms devoted to parameter optimization, especially when the objective function is the only available information on the problem. This work points out the applicability of GA in solving a system of fuzzy relational equations. Besides the theoretical method, we also present some encouraging results on an identification of a dynamic process.

Acknowledgements

This research has been performed under the cooperation contract RUM X053.4 between the Institute of Microtechnology and Dortmund University. The authors express the gratitude to all the people from Chair Informatics I - Dortmund University for their substantial support.

References

1. E. Sanchez, "Resolution of composite relation equations", Information and Control, 30, 1976, 38-48.
2. W.Pedrycz, "Numerical and applicational aspects of fuzzy relational equations", Fuzzy Sets and Systems, 11, 1983, 1-18.
3. K.Hirota , W.Pedrycz, "Analysis and synthesis of fuzzy systems by the use of probabilistic sets", Fuzzy Sets and Systems, 10, 1983, 1-13.
4. D.E. Goldberg, "Genetic algorithms in search, optimization and machine learning", Addison-Wesley Publishing Company, 1989.
5. W. Pedrycz, "Algorithms for solving fuzzy relational equations in a probabilistic setting", Fuzzy Sets and Systems, 38, 1990, 313-327.
6. E. Sanchez, "Genetic Algorithms and Soft Computing", A Tutorial Presentation , EUFIT '93, Aachen, Germany, September 1993.
7. M.Gh.Negoita, F.Fagarasan, A.Agapie, "Genetic algorithms in fuzzy processing of information", 4.Dortmunder Fuzzy Tage, Dortmund, Germany, June 1994
8. M.Gh.Negoita, A.Agapie, F.Fagarasan, "Genetic Algorithms for solving fuzzy relational equations", EUFIT 94, Aachen, Germany, September 1994.
9. M.Giuclea, A.Agapie, F.Fagarasan "A GA Approach for Closed-Loop System Identification", in Proceedings of EUFIT 96, Aachen, Germany, September 1996, 410-414.
10. Y.- C. Lee, C. Hwang, Y.-P. Shih, "A combined approach to fuzzy model identification", IEEE Transaction on Systems, Man and Cybernetics, vol. 24, nr. 5, 1994, 736-744.
11. J. Q. Chen, J. H. Lu and L. J. Chen, "Analysis and syntesis of fuzzy closed-loop control systems", IEEE Transaction on Systems, Man and Cybernetics, vol. 25, nr. 5, 1995, 881-888.
12. D. Park, A. Kandel and G. Langholz, "Genetic-Based New Fuzzy Reasoning Models with Application to Fuzzy Control", IEEE Transaction on Systems, Man and Cybernetics, vol. 24, nr.1, 1994, 39-47.

On the Chaining of Fuzzy IF-THEN Rule Bases Interpreted by the Principle FATI II. The Case of Non-singleton Inputs*

Karl-Heinz Temme

temme@ls1.informatik.uni-dortmund.de
Tel.: +49 231 755 6373, Fax: +49 231 755 6555

Helmut Thiele

thiele@ls1.informatik.uni-dortmund.de
Tel.: +49 231 755 6152, Fax: +49 231 755 6555

University of Dortmund, Dept. of Computer Science I, D-44221 Dortmund, Germany

1. Introduction

In the papers [19, 20] we have studied the chaining problem for fuzzy IF-THEN rules and for fuzzy IF-THEN rule bases, but only for the restricted case of singleton inputs. Recently, non-singleton inputs became more and more importance, in particular, w.r.t. applications [1,11]. Therefore, in the following we will study the chainability of fuzzy IF-THEN rules and of fuzzy IF-THEN rule bases for arbitrary inputs. Because of space restrictions, we consider fuzzy IF-THEN rule bases only under interpretations made by the principle FATI.

2. The Case of Single Rules

We denote the set of all real numbers r with $0 \leq r \leq 1$ by $<0,1>$. Let U be a fixed arbitrary non-empty set called universe. Fuzzy sets on U are mappings $F:U \to <0,1>$.
Fuzzy IF-THEN rules have the form R: IF F THEN G with $F,G \in FP(U)$, the set of all fuzzy sets on U. We underline that by the well known principle of cylindrical extension and suitable operations with fuzzy sets we can restrict the considerations to rules in the given "normal form", without loss of generality. For definiteness we recall the following definitions.

Definition 2.1

$I = [\pi, \kappa, Q]$ is said to be an interpretation (of an arbitrary fuzzy IF-THEN rule) $=_{def}$
    1. $\pi, \kappa: <0,1>^2 \to <0,1>$
    2. $Q: P(<0,1>) \to <0,1>$.

Now for a given interpretation $I = [\pi, \kappa, Q]$ and a given fuzzy IF-THEN rule R: IF F THEN G we define the functional operator $\Phi_I^{F,G}$ on U where $F':U \to <0,1>$ and $y \in U$.

* This research was supported by the Deutsche Forschungsgemeinschaft as part of the Collaborative Research Center "Computational Intelligence" (531).

Definition 2.2

$$\Phi_I^{F,G}(F')(y) =_{def} Q\{\kappa(F'(x), \pi(F(x),G(y))) \mid x \in U\}$$

Now we consider the fuzzy IF-THEN rules

R_1: IF F THEN G
R_2: IF G THEN H
R_3: IF F THEN H

and define for $\mathcal{F} \subseteq FP(U)$:

Definition 2.3

The rule R_3 is the result of chaining the rules R_1 and R_2 w.r.t. I and $\mathcal{F}$ $=_{def}$

$$\forall F' (F' \in \mathcal{F} \rightarrow (\Phi_I^{F,G} \circ \Phi_I^{G,H})(F') = \Phi_I^{F,H}(F')).$$

On the basis of this definition we can prove the following theorems.

THEOREM 2.1

If 1. π is κ-transitive, i.e. $\forall r \forall s \forall t$ $(r,s,t \in <0,1> \rightarrow \kappa(\pi(r,s), \pi(s,t)) \leq \pi(r,t))$
2. κ is monotone
3. κ is associative
4. κ is continuous w.r.t. the first argument
5. $Q = \text{Sup}$
then for every $F' : U \rightarrow <0,1>$ and every $y \in U$,

$$\Phi_I^{G,H}(\Phi_I^{F,G}(F'))(y) \leq \Phi_I^{F,H}(F')(y).$$

Example to theorem 2.1:
$\kappa = \pi$ where κ is an arbitrary continuous T-norm, i.e. we have the so-called generalized Mamdani case.

THEOREM 2.2

If 1. π is reflexive, i.e. $\forall r(r \in <0,1> \rightarrow \pi(r,r)=1)$
2. κ is monotone
3. κ is associative
4. κ is continuous w.r.t. the first argument
5. $\forall s(s \in <0,1> \rightarrow \kappa(1,s)=s)$
6. $Q = \text{Sup}$
7. $\forall x (x \in U \rightarrow \exists y(y \in U \wedge G(y)=F(x)))$
then for every $F' : U \rightarrow <0,1>$ and every $y \in U$,

$$\Phi_I^{F,H}(F')(y) \leq \Phi_I^{G,H}(\Phi_I^{F,G}(F'))(y).$$

Example to theorem 2.2:

$\pi(r,s) =_{def}$ min $(1,1-r+s)$ and κ is an continuous T-norm, where $\kappa(r,s) \leq$ max$(0,r+s-1)$ for every $r,s \in <0,1>$.

From the theorems 2.1 and 2.2 we obtain the following "chaining" theorem:

THEOREM 2.3

If 1. π is reflexive, e.g. $\forall r(r \in <0,1> \rightarrow \pi(r,r)=1)$
   2. π is κ-transitive, i.e. $\forall r \forall s \forall t$ $(r,s,t \in <0,1> \rightarrow \kappa(\pi(r,s), \pi(s,t)) \leq \pi(r,t))$
   3. κ is monotone
   4. κ is associative
   5. κ is continuous w.r.t. the first argument
   6. $\forall s(s \in <0,1> \rightarrow \kappa(1,s)=s)$
   7. $Q = $ Sup
   8. $\forall x$ $(x \in U \rightarrow \exists y(y \in U \wedge F(x)=G(y)))$
then for every $F' : U \rightarrow <0,1>$,

$$\Phi_I^{G,H}(\Phi_I^{F,G}(F')) = \Phi_I^{F,H}(F').$$

Example to theorem 2.3: as for theorem 2.2

If one is of opinion that the assumption 7 and 8 of theorem 2.2 and 2.3. respectively, is not suitable then we can derive and use the following theorem:

THEOREM 2.4

If 1. $\forall s(s \in <0,1> \rightarrow \pi(1,s)=s)$
   2. $\forall r \forall s(r,s \in <0,1> \rightarrow \pi(r,s) \leq \kappa(\pi(r,1),s))$
   3. κ is monotone w.r.t. the first argument
   4. κ is associative
   5. κ is continuous w.r.t. the first argument
   6. $Q = $ Sup
   7. G is strongly normal, i.e. $\exists x$ $(x \in U \wedge G(x)=1)$
then for every $F' : U \rightarrow <0,1>$ and every $y \in U$,

$$\Phi_I^{F,H}(F')(y) \leq \Phi_I^{G,H}(\Phi_I^{F,G}(F'))(y).$$

Example to theorem 2.4:
$\kappa = \pi$ where κ is an arbitrary continuous T-norm, i.e. we have the so-called generalized Mamdani case, as to theorem 2.1.

By combining the theorems 2.1 and 2.4 we obtain the following "chaining" theorem:

THEOREM 2.5

If 1. $\forall s(s \in <0,1> \to \pi(1,s)=s)$
    2. $\forall r \forall s(r,s \in <0,1> \to \pi(r,s) \le \kappa(\pi(r,1),s))$
    3. π is κ-transitive, i.e. $\forall r \forall s \forall t$ $(r,s,t \in <0,1> \to \kappa(\pi(r,s), \pi(s,t)) \le \pi(r,t))$
    4. κ is monotone
    5. κ is associative
    6. κ is continuous w.r.t. the first argument
    7. $Q = \text{Sup}$
    8. G is strongly normal, i.e. $\exists x\ (x \in U \wedge G(x)=1)$
then for every $F' : U \to <0,1>$,

$$\Phi_I^{G,H}(\Phi_I^{F,G}(F')) = \Phi_I^{F,H}(F').$$

Example to theorem 2.5: like to theorem 2.4

3. The Case of Fuzzy IF-THEN Rule Bases

We consider an IF-THEN rule base of the form:

$$\text{IF } F_1 \text{ THEN } G_1$$
$$RB_1 : \quad \quad ...$$
$$\text{IF } F_n \text{ THEN } G_n$$

where n is an integer with $n \ge 1$ and $F_1,...,F_n,...,G_1,...,G_n$ are fuzzy sets on U. For shortness we put: $\mathbf{F} =_{def} [F_1,...,F_n]$, $\mathbf{G} =_{def} [G_1,...,G_n]$ and RB_1 : IF $\mathbf{F}$ THEN $\mathbf{G}$. For interpreting RB_1 we fix a (3n+4)-tupel I called interpretation of the form:

$I = [\pi_1,...,\pi_n,\kappa_0,\kappa_1,...,\kappa_n,Q_0,Q_1,...,Q_n,\alpha,\beta]$ where
    1. $\pi_1,...,\pi_n,\kappa_0,\kappa_1,...,\kappa_n$: $<0,1>^2 \to <0,1>$
    2. $Q_0,Q_1,...,Q_n$: $P(<0,1>) \to <0,1>$
    3. α,β: $<0,1>^n \to <0,1>$.

Definition 3.1

Let F be a fuzzy set on U and assume $y \in U$,

$$\text{FATI}_I^{\mathbf{F,G}}(F)(y) =_{def} \quad Q_0\{\kappa_0(F(x),\alpha(\pi_1(F_1(x),G_1(y)),...,\pi_n(F_n(x),G_n(y))))|x \in U\}$$

$$\text{FITA}_I^{\mathbf{F,G}}(F)(y) =_{def} \quad \beta\ (Q_1\{\kappa_1(F(x),\pi_1(F_1(x),G_1(y)))|x \in U\}$$
$$,...,Q_n\{\kappa_n(F(x),\pi_n(F_n(x),G_n(y)))|x \in U\})$$

For formulating the fundamental definitions we consider two further IF-THEN rule bases, namely:

$$\text{RB}_2: \quad \begin{matrix} \text{IF } G_1 \text{ THEN } H_1 \\ \dots \\ \text{IF } G_n \text{ THEN } H_n \end{matrix} \quad \text{and} \quad \text{RB}_3: \quad \begin{matrix} \text{IF } F_1 \text{ THEN } H_1 \\ \dots \\ \text{IF } F_n \text{ THEN } H_n \end{matrix}$$

where $H_1,...,H_n$ are also fuzzy sets on U. Again we introduce the abbreviations $\mathbf{H} =_{def} [H_1,...,H_n]$, $\text{RB}_2: \text{IF } \mathbf{G} \text{ THEN } \mathbf{H}$ and $\text{RB}_3: \text{IF } \mathbf{F} \text{ THEN } \mathbf{H}$. Furthermore we put $\Phi_{I,1}^{\mathbf{F},\mathbf{G}} =_{def} \text{FATI}_I^{\mathbf{F},\mathbf{G}}$ and $\Phi_{I,2}^{\mathbf{F},\mathbf{G}} =_{def} \text{FITA}_I^{\mathbf{F},\mathbf{G}}$.

The same holds for $\text{FATI}_I^{\mathbf{G},\mathbf{H}}$, $\text{FATI}_I^{\mathbf{F},\mathbf{H}}$ and $\text{FITA}_I^{\mathbf{F},\mathbf{H}}$.

Then by generalizing definition 2.3 we define, where $i,j,k \in \{1,2\}$ and $\mathcal{F} \subseteq \text{FP}(\text{U})$:

Definition 3.2:

1. The rule base IF **F** THEN **H** is the result of [i,j,k]-chaining of the rule bases IF **F** THEN **G** and IF **G** THEN **H** w.r.t. I and $\mathcal{F}$ $=_{def}$

$$\forall F \ (F \in \mathcal{F} \ \rightarrow \ (\Phi_{I,i}^{\mathbf{F},\mathbf{G}} \circ \Phi_{I,j}^{\mathbf{G},\mathbf{H}})\ (F) = \Phi_{I,k}^{\mathbf{F},\mathbf{H}}\ (F)\ )$$

2. The rule base IF **F** THEN **H** is the result of weakly [i,j,k]-chaining of the rule bases IF **F** THEN **G** and IF **G** THEN **H** w.r.t. I and $\mathcal{F}$ $=_{def}$

$$\forall F \ (F \in \mathcal{F} \ \rightarrow \ (\Phi_{I,i}^{\mathbf{F},\mathbf{G}} \circ \Phi_{I,j}^{\mathbf{G},\mathbf{H}})\ (F) \subseteq \Phi_{I,k}^{\mathbf{F},\mathbf{H}}\ (F)\ )$$

Each case in definition 3.2 contains eight different versions of a chaining procedure. Because of space restrictions we consider only the [1,1,1]-chaining, but for arbitrary inputs, i.e. we study the question whether

$$\forall F \ (F \in \mathcal{F} \rightarrow \text{FATI}_I^{\mathbf{G},\mathbf{H}}(\text{FATI}_I^{\mathbf{F},\mathbf{G}}(F)) = \text{FATI}_I^{\mathbf{F},\mathbf{H}}\ (F))$$

holds.

Definition 3.3

1. κ_0 is said to be right-side strongly α-distributive $=_{def}$
$\forall r_1,...,\forall r_n, \forall s_1,...,\forall s_n \ (r_1,...,r_n,s_1,...,s_n \in \ <0,1> \ \rightarrow$
$\kappa_0(\alpha(r_1,...,r_n),\alpha(s_1,...,s_n)) \leq \alpha(\kappa_0(r_1,\ s_1),...,\kappa_0(r_n,\ s_n)))$

2. κ_0 is said to be left-side weakly α-distributive $=_{def}$
$\forall r_1,...,\forall r_n, \forall s_1,...,\forall s_n \ (r_1,...,r_n,s_1,...,s_n \in \ <0,1> \ \rightarrow$
$\alpha(\kappa_0(r_1,\ s_1),...,\kappa_0(r_n,\ s_n)) \leq \kappa_0(\alpha(r_1,...,r_n),\alpha(s_1,...,s_n)))$

THEOREM 3.1

If   1. for every $i \in \{1,...,n\}$, π_i is κ_0-transitive, e.g.
$$\forall r \forall s \forall t \ (r,s,t \in <0,1> \ \rightarrow \ \kappa_0(\pi_i(r,s),\pi_i(s,t)) \leq \pi_i(r,t))$$
    2. κ_0 is monotone w.r.t. the first argument
    3. κ_0 is associative
    4. κ_0 is continuous w.r.t. the first argument
    5. $Q = \text{Sup}$
    6. α is monotone
    7. κ_0 is right-side strongly α-distributive, i.e.
$$\forall r_1,...,\forall r_n, \forall s_1,...,\forall s_n \ (r_1,...,r_n,s_1,...,s_n \in <0,1> \rightarrow$$
$$\kappa_0(\alpha(r_1,...,r_n),\alpha(s_1,...,s_n)) \leq \alpha(\kappa_0(r_1,s_1),...,\kappa_0(r_n,s_n)))$$
then for every $F : U \rightarrow <0,1>$ and every $y \in U$,
$$\text{FATI}_I^{G,H}(\text{FATI}_I^{F,G}(F))(y) \leq \text{FATI}_I^{F,H}(F)(y).$$

Example to theorem 3.1:
1. $\pi_i(r,s) =_{\text{def}} \min(1,1-r+s)$
2. κ_0 is an arbitrary continuous T-norm
3. $\alpha(r_1,...,r_n) =_{\text{def}} \kappa_0(r_1,\kappa_0(r_2,\kappa_0(r_3,...,\kappa_0(r_{n-1},r_n)...).$

For shortness we define the following "aggregated" function π and "aggregated" binary fuzzy relation S:

Definition 3.4

1. $\pi(r,s) =_{\text{def}} \alpha(\pi_1(r,\ s),...,\pi_n(r,\ s))$ $(r,s \in <0,1>)$
2. $S(x,y) =_{\text{def}} \alpha(\pi_1(F_1(x),G_1(y)),...,\pi_n(F_n(x),G_n(y)))$ $(x,y \in <0,1>)$

Then we can reformulate the definition of $\text{FATI}_I^{F,G}$ as follows where $F : U \rightarrow <0,1>$ and $y \in U$,
$$\text{FATI}_I^{F,G}(F)(y) =_{\text{def}} Q_0 \ \{\kappa_0(F(x),S(x,y)) \mid x \in U\}.$$

Using the function π we obtain the following modification of theorem 3.1:

THEOREM 3.2

If   1. π is κ_0-transitive, e.g. $\forall r \forall s \forall t \ (r,s,t \in <0,1> \ \rightarrow \ \kappa_0(\pi(r,s),\pi(s,t)) \leq \pi(r,t))$
    2. κ_0 is monotone w.r.t. the first argument
    3. κ_0 is associative
    4. κ_0 is continuous w.r.t. the first argument
    5. $Q = \text{Sup}$
then for every $F : U \rightarrow <0,1>$ and every $y \in U$,
$$\text{FATI}_I^{G,H}(\text{FATI}_I^{F,G}(F))(y) \leq \text{FATI}_I^{F,H}(F)(y).$$

Example to theorem 3.2: as to theorem 3.1.

REMARK: Theorem 3.2 can be interpreted as a "vectorial" version of theorem 2.1.

THEOREM 3.3

If 1. π_i is reflexive for every $i \in \{1,...,n\}$, e.g. $\forall r \ (r \in <0,1> \rightarrow \pi_i(r,r)=1)$
    2. $\forall s \ (s \in <0,1> \rightarrow \kappa_0(1,s)=s)$
    3. κ_0 is monotone w.r.t. the first argument
    4. κ_0 is associative
    5. κ_0 is continuous w.r.t. the first argument
    6. $Q = \text{Sup}$
    7. α is monotone
    8. κ_0 is left-side weakly α-distributive, e.g.
       $\forall r_1,...,\forall r_n, \forall s_1,...,\forall s_n \ (r_1,...,r_n,s_1,...,s_n \in <0,1> \rightarrow$
       $\alpha(\kappa_0(r_1,s_1),...,\kappa_0(r_n,s_n)) \leq \kappa_0(\alpha(r_1,...,r_n),\alpha(s_1,...,s_n)))$
    9. $\forall x \ (x \in U \rightarrow \exists y(F_1(x)=G_1(y) \wedge...\wedge F_n(x)=G_n(y)))$
then for every $F : U \rightarrow <0,1>$ and every $y \in U$,
$$\text{FATI}_I^{F,H}(F)(y) \leq \text{FATI}_I^{G,H}(\text{FATI}_I^{F,G}(F))(y).$$

Example to theorem 3.3:
1. $\pi_i(r,s) =_{\text{def}} \min(1,1-r+s)$ 2. $\kappa_0 =_{\text{def}} \min$ 3. $\alpha =_{\text{def}} \max$.

By combining the theorems 3.2 and 3.3 we obtain the following "chaining" theorem:

THEOREM 3.4

If 1. π_i is reflexive for every $i \in \{1,...,n\}$, e.g. $\forall r \ (r \in <0,1> \rightarrow \pi_i(r,r)=1)$
    2. $\pi(r,s) = \alpha(\pi_1(r, s),...,\pi_n(r, s))$ is κ_0-transitive
    3. $\forall s \ (s \in <0,1> \rightarrow \kappa_0(1,s)=s)$
    4. κ_0 is monotone w.r.t. the first argument
    5. κ_0 is associative
    6. κ_0 is continuous w.r.t. the first argument
    7. $Q = \text{Sup}$
    8. α is monotone
    9. κ_0 is left-side weakly α-distributive, e.g.
       $\forall r_1,...,\forall r_n, \forall s_1,...,\forall s_n \ (r_1,...,r_n,s_1,...,s_n \in <0,1> \rightarrow$
       $\alpha(\kappa_0(r_1,s_1),...,\kappa_0(r_n,s_n)) \leq \kappa_0(\alpha(r_1,...,r_n),\alpha(s_1,...,s_n)))$
    10. $\forall x (x \in U \rightarrow \exists y \in U(F_1(x)=G_1(y) \wedge...\wedge F_n(x)=G_n(y)))$
then for every $F : U \rightarrow <0,1>$ and every $y \in U$,
$$\text{FATI}_I^{G,H}(\text{FATI}_I^{F,G}(F))(y) = \text{FATI}_I^{F,H}(F)(y).$$

Example to theorem 3.4:

1. $\pi_i(r,s) =_{def} \begin{cases} 1 \text{ if } r \leq s \\ s \text{ if } r > s \end{cases}$ 2. $\kappa_0 =_{def} \min$ 3. $\alpha =_{def} \max.$

REMARKS
1. Like the combining of theorems 3.2 and 3.3 to the "chaining" theorem 3.4 one can obtain a further "chaining" theorem from the theorems 3.1 and 3.3.
2. Maybe one is of opinion that the assumption 9 and 10 in theorem 3.3 and 3.4, respectively, is too strong. Therefore we formulate the following theorem:

THEOREM 3.5

If 1. $\forall s \ (s \in <0,1> \rightarrow \pi_i(1,s)=s)$ for every $i \in \{1,...,n\}$
   2. $\forall r \forall s \ (r,s \in <0,1> \rightarrow \pi_i(r,s) \leq \kappa_0(\pi_i(r,1),s))$ for every $i \in \{1,...,n\}$
   3. κ_0 is monotone w.r.t. the first argument
   4. κ_0 is associative
   5. κ_0 is continuous w.r.t. the first argument
   6. $Q = \mathrm{Sup}$
   7. α is monotone
   8. κ_0 is left-side weakly α-distributive
   9. $\exists y \ (y \in U \wedge G_1(y)=...=G_n(y)=1)$
then for every $F : U \rightarrow <0,1>$ and every $y \in U$,

$$\mathrm{FATI}_I^{F,H}(F)(y) \leq \mathrm{FATI}_I^{G,H}(\mathrm{FATI}_I^{F,G}(F))(y).$$

Example to theorem 3.5:
1. π_i is an arbitrary T-norm   2. $\kappa_0 =_{def} \min$ 3. $\alpha =_{def} \max.$

By combining theorems 3.2 and 3.5 we obtain the following "chaining" theorem:

THEOREM 3.6

If 1. $\forall s \ (s \in <0,1> \rightarrow \pi_i(1,s)=s)$ for every $i \in \{1,...,n\}$
   2. $\forall r \forall s \ (r,s \in <0,1> \rightarrow \pi_i(r,s) \leq \kappa_0(\pi_i(r,1),s))$ for every $i \in \{1,...,n\}$
   3. the aggregated function $\pi(r,s) =_{def} \alpha(\pi_1(r, s),...,\pi_n(r, s))$ is κ_0-transitive
   4. κ_0 is monotone w.r.t. the first argument
   5. κ_0 is associative
   6. κ_0 is continuous w.r.t. the first argument
   7. $Q = \mathrm{Sup}$
   8. α is monotone
   9. κ_0 is left-side weakly α-distributive
   10. $\exists y \ (y \in U \wedge G_1(y)=...=G_n(y)=1)$
then for every $F : U \rightarrow <0,1>$ and every $y \in U$,

$$\mathrm{FATI}_I^{G,H}(\mathrm{FATI}_I^{F,G}(F))(y) = \mathrm{FATI}_I^{F,H}(F)(y).$$

Example to theorem 3.6:

1. $\pi_1 = ... = \pi_n = \kappa$ where κ is an arbitrary T-norm
2. $\kappa_0 =_{def} \min$
3. $\alpha =_{def} \max$.

REMARK: If one has the feeling that the assumption 9 and the assumption 10 of theorem 3.5 and theorem 3.6, respectively, is too strong then in order to formulate weaker assumptions, we define for the given RB_1, RB_2 and RB_3, where x, y, z$\in$ U:

$$P_1(x,y) =_{def} \alpha(\pi_1(F_1(x),G_1(y)),...,\pi_n(F_n(x),G_n(y)))$$
$$P_2(y,z) =_{def} \alpha(\pi_1(G_1(y),H_1(z)),...,\pi_n(G_n(y),H_n(z)))$$
$$P_3(x,z) =_{def} \alpha(\pi_1(F_1(x),H_1(z)),...,\pi_n(F_n(x),H_n(z)))$$

THEOREM 3.7

If 1. κ_0 is monotone w.r.t. the first argument
   2. κ_0 is associative
   3. κ_0 is continuous w.r.t. the first argument
   4. $\forall x \forall z (x,z \in U \rightarrow \exists y (y \in U \wedge P_3(x,z) \leq \kappa_0(P_1(x,y),P_2(y,z)))$
then for every F : U $\rightarrow$ <0,1> and every y$\in$ U,
$$\text{FATI}_I^{F,H}(F)(y) \leq \text{FATI}_I^{G,H}(\text{FATI}_I^{F,G}(F))(y).$$

If we recall the definition of π as $\pi(r,s) =_{def} \alpha(\pi_1(r, s),...,\pi_n(r, s))$, then we can combine theorem 3.2 and theorem 3.7 to the following "chaining" theorem:

THEOREM 3.8

If 1. π is κ_0-transitive, e.g. $\forall r \forall s \forall t (r,s,t \in <0,1> \rightarrow \kappa_0(\pi(r,s),\pi(s,t)) \leq \pi(r,t))$
   2. κ_0 is monotone w.r.t. the first argument
   3. κ_0 is associative
   4. κ_0 is continuous w.r.t. the first argument
   5. $\forall x \forall z (x,z \in U \rightarrow \exists y (y \in U \wedge P_3(x,z) \leq \kappa_0 (P_1(x,y),P_2(y,z)))$
then for every F : U $\rightarrow$ <0,1> and every y$\in$ U,
$$\text{FATI}_I^{G,H}(\text{FATI}_I^{F,G}(F))(y) = \text{FATI}_I^{F,H}(F)(y).$$

References

[1]   DRIANKOV, D., R. PALM, AND H. HELLENDORN: Fuzzy control with fuzzy inputs: The need for new rule semantics. In FUZZ-IEEE '94 - Third IEEE International Conference on Fuzzy Systems, pages 13 - 21, Orlando, Florida, USA, June 26 - July 1, 1994.

[2]   DIMITER DRIANKOV and HANS HELLENDOORN: Chaining of fuzzy if-then rules in Mamdani-controllers. In: FUZZ-IEEE '95 - Fourth IEEE International Conference on Fuzzy Systems, volume 1, pages 103 - 108, Yokohama, Japan, March 1995.

[3]   DIDIER DUBOIS and HENRI PRADE: On fuzzy syllogism. Computational Intelligence 4 (2), pages 171 - 179, 1988.

[4]   DIDIER DUBOIS and HENRI PRADE: Fuzzy sets in approximate reasoning, Part 1: Inference with possibility distributions: Fuzzy Sets and Systems 40, pages 142 - 202, 1991. Special Memorial Volume: 25 years of fuzzy sets.

[5]   J. FODOR, M. ROUBENS: Fuzzy Preference Modelling and Multicriteria Decision Support. Theory and Decision Libr., Ser. D. vol. 14, Kluwer Academic Publications, Dordrecht 1994.

[6]   S. GOTTWALD: Fuzzy Sets and Fuzzy Logic. Vieweg Verlag, Braunschweig, 1993.

[7]   SIEGRFIED GOTTWALD: On the Chainability of Fuzzy if-then Rules. IPMU '96. Information Processing and Management of Uncertainty in Knowledge-Based Systems, Granada, Spain, July 1 - 5, 1996. Vol. 2, pages 997 - 1001.

[8]   M. M. GUPTA, J. QI: Theory of t-norms and fuzzy inference methods. Fuzzy Sets and Systems 40 (1991), pages 431 - 450.

[9]   E. H. MAMDANI, S. ASSILIAN: An experiment in linguistic synthesis with a fuzzy logic controller. International Journal Man - Machine Studies 7 (1975), pages 1 - 13.

[10]   M. MIZUMOTO and HANS-JÜRGEN ZIMMERMANN: Comparison of fuzzy reasoning methods. Fuzzy Sets and Systems 8, pages 253 - 283, 1982.

[11]   MOUZOURIS, GEORGE C. and JERRY M. MENDEL: Non-singleton fuzzy logic systems. In FUZZ-IEEE '94 - Third IEEE International Conference on Fuzzy Systems, pages 456 - 461, Orlando, Florida, USA, June 26 - July 1, 1994.

[12]   DA RUAN: A critical study of widely used fuzzy implication operators and their influence on the inference rules in fuzzy expert systems. Ph. D. thesis, Gent University, Belgium, 1990.

[13]   DA RUAN et al.: Influence of the fuzzy implication operator on the method of cases inference rule. International Journal of Approximate Reasoning 4, pages 307 - 318, 1990.

[14]   DA RUAN and ETIENNE E. KERRE: Fuzzy implication operators and generalized fuzzy method of cases. Fuzzy Sets and Systems 54, pages 23 - 37, 1993.

[15]   DA RUAN and ETIENNE E. KERRE: On the extension of the compositional rule of inference. International Journal of Intelligent Systems 8, pages 807 - 817, 1993.

[16]   DA RUAN and ETIENNE E. KERRE: Extended Fuzzy Chaining Syllogism. In: IFSA '95 - Sixth International Fuzzy Systems Association World Congress, volume I, pages 145 - 148, São Paulo, Brazil, July 22 - 28, 1995.

[17]   PH. SMETS, P. MAGREZ: Implication in fuzzy logic. International Journal on Approximate Reasoning 1 (1987), pages 327 - 347.

[18]   KARL-HEINZ TEMME and HELMUT THIELE: On the Correctness of the Principles FATI and FITA and their Equivalence. In: IFSA '95 - Sixth International Fuzzy Systems Association World Congress, volume I, pages 475 - 478, São Paulo, Brazil, July 22 - 28, 1995.

[19]   KARL-HEINZ TEMME and HELMUT THIELE: On the Chaining of IF-THEN Rule Bases Interpreted by the Principle FATI. In: Proceedings of the Symposium on Qualitative System Modelling, Qualitative Fault Diagnosis and Fuzzy Logic and Control, Budapest and Balatonfüred, Hungary, April 17 - 20, 1996.

[20]   KARL-HEINZ TEMME and HELMUT THIELE: On the Chaining of IF-THEN Rule Bases interpreted by the principle FITA - Extended Abstract. EUFIT '96. Fourth European Congress on Intelligent Techniques and Soft Computing, Aachen, Germany, September 2 - 5, 1996. Proceedings, vol. 2, pages 939-945.

[21]   HELMUT THIELE: On the Calculus of IF-THEN Rules. In: Real World Applications of Intelligent Technologies, Edited by Hans Jürgen Zimmermann, Mircea Gh. Negoita, Dan Dascalu. Editura Academiei Romane, Bucharest, Romania, pages 91 - 111, 1995.

[22]   I. B. TÜRKSEN and Y. TIAN: Combination of rules or their consequences in fuzzy expert systems. Fuzzy Sets and Systems 58, pages 3 - 40, 1993.

[23]   LOTFI A. ZADEH: Outline of a new approach to the analysis of complex systems and decision processes. IEEE Transactions on Systems, Man and Cybernetics 3 (1), pages 28 - 44, 1973.

[24]   LOTFI A. ZADEH: Calculus of fuzzy restrictions. In: Lotfi A. Zadeh et al. (editors), Fuzzy Sets and Their Applications to Cognitive and Decision Processes, pages 1 - 39, Academic Press, New York, 1975.

[25]   LOTFI A. ZADEH: Syllogistic reasoning in fuzzy logic and its application to reasoning with dispositions. IEEE Transactions on Systems, Man and Cybernetics SMC-15, pages 754 - 763, 1985.

Chaining Syllogism Applied to Fuzzy IF-THEN Rules and Rule Bases

Christian Igel and Karl-Heinz Temme

University of Dortmund, Department of Computer Science I,
D-44221 Dortmund, Germany

Abstract. This paper presents a new approach to show the validity of the chaining syllogism for fuzzy IF-THEN rules and rule bases. Based on this approach new conditions are given on which the investigated deduction scheme holds.

1   Introduction

In the two-valued propositional calculus the chaining syllogism

$$(p \to q) \wedge (q \to r) \to (p \to r),$$

where p, q and r are propositional variables, is a tautology. The validity of this deduction scheme applied to fuzzy IF-THEN rules and rule bases is studied in this paper, a problem investigated e.g. in [3,6,11] and recently in [2,5,8]. The chaining syllogism provides a way to combine blocks of fuzzy IF-THEN rules e.g. to reduce the complexity of a fuzzy controller or expert system.

Let U be an arbitrary non-empty set called universe. Fuzzy sets on U are mappings $F : U \to [0,1]$. The power set of $[0,1]$ is denoted by $\mathcal{P}([0,1])$ and the set of all fuzzy sets on U by $\mathcal{F}(U)$. Let $F, G, H \in \mathcal{F}(U)$. The range of a fuzzy set is defined as $RANGE(G) =_{def} \{G(x) | x \in U\}$.

2   Chainability of single rules

We consider rules of the form IF F THEN G. The case of multiple antecedents or several universes can be covered by using boolean expressions of fuzzy sets resp. the construction of cylindrical extension.

Definition 2.1
$\Im = [\pi, \kappa, Q]$ *is an interpretation of an IF-THEN rule*
$=_{def}$ *1. $\pi, \kappa : [0,1]^2 \to [0,1]$*
        2. $Q : \mathcal{P}([0,1]) \to [0,1]$.

For a given interpretation $\Im$ the "implication function" π interprets the rule IF F THEN G by defining the binary fuzzy relation

$$\Theta_\pi^{F,G}(x,y) =_{def} \pi(F(x), G(y)).$$

For an arbitrary $F' \in \mathcal{F}(U)$ the image $G' \in \mathcal{F}(U)$ is inferred by a generalized form of the compositional rule of inference (see [10]) as follows

$$G'(y) =_{def} (F' \circ_{Q,\kappa} \Theta_\pi^{F,G})(y),$$

where the functions κ and Q define the Q-κ-composition $\circ_{Q,\kappa}$ of fuzzy relations, e.g.

$$(F' \circ_{Q,\kappa} \Theta_\pi^{F,G})(y) = Q\{\kappa(F'(x), \Theta_\pi^{F,G}(x,y)) | x \in U\}.$$

Two rules IF F THEN G and IF G THEN H are called chainable iff the deduction scheme

$$\frac{\begin{array}{c} \text{IF } F \text{ THEN } G \\ \text{IF } G \text{ THEN } H \end{array}}{\text{IF } F \text{ THEN } H}$$

is valid. An intuitive understanding of the validity of this "chaining scheme" leads to the following definition.

Definition 2.2
Two fuzzy rules IF F THEN G and IF G THEN H are called chainable with respect to $\Im$ and $\mathcal{F}' \subseteq \mathcal{F}(U)$ iff

$$\forall F'(F' \in \mathcal{F}' \to (F' \circ_{Q,\kappa} \Theta_\pi^{F,G}) \circ_{Q,\kappa} \Theta_\pi^{G,H} = F' \circ_{Q,\kappa} \Theta_\pi^{F,H}).$$

On which conditions are two rules chainable? The main problem is to show the chainability property for every input fuzzy set $F' \in \mathcal{F}'$. This leads to the following new approach.

Definition 2.3
Two fuzzy rules IF F THEN G and IF G THEN H are called loosely chainable with respect to $\Im$ iff

$$\Theta_\pi^{F,G} \circ_{Q,\kappa} \Theta_\pi^{G,H} = \Theta_\pi^{F,H}.$$

We can split the problem of chainability into two separate questions.

① On which conditions does loose chainability imply chainability? The following section describes some sufficient conditions.
② On which conditions are rules loosely chainable? Examples are given in the third section.

Different answers to ① and ② can be combined and yield various "chainability theorems".

3 Chainability of single rules implied by loose chainability

There is a close relation between the associativity of the Q-κ-composition, chainability and loose chainability.

Lemma 3.1
If the Q-κ-composition is associative, then two fuzzy rules IF F THEN G and IF G THEN H are chainable if they are loosely chainable.

To characterize an important class of associative compositions we need the following property.

Definition 3.1
A mapping $\lambda : [0,1] \to [0,1]$ is lower-semicontinuous
$=_{def} \forall x_0(x_0 \in [0,1] \to$
$$\forall \epsilon(\epsilon > 0 \to \exists \delta(\delta > 0 \wedge \forall x(|x - x_0| < \delta \to \lambda(x_0) - \lambda(x) < \epsilon))))$$

Lemma 3.2
Every nondecreasing mapping $\lambda : [0,1] \to [0,1]$ is lower-semicontinuous iff it is left continuous.

Based on [1,4] one can prove

Lemma 3.3
If 1. κ is nondecreasing and associative
   2. every partial mapping of κ is lower-semicontinuous
   3. $Q = Sup$
then the Q-κ-composition is associative.

Lemma 3.1 and 3.3 lead to the following theorem.

Theorem 3.4
If 1. κ nondecreasing and associative
   2. every partial mapping of κ is lower-semicontinuous
   3. $Q = Sup$
then two fuzzy rules IF F THEN G and IF G THEN H are chainable if they are loosely chainable.

If a partial mapping of κ is not lower-semicontinuous, one can restrict the arguments of Q to finite sets to achieve the associativity of the Q-κ-composition.

Lemma 3.5
If 1. κ is nondecreasing and associative
   2. $Q = Sup$
   3. the arguments of Q are always finite sets
then the Q-κ-composition is associative.

With Lemma 3.1 we get

Theorem 3.6
If 1. κ is nondecreasing and associative
  2. $Q = Sup$
  3. for every $F' \in \mathcal{F}'$ the union $RANGE(F') \cup RANGE(F) \cup RANGE(G)$ a finite set

then two fuzzy rules IF F THEN G and IF G THEN H are chainable with respect to $\mathcal{F}'$ if they are loosely chainable.

If U is a finite set, then assumption 3 of the previous theorem is fulfilled.

For many (fuzzy control) applications the input can be restricted to fuzzy singletons of the form

$$F'_{x_0}(x) = \begin{cases} 1 & \text{; if } x = x_0 \\ 0 & \text{; else} \end{cases}$$

where $x, x_0 \in U$. From [8] we quote the following lemma.

Lemma 3.7
If 1. $\forall r(r \in [0,1] \rightarrow \kappa(0,r) = 0 \wedge \kappa(1,r) = r)$
  2. $\forall r(r \in [0,1] \rightarrow Q\{0,r\} = r)$

then for every $x_0, y \in U$

$$(F'_{x_0} \circ_{Q,\kappa} \Theta^{F,G}_\pi)(y) = \Theta^{F,G}_\pi(x_0, y).$$

This leads to the following theorem, where the associativity of the Q-κ-composition is not necessarily assumed.

Theorem 3.8
If 1. $\forall r(r \in [0,1] \rightarrow \kappa(0,r) = 0 \wedge \kappa(1,r) = r)$
  2. $\forall r(r \in [0,1] \rightarrow Q\{0,r\} = r)$

then two fuzzy rules IF F THEN G and IF G THEN H are chainable with respect to singleton inputs if they are loosely chainable.

4   Loose chainability of single rules

We give three examples for the loose chainability of single IF-THEN rules, which merge with the theorems of the previous section into different "chainability theorems".

Theorem 4.1
If 1. $\forall r \forall t(r, t \in [0,1] \rightarrow Q\{\kappa(\pi(r,s), \pi(s,t)) | s \in I\} = \pi(r,t))$
  2. $RANGE(G) = I$

then

$$\Theta^{F,H}_\pi = \Theta^{F,G}_\pi \circ_{Q,\kappa} \Theta^{G,H}_\pi.$$

In [2] property 1 of Theo. 4.1 with $I = [0, 1]$ and $Q = Sup$ is investigated for several "implication functions" and t-norms. These results can also serve as sufficient conditions for the transitivity of the "implication functions" with respect to the t-norms.

Definition 4.1

$\pi : [0, 1]^2 \rightarrow [0, 1]$ *is called* κ-*transitive*
$=_{def} \forall r \forall s \forall t(r, s, t \in [0, 1] \rightarrow \kappa(\pi(r, s), \pi(s, t)) \leq \pi(r, t))$.

Theorem 4.2

If 1. $\forall r(r \in [0, 1] \rightarrow \pi(1, r) = r)$
   2. $\forall r \forall s(r, s \in [0, 1] \rightarrow \pi(r, s) \leq \kappa(\pi(r, 1), s))$
   3. π *is* κ-*transitive*
   4. $Q = Sup$
   5. G *is a normal fuzzy set on* U, *i.e.* $\exists y(y \in U \wedge G(y) = 1)$

then

$$\Theta_\pi^{F,H} = \Theta_\pi^{F,G} \circ_{Q,\kappa} \Theta_\pi^{G,H}.$$

If π is a t-norm and $\kappa = \pi$, assumptions 1, 2 and 3 of Theo. 4.2 are fulfilled. In fact if $Q = Sup$ and π is a t-norm, then $\kappa = \pi$ is a necessary condition for the loose chainability.

Theorem 4.3

If 1. π *is reflexive, i.e.* $\forall r(r \in [0, 1] \rightarrow \pi(r, r) = 1)$
   2. $\forall r(r \in [0, 1] \rightarrow \kappa(1, r) = r)$
   3. π *is* κ-*transitive*
   4. $Q = Sup$
   5. $\forall x(x \in U \rightarrow \exists y(y \in U \wedge F(x) = G(y)))$

then

$$\Theta_\pi^{F,H} = \Theta_\pi^{F,G} \circ_{Q,\kappa} \Theta_\pi^{G,H}.$$

If κ is a lower-semicontinuous t-norm and π is the r-implication generated by κ, i.e. for all $r, s \in [0, 1]$

$$\pi(r, s) = Sup\{t | \kappa(r, t) \leq s \text{ and } t \in [0, 1]\},$$

then the assumptions 1-3 of Theo. 4.3 are fulfilled (see [4,5]).

5 Chainability of rule bases

For a positive natural number n, let $F_1, \ldots, F_n, G_1, \ldots, G_n, H_1, \ldots, H_n \in \mathcal{F}(U)$. We consider fuzzy IF-THEN rule bases of the form

$$
RB_1 : \quad
\begin{array}{l}
\text{IF } F_1 \text{ THEN } G_1 \\
\quad\quad \vdots \\
\text{IF } F_n \text{ THEN } G_n
\end{array}
$$

and introduce the abbreviations $\mathfrak{F} =_{def} [F_1, \ldots, F_n]$, $\mathfrak{G} =_{def} [G_1, \ldots, G_n]$ and RB_1 : IF $\mathfrak{F}$ THEN $\mathfrak{G}$.

Definition 5.1

$\mathfrak{S} = [\pi_1, \ldots, \pi_n, \kappa_0, \kappa_1, \ldots, \kappa_n, Q_0, Q_1 \ldots, Q_n, \alpha, \beta]$ *is called an interpretation of an IF-THEN rule base*

$=_{def}$ 1. $\pi_1, \ldots, \pi_n, \kappa_0, \kappa_1, \ldots, \kappa_n : [0,1]^2 \to [0,1]$
      2. $Q_0, Q_1 \ldots, Q_n : \mathcal{P}([0,1]) \to [0,1]$
      3. $\alpha, \beta : [0,1]^n \to [0,1]$

For a given interpretation $\mathfrak{S}$ we define the "superrelation" $\Theta_{\mathfrak{S}}^{\mathfrak{F},\mathfrak{G}}$ and the operators $FATI_{\mathfrak{S}}^{\mathfrak{F},\mathfrak{G}}$ and $FITA_{\mathfrak{S}}^{\mathfrak{F},\mathfrak{G}}$ as follows.

Definition 5.2

1. $\Theta_{\mathfrak{S}}^{\mathfrak{F},\mathfrak{G}}(x,y) =_{def} \alpha(\pi_1(F_1(x), G_1(y)), \ldots, \pi_n(F_n(x), G_n(y)))$

2. $FATI_{\mathfrak{S}}^{\mathfrak{F},\mathfrak{G}}(F')(y) =_{def} (F' \circ_{\kappa_0, Q_0} \Theta_{\mathfrak{S}}^{\mathfrak{F},\mathfrak{G}})(y) =$
         $Q_0\{\kappa_0(F'(x), \alpha(\pi_1(F_1(x), G_1(y)), \ldots, \pi_n(F_n(x), G_n(y))))|x \in U\}$

3. $FITA_{\mathfrak{S}}^{\mathfrak{F},\mathfrak{G}}(F')(y) =_{def} \beta\{Q_1\{\kappa_1(F'(x), \pi_1(F_1(x), G_1(y)))|x \in U\}, \ldots,$
         $Q_n\{\kappa_n(F'(x), \pi_n(F_n(x), G_n(y)))|x \in U\}\}$

We need two further rule bases

$$RB_2 : \begin{array}{c} \text{IF } G_1 \text{ THEN } H_1 \\ \vdots \\ \text{IF } G_n \text{ THEN } H_n \end{array} \quad \text{and } RB_3 : \begin{array}{c} \text{IF } F_1 \text{ THEN } H_1 \\ \vdots \\ \text{IF } F_n \text{ THEN } H_n \end{array}$$

or for shortness RB_2 : IF $\mathfrak{G}$ THEN $\mathfrak{H}$ and RB_3 : IF $\mathfrak{F}$ THEN $\mathfrak{H}$ with $\mathfrak{H} =_{def} [H_1, \ldots, H_n]$.

By generalizing Def. 2.2 we obtain two versions of chainability of fuzzy rule bases.

Definition 5.3

1. *Two rule bases IF $\mathfrak{F}$ THEN $\mathfrak{G}$ and IF $\mathfrak{G}$ THEN $\mathfrak{H}$ interpreted by the principle FATI are called chainable with respect to $\mathfrak{S}$ and $\mathcal{F}'$, iff*

$$\forall F'(F' \in \mathcal{F}' \to FATI_{\mathfrak{S}}^{\mathfrak{G},\mathfrak{H}}(FATI_{\mathfrak{S}}^{\mathfrak{F},\mathfrak{G}}(F')) = FATI_{\mathfrak{S}}^{\mathfrak{F},\mathfrak{H}}(F')).$$

2. *Two rule bases IF $\mathfrak{F}$ THEN $\mathfrak{G}$ and IF $\mathfrak{G}$ THEN $\mathfrak{H}$ interpreted by the principle FITA are called chainable with respect to $\mathfrak{S}$ and $\mathcal{F}'$, iff*

$$\forall F'(F' \in \mathcal{F}' \to FITA_{\mathfrak{S}}^{\mathfrak{G},\mathfrak{H}}(FITA_{\mathfrak{S}}^{\mathfrak{F},\mathfrak{G}}(F')) = FITA_{\mathfrak{S}}^{\mathfrak{F},\mathfrak{H}}(F')).$$

The considerations of the previous sections also hold for the chaining of IF-THEN rule bases interpreted by the principle FATI, where the rules of each rule base are aggregated to a single "superrelation".

Definition 5.4
Two rule bases IF $\mathfrak{F}$ THEN $\mathfrak{G}$ and IF $\mathfrak{G}$ THEN $\mathfrak{H}$ are called loosely chainable *with respect to $\mathfrak{S}$, iff*

$$\Theta_{\mathfrak{S}}^{\mathfrak{F},\mathfrak{G}} \circ_{\kappa_0, Q_0} \Theta_{\mathfrak{S}}^{\mathfrak{G},\mathfrak{H}} = \Theta_{\mathfrak{S}}^{\mathfrak{F},\mathfrak{H}}.$$

Again we split the problem of chainability into two questions.

① On which conditions does loose chainability imply chainability of rule bases interpreted by the principle FATI? Answers are given in the next section.
② On which conditions are rule bases loosely chainable? Examples are given in Sect. 7.

The results also hold for FITA, if the principles FATI and FITA coincide. The equivalence of the two principles is investigated in [7,10].

6 Chainability of rule bases interpreted by the principle FATI implied by loose chainability

The basic ideas are the same as in Sect. 3.

Theorem 6.1
If 1. κ_0 is nondecreasing and associative
2. every partial mapping of κ_0 is lower-semicontinuous
3. $Q_0 = Sup$
then two rule bases IF $\mathfrak{F}$ THEN $\mathfrak{G}$ and IF $\mathfrak{G}$ THEN $\mathfrak{H}$ interpreted by the principle FATI are chainable if they are loosely chainable.

Theorem 6.2
If 1. κ_0 is nondecreasing and associative
2. $Q_0 = Sup$
3. for every $F' \in \mathcal{F}'$ the union $RANGE(F') \cup RANGE(F_1) \cup \cdots \cup RANGE(F_n) \cup RANGE(G_1) \cup \cdots \cup RANGE(G_n)$ is a finite set
then two rule bases IF $\mathfrak{F}$ THEN $\mathfrak{G}$ and IF $\mathfrak{G}$ THEN $\mathfrak{H}$ interpreted by the principle FATI are chainable with respect to $\mathcal{F}$ if they are loosely chainable.

If U is a finite set, then assumption 3 of the previous theorem is fulfilled.
From [8] we quote the next lemma.

Lemma 6.3
If 1. $\forall r(r \in [0,1] \to \kappa_0(0,r) = 0 \wedge \kappa_0(1,r) = r)$
2. $\forall r(r \in [0,1] \to Q_0\{0,r\} = r)$
then for every $x_0, y \in U$

$$FATI_{\mathfrak{S}}^{\mathfrak{G},\mathfrak{H}}(F'_{x_0})(y) = \alpha(\pi_1(F_1(x_0), G_1(y)), \ldots, \pi_n(F_n(x_0), G_n(y)))$$

Theorem 6.4

If 1. $\forall r(r \in [0,1] \to \kappa_0(0,r) = 0 \land \kappa_0(1,r) = r)$
  2. $\forall r(r \in [0,1] \to Q_0\{0,r\} = r)$

then two rule bases IF $\mathfrak{F}$ THEN $\mathfrak{G}$ and IF $\mathfrak{G}$ THEN $\mathfrak{H}$ interpreted by the principle FATI are chainable with respect to singleton inputs if they are loosely chainable.

7 Loose chainability of rule bases

Based on [8] we give two examples for the loose chainability of rule bases, which can be combined with the theorems of the previous section.

Definition 7.1

1. α is monotone
$$=_{def} \forall r_1 \ldots \forall r_n \forall s_1 \ldots \forall s_n (r_1, \ldots, r_n, s_1, \ldots, s_n \in [0,1] \land$$
$$r_1 \leq s_1 \land \cdots \land r_n \leq s_n \to \alpha(r_1, \ldots, r_n) \leq \alpha(s_1, \ldots, s_n))$$

2. κ_0 is right-side strongly α-distributive
$$=_{def} \forall r_1 \ldots \forall r_n \forall s_1 \ldots \forall s_n (r_1, \ldots, r_n, s_1, \ldots, s_n \in [0,1]$$
$$\to \kappa_0(\alpha(r_1, \ldots, r_n), \alpha(s_1, \ldots, s_n)) \leq \alpha(\kappa_0(r_1, s_1), \ldots, \kappa_0(r_n, s_n)))$$

3. κ_0 is left-side weakly α-distributive
$$=_{def} \forall r_1 \ldots \forall r_n \forall s_1 \ldots \forall s_n (r_1, \ldots, r_n, s_1, \ldots, s_n \in [0,1]$$
$$\to \alpha(\kappa_0(r_1, s_1), \ldots, \kappa_0(r_n, s_n))) \leq \kappa_0(\alpha(r_1, \ldots, r_n), \alpha(s_1, \ldots, s_n)))$$

Theorem 7.1

If 1. for every $i \in \{1, \ldots, n\}$, π_i is reflexive
  2. for every $i \in \{1, \ldots, n\}$, π_i is κ_0-transitive
  3. $\forall r(r \in [0,1] \to \kappa_0(1,r) = r)$
  4. $Q_0 = Sup$
  5. α is monotone
  6. κ_0 is right-side strongly α-distributive
  7. (a) κ_0 is left-side weakly α-distributive **or**
    (b) $\alpha(1, \ldots, 1) = 1$
  8. $\forall x(x \in U \to \exists y(y \in U \land F_1(x) = G_1(y) \land \cdots \land F_n(x) = G_n(y)))$,

then

$$\Theta_{\mathfrak{F}}^{\mathfrak{F},\mathfrak{G}} \circ_{\kappa_0, Q_0} \Theta_{\mathfrak{F}}^{\mathfrak{G},\mathfrak{H}} = \Theta_{\mathfrak{F}}^{\mathfrak{F},\mathfrak{H}}.$$

If e.g. 1. $\forall r \forall s(r, s \in [0,1] \to \pi_1(r,s) = \cdots = \pi_n(r,s) = min(1, 1-r+s))$
  2. $\forall r \forall s(r, s \in [0,1] \to \kappa_0(r,s) = max(0, r+s-1))$
  3. $Q = Sup$
  4. $\forall r_1 \ldots \forall r_n (r_1, \ldots, r_n \in [0,1] \to$
$$\alpha(r_1, \ldots, r_n) = \kappa_0(r_1, \kappa_0(r_2, \ldots \kappa_0(r_{n-1}, r_n) \ldots)))$$

then the assumptions 1-7 of the previous theorem are fulfilled, hence the two fuzzy rule bases are loosely chainable if assumption 8 is satisfied.

187

Theorem 7.2

If 1. *for every* $i \in \{1, \dots, n\}$, $\forall r(r \in [0,1] \to \pi_i(1,r) = r)$,
    2. *for every* $i \in \{1, \dots, n\}$, $\forall r \forall s(r,s \in [0,1] \to \pi_i(r,s) \leq \kappa_0(\pi_i(r,1),s))$
    3. *for every* $i \in \{1, \dots, n\}$, π_i *is* κ_0-*transitive*
    4. $Q_0 = Sup$
    5. α *is monotone*
    6. κ_0 *is right-side strongly* α-*distributive*
    7. κ_0 *is left-side weakly* α-*distributive*
    8. $\exists y(y \in U \wedge G_1(y) = \cdots = G_n(y) = 1)$

then

$$\Theta_{\mathfrak{F}}^{\mathfrak{F},\mathfrak{G}} \circ_{\kappa_0,Q_0} \Theta_{\mathfrak{F}}^{\mathfrak{G},\mathfrak{H}} = \Theta_{\mathfrak{F}}^{\mathfrak{F},\mathfrak{H}}.$$

If e.g. 1. $\pi_1 = \cdots = \pi_n = \kappa_0 = \kappa_1 = \cdots = \kappa_n = min$
       2. $Q_0, Q_1, \dots, Q_n = Sup$
       3. $\alpha = \beta = Min$

then the assumptions 1-7 of the previous theorem are satisfied and FITA and FATI coincide.

8 Conclusion

The proposed systematic approach combines new considerations with the results of former investigations and leads to various new conditions on which fuzzy rules and rule bases are chainable.

References

1. B. Cappelle, Etienne E. Kerre, Ruan Da, and F. R. Vanmassenhove. Characterization of binary operations on the unit interval satisfying the general modus ponens inference rule. *Mathematica Pannonica*, 2(1):105–121, 1991.
2. Ruan Da and Etienne E. Kerre. Extended Fuzzy Chaining Syllogismus. In *ISFA '95–Sixth International Fuzzy System Assocition World Congress*, volume 1, pages 145–148, São Paulo, Brazil, Juli 1995.
3. Didier Dubois and Henri Prade. Fuzzy sets in approximate reasoning, Part 1: Inference with possibility distributions. *Fuzzy Sets and Systems*, 40:143–202, 1991.
4. Siegfried Gottwald. *Fuzzy Sets and Fuzzy Logic*. Vieweg Verlag, Braunschweig, 1993.
5. Siegfried Gottwald. On the Chainability of Fuzzy if-then Rules. In *IPMU '96–Sixth International Conference on Information Processing and Management of Uncertainty in Knowledge-Based Systems*, volume 2, pages 997–1001, Granada, Spain, Juli 1996.
6. Masaharu Mizumoto and Hans-Jürgen Zimmermann. Comparison of fuzzy reasoning methods. *Fuzzy Sets and Systems*, 8:253–283, 1982.
7. Karl-Heinz Temme and Helmut Thiele. On the Correctness of the Priciples FATI and FITA and their Equivalence. In *IFSA '95—Sixth International Fuzzy System Assocition World Congress*, volume 2, pages 475–478, São Paulo, Brasil, Juli 1995.

8. Karl-Heinz Temme and Helmut Thiele. On the Chaining of If-Then Rule Bases Interpreted by the Principle FATI. In *Proceedings of Symposium on Qualitative System Modelling, Qualitative Fault Diagnosis and Fuzzy Logic and Control*, pages 143–152, Budapest and Balantonfüret, April 1996. Further references may be found in the bibliography of this paper.

9. Karl-Heinz Temme and Helmut Thiele. On the Chaining of If-Then Rule Bases Interpreted by the Principle FITA. In *EUFIT '96—Fourth European Congress on Intelligent Techniques and Soft Computing*, Aachen, Germany, September 1996.

10. Helmut Thiele. On the calculus of fuzzy if-then rules. In *Intelligent Technologies and Soft Computing*, Mangalia, Romania, September 1995.

11. Lotfi A. Zadeh. Syllogistic reasoning in fuzzy logic and its application to reasoning with dispositions. *IEEE Trans. on Systems, Man and Cybernetics*, pages 754–763, 1985.

12. Lotfi A. Zadeh. The Calculus of Fuzzy If-Then Rules. In T. Terano, M. Sugeno, M. Mukaidono, and K. Shigemasu, editors, *Fuzzy Engineering toward Human Friendly Systems–Proceedings of the International Fuzzy Engineering Symposium '91*, volume 1, pages 11–12, Yokohama, Japan, November 1991.

Computational Properties of Fuzzy Logic Deduction

Antonín Dvořák

University of Ostrava
Institute for Research and Applications of Fuzzy Modeling
Bráfova 7, 701 03 Ostrava, Czech Republic

Abstract. This paper studies computational properties of fuzzy logic deduction and compares them with the standard inference methods. The principles of deduction in fuzzy logic are explained and algorithms for its computer realization are described. Basic algorithm has exponential complexity with respect to the number of antecedent variables, and in case of fuzzy observations we are able to improve its performance only by constant factor.

1 Introduction

In this contribution, we study computational and algorithmic aspects of fuzzy logic deduction. This type of fuzzy inference mechanism was proposed and investigated by V. Novák in, e.g. [6] and it has been demonstrated, that its properties make it suitable for applications in control systems as well as in decision making. Algorithmic aspects of the most commonly used Mamdani-Zadeh inference mechanism were extensively studied [5], where was shown that computational complexity grows exponentially with the number of antecedent variables. However, algorithmic properties of fuzzy logic deduction have not been studied so far. Some results on these problems and on the new method of linguistic approximation were published in [1].

This paper is organized as follows: In Section 2 we present a brief introduction to the theory of the deduction in fuzzy logic, Section 3 contains description of inference in situation when observations are crisp numbers or fuzzy singletons. Section 4 presents situation when observations are fuzzy sets. This is the main part of this paper and we describe here some methods for speeding-up the inference. Section 5 contains some conclusions and prospects of further research.

2 Fuzzy Logic Deduction

There are several approaches to fuzzy inference, which is the implementation of the modus ponens inference rule in which the implication and possibly also premise are given vaguely. They differ mainly in the interpretation of implication. One approach, known as *Mamdani-Zadeh inference* or fuzzy interpolation or Max-t-norm inference, is essentially approximation of unknown function [2].

The second approach, which we call *fuzzy logic deduction*, uses Łukasiewicz implication operator as a basis for inference mechanism. The basic scheme known as *generalized modus ponens* is the following:

Condition: $\mathscr{R} :=$ IF X_1 is $\mathscr{A}_1$ AND ... AND X_n is $\mathscr{A}_n$ THEN Y is $\mathscr{B}$
Observation: X_1 is $\mathscr{A}'_1$ AND ... AND X_n is $\mathscr{A}'_n$

Conclusion: Y is $\mathscr{B}'$.

The $\mathscr{A}'_1, \ldots, \mathscr{A}'_n$ are expressions of natural language, which may be slightly different from $\mathscr{A}_1, \ldots, \mathscr{A}_n$. It follows, that also the conclusion $\mathscr{B}'$ can be slightly different from $\mathscr{B}$.

The *condition* in the scheme presented above is translated into the fuzzy relation

$$R \underset{\sim}{\subseteq} U_1 \times \ldots \times U_n \times V$$

and *observation* into the fuzzy set

$$A' \underset{\sim}{\subseteq} U_1 \times \ldots \times U_n.$$

For practical reasons, we assume that the universes of discourse of all the variables are discrete, i.e.

$$U_i = \{x_1, x_2, \ldots, x_P\}, \quad i = 1, \ldots, n,$$
$$V = \{y_1, y_2, \ldots, y_P\},$$

where the number P of elements is the same for all $U_i, i = 1, \ldots, n$ and V.

If more independent variables are present, then the linguistic conjunction AND is usually interpreted as minimum operation on the corresponding membership degrees $A'_i x_i$, $i = 1, \ldots, n$ and $x_i \in U_i$.

The *conclusion* is computed using a certain operation $\circ^*$

$$B' = (A'_1 \cap \cdots \cap A'_n) \circ^* R. \tag{1}$$

The operation $\circ^*$ depends on the knowledge about dependence among considered phenomena. Formula (1) is the interpretation of the inference scheme.

We may rewrite (1) into formula

$$B'y = \bigvee_{x_1 \in U_1, \ldots, x_n \in U_n} ((A'_1 x_1 \wedge \cdots \wedge A'_n x_n) \otimes R(\langle x_1, \ldots, x_n \rangle, y)), \tag{2}$$

where $y \in V$. In this formula, R denotes the interpretation of one IF–THEN rule $\mathscr{R}$. The operation $\otimes$ is the Łukasiewicz conjunction

$$a \otimes b = 0 \vee (a + b - 1), \quad a, b \in [0, 1].$$

If a set of linguistic IF–THEN rules is given, then R should be replaced by an appropriate fuzzy relation. Then the formula (2) turns into

$$B'y = \bigvee_{x_1 \in U_1, \ldots, x_n \in U_n} ((A'_1 x_1 \wedge \cdots \wedge A'_n x_n) \otimes \bigwedge_{j=1}^{m} ((A_{1j} x_1 \wedge \cdots \wedge A_{nj} x_n) \rightarrow B_j y))),$$

$$(3)$$

where the operation $\rightarrow$ is the Łukasiewicz implication

$$a \rightarrow b = 1 \wedge (1 - a + b), \qquad a, b \in [0, 1].$$

Two cases should be distinguished:

– A' are crisp numbers or fuzzy singletons, or
– A' are fuzzy sets.

In the first case, there is a possibility to perform the inference for individual rules first, and then combine results to one fuzzy set. This approach is similar to the well-known one of Mamdani. In the second case, however, such "local" inference is not possible, and we are forced to proceed in accordance with the formula (3). In the following section, we elaborate two above mentioned cases in more details.

3   Local Inference

If observations which enter inference engine are fuzzy singletons

$$\{ \beta_1 / x_1 \}, \ldots, \{ \beta_n / x_n \}, \qquad 0 < \beta_i \leq 1, x_i \in U_i$$

then the formula (3) may be rewritten to

$$B'y = (\beta_1 \wedge \cdots \wedge \beta_n) \otimes \bigwedge_{j=1}^{m} ((A_{1j} x_1 \wedge \cdots \wedge A_{nj} x_n) \rightarrow B_j y)),$$

because there is only one combination of $x_1, \cdots, x_n$ for which the expression $A'_1 x_1 \wedge \cdots \wedge A'_n x_n$ in (3) has a non-zero value. This value should be 1 for crisp observations. In the case of fuzzy singletons we denote it as β', i.e.

$$\beta' = \beta_1 \wedge \cdots \wedge \beta_n.$$

It is easy to show that the former formula is equivalent with

$$B'y = \bigwedge_{j=1}^{m} (\beta' \otimes ((A_{1j} x_1 \wedge \cdots \wedge A_{nj} x_n) \rightarrow B_j y)),$$

which allows us to perform the inference step separately for individual rules. In this situation we can introduce the notion of *degree of matching* γ_j as the degree in which individual j-th rule fires for a given (crisp) observation $x_1, \ldots, x_n$:

$$\gamma_j = A_{1j} x_1 \wedge \cdots \wedge A_{nj} x_n.$$

Then the algorithm of computation B' can be described as follows:

1. For each rule $\mathscr{R}_j$ we compute the degree of matching γ_j.
2. Partial outputs B'_j are determined as

$$B'_jy = \beta' \otimes (\gamma_j \to B_jy).$$

3. The aggregated fuzzy set B' is determined as an intersection of partial outputs

$$B'y = \bigwedge_{j=1}^{m} B'_jy.$$

This type of inference is analogous with the well-known Mamdani inference method. The complexity of this algorithm is ([4])

$$\mathscr{C} = \mathscr{O}(m(n+1)P), \tag{4}$$

where P is precision of the discretization introduced in Section 2.

4 Global Inference

4.1 General Properties

If the observations A' are fuzzy sets, it is not possible to perform local inference, and we are forced to use formula (3) directly. There is a possibility to adapt it in the following manner:

$$B'y = \bigvee_{x_1 \in U_1, \ldots, x_n \in U_n} \bigwedge_{j=1}^{m} ((A'_1x_1 \wedge \cdots \wedge A'_nx_n) \otimes ((A_{1j}x_1 \wedge \cdots \wedge A_{nj}x_n) \to B_jy)), \tag{5}$$

because operations $\wedge$ and $\otimes$ are infinitely distributive. If we exchange the supremum and the infimum, we obtain only upper estimate of B.

Theorem 1. *If B' is the fuzzy set computed according to formula (5) and B'' the fuzzy set computed according to formula*

$$B''y = \bigwedge_{j=1}^{m} \bigvee_{x_1 \in U_1, \ldots, x_n \in U_n} ((A'_1x_1 \wedge \cdots \wedge A'_nx_n) \otimes ((A_{1j}x_1 \wedge \cdots \wedge A_{nj}x_n) \to B_jy)),$$

then $B' \subseteq B''$.

Proof. See, for example, [3, p. 320].

The basic algorithm which uses the formulas (3) or (5) has very high complexity

$$\mathscr{C} = \mathscr{O}(mP^{n+1}). \tag{6}$$

There are no methods known so far which allow us to get rid of inherent exponential complexity with respect to number of antecedent variables. If we use this inference mechanism in practice, then it is important to speed up computations also with respect to other parts of the expression (6). The number of antecedent variables n is usually constant for the given application or problem.

We can use several techniques to speed up the computations:

1. We can compute only some interesting part of B',
2. we can use only several rules from rule base, and
3. we can do some computations beforehand.

In the following, we begin with formula (3) and study formerly mentioned methods in detail. Notice, that for individual $x_1, \ldots, x_n$ the supremum in (3) can be non-zero only if the expression $A'_1 x_1 \wedge \ldots \wedge A'_n x_n$ is greater than zero. This follows from the properties of the operation $\otimes$ (and, in fact, any t-norm). Hence we compute the remaining part of the supremum only if $A'_1 x_1 \wedge \ldots \wedge A'_n x_n > 0$.

The ratio of improving of the performance is dependent on the width of fuzzy sets in the observation. The closer are fuzzy sets to singletons, the faster is the computation. The speeding-up factor is equal to

$$f_0 = \frac{P^n}{\prod\limits_{i=1}^{n} \mathrm{Card}(\mathrm{Supp}(A'_i))}.$$

4.2 The First Method

Sometimes it is sufficient to compute only important part of the conclusion B', e.g. the part where membership degrees are greater than some α. In this case we can use the well-known property of t-norm $\otimes$, namely that

$$a \otimes b \leq a \qquad a, b \in [0, 1].$$

Hence, if we are interested only in that part of B', which has membership degrees greater than $\alpha \in [0, 1]$ we compute the implication part of the supremum in (3) only if $A'_1 x_1 \wedge \ldots \wedge A'_n x_n > \alpha$. The speeding-up factor is

$$f_1 = \frac{P^n}{\prod\limits_{i=1}^{n} \mathrm{Card}(A'_{\alpha i})},$$

where A'_α is the α-cut of the fuzzy set A'.

However, it can happen than this part of B' is empty or too small for performing defuzzification or linguistic approximation procedure. But one of the properties of fuzzy logic deduction is, that the obtained conclusions are usually normal fuzzy sets, i.e. fuzzy sets with non-empty kernel. If, for example, there is no rule which fires for a given input, then the result is fuzzy set with membership function identically equal to 1. The subnormal fuzzy sets are obtained in cases when the rulebase contains a contradiction and observation is such that two or more rules with strongly different consequent fire. Even in these situations, which can be simply tested, we can use similar method with the addition, that if the conclusion B' is empty fuzzy set, we determine a new smaller threshold α.

4.3 The Second Method

The second method is based on the idea, that it is not necessary to compute conclusion over all the rules, and that we can determine the most appropriate rules in the rulebase for the given observation. The idea is similar to *fuzzy analogical reasoning* introduced in [8]. Our approach is different in the following sense: we choose more than one rule from rulebase with the highest similarity to observation and then perform standard fuzzy logic inference. For determining the most suitable rules we need some *similarity measure* between fuzzy sets ([9, 7]).

Unfortunately, the commonly used similarity measures are not suitable for our purposes, because even very dissimilar fuzzy sets can have crucial importance for inference. Hence, we propose to determine the most important rules from the area of intersection between observation and antecedent fuzzy sets.

The area of intersection is determined by the formula

$$M(A, B) = \int\limits_{x \in U} (A \cap B) x dx,$$

provided that membership functions of fuzzy sets A, B are measurable.

The overall measure should be computed with respect to the area of individual observations. This is necessary because otherwise wider observations can influence conclusion more than narrower ones. The overall measure for one rule is defined by

$$M^\star = \sum_{i=1}^{n} \frac{\int\limits_{x \in U_i} (A_i' \cap A_i) x dx}{\int\limits_{x \in U_i} A_i' x dx}. \tag{7}$$

In practice, when we work with discrete fuzzy sets, the formula (7) turns into

$$M^\star = \sum_{i=1}^{n} \frac{\sum_{k=1}^{P} min(A_i' x_k, A_i x_k)}{\sum_{k=1}^{P} A_i' x_k}.$$

The next step of this method is determination of the number of rules which we use for inference. We can fix this number l beforehand, or we can determine it from the results of the preselection procedure.

The algorithm of computation B' is:

1. For all rules $\mathscr{R}_j$ we determine $M_j^\star$.
2. Compute $B'y$ by (3) for l rules with the highest $M_j^\star$.

The speeding-up factor is approximately equal to

$$f_2 \approx \frac{m}{l},$$

where m is the number of rules in the rulebase.

4.4 The Third Method

Two methods described above generally do not give the same conclusion B' in comparison with the use of formula (3). The third method can compute precise conclusion, but we should do some calculations beforehand.

The first possibility is to compute the fuzzy relation R. However, as this relation is $(n+1)$-dimensional, the amount of required memory is intractable for more than three antecedent variables.

The second possibility is based on observation, that for one given y the expression $(A_{1j}x_1 \wedge \cdots \wedge A_{nj}x_n) \to B_j y$ is independent on B_j. Furthermore, we can use some properties of the operation $\to$ and also of the fact, the we search the infimum. The algorithm for computation B' can be described as follows:

1. For each $y \in V$ make the ascending order of the rules $\mathscr{R}_j$ by the value of By.
2. Compute $B'y$ by means of formula (3), but when searching the infimum

$$\bigwedge_{j=1}^{m} ((A_{1j}x_1 \wedge \cdots \wedge A_{nj}x_n) \to B_j y)$$

use the following:
   (a) compute $\gamma_j = (A_{1j}x_1 \wedge \cdots \wedge A_{nj}x_n) \to B_j y$, where $j = 1, \ldots, m$ in order determined in Step 1,
   (b) if $\gamma_j = 0$, then the infimum $\bigwedge_{j=1}^{m}((A_{1j}x_1 \wedge \cdots \wedge A_{nj}x_n) \to B_j y) = 0$,
   (c) if $\gamma_j \leq B_j y$, then the infimum $\bigwedge_{j=1}^{m}((A_{1j}x_1 \wedge \cdots \wedge A_{nj}x_n) \to B_j y) = B_j y$.

The speeding-up factor is difficult to estimate beforehand, because there is always the possibility that By which gives the infimum is hidden in the end of the sequence of rules ordered in accordance with the first step of the algorithm. However, by this method we find the infimum as soon as possible.

4.5 Comparison of Methods

If we are interested only in the most important parts of fuzzy conclusion, we can use The First Method. However, some important information may be hidden in that part of conclusion which is not computed. The Second method is appropriate in cases when the number of rules and the number of variables in antecedent is high, because the time which is needed for computation of important rules from, say, 500 rules is much smaller than the time for inference with all of them using directly formula (3). Experimental results show that the performance of this method is good, i.e. the algorithm is able to find the most appropriate rules. The Third Method is proposed for cases when we want to compute the conclusion precisely.

As a matter of course, there are possibilities of combining above mentioned methods. In particular, if we want speed up the computation of conclusions maximally, it is advantageous to combine The First and The Second methods.

5 Conclusion

All the described methods make possible to speed-up computation of conclusions in fuzzy logic deduction. This is important in practical realization of these inference method. The computational properties of fuzzy logic deduction are equivalent with Mamdani inference in situation when observations are crisp numbers or fuzzy singletons. If observations are fuzzy sets, then the computational complexity is basically described by (6). The methods from Section 4 can improve the performance of inference engine and allow us to use fuzzy logic deduction in real-time systems.

Applications of the above described methods are expected in the field of fuzzy expert and decision support systems, where observations are fuzzy sets. Fuzzy observations can occur also in complex fuzzy systems, when conclusion deduced by some subsystem is observation for the subsequent one. In further research we intend to concentrate on the use of sparse rulebases based on principles described in [5].

Acknowledgements

This paper has been supported by the grant A1086501 of the GA AV ČR.

The author wishes to thank to Prof. Laszló T. Kóczy for valuable discussions during author's CEEPUS stay at Technical University of Budapest.

References

1. Dvořák, A.: Properties of the Generalized Fuzzy logic Inference. Proceeding of SIC'96 Budapest, 1996, 75–80.
2. Klawonn, F. and V. Novák: The Relation between Inference and Interpolation in the Framework of Fuzzy Systems. Fuzzy Sets and Systems 81(1995), 331–354.
3. Klir, G.J. and Bo Yuan: Fuzzy Sets and Fuzzy Logic. Theory and Apllications. Prentice Hall 1995.
4. Kóczy, L.T.: Computational Complexity of Various Fuzzy Inference Algorithms. Annales Univ. Sci. Budapest., Sect. Comp. 12(1991), 151–158.
5. Kóczy, L.T.: Algorithmic Aspects of Fuzzy Control. Int. J. of Approximate Reasoning, 12(1995), 159–219.
6. Novák, V.: Linguistically Oriented Fuzzy Logic Controller and Its Design. Int. J. of Approximate Reasoning, 12(1995), 263–277.
7. Lee, E.S. and Q. Zhu: Fuzzy and Evidence Reasoning. Physica–Verlag 1995.
8. Turksen, I.B. and Z. Zhong: An Approximate Analogical Reasoning Schema Based on Similarity Measures and Interval Valued Fuzzy Sets. Fuzzy Sets and Systems 34(1990), 323–346.
9. Zwick, R., E. Carlstein and D.V. Budescu: Measures of Similarity Among Fuzzy Concepts: A Comparative Analysis. International Journal of Approximate Reasoning, 1(1987), 221–242.

Fuzzy Diagnosis of Float-Glass Production Furnace

L. Spaanenburg, H. TerHaseborg and J.A.G. Nijhuis

Rijksuniversiteit Groningen, Dept. Computing Science
P.O.Box 800, 9700 AV Groningen (The Netherlands)

Abstract. The industrial production of high–quality float–glass is usually super-vised by the single human expert. It is of interest to formalize his empirical knowl-edge to support the furnace operator at all times during the day. The paper describes the systematic development of a fuzzy expert with 6 blocks of 297 knowledge rules and 85 fuzzy terms for verified production quality.

1 Introduction

Glass is a famous culture carrier. For centuries now a characteristic mark of civilization has been its use of glass for a variety of purposes. Glass was already known in 400 B.C.. The legend tells how merchants on the Lebanese coast cleaned their pots and pans with potassium sulfate and accidentally created glass. In the course of history, glass was first used for very prestigious items and later became more common place. As long as the fabrication was in the hands of artisans, each product was an individual piece of art and quality was not at stake.

With the advent of industrialization came also the industrial production of glass: mass production with a guaranteed and constant quality. Glass results from the heating of a material mix under well–defined conditions. The harsh environment within the furnace makes it difficult to maintain these conditions, as only external observation is feasible. It seems likely that, where over years a wealth of experience is assembled, insight has been gained that may deepen the understanding and interpretation of external measure-ments. This insight is presently available through the single human expert, when he is available. It should be replaced by the always present fuzzy rule–set, providing diagno-sis of the current operation and sound advice on a preferred control strategy.

This paper starts by introducing industrial glass production and focuses on the float–glass technique. Then knowledge acquisition based on multiple rule–blocks is introduced through a small example. Ensuing the use of the fuzzy neural network is dis-cussed to facilitate personalization of the model and subsequently applied to the float–glass case.

2 The current system

The oldest fabrication styles (or processes) are discontinuous, as the shaping of glass objects is performed off–line. In the Bicheroux process this is only partly continuous, while in modern times the process is fully automated. In coarse division, there are two type of glass objects: hollow and flat. Here, we will pay only attention to the flat glass as commonly used in windows.

2.1 Float–glass production

Glass is industrially produced in a furnace of minimal 500 ton capacity. The materials are supplied at one side of the furnace and go through four phases: (a) melting the raw material, (b) refining the melt for specified characteristics, (c) cooling–down to 1100 degrees Celsius in order to (d) cast the melt into shape. Different chemical receipts (called *processes*) require different architectures of the furnace. They can be seen as constant improvements with respect to the way the thickness is determined and to the smoothness of the surface. The process of interest here (the *Pinkington process*) is based on the concept of floating transport of the material: the *float–glass*. The glass melt is spawned on a tin bedding, torn for thickness, cooled and finally cut to measure.

By means of heat converters the glass melt is created at 1800 degrees Celsius. At refinement, oxygen is produced and will partly escape the furnace. However, some oxygen may be enclosed as small bulbs within the glass. For window–glass the amount of oxygen bulbs indicates the isolation quality and robustness of the panel. As the occurrence of those bulbs is further raised by sudden changes in temperature, a prerequisite for production quality is a constant temperature throughout the oven. Unfortunately, the local temperature is strongly dependent on local production parameters; moreover, the temperature can not be measured at every spot within the oven.

The oven conditions are set by the operator through the use of a PID–controller; the settings are changed according to quality checks on the final product on basis of the judgement of the human specialist. However, this specialist is not always available and different specialists make different judgements. A further formalization is required to provide for a constant and consistent quality of problem diagnosis and repair [1].

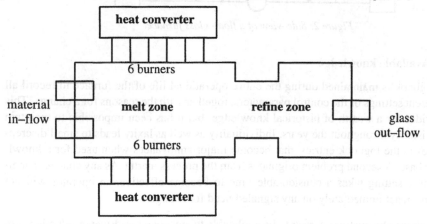

Figure 1: Top–view of a float–glass furnace.

2.2 Need for diagnosis

There are two fundamental reasons why diagnosis is required. As indicated in Figure 1, the glass can hardly be observed during the production process. Therefore it will be difficult to relate production failures to failing settings for the different production parts.

This states the need for *deep knowledge* (i.e. educated guesses about the internal operation) based on a black–box model (having only external access and therefore *shallow knowledge*)

The problem seems even more serious when looking at Figure 2, the cross–section of a furnace. Because of the environmental conditions, measurements of important variables can not be made directly. For instance, the temperature at the sensor will not be identical to the glass temperature. Because of the physical phenomena involved, there is not even a linear relationship. In other words, the characteristics involve a difficult, non–linear correspondence with large time constants. Despite that, a direct action is required to reduce fabrication waste and this action involves the knowledge of an expert, that as a human being can not always be available.

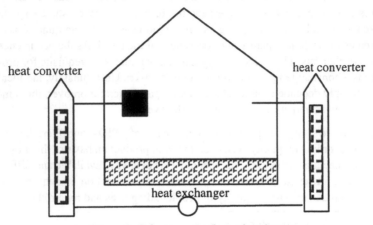

Figure 2: Side–view of a float–glass furnace.

2.3 Available knowledge

A logbook is maintained during the entire operational life of the furnace to record all different settings of the control parameters, together with the reasons for a change. This provides for a wealth of historical knowledge, but it has been impossible to maintain consistency throughout the years. Individuality as well as laxity leads to small discrepancies in the logbook entries, that become major annoyances when used for a knowledge base. A second problem originates from the process inertia. As any change in temperature setting takes a considerable time to become effective, an operator will not always react immediately on any signaled need for change.

The above observations appear to be applicable to any operator–driven production process and lead to a increased amount of *data noise* in the maintained records. Therefore, we have initially attempted to improve the written records by oral interviews. This is of course a tedious procedure, but also very rewarding for the novice interviewer. However, it is hardly possible and definitely not necessary to remove all data noise. The physical process to be controlled will always lead to an amount of noise that changes over periods.

Therefore we have cast the knowledge into fuzzy rules. From the human expert, the main features (or principal components) are determined. To a large degree, cross–correlations are used to reveal the interesting parameters. The overall architecture of the fuzzy system is then visible from these features and their mutual relations. This is important in order to keep into grip with the ongoing design as the fully expanded system is too large to be easily understood [2], [3].

3 Development method

In order to introduce the knowledge acquisition scheme, we will describe a simplified example of the traffic light regulation on one side of a street crossing. The problem is here to control the green period depending on the local traffic density, assuming only nearby detectors. As a consequence of not using a remote detection, there is no direct knowledge about the traffic running towards the light. So the decision on whether or not to extend the green period from its minimal value has to be based on a prediction of the traffic density.

3.1 Feature determination

The traffic density can in the first instance be estimated from the vehicle arrival times at the crossing. When there are no cars arriving, the density will be presumably zero; when all the time cars are passing by the density will be one. But when the situation is not extreme, the density will harder to guess. The difference between the density on the left and on the right lane may play a role, while further also the time of fay will be of importance.

From a reasonable estimate of the traffic density one may decide on the need to provide a green phase. The weather conditions may play a role in the evaluation of the density. Further, if the light has just turned on to green, we have to provide for a minimum green time. There will also be an upper limit to the green phase. Whether to extend the minimum green phase to ultimately the maximum phase, is dependent on the density on this road and on others.

3.2 Rule determination

For the feature green_phase we can easily derive three rules:
> IF (time – green_start is LOW) THEN (green_phase is FORCE)
> IF (time – green_start is MEDIUM) THEN (green_phase is EXTEND)
> IF (time – green_start is HIGH) THEN (green_phase is STOP)

Upon considering the traffic density with this rules we find a small explosion in the number of rules. For instance, the rule to extend the green phase may explode to
> IF ((time – green_start is MEDIUM) AND (density is LOW))
> > THEN (green_phase is STOP)
> IF ((time – green_start is MEDIUM) AND (density is MEDIUM))
> > THEN (green_phase is EXTEND)
> IF ((time – green_start is MEDIUM) AND (density is HIGH))
> > THEN (green_phase is FORCE)

Such an explosion in the number of rules does not make the set easily accessible and therefore one may opt not to merge the difference contribuants during the initial specification of the fuzzy controller. This will give a description based on the interconnection of many blocks, each with only a small amount of rules. An example is shown in Figure 3.

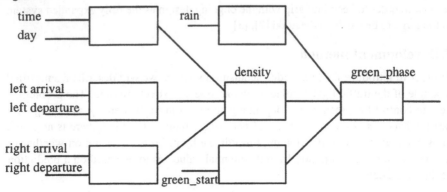

Figure 3: Multiple rule–blocks for traffic light control.

Roughly one can distinguish in a prediction part that provides an educated guess on the overall traffic density and a decision–making part that concludes on the need for the green phase extension. This composition appears very often when using this specification style and can be taken as a reasonable guideline [4].

3.3 Technology mapping

In order to derive an efficient fuzzy logic controller from this hierarchical specification, the set of rule blocks must first be merged. The next step is then to re–structure this flat description on basis of the time/area balance set by the target platform. The flat specification will be fast to execute but at the expense of a large memory consumption. On the other hand, the structured description will be slower to execute but more efficient in its storage requirements. This process is automated by the FASCOGEN tool, taking a FuzzyTech description on input and after an interactive trial–and–error provide C–code for the target controller platform [5].

4 Personalization

In case of a natural process, that defies mere simulation but has no mathematical description, it will be hard to render a fuzzy description for which proper values have been assigned to the linguistic variables [6]. In this case, an optimization of the fuzzy description is required either by statistical or by neural methods. This first requires a transformation to a neural network, then an optimization by learning through the neural network, followed by a transformation back to a fuzzy description.

4.1 Creating a fuzzy neural network

One of the early architectures to initialize a neural network according to fuzzy rules is shown in Figure 4 [7]. It is a five layer structure, wherein the input neurons represent the linguistic variables. The first hidden layer is composed of a set of neurons to fuzzify each linguistic variable. Then the input terms are aggregated to give the rules represented in the second hidden layer. After combination of the rules, the output terms are constructed in the third hidden layer, whereupon the fuzzification results in the linguistic outputs. Such a network can be precisely initialized and is a 1–to–1 representation of the fuzzy rules. Combining such networks into a single modular one can then reflect the multiple rule–block specification.

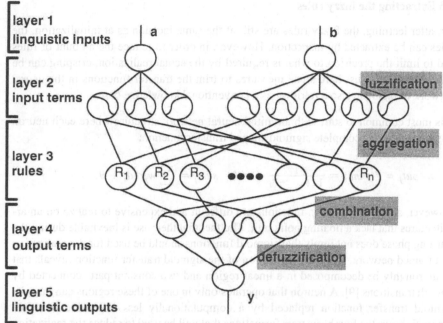

Figure 4: Structure of a fuzzy neural network.

4.2 Optimization

The initialized fuzzy neural network will contain the operator knowledge as collected by inspection of the logbook. This is the so–called *domain knowledge*, that may represent the fabrication process but not necessarily the specific furnace. To personalize for the individual characteristics, a small set of actual measurements can be used for learning. During learning, each set of neurons involved in (de)fuzzification must be guarded to ensure that the fuzzy composition is maintained. Though the advantage of having an exact replica of a fuzzy rule is clear, the drawback of increased learntimes has often led to alternative approaches.

Consider a Fuzzy system with one output variable and m input variables with n terms each. The m input variables create an m–dimensional hypercube that can be divided in n^m sub–hypercubes representing the n^m rules. Then, the data are clustered into the sub–hypercubes: the input data activate one rule, the desired output is written into the sub–hypercube. If a sub–hypercube is activated more than once, the mean and standard deviation of the desired output data are calculated. After the clustering is done, the output terms can be assigned to the rules [8].

If we interpret this general procedure in terms of our multiple rule–block description or the modular fuzzy neural network, each rule–block will be optimized as a sub–hypercube. The advantage of the hypercube notation is that it lends itself better for evolutionary programming.

4.3 Extracting the fuzzy rules

As, after learning, the fuzzy rules are still at the same location as at initialization, the rules can be extracted by inspection. However, in order to reduce the amount of rules and to limit the precision to what is required by the actual realization, crisping can be used. We will discuss here some measures to trim the transfer functions in the neural network to facilitate also a retrieval as a conventional knowledge base.

It is most common to start with an initial neural network structure where each neuron is equipped with a complete sigmoid shaped transfer function:

$$out_j = f(sum_j) = \frac{1}{1 + e^{-sum_j}} \quad with \quad sum_j = \sum_{i=0}^{N} w_{ji} out_i + \theta_j$$

However, sigmoid activation functions are difficult and expensive to realize on an architectures that lack a floating point unit. The fact that their use is inevitable during the learning phase does not imply that sigmoid functions should be used for the realization of a trained network. A closer inspection of the sigmoid transfer function reveals that it can roughly be decomposed in a linear region and two constant parts connected by smooth transitions [9]. A neuron that operates only in one of these regions can have its sigmoid transfer function replaced by a computationally less expensive function. Figure 5 shows the four basis transformations that will be used to reduce the realization costs.

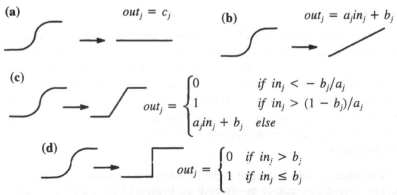

(a) $out_j = c_j$ **(b)** $out_j = a_j in_j + b_j$

(c) $out_j = \begin{cases} 0 & if\ in_j < -b_j/a_j \\ 1 & if\ in_j > (1 - b_j)/a_j \\ a_j in_j + b_j & else \end{cases}$

(d) $out_j = \begin{cases} 0 & if\ in_j > b_j \\ 1 & if\ in_j \le b_j \end{cases}$

Figure 5: The behavioral transformations used to reduce the computational complexity of the neuron transfer function: (a) constant replace, (b) linear replace, (c) threshold replace, and (d) hardlimiter replace.

- **Constant replace**: This transformation implies that the neuron and all its connections can be removed from the network. The constant output level c_j of the neuron will be added to the bias level of the neurons that received input from this neuron.

- **Linear replace**: Neurons that use only the linear region or only a small part of the transition region can use a linear transfer function. The slope a_j and offset b_j can be determined by means of a simple least means square algorithm.

- **Threshold replace**: After learning many neurons operate partly in one of the saturated regions and partly in the transition region (only a piece of the lower or upper part of the sigmoid is used). This behaviour can be accomplished for by using a (part of) threshold transfer function.

- **Hardlimiter replace**: For neurons with large weighted inputs sums the sigmoid transfer function behaves like a hardlimiter, i.e. the neuron is only used in saturation. The threshold value b_j can be copied from the threshold value of the original sigmoid function.

Suppose a neural network is trained to interpolate a given function. Usually not all domains covered by the learned function have equal importance. Because of the monotonically increasing activation function, a neuron is only active for a part of the trained function. A fit with the selected transfer function is made on basis of a RMS criteria in which only the used part of the sigmoid is considered. The actual slope of a linear or threshold replace can therefor differ from the slope of the original transfer function is illustrated in Figure 6.

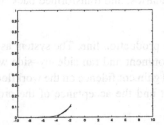 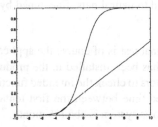

Figure 6: An example of a threshold replace. The actual slope of the threshold function can differ from the initial slope of the sigmoid transfer function as a"best" fit is made only with the part of the sigmoid function that is actually used by the neuron. Plot A shows the neuron response in the used area only. Plot B covers the complete output range of the sigmoid. A large difference between the threshold and the sigmoid occurs only for unused input sums.

A designer knows which parts of the function are important, and with this information he can roughly determine the required number of bits and an appropriate activation function. However, because of the interdependencies in the network it is awkward to go directly into detail. Therefore, we suggest here to first chart the available redundancy per computational node as well as the interdependencies, before moving to the actual optimization. To this purpose of charting we introduce the *redundancy index* [10].

5 Simulation and Application

The diagnosis system consists of two parts: (a) the process predictor, that acts largely on temperature gradients according to knowledge from the expert interviews, and (b) the decision maker, that interprets predicted changes within the context of production requirements according to knowledge from the logbooks. It is built using FuzzyTech and structured into many small rule blocks. Each rule block can be visually inspected and discussed with an expert.

The FuzzyTech model is used to produce C–code and integrated with a dedicated graphical user interface to provide for a direct interaction between modeler and model. In this way, a number of situations from the logbook can be re–run as a first coarse check on the validity of the model. It stands to reason that it is impossible to model, simulate and verify all situations from the logbook.

On the other hand, it is always necessary to check and alter the numeric details within rule blocks. In other words, there is a need to *autonomously* learn from the logbook. This can be achieved by re–modeling the rule blocks as a neural network. In general, the neural network will distribute its knowledge over the hidden nodes and an automatic transfer between the fuzzy and the neural domain and vice versa will be hard, unless dedicated fuzzy neural networks are used. However, where the learning is almost complete, it suffices to provide for adaptation (or *on–line learning* [11]). Under such circumstances, the location of the knowledge will not change anymore and the bi–directional transformation can be accomplished. In short, the coarse–tuned fuzzy rules are transformed to neural nets, fine–tuned by typical scenarios, and transformed back to fuzzy rules.

The final test is of course the application on the production line. The system as developed has been installed in the production environment and ran side–by–side with the operators to check the intended functionality and gain confidence on the workfloor. The elapsed time between the first tentative contact and the acceptance of the prototype diagnosis system is about 5 month.

The system consists of 6 fuzzy rule blocks with 19 linguistic variables for a total of 85 terms. Of the variables, 11 are used for input and 8 for output. The operator knowledge is stored in the 6 rule block by means of 297 knowledge rules. In comparison with similar projects using conventional A.I. in the chemical process industry, the efficiency in rule generation for a first prototype is a factor 4 higher and moreover it did work!

Acknowledgements

The here reported development has been commissioned by Ir. R. Oosterhaven of Maas-Glas b.v.. Furthermore, intensive use has been made of FuzzyTech in combination with InterAct. Lastly, the experience of W. Jansen with relevance to hierarchical design in neural nets has been applied with gratitude.

References

1. K. Matsushima and H. Sugiyama: Human operator's fuzzy model Man–Machine systems with a nonlinear controled object. Digest Industrial Applications in Fuzzy Control, 175–185 (1985)

2. Th. Froese, C. van Altrock and St. Franke: Optimization of water–treatment system with fuzzy logic control. Proceedings 3rd IEEE Int. Conf. on Fuzzy Systems III, pp. 1614–1619 (1994)

3. T. Furuhshi, K. Nakaoka and Y. Uchikawa: Suppression of excessive fuzziness using multiple fuzzy classifier systems. Proceedings 3rd IEEE Int. Conf. on Fuzzy Systems I, pp. 411–414 (1994)

4. S. Yasurobu and Y. Murai: Parking control based on predictive fuzzy control. Proceedings 3rd IEEE Int. Conf. on Fuzzy Systems II, pp. 1338–1341 (1994)

5. J.A.G. Nijhuis, H. vanAartsen, E.I. Barakova, W.J. Jansen and L. Spaanenburg: On the optimal mapping of fuzzy rules on standard micro–controlers. Microprocessing and Microprogramming 40, 697–700 (1994)

6. N. Funke and Ch. Grallen: Optimierung einer Hydraulikregelung für Personenaufzüge. Tagungsband 5. Aachener Fuzzy Symposium, pp. 33–41 (April 1994)

7. C.–T. Lin and C.S.G. Lee: Neural–Network–Based Fuzzy Logic Control and Decision System. IEEE Transactions on Computers 40, No. 12 (1991)

8. J. Blattner, S. Neußer, L. Spaanenburg and J.A.G. Nijhuis: Optimizing Fuzzy rules by neural learning. Digest Int. Conf. on Mathematical and Intelligent Models in System Simulation MISS'93, pp. 238–246 (1993)

9. A. Siggelkow, J. Nijhuis, S. Neusser and L. Spaanenburg: Influence of Hardware Characteristics on the Performance of a Neural System. Proceedings of the 1991 International Conference on Artificial Neural Networks, pp. 697–702 (1991)

10. H. Keegstra, W.J. Jansen, J.A.G. Nijhuis, L. Spaanenburg, J.H. Stevens and J.T. Udding: Exploiting Network Redundancy for Lowest–Cost Neural Network Realizations. Digest ICNN'96, pp. 951–955 (June 1996)

11. V. Ruiz de Angulo and C. Torres: On–line learning with minimal degradation in feedforward networks. IEEE Transactions on Neural Networks 6, No. 3. pp. 657–668 (May 1995)

Application of Fuzzy Control to Fed-Batch Yeast Fermentation

N Besli[1], Ensar Gul[2,*], M. Türker[3]

[1]Department of Electronic Engineering, University of Southern California
California, USA, email: besli@chaph.usc.edu

[2]Department of Computer Engineering, Faculty of Engineering, Marmara University
Goztepe, 81040 Istanbul, Turkey, email: ensar@marun.edu.tr

[3]Pakmaya P.O. Box 149, 41001 Izmit, Turkey

Address for correspondence

Keywords: Fuzzy control, fed-batch process, baker's yeast

Abstract: A new fuzzy controller has been designed and compared with the one developed previously for the control of fed-batch yeast fermentation. The respiratory quotient (RQ) was used as controller input and flow rate of glucose solution was controlled to maximize conversion of glucose to biomass. The new controller was found to be stable throughout the production period, keeping the glucose flow rate at the desired value, whereas the old controller was unstable towards the end of the fermentation and was unable to control and minimize the ethanol concentration in the fermentation broth.

1 Introduction

Baker's yeast is produced by fed-batch technique in which one ore more nutrients are supplied to the fermenter during cultivation and the products remain in the containment until the end of the run. This technique may be regarded as a modification of batch operation[1]. Essential nutrients are fed in a predetermined rate to the fermenter during the growth period. This type of operation does not take into account the need of organism at particular time. If there is variation in the quality and concentration of the carbon source in the feed, the fermentation behavior changes, but no corrective action is taken due to the predetermined nature of the feeding scheme. Excess feeding of carbon source results in the Crabtree effect in which even in the presence of oxygen ethanol production occurs thus reducing the cellular yield on carbon source whereas underfeeding results in underusage of the fermenter capacity.

Such deficiencies of predetermined feeding strategy could be overcome by appropriate control strategy such as fuzzy control. Therefore, it could be appropriate to use fuzzy control techniques to maintain optimal feeding of carbon source. Fuzzy control techniques applied successfully in many fermentation processes[2-7]. In this paper we will present an improved version of the fuzzy controller presented in our previous paper [8]. In section 2, the design of this controller is explained. The simulation results and comparisons with the previous design are presented in section 3

2 Design of New Fuzzy Controller for Fed batch Fermentation

One of the reason of reduced yield in the yeast fermentation is the formation of ethanol. When there is an oxygen starvation and/or high glucose concentration in the medium, ethanol production occurs. The production of ethanol can be detected either by directly measuring ethanol in gas or liquid phase or by indirectly analyzing exit gas composition. The formation or consumption of ethanol can be related to exit gas composition by mathematical model which is explained in detail in previous paper (8). The magnitude of respiratory quotient, RQ, which is defined as the ratio of the carbon dioxide evolution rate, Q_c, to the oxygen uptake rate, Q_o, ($RQ = Q_c / Q_o$) indicates four types of yeast metabolism[1]:

- oxidative assimilation of glucose
- oxidative assimilation of glucose and ethanol formation
- oxidative assimilation of both glucose and ethanol
- oxidative assimilation of ethanol

Experimental results show that to get maximum cellular yield, the RQ should be constant and equal to RQ_o. at which no ethanol occurs. Thus RQ provides a criteria that is a quantitative indicator of ethanol formation[4].

Therefore, the input variables of fuzzy logic controller (FLC) are chosen as the error in RQ and the change in this error, which is defined as:

$$e_k = RQ - RQ_0$$

$$\Delta e_k = e_k - e_{k-1}$$

The glucose feeding rate F(t) holds RQ at its preset value RQ_o. The values of F(t) are predetermined if there is no controller in the system. By applying a correction factor to the glucose feeding rate, the RQ can be kept constant, thus minimizing the ethanol production in the system. Therefore, the output of he fuzzy controller is chosen as ΔF to adjust the glucose feeding rate F(t). The block diagram of the input conditioning interface and fuzzy controller are shown in Figure 1.

The difficulty in this problem is to control the varying set point of glucose flow rate, F(t), which is given as

$$F(t) = (\frac{\mu_c}{Y_{x/s}} + m)\frac{V_0 X_0}{S_0 - C_{s,c}}\exp(\mu_c t)$$

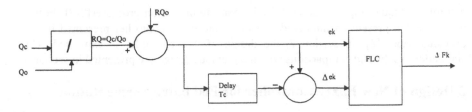

Figure 1: The input and output variables of FLC

where μ_c, $C_{s,c}$, X_0, V_0, m and $Y_{x/s}$ denote the maximum specific grow rate, the maximum concentrations of the glucose in the fermenter, initial concentrations of cell, initial culture volume, concentrations of feed glucose, the maintenance energy coefficient and yield of biomass on glucose respectively. The controller designed before [8], accumulates ΔF's at each sampling instants and adds these to F(t). However, even these accumulated ΔF's can not produce the sufficient control action near to the end of fermentation process. To solve this problem, a new controller is proposed in which the accumulated outputs of the controller is multiplied by an exponential term. This term depends on the specific grow rate μ_c. As a result better control action has been obtained because of the exponontial curve of feeding rate until the process runs into the oxygen starvation. The block diagram of this controller is shown in Figure 2.

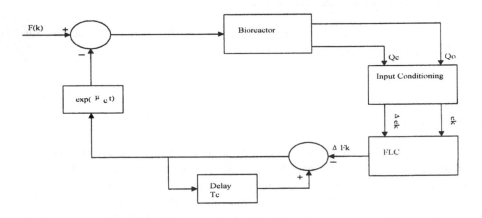

Figure 2: The block diagram of the new controller

The range of input and output variables are divided into seven linguistic sets. These are NB(Negative big), NM(negative medium), NS (Negative small), ZE(zero), PS(positive small), PM (positive medium) and PB(positive big). The membership functions have triangular shapes that are either symmetric or asymmetric depending on the range of variables in each set as shown in Figure 3.

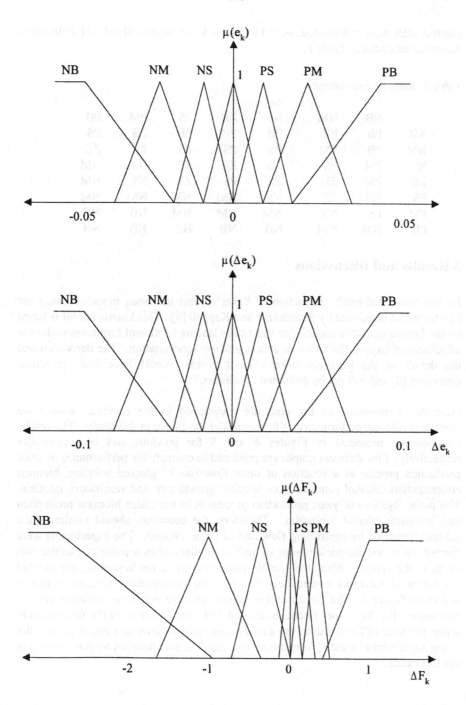

Figure 3. Fuzzy set values of fuzzy variables e_k, Δe_k and ΔF_k .

Control rules were constructed using the expert knowledge as IF-THEN statements. These are tabulated in Table 1.

Table 1. Rules for new controller

		Δe_k						
		NB	NM	NS	ZE	PS	PM	PB
	NB	PB	PB	PB	PB	PB	PB	PS
	NM	PB	PM	PS	PS	PS	ZE	ZE
e_k	NS	PM	PS	ZE	ZE	ZE	NS	NM
	ZE	PM	ZE	ZE	PS	ZE	NS	NM
	PS	PM	ZE	NS	NM	NM	NM	NM
	PM	PS	NS	NM	NM	NM	NB	NB
	PB	NM	NM	NB	NB	NB	NB	NB

3 Results and Discussions

In this study, fed-batch production of baker's yeast has been modeled using the kinetic model developed by Sonnleitner and Kappeli [9]. This kinetic model is based on the limited oxidation capacity of yeast cells leading to ethanol formation under the conditions of oxygen limitation or excess glucose concentration. The derivation and the details of the fed-batch baker's yeast process model have been presented elsewhere [8] and will not be discussed here again.

Here the performance of the controller developed in the previous work were compared with the performance of the controller developed in this study. The results obtained are presented in Figures 4 and 5 for previous and new controller respectively. Five different graphs are produced to compare the performance of yeast production process as a function of time: flow rate of glucose solution, biomass concentration, alcohol concentration, specific growth rate and respiratory quotient. The main objective of yeast production process is to maximize biomass production and minimize ethanol formation. Therefore, the controller should minimize the ethanol formation by controlling flow rate of sugar solution. The experiments were carried out to test the performance of both controllers when a pulse of glucose was given to the system. When the glucose pulse is given at the beginning, the alcohol concentration increases immediately and with the concomitant increase in RQ as shown in Figure 4. The old controller takes corrective action to minimize ethanol formation. The RQ values fluctuates around 1.08 until 10th hour of the fermentation. After 10th hour of fermentation the alcohol concentration increases and the controller is unable to control alcohol formation. This result is also justified by the increase in the RQ values.

212

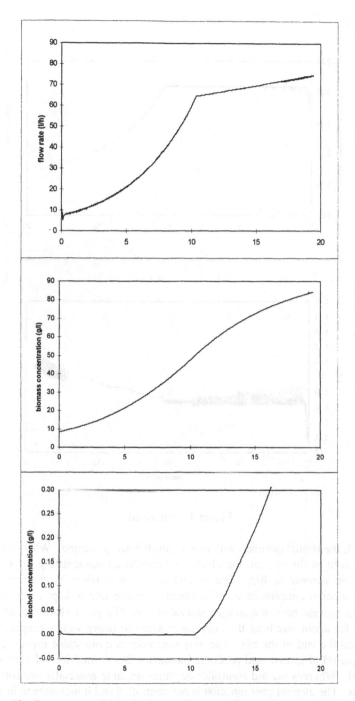

Figure 4: The Process variables obtained with controller 1 when a pulse of glucose is given at the beginning of fermentation

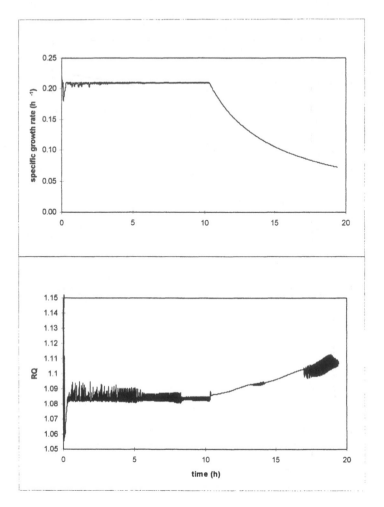

Figure 4: continued

In figure 5, the results obtained with new controller are presented. When the glucose pulse is given to the system, the alcohol concentration immediately increases with concomitant increase in RQ. The corrective action is taken by the controller to minimize alcohol concentration. The specific growth rate is kept at its maximum value until process runs into oxygen starvation. At this point alcohol concentration increases for about one hour then decreases again to lower values keeping at these values until the end of the run. The new controller is quite stable especially during the first period of growth and produces biomass with minimal loss of carbon source as alcohol. Whereas the old controller becomes unstable especially towards the end of process. The alcohol concentration is not controlled and it increases to high values which is unacceptable from economic point of view.

214

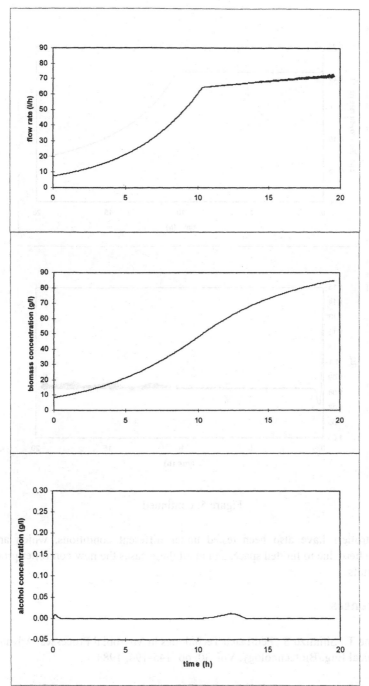

Figure 5:The process variables obtained with controller 2 when a pulse of glucose is
given at the beginning of fermentation

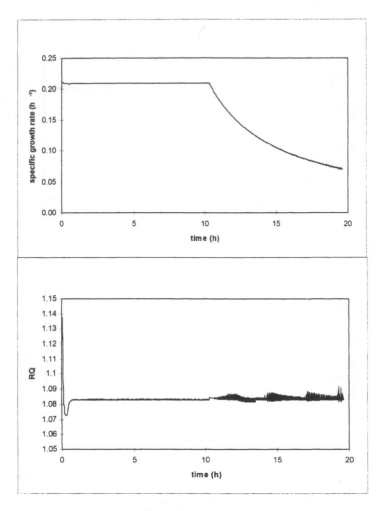

Figure 5: continued

The controllers have also been tested under different conditions, which are not presented here due to limited space, and in all these cases the new controller produced better results.

4 References

1. Yamane T., Shimizu S., Fed-batch Techniques in Microbial Processes, Advances in Biochemical Eng./Biotechnology, Vol. 30, pp. 145-194, 1984.

2. M. Kishimoto, Y. Kitta, S. Takeuchi, M. Nakajima, T. Yoshida, "Computer Control of Glutamic Acid Production Based on Fuzzy Clusterization of Culture Phases", J. Ferment. Bioeng., Vol. 72,pp.110, 1991.

3. T. Nakamura, T. Kuratani, Y. Morita, "Fuzzy Control Application to Glutamic Acid Fermentation", in Proc. 1st IFAC Symposium on Modelling and Control of Biotechnological Processes, Noordwijkerhout, The Netherlands, December 1985.

4. K. Oishi, M. Tominaga, A. Kawato, Y. Abe, S. Imayasu, A. Nanba, "Application of Fuzzy Control to Sake Brewing Process", J. Ferment. Bioeng., Vol. 72, pp. 115, 1991.

5. C.G. Alfafara, K. Miura, H. Shimizu, S. Shioya, K. Suga, and K. Suzuki, "Fuzzy Control of Ethanol Concentration and Its Application to Maximum Glutathione Production in Yeast Fed-Batch Culture", Biotechnol. Bioeng., Vol. 41, pp. 493-501, 1993.

6. Y.S. Park, Z.P. Shi, S. Shiba, C. Chantal, S. Iijima, T. Kobayashi, "Application Of Fuzzy Reasoning To Control Of Glucose And Ethanol Concentrations In Baker's Yeast Culture", Appl. Microbial. Biotechnol., Vol. 38, pp. 649-655, 1993.

7. B.E. Postlethwaite, "A Fuzzy State Estimator For Fed-Batch Fermentation", Chem. Eng. Res. Des.,Vol. 67,pp. 267-272, 1989.

8. N. Besli, M. Turker, E. Gul, Design and Simulation of a Fuzzy Controller for Fed-Batch Yeast Fermentation, Bioprocess Engineering, Vol 13, pp. 141-148, 1995.

9. B. Sonnleitner, O. Kappeli, "Growth of *Saccharomyces cerevisiae* is Controlled by Its Limited Respiration Capacity: Formulation and Verification of a Hypothesis", Biotechnol. Bioeng.,Vol. 28, pp. 927-937, 1986.

Adaptation, Learning, and Evolutionary Computing for Intelligent Robots

Toshio FUKUDA*, Koji SHIMOJIMA**

*Dept. of Micro System Engineering, Nagoya University
1 Furo-cho, Chikusa-ku, Nagoya 464-01, Japan
**Material Processing Dept., National Research Institute of Nagoya, AIST, MITI
Hirate-cho, Kita-ku, Nagoya 462, Japan
Phone: +81-52-911-2111, Fax: +81-52-916-2802

Abstract. There have been growing demands for the intelligent systems for many areas. In this lecture, the methodologies for the adaptation, learning and evolutionary computing will be shown to make robotic system more intelligent through Fuzzy, Neuro and Genetic Algorithm basisses. Robotic manipulators can generate the optimal trajectory automatically. Mobile robots can find the path and work cooperatively, by sensing the environments, scheduling the optional path and actuating properly. Some of the examples are also shown in this presentation.

1 Introduction

Living things can hear, see, smell, and so on, for perceiving their environments. The information acquired by sensing from their external environments is very important and useful for surviving under their dynamic environments. By using the information effectively, living things adapt to their dynamic environments. Here adaptation means a adjustment in structure or function to a dynamic environment. The driving force for adaptation is mainly genetic inheritance and learning. Inheritance is a process of transmission of genetic information from parents to offspring. As transmitting errors, mutation rarely changes genetic information and therefore phenotype of a living thing also changes. Leaning is a process or experience of gaining knowledge or skill. Learning includes learning by discovery, learning by show from observers, and reinforced learning through the interaction with the environments.

In order to archive these function in robots, this paper introduces a *hierarchical intelligent control* architecture. The hierarchical intelligent control consists of three levels: 1) *adaptation* level, 2) *skill* level, and 3) *learning* level. This scheme has two characteristics with respect to learning process: top-down approach and bottom-up approach. To link three levels and have such characteristics for knowledge acquisition, the scheme uses artificial intelligence (AI), fuzzy logic, neural networks

(NN) and genetic algorithm (GA) [1-3]. Each technique has advantages and disadvantages as shown in table 1. In order to overcome the disadvantages and enhance the advantages, this paper introduces integration and synthesis techniques of these intelligent methods as shown in Fig.1. This is one of key techniques for intelligent systems and robots. The next section explains the hierarchical intelligent control architecture. The third section explains a new approaches on the learning level by the Virus Evolutionary Genetic Algorithm and, shows the effectiveness on the learning level through trajectory planning of a robot manipulators [10]. The fourth section, fuzzy and neural network based navigation system for autonomous mobile robotic system is described.

Table 1 Comparison of Neural Network, Fuzzy Logic, AI, and Genetic Algorithm

	Math Model	Learning Data	Operator Knowledge	Real Time	Knowledge Representation	Nonlinearity	Optimization
Control Theory	O	X	△	O	X	X	X
Neural Network	X	O	X	O	X	O	◐
Fuzzy	◐	X	O	O	△	O	X
AI	△	X	O	X	O	△	X
GA	X	O	X	△	X	O	O

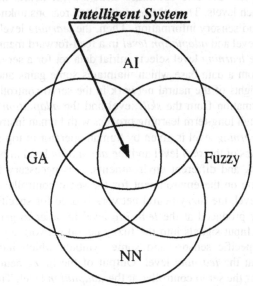

Intelligent System

AI

GA

Fuzzy

NN

Fig.1 Synthesis of Fuzzy, Neural Network, AI and Genetic Algorithms

2 Hierarchical Intelligent Control

The hierarchical intelligent control architecture comprises three levels: a) *learning* level, b) *skill* level, and c) *adaptation* level as shown in Fig. 2 [4], and there are three feed-back loops. The *learning* level is based on the expert system for a reasoning mechanism and has a hierarchical structure: recognition and planning to develop control strategies. The recognition level uses neural networks or fuzzy neural network as nodes of a decision tree. In the case of the neural network, inputs are numeric quantity sensed by some sensors, while outputs are symbolic quality which indicates process states. The structured neural network for incremental learning is effective to memorize new patterns [5]. In the case of the fuzzy neural network, inputs and outputs are numeric quantities and the fuzzy neural network clusters input signals by using membership functions. That is, the fuzzy neural network transforms numerical quantity into symbolic quality by membership functions. Both the neural network and the fuzzy neural network are trained with the training data sets of a-priori knowledge obtained from human experts. As a result, these networks can transform various sensed data from numerical quantities to symbolic qualities, and perform *sensor fusion* and production of *meta-knowledge* at the *learning* level. The important information is sensed actively on using the knowledge base. The sensors of vision, weight, force, touch, acoustic, and others can be used as nodes of decision tree for recognition of the environment.

Then, the planning level reasons symbolically for strategic plans or schedules of robotic motion, such as task, path, trajectory, force, and other planning in conjunction with the knowledge base. The system can include another *common sense* for robotic motion. The GA optimizes control strategies for robotic motion heuristically [6,7]. The GA also optimizes structures of neural network and fuzzy logic connecting each levels. Thus, the *learning* level reasons unknown facts from a-priori knowledge and sensory information. Then, the *learning* level produces control strategies for *skill* level and *adaptation level* in a feed-forward manner. Following the control strategy, the *learning* level selects initial data set for a servo controller at the *adaptation* level from a data base which maintains some gains and initial values of interconnection weights of the neural network in the servo controller. Moreover, the recent sensed information from the *skill* level and the *adaptation* level updates the *learning* level through long-term learning process with human instruction. Therefore, knowledge at the *learning* level is given by human operator in top-down manner and acquired by heuristics of the *skill* level and the *adaptation* level in bottom-up manner.

In the same task and different environments, it is necessary to change control references depending on the environment for the servo controller at the *adaptation* level. At the *skill* level, the fuzzy neural network is used for specific tasks following the control strategy produced at the *learning* level in order to generate appropriate control references. Input signals into the fuzzy neural network are numerical values sensed by some specific sensors and some symbols which indicate the control strategy produced at the *learning* level. Output of the fuzzy neural network is the control reference for the servo controller at the *adaptation* level. This output is based on the skill extracted from human experts through learning training sets obtained

220

from them. At the same moment, the fuzzy neural network clusters the input signals in the shape of membership functions. These membership functions are used as the symbolic information for the *learning* level.

In the *adaptation* level, a neural network in the servo controller adjusts control law to current status of dynamic process [8]. Particularly, compensation for non-linearity of the system and uncertainties included in the environment must be dealt with by the neural network. Thus, the neural network in the adaptation process works more rapidly than that in the learning process. It is shown that the neural network-based controller, the Neural Servo Controller, is effective to the nonlinear dynamic control with uncertainties such as force control of a robotic manipulator. Eventually, the neural networks and the fuzzy neural networks connect neuromorphic control with symbolic control for hierarchical intelligent control while combining human skills.

The hierarchical intelligent control is applied not only to a single robot, but also to multi-agent robot system. If there is no interaction between robots, each robot has to work optimally for its purpose, so that the total task should be achieved optimally. That is, each robot should work selfishly. Or else conflicts among the robots might occur when using a public resource. The competition may cause collisions and deadlock states among the robots in a local area. In order to avoid competition, it is necessary for the robots to communicate and to coordinate among themselves. The coordination among the robots is as important as selfishness. The GAs are applied hierarchically to balance selfishness with coordination for efficient motion planning [6]. When multiple robots works independently as decentralized system, the learning capability of the robots is indispensable for evolution of the system [9].

As results, integration and synthesis of AI, Fuzzy Logic, Neural Network and GA are important for intelligent system, depending on their characteristics. Hierarchical intelligent control using these techniques is effective to control intelligent systems in robotics.

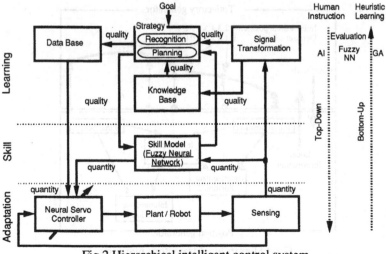

Fig.2 Hierarchical intelligent control system

3 Hierarchical Trajectory Planning with Virus Evolutionary Genetic Algorithm

Trajectory planning is one of the most important and difficult tasks required to robot manipulators. To achieve a given task, the robot manipulator, in, general, performs the follow steps: 1) find obstacles (perception), 2) generate a collision-free trajectory (decision making), and 3) trace the trajectory actually (action). In this section, we focus on the decision making and discuss how to generate an optimal trajectory of a manipulator.

3.1 Hierarchical Trajectory Planning

We have proposed a hierarchical trajectory planning method [10], which is composed of a position generator as local search and a trajectory generator as global search (Fig.3). The position generator generates some intermediate positions of the manipulator between the given initial and final positions. A position   is expressed by a set of joint angles of the manipulator. All intermediate positions are generated based on their before and after intermediate positions simultaneously. An intermediate position satisfied the aspiration level is sent to the trajectory generator as local information. An appropriate intermediate position is generated based on the distance between the manipulator and obstacles. To measure the distance between the manipulator and obstacles, we apply the concept of pseudo-potential [11].

The trajectory generator generates a collision-free trajectory combining some intermediate positions generated in the position generator. Here the intermediate positions of the best candidate solution are constraint for generating intermediate positions, that is, the position generator generates other intermediate positions based on the intermediate positions of the best candidate solution. Therefore, the hierarchical trajectory planning results in a co-optimization problem of the trajectory generation and the intermediate position generation.

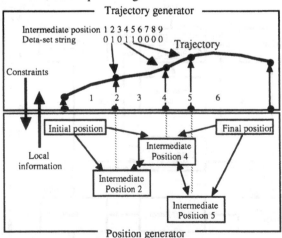

Fig.3 Hierarchical trajectory planning

3.2 Virus-Evolutionary Genetic Algorithm for Hierarchical Trajectory Planning

A virus-evolutionary genetic algorithm (VEGA) simulates the evolution with both horizontal propagation and vertical inheritance of genetic information (Fig.4). The VEGA is composed of two populations, a host population and a virus population. Here the host and virus population are defined as a set of candidate solutions and a substring set of the host population, respectively. Genetic operators are performed between the host population. In the VEGA, virus infection operators are introduced into GA. The VEGA has two virus infection operators as follows:

• Reverse transcription operator: A virus overwrites its substring on the string of a host individual to evolve host population.

• Transduction operator: A virus takes out a substring from a host individual to evolve virus population.

In this way, the VEGA performs genetic operators and virus infection operators. The procedure of the VEGA is as follows,

```
Initialization
    repeat
        Selection
        Crossover
        Mutation
        Virus _infection
        Replacement
    until Termination _condition = True
end.
```

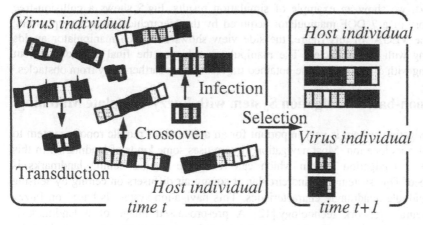

Fig.4 Virus-evolutionary genetic algorithm

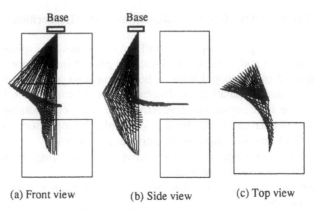

(a) Front view   (b) Side view   (c) Top view

Fig.5 Collision-free trajectory

The objective of trajectory planning is to generate a trajectory realizing minimum distance from the initial point to the final point and farther from the obstacles. To achieve the objective, we use the fitness function as follows,

$$fitness = w_1 f_p + w_2 f_d + w_3 f_{fr}$$
$$+ w_4 \max_{pot}^2 + w_5 \operatorname{sum}_{pot} \tag{1}$$

where $w_1, \cdots, w_5$ are weight coefficients. The first and second terms denote the sum of squares of the distance and joint angle, respectively. The third term makes each joint be within an available range. The fourth term denotes the maximum value in the pseudo-potential values on the sampling points. The last term denotes the sum of pseudo-potential values on all sampling points.

Next, we show an example of simulation results. Fig.5 shows a collision-free trajectory of a 7 DOF manipulator acquired by the hierarchical trajectory planning with the VEGA. In this figure, the side view shows that each manipulator avoids colliding with the obstacles. The manipulator achieves the final position without colliding with obstacles and the obtained trajectories are farther away from obstacles.

4. Vision-based Navigation System with Fuzzy Template Matching

Navigation system is very important for an autonomous mobile robotic system to recognize its location. Most navigation system uses some kinds of landmarks. In this section, a navigation system which can recognize regular ceiling landmarks is described. This system can find circular or square air diffusers on ceiling by sensing their multiple-quadrangle characteristics. This navigation system is based on Fuzzy and Neural Network technology.[12] A pre-processed image of a landmark is compared with a fuzzy template made from the fuzzy membership function. In order to find specific landmarks, the system determines the degrees of similarity by comparing the sited object with a variety of templates stored in its memory. Then a Neural Network integrates calculation results to determine its search direction. Figure 6 shows the experimental autonomous mobile robotic system[12-14]. Figure.7(a), (b), (c), and (d) are shown original images of landmarks and these pre-processed images.

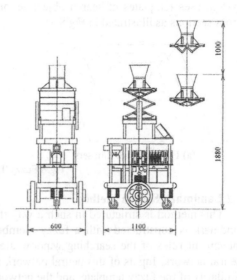

Fig.6 Experimental Mobile Robotic System with Fuzzy Template Navigation System

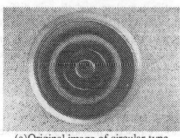

(a)Original image of circular type

(b) Image Processing Result of circular type

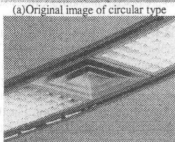

(c)Original image of square type

(d)Image Processing Result of square type

Fig.7 Pretreatment results

4.1 Fuzzy Templates

Air diffuser as the recognition object in this paper often accompany shadows because they are installed on ceilings. For this reason, an algorithm is required which can accurately search out an air diffuser even through the lines composing the air diffuser are thick to a certain extent. Under these circumstances, this navigation system uses templates of search objects expressed by fuzzy sets having triangular cross sections as illustrated in Fig.8.

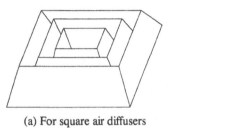 

(a) For square air diffusers   (b) For circular air diffusers

Fig.8 Fuzzy Templates

4.2 Landmark Search Method

This method is structured in such a way that the template of an air diffuser as a landmark is constructed with a fuzzy membership function as Fig.8 and then the movement rules of the searching window are created automatically by means of a neural network. Inputs of this neural network are the center of gravity and degree of similarity of the fuzzy template, and the network is taught beforehand so as to be able to determine in which direction the air diffuser would be located. With this structure, it is possible to move the searching window in a direction that increases the grade (degree of similarity) of the membership function of a component for the air diffuser, and determine the place where the place where the template and the air diffuser precisely coincide, namely the place where the air diffuser is found, thereby discovering and recognizing the object. This method can work both for circular and square types. In the following, methods of searching for circular air diffusers as well as square air diffusers will be described.

Search for Circular Air Diffusers

Suppose that the number of circles comprising an air diffuser is a. Then a different concentric circular fuzzy sets $A_k(k=1,...,a)$ are placed; in addition, fuzzy sets $B_j(j=1,...,a-1)$ between templates and fuzzy sets between the center of the circles C and the outer sides O of the templates are provided so as to detect shifting of the templates. Figure 9 shows the structures of the templates and the neural network. The fuzzy sets concerned are represented by triangular columns with height 1 and with a triangular cross section. However, unless these sets were modified, they would output the same degree of similarity numerically when the shifts in the left direction and the right direction are the same, and the direction of the shift cannot be learnt. For this reason, it was decided to detect the shifting direction of the templates by finding the center of gravity (g_{xi}, g_{yi}), of each fuzzy set. In this case, let the membership grade of a certain fuzzy set i (i is any one of A_k, B_j, C, or O) be expressed as μ_i

$$V_i = \sum_x \sum_y \mu_i(x, y) \tag{2}$$

where V_i are the volumes of the templates.

$$g_{xi} = \sum_x \sum_y \frac{x \cdot \min(G(x,y), \mu_i(x,y))}{V_i} \tag{3}$$

$$g_{yi} = \sum_x \sum_y \frac{y \cdot \min(G(x,y), \mu_i(x,y))}{V_i} \tag{4}$$

where min( ) is the common set of two fuzzy sets and $G(x,y)$ is the value of pixel (x,y) (which is 0 or 1 here).

By using these equations, it now becomes possible to determine the centers of gravity inside the templates, and the total sum of degrees of similarity, U, can be obtained as follows:

$$u_{AK} = \sum_A \sum_K \frac{\min(G(x,y), \mu_{AK}(x,y))}{V_{AK}} \tag{5}$$

where

u_{AK} : degree of similarity of fuzzy set A_K

μ_{AK} : membership grade of fuzzy set A_K

V_{AK} : Volume of fuzzy set A_K

Hence

$$U = \sum_{K=1}^{a} u_{AK} \tag{6}$$

Here, search is carried out by putting the center of gravity, (g_{xi}, g_{yi}), of each fuzzy set (altogether 4a+2 values) and the total sum of the degrees of similarity of the templates (4a+3 values in total) into the neural network.

The neural network, which uses multi-layered perceptrons for learning by the back provocation method, is constructed with 4a+3 units for its input layer, ten units for each of two intermediate layers, and 2 units for output layer. The each value is put into the input layer, while displacements Δx and Δy in the x direction and the y direction are output from the output layer, and when the template and the air diffuser are in perfect agreement, then both Δx and Δy produce 0, thereby completing the search.

Incidentally, when landmarks composed of diagrams other than the circular or square type, search can be carried out by the same algorithm as the one used for the circular type.

Search for Square Air Diffusers

In the case of the square type, the positions of the templates themselves contain information on their left and right shifts. For this reason, center of gravity is not input in the case of a square air diffuser, but instead a fuzzy set is established for one side of each of the squares involved and changed to the degree of similarity of each side, thereby achieving more accurate search.

Figure 10 shows the structures of the templates and the neural network. Suppose that a denotes the number of squares composing a square air diffuser. Then 4a+1 fuzzy sets are prepared for each of the vertical and horizontal directions, where 2a sets out of 4a+1 fuzzy sets are for templates A_k(k=1,...,2a); in addition, fuzzy sets B_j(j=1,...,2a-2) and fuzzy sets for the true center C, for the two ends L and R of templates, and for the top and bottom, U and D, of templates are provided so as to detect the shifting of the templates. Each fuzzy set has a triangular cross section with a height of 1, as in the case of the circular type, and the degree of similarity, u_{xi}, for a certain fuzzy set x_i in the horizontal direction, for example, is determined as follows (i is any one of A_1 to A_{2a}, B_1 to B_{2a-2}, C, L, and R):

$$u_{xi} = \sum_x \sum_y \frac{x \cdot \min(G(x,y), \mu_{xi}(x,y))}{V_i} \tag{7}$$

μ_{xi} : degree of similarity of fuzzy set xi.

This calculation is carried out for each of the 4a+1 fuzzy sets in each of the horizontal and vertical directions, and the resulting 8a+2 degrees of similarity are put into the neural network, thereby prompting the neural network to carry out search, as in the case of the circular type, to eventually discover and recognize a square air diffuser.

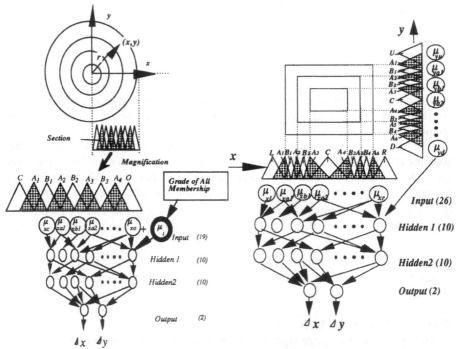

Fig.9 Search for circular air diffuser   Fig.10 Search for square air diffuser

5 Summary

This paper describes a hierarchical intelligent control architecture and proposes a new approaches on the *learning level* by the Virus Evolutionary Genetic Algorithm. The simulation result through the trajectory planning of a robot manipulator, shows the effectiveness on the *learning level*. The fuzzy-neuro based navigation system for the autonomous mobile robotic system is also described. This navigation system can adapt for its environment by *learning level*.

References

[1] T. Fukuda and T. Shibata, Theory and Applications for Neural Networks for Industrial Control Systems, IEEE Trans. on Industrial Electronics, Vol. 39, No. 6, pp. 472-489 (1992)

[2] L. A. Zadeh, Fuzzy Sets, Information and Control, Vol. 8, pp. 228, (1965)

[3] D. E. Goldberg, Genetic Algorithms in Search, Optimization, and Machine Learning, Addison Welsey (1989)

[4] T. Shibata and T. Fukuda, Hierarchical Intelligent Control of Robotic Motion, Trans. on NN (1992)

[5] T. Fukuda, S. Shiotani, F. Arai, A New Neuron Model for Additional Learning, Proc. of IJCNN92-Baltimore, Vol. 1, pp. 938-943, (1992)

[6] T. Shibata, T. Fukuda, K. Kosuge, F. Arai, Selfish and Coordinative Planning for Multiple Mobile Robots by Genetic Algorithm, Proc. of the 31st IEEE Conf. on Decision and Control, Tucson, Vol. 3, pp. 2686-2691 (1992)

[7] J. Koza, Genetic Programming on the Programming of Computers by means of Natural Selection, MIT Press (1992)

[8] T. Fukuda, T. Shibata, M. Tokita, T. Mitsuoka, Neuromorphic Control - Adaptation and Learning, IEEE Trans. on Industrial Electronics, Vol. 39, No. 6, pp. 497-503 (1992)

[9] T. Shibata and T. Fukuda, Coordinative Behavior by Genetic Algorithm and Fuzzy in Evolutionary Multi-Agent System, Proc. of IEEE Int'l Conf. on Robotics and Automation, Vol. 1, pp. 760-765 (1993)

[10] N. Kubota, T. Fukuda, K. Shimojima, Trajectory Planning of Reconfigurable Redundant Manipulator Using Virus-Evolutionary Genetic Algorithm, Proc. of The 22nd International Conference on Industrial Electronics, Control, and Instrumentation, pp. 836-841 (1996).

[11] O. Khatib, Real-Time Obstacle Avoidance for Manipulators and Mobile Robots, Robotics and Research, Vol.5, No.1, pp. 90-98 (1986).

[12] Y.Abe, K.Tanaka Y. Tanaka, T. Fukuda, F. arai, and S. Ito, Development of Air Conditioning Equipment Inspection Robot with Vision Based Navigation System, 11th International Symposium on Automation and Robotics in Construction, 665-674 (1994).

[13] T. Fukuda, S. Ito, F. Arai, Y. Abe, K. Tanaka, and Y. Tanaka, Navigation System Based on Ceiling Landmark Recognition for Autonomous Mobile Robot, Proceedings of the IECON '93 , Vol.3, 1466-1471(1993).

[14] Y. Abe, T. Fukuda, K. Tanaka, Y. Tanaka, F. Arai, K. Shimojima, S. Ito, Navigation System for Air Conditioning Equipment Inspection Robot, J. of Robotics and Mechatronics, Vol.7, No.5, 354-366 (1995)

About Linguistic Negation of Nuanced Property in Declarative Modeling in Image Synthesis

Daniel Pacholczyk

LERIA, University of Angers
UFR Sciences, 2 Bd Lavoisier, 49045 Angers CEDEX 01, France

Abstract. In this paper, we focus our attention on the explicit Representation of Linguistic Negation of Nuanced Properties. Our Approach is based upon a Similarity Relation between Nuanced Properties through their corresponding Fuzzy Sets. We propose a Choice Strategy improving the abilities in Declarative Modeling in Image Synthesis. The Designer can refer to Linguistic Negations both in Scene Descriptions and in Validation Rules of potential Solutions, since the Interactive Choice Strategy allows him to explain their implicit meanings.

1 Introduction

In Image Synthesis, the work of the Scene Designer becomes simplified with Declarative Modeling [7]. Indeed, he only has to provide the Scene Description whose translation generally leads to a set of Properties. Then, the Designer explores the Universe of potential Scenes and makes his choice of the most appropriate Scene. So, the fundamental notion in Declarative Modeling is the Property concept. We can find in [2] a Model dealing with *Affirmative assertions* using Fuzzy Information evaluated in a Numeric or Symbolic way. So, its construction is based upon Zadeh's Fuzzy Sets Theory [16]. In other words, the Designer can introduce in a Scene Description an assertion like « the menhirs are *really very* tall ». In order to make the basic terminology clear, this point is briefly presented in Section 2.

Our purpose here is to improve the abilities of the previous Model in such a way that the Designer can also refer to *Linguistic Negations* in Scene Descriptions. A lot of work on Linguistic Negation has already been proposed. Some authors [8, 5, 3, 4, 6, 9, 10, 14] have developed *pragmatic* or *formal* methods dealing with Negation. We can note that in the recent paper by Torra [14], some similarities could be found with our initial Approach.

Our main objective here is to express *Linguistic Negations with Fuzzy Properties*. Indeed, the Scene Designer can also assert that « the number of menhirs is *not* small ». In such a case, he may refer to the Fuzzy Complement Property « not small », but very often he intends to mean another Fuzzy Property like « *very* great » or « *really extremely* small ». So, « x is not A » seems to have, in a Fuzzy Context, an equivalent affirmative translation of the form « x is P », where the Property P is defined in the same domain as A, but is in « weak agreement » with A, or has a « weak similarity » to A.

Some authors have studied *Similarity Relations* [17, 15, 1, 12, 11] conceived as a Reflexive and Symmetric, but not Transitive Relation. Our main idea has been to approach symbolically the Linguistic Negation by using the weakly transitive Similarity Relation proposed in [11].

The basic notions leading to the symbolic concept of θ_i-*similarity of Fuzzy Sets* are defined in Sections 3.2 and 3.3. Thus, we generate (§ 3.4) a set of ρ-*plausible Linguistic Negations* in the domain. In Section 4, we propose to the Designer a *Choice Strategy* of « x is P » as its interpretation of « x is not A ». Section 5 is devoted to some aspects of the Validation of a potential Scene. In Section 6, we finally point out the fact that this Linguistic Negation satisfies some Common sense properties of the Negation.

2 Brief Description of Nuanced Properties in terms of Fuzzy Modifiers and Operators [2]

A *Scene* is characterized by some *Concepts* C_i. A set of Properties P_{ik} is associated with each C_i, whose *Description Domain* is denoted as D_i. A finite set of *Fuzzy Modifiers* m_α allows us to define Fuzzy Properties, denoted as « $m_\alpha P_{ik}$ », whose membership L-R function simply results from P_{ik} by using a translation and a contraction. We can select the following Modifiers set (*Cf.* Fig. 1):

M_7={extremely little, very little, rather little, moderately ($\varnothing$), rather, very, extremely}.

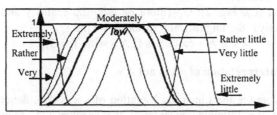

Fig. 1. A set of Fuzzy Modifiers

In order to modify the Precision or Imprecision of each « $m_\alpha P_{ik}$ », we use a finite set of *Fuzzy Operators* f_α defining new Fuzzy Properties « $f_\alpha m_\beta P_{ik}$ ». Their membership L-R functions simply result from the ones of « $m_\beta P_{ik}$ ». The following set F_6 gives us a possible choice of Fuzzy Operators (*Cf.* Fig. 2):

F_6={vaguely, neighboring, more or less, moderately ($\varnothing$), really, exactly }.

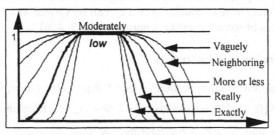

Fig. 2. A set of Fuzzy Operators

Definition 1. Given a Property P, a property such as «$f_\alpha\, m_\beta\, P$» which requires for its expression the list of linguistic terms (f_α, m_β) can be called a *Nuanced Property*.

As an example, asserting that « the menhirs are *really very* tall », the Scene Designer refers to the Nuanced Property « *really very* tall ».

Remark: It is obvious that more than one Concept can be connected with a particular Domain. As an example, « height » concerns either « humans » or « buildings ». So, strictly speaking, it must be necessary to define for any concept its Dependence on other Concepts in order to express its particular Graduation Scale. This point not being studied in this paper, we simply denote as D_i the Domain of the Concept C_i.

3 Towards an Interpretation of Linguistic Negation through Similarity of Fuzzy Sets

3.1 Linguistic Interpretation of « x is A »

As already pointed out in [13], linguistically speaking, the statement « x is A » implicitly covers some statements. So, applying this result to Nuanced Properties, « x is A » may be interpreted as one of the following statements : « x is $\varnothing_f$ A », « x is really A » or « x is more or less A ». As an example, the statement « x is low » corresponds to one of the following statements : « x is $\varnothing_f$ low », « x is really low » and « x is more or less low ».

In F_6, we can put (*Cf.* Fig. 2): $G_1 = \{$more or less, $\varnothing_f$, really $\}$ and $G_2 = \{$vaguely, neighboring, exactly$\}$. So, linguistically speaking we put :

$$\text{« x is A »} \Leftrightarrow \{\text{« x is } f_\alpha \text{ A » with } f_\alpha \in G_1\}.$$

3.2 Linguistic Interpretation of « x is not A »

Within a given Domain, the Linguistic Negation of « x is A », denoted as « x is not A », can receive the following meaning : the speaker (1), *rejects* all previous interpretations of « x is A », and (2), *refers to* a Nuanced Property P having a « weak agreement » with A, and, in such a way that « x is P » is equivalent to the statement « x is not A ». By using our previous « Nuances », P is « f_β A » with $f_\beta \in G_2$, *or* P is « m_δ A » with $m_\delta \neq \varnothing$, *or* P is « $f_\chi B$ » with $B \neq A$ defined in the same domain and $f_\chi \in F_6$. The main difficulty is then due to the fact that P is not explicitly given.

In the following, we propose to the Scene Designer a Choice Method based upon the notion of Fuzzy Sets Nuanced Similarity, which can be viewed as a Generalization to Nuanced Properties of some Linguistic Approaches [8, 3, 4, 6].

3.3 Nuanced Similarity of Fuzzy Sets

Let us recall Lukasiewicz's Implication : $u \rightarrow v = 1$ if $u \leq v$ else $1 - u + v$, where the values u and v belong to [0, 1]. The *neighborhood relation* $\mho_\alpha$ is defined in [0, 1] as follows.

Definition 2. u and v are α-*neighboring*, denoted as $u \mho_\alpha v$, if and only if, Min $\{u \rightarrow v, v \rightarrow u\} \geq \alpha$.

Given two Fuzzy Sets defined in the same domain, we can introduce the following *neighborhood relation* $\approx_\alpha$.

Definition 3. The Fuzzy sets A and B are said to be α-*neighboring*, and we denote this as A$\approx_\alpha$B, if and only if $\forall x$, $\mu_A(x)\, U_\alpha\, \mu_B(x)$.

In order to define Linguistic *Nuanced Similarity of Fuzzy Sets*, we have introduced a totally ordered partition of $[0, 1]$: $\{I_1, I_2,..., I_7\}=[0]\cup]0, 0.25]\cup]0.25, 0.33]\cup$ $]0.33, 0.67[\cup[0.67, 0.75[\cup[0.75, 1[\cup[1]$.

We have defined a one to one correspondence between these intervals and the following totally ordered set:

$\{\theta_1,..., \theta_7\}=\{$not at all, very little, rather little, moderately (or $\varnothing$), rather, very, entirely$\}$.

Finally, we put the following definition of θ_i-*similar Fuzzy Sets*.

Definition 4. A and B are θ_i-*similar* if and only if (θ_i is such that $\alpha\in I_i$, knowing that α=Max $\{\delta\,|\, A\approx_\delta B\}$).

Remark: It is obvious that these boundary choices are arbitrary. But, note that they lead to results corresponding in a satisfactory way to those intuitively expected.

3.4 ρ-plausible Linguistic Negations

Let ρ be a real such that $0.33\geq\rho\geq0$ and P a property defined in the same domain as A.

Definition 5. If P satisfies the following conditions :
[C1] : P and A are θ_i-*similar* with θ_i < moderately (or $\varnothing$),   (Global Condition)
[C2] : $\forall x$, (($\mu_A(x) = \xi \geq 0.67 + \rho$) $\Rightarrow$ ($\mu_P(x)\leq\xi - 0.67$)),   (Local Condition)
[C3] : $\forall x$, (($\mu_P(x) = \xi \geq 0.67 + \rho$) $\Rightarrow$ ($\mu_A(x) \leq \xi - 0.67$)),   (Local Condition)
then « x is P » is said to be a ρ-*plausible Linguistic Negation* of « x is A ».

Property 1. The Fuzzy Operators and Modifiers have been defined in such a way that, for any property A :
- A and m_α A are at least $\varnothing$-similar iff $m_\alpha\in\{$rather little, moderately, rather$\}$.
- A and f_α A are at least $\varnothing$-similar iff $f_\alpha\in\{$more or less, moderately, really$\}$.

We can verify the Adequation of this definition with the intuitive properties. The Similarity of Fuzzy Sets (§ 3.3) has been conceived in such a way that a *weak global Similarity* leads to a degree less than moderately ([C1]). Conditions [C2] and [C3] translate a *weak local Neighborhood* of membership degrees. Indeed, [C2] gives us: $\mu_P(x)\rightarrow\mu_A(x)=1$. So, we obtain: $\mu_A(x)\rightarrow\mu_P(x)\leq\xi\rightarrow(\xi-0.67)=1-\xi+(\xi-0.67)=0.33$. In other words, $\mu_A(x)U_\alpha\mu_P(x)$ with $\alpha\leq0.33$. These conditions define a weak global Similarity and a weak Neighborhood for the more significant values. We can note that: $\mu_A(x)=1\Rightarrow\mu_P(x)\leq0.33$. So, we can state the following Property.

- **Property 2.** Given ρ such that $0.33 \geq \rho \geq 0$, each ρ-plausible Linguistic Negation P is less than moderately-similar to A and less than 0.33-neighboring with A.

- **Property 3.** For any Property A, there exists a value of ρ such that the set of ρ-plausible Linguistic Negations is not empty.

We can distinguish two cases :

a : A is the only property defined in the domain. As noted before, Fuzzy Operators and Modifiers have been defined in such a way that some nuanced properties based upon A satisfy previous conditions. More precisely, it is the case when ρ ≥ 0.3.

b : At least two different Fuzzy Properties has been defined in the domain. Then, the translations and contractions have been defined in such a way that some Nuanced Properties based upon them fulfil all the conditions for any ρ ≥ 0.1.

Example: The concept being « the number of intersection points », its associated properties will be « low », « average » and « high ». Then, a plausible interpretation of these properties is given in Figure 3.

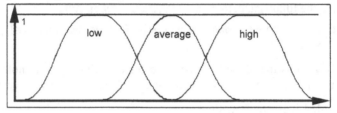

Fig. 3. A set of Basic Properties

The following Figures 4 and 5 give us illustrations of ρ-plausible negations with the Basic Property « low ».

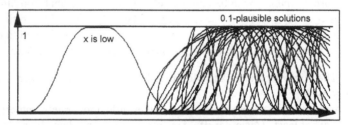

Fig. 4. 0.1-plausible Negations of « x is low »

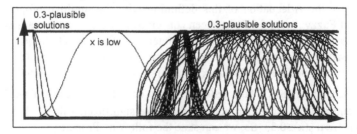

Fig. 5. 0.3-plausible Negations of « x is low »

Remark: We can note that, for any Fuzzy Property A, the function of ρ is to increase (or not) the number of Nuanced Negations and to accept (or not) some Negations based upon A.

4 The Choice Strategy

Resulting from the fact that our model accepts a great number of Operator and Modifier combinations, the number of ρ-plausible Negations can be high. In order to allow the Scene Designer to make his Choice from a limited number of solutions, we have constructed a Strategy essentially based upon the following *Simplicity Rule : a Speaker generally uses a very small number of expressions to define a Nuanced Property.* So, the ρ-plausible Solutions far too Nuanced will be rejected.

In order to develop our Choice Strategy, we firstly describe the Linguistic Analysis of Negation proposed by the previous Linguists, an *analysis being restricted here* to a statement having the form « x is not A ».

4.1 Negation Characterization in Linguistics

Linguists [8, 3, 4] have pointed out that Negation must be defined within a *Pragmatic Context.* More precisely, saying that « x is not A », the *Speaker characterizes as Negation i)* the *judgement of rejection* and *2)* the *semiologic means* exclusively used to notify this rejection. So, « x is A » is not in Adequation with Discourse Universe, but « x is not A » does not necessarily imply its Adequation with this Universe. It can only be a step in the outcome of this precise Adequation by the Speaker. Another point has also been pointed out by Linguists. Generally, Common sense Reasoning prefers affirmative statements to negative ones. In other words, Reasoning Process is based upon Rules consisting of Affirmative statements. So, a statement like « x is not A » has to be translated as an equivalent affirmative statement « x is P ».

4.2 Different Interpretations of « x is not A »

We have chosen some characteristic cases to present the argumentation of linguists leading to the interpretations of « x is not A ». But, note that we have extended these interpretations by using Operators or Modifiers applied to Fuzzy Properties.

1 : *Linguistic Negation based upon A.* Saying that « John is not tall », the Speaker does not deny a certain height ; he simply denies that his height can be high. So, his precise Adequation to Reality can be « John is extremely little tall ». In this case, the Speaker refers to the same Property and expresses a weak agreement between « tall » and « extremely little tall ».

2 : *Marked and not marked Properties*
a - Let us suppose that three Properties « thin », « big » and « enormous » are associated with the basic concept « weight » (*Cf.* Fig. 6). Then, « x is not thin » can mean that the speaker 1) rejects « x is thin » and 2) refers to « x is enormous » or « x is *really* big », but not to « x is *vaguely* big ». It can be noted that « big » and « enormous » have a weak agreement with « thin ». On the other hand, asserting that « x is not big » generally means that « x is thin », that is to say, the affirmative Interpretation is precise and unique. So, linguists distinguish a *marked* property like « thin » from a *not marked* property such as « big ». This distinction is important *since the Negation of a not marked Property is explicitly defined.*

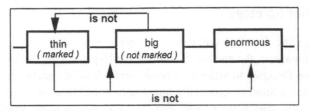

Fig. 6. Negations of Properties associated with the concept « weight »

b - The same analysis with two Properties « unkind » and « nice » associated with the basic concept « personality », leads to results collected in Figure 7.

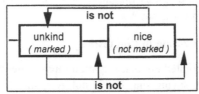

Fig. 7. Negations of Properties associated with the concept « personality »

- 3 : *A new property denoted as « not-A ».* When the Speaker asserts that « Mary is not ugly », in this particular case, this semblance of Negation is in fact an affirmative reference to the new Property « not-ugly ». So, in some cases, Negation of « A » introduces *a new Basic Property* denoted as « not-A ».

- 4 : *Reference to logical ¬ A.* The interpretation of the assertion « this head-scarf is not green » is « this head-scarf has another colour than green ». So, in this case, « x is not A » means that « x is P (colour) » with P ≠ A. The Property P is exactly ¬ A. We can note that between P and A, no similarity can exist.

- 5 : *A simple rejection of « x is A ».* Saying that « John is not guilty », the Speaker only rejects the fact « John is guilty ». So, we progress weakly with the Adequation of « x is not guilty » to Reality.

4.3 Construction of ρ-plausible solution Sets

Let us now go into all the details of the Choice Strategy. Firstly, we ask the Speaker for Possibility or not of Negation based upon A. So, we can determine a value of ρ satisfying this condition. If he does not make a Choice, we include Possibility of Negation based upon A by putting ρ=0.3. This being so, we can define the sets of ρ-plausible solutions by using successively the following Rules.

[R1] : *Simplicity Principle.* Among the previous ρ-plausible solutions, we define the set S of Nuanced Properties P based upon *at most two Nuances of a Basic Property.*

[R2] : *Increasing Similarity.* For each degree θ_i with θ_i<moderately, we define the subset S_i of S whose elements P are θ_i-*similar* to A.

[R3] : *Increasing Complexity.* We constitute a partition of each S_i in subsets $S_{i\,P}$ where P is a Basic Property defined in the same Domain as A. The last subset will be $S_{i\,A}$. Moreover, each $S_{i\,P}$ is reorganized in such a way that its elements appear ordered to an increasing *Complexity* extent, that is to say, the number of Nuances (different from ∅) required in their construction.

Example: The concept being « the number of intersection points », and the properties « low », « average » and « high » (*Cf.* Fig. 3), the Choice Strategy suggests the set S of 0.3-plausible Linguistic Negations of « x is low » collected in Figure 8.

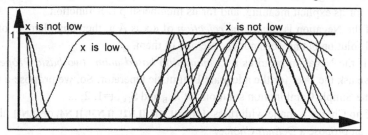

Fig. 8. The set S of 0.3-plausible Linguistic Negations of « x is low »

As noted before, for any Property A, the function of ρ is to increase the number of Nuanced Negations and to accept some Negations based upon A. As an example, a Choice is made among 20 interpretations of « x is not low » when $\rho = 0.3$, and among 10 interpretations when $\rho = 0.1$. So, the Scene Designer can choose, for any ρ, solutions such as « high », « extremely high », « exactly average », « really average », and in addition, « extremely low », « really extremely low » and « exactly extremely low » for $\rho = 0.3$. The function of the *mark* i, defined in the following section, will be then the selection of the most appropriate solutions in the previous set.

4.4 The Choice of the Linguistic Negation

The Choice Strategy, based upon the following Rules denoted [LNi], allows the Scene Designer to explain his intended interpretation of a *particular occurrence* of the assertion « x is not A ». In other words, the procedure must be applied to each previous occurrence appearing in a Scene Description. So, each occurrence of « x is not A » being explained with the aid of a particular Rule [LNi], we associate the *mark* i to this particular occurrence. Indeed, we have generalized the notion of the *marked or not marked* Property proposed in linguistics (*Cf.* § 4.2), in such a way that we can manage the Fuzzy Property in a Scene Description.

An interactive Process proposes to the User the following ordered Rules.

[LN0] : If the Negation is in fact the *logical Fuzzy Negation* denoted as ¬A, then :

- if ¬A exists as a Property defined in the same domain as A, we propose ¬A,

- if not, then for any P such that P⊂(¬A), we propose a Choice among the solutions of $S_{i\,P}$, i=1,2, ...

[LN1] : If the Negation *is based upon* A, we propose a Choice among the plausible solutions of $S_{i\,A}$, i=1,2, ...

[LN2] : If the Negation *is based upon only one P≠A*, we ask the User for its basic Negation P. Then, his Choice has to be made among the plausible solutions of the previous sets $S_{i\,P}$, i=1,2, ...

[LN3] : If the Negation can be *based upon B* and C *being different from A*, then the User has to retain his interpretation among the solutions of $S_{i\,B}$ and $S_{i\,C}$ where i=1,2, ...

|LN4| : If the Negation can be based upon *one of all the basic properties*, his interpretation is one of the plausible solutions of $S_{i\,P}$ where i=1,2, ... , for any P.

|LN5| : If the Negation *is in fact a New basic Property denoted as not-A*, we must ask the User for its explicit meaning, and for its membership L-R function.

|LN6| : If the Negation is *simply the rejection* of « x is A », then any element of S is a potential solution but he does not choose one of them.

|LN7| : If the Negation requires a *Combination based upon two basic Properties B and C*, we ask the User for the adequate Linguistic Operator. So, we propose a Choice among the Simple Combination solutions of $S_{i\,B}$ and $S_{i\,C}$, i=1, 2, ...

|LN8| : If no Choice is made with rules |LN0|, |LN1|, |LN2|, |LN3|, |LN4| or |LN7| then the System can propose a *Default Choice*.

Remark: The solutions are always proposed by increasing Complexity

4.5 Examples

We present here some results obtained by using the previous Choice Strategy.

Example 1 (Cf. Fig. 3) *:* The concept and its associated Properties being the ones presented in § 3.4, we search for the interpretation of Negation « x is not low ».

a : If its mark is 1, then 3 solutions are proposed, that is to say, « x is extremely low », « x is really extremely low » and « x is exactly extremely low ».

b : If its mark is 4, then the Choice is made among the following solutions :

- Complexity = 1 : « x is m_α high » with α=2..7 ; « x is f_β high » with β=1..3, 5, 6 ; « x is exactly average » and « x is extremely low ».

- Complexity = 2 : « x is really m_α average » with α=1, 2, 6, 7 ;« x is exactly m_α average » with α=1..3, 5.. 7 ; « x is really extremely low » and « x is exactly extremely low » ; « x is f_β m_α high » with α=1..3, 5..7 and β=1..3, 5, 6.

Example 2: Properties « small », «tall » and « gigantic » are associated with the concept « height » *(Cf.* Fig. 9). We study now the following Negations.

a : « x is not tall» has received the mark 2. So, the User constructs its negation with « small ». We first propose the solutions having a Complexity equal to 1 : « x is m_α small» with α=4, 5, 6, 7; Solutions having 2 as a complexity are obtained as before for Property « low».

b : « x is not small » has received the mark 3. The Linguistic Negation is then based upon Properties « tall » and « gigantic ». We can propose the successive solutions :

- Complexity = 1 : « x is m_α gigantic » with α=2..7 ; « x is f_β gigantic » with β=1..3, 5, 6 ; « x is exactly tall».

- Complexity = 2 : « x is really m_α tall» with α=1, 2, 6, 7 ; « x is exactly m_α tall» with α=1..3, 5..7 ; « x is f_β m_α gigantic » with α=1..3, 5..7 and β=1..3, 5, 6.

Remark: The implementation of this Choice Strategy is actually in progress within Project « Filiforms » in Image Synthesis. *We can point out that Designers generally refer to Negations having 1 as complexity.*

Remark: It is obvious that previous results must be applied to any instance of « x is not A », denoted as « a is not A ».

Indeed, two different instances require two different applications of the Choice Strategy. As an example, saying that « the menhir A is not tall » can be translated « the menhir A is very small », and, on the other hand, the translation of « the menhir B is not tall » can be « the menhir B is really very tall ».

5 Validation of a Potential Scene

It results from previous analysis that our interactive Choice Strategy allows the User to explain affirmatively the intended meanings of Linguistic Negations. Moreover, he can modify a Scene Description in order to obtain a more appropriate Scene, and this, by using Validation Rules containing Negative Information.

Example: In the following Figures 9 and 10, we have collected the Properties associated with the concepts « height » and « appearance ». The arrows specifying the directions of the plausible Negations result from the marks associated with the basic Properties. The Nuances of the intended Meanings of Linguistic Negations appear in brackets.

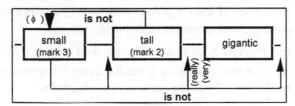

Fig. 9. The concept « height »

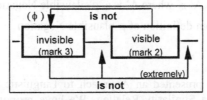

Fig. 10. The concept « appearance »

Let us now analyse the Validation Process of some Potential Scenes.

1 - Validation Rule : if « x is not tall » then «x is not visible in a crowd ». It results from Designer interpretation that we obtain the following equivalent Validation Rule: if « x is small » then « x is invisible in a crowd ». So, any small human must be invisible in the Scene describing a crowd. If it appears that Jack, who is small, is extremely visible in the proposed Scene, then the Designer must reject such a potential Scene. Moreover, he has to modify his previous Scene Description in order to avoid such inconsistencies.

2 - Validation Rule : if « x is not small » then « x is visible in a crowd». Let us suppose that the Designer has selected « x is really very tall » as the intended meaning of « x is not small ». So, the initial Rule is equivalent to : if « x is really very tall » then « x is visible in a crowd ». In this case, John being really very tall must be visible in a crowd. If so, he does not reject the proposed Scene and he can continue his Validation Process. If not, he modifies his Scene Description.

3 - Validation Rule : if « x is not small » then « x is not invisible in a crowd ».The previous Choice gives us : if « x is really very tall » then « x is not invisible in a crowd ». If « x is not invisible in a crowd » receives as a meaning «x is extremely visible in a crowd », we obtain : if « x is really very tall » then « x is extremely visible in a crowd ». Knowing that Ann is really very tall and showing that she clearly appears in the proposed Scene, the Designer continues his validation of this Scene.

6 Some Properties of Linguistic Negation

We can point out the fact that this Linguistic Negation satisfies some Common sense properties of Negation.

Property 4. For many Fuzzy Properties A, «x is A» does not automatically define the Knowledge about «x is not A ».

This property results directly from our construction Process of Linguistic Negation. Knowing exactly A does not imply, as does the Logical Negation, precise Knowledge of its Negation, since most of them require complementary information, as the mark of the property, and the Choice among possible interpretations.

Property 5. Given a Property A, its double Negation does not generally lead to A.

Using Figure 9, the User has selected « x is small » as the interpretation of « x is not tall », and « x is really very tall » as the negation of « x is small».

Property 6. Given the Rule « if « x is A » then « y is B » », we can deduce that if « y is not B » then « z is A' » where A' is a ρ-plausible Negation of A.

This property results from definition of ρ-plausible Negation

7 Conclusion

In this paper, we have presented an Approach to Linguistic Negation of Nuanced Properties based upon a Similarity Relation. We have proposed a Choice Strategy improving the abilities in Declarative Modeling in Image Synthesis. Indeed, a Scene Designer can refer to Linguistic Negations both in Scene Descriptions and in Validation Rules of potential Solutions, since this Interactive Strategy allows him to explain their intended meanings.

References

1.   J. F. Baldwin, B. W. Pilsworth: Axiomatic approach to implication for approximate reasoning with fuzzy logic. in: Fuzzy Sets and Systems 3, 193 -219 (1980)
2.   E. Desmontils, D. Pacholczyk: Modélisation déclarative en Synthèse d'images: traitement semi-qualitatif des propriétés imprécises ou vagues. in: Proc. AFIG'96, 173-181 (1996)
3.   O. Ducrot, J.-M. Schaeffer et al.: Nouveau dictionnaire encyclopédique des sciences du langage. Seuil, Paris (1995)

240

4.   L.R. Horn: A Natural History of Negation. The University of Chicago Press, Chicago (1989)
5.   W.A. Ladusaw, Negative Concord and « Made of Judgement », in: Negation, a notion in: Focus, H. Wansing, W. de Gruyter eds., Berlin, 127 - 144, 1996
6.   W. Lenzen, Necessary Conditions for Negation Operators, in: Negation, a notion in Focus, H. Wansing, W. de Gruyter eds., Berlin, 37 - 58, 1996
7.   M. Lucas, D. Martin, P. Martin, D. Plemenos: Le projet ExploFormes: quelques pas vers la modélistion déclarative de formes. Journées AFCET-GROPLAN, Strasbourg, 1989, publié dans BIGRE n°67, 35-49 (1990)
8.   C. Muller: La négation en français, Publications romanes et françaises. Genève, (1991)
9.   D. Pearce, Reasoning with negative information II: Hard Negation, Strong Negation and Logic Programs, in: LNAI 619, Berlin, 63 - 79, 1992
10. D. Pearce & G. Wagner, Reasoning with negative information I: Hard Negation, Strong Negation and Logic Programs, in: Language, Knowledge, and Intentionality : Perspectives on the Philosophy of J. Hintikka, Acta Philosophica Fennica 49, Helsinki, 430 - 453, 1990
11. D. Pacholczyk: Contribution au traitement logico-symbolique de la connaissance. Thèse d'état, Part C, Paris VI (1992)
12. E. H. Ruspini: The Semantics of vague knowledge. in: Rev. int. de Systémique, Vol. 3, n 4, 387 - 420 (1989)
13. P. Scheffe: On foundations of reasoning with uncertain facts and vague concepts. in: fuzzy reasoning and its applications. 189 - 216 (1981)
14. V. Torra, Negation Functions Based Semantics for Ordered Linguistic Labels, in: Int. Jour. Intelligent Systems, 975-988, 1996
15. A. Tversky: Features of Similarity. Psychological Review, 4, 84 (1977)
16. L.A. Zadeh: Fuzzy Sets. Information and Control, vol. 8, 338-353 (1965)
17. L. A. Zadeh: Similarity relations and Fuzzy orderings. in: Selected Papers of A. Zadeh (1987)

Fuzzy Clustering and Images Reduction *

A. F. Gómez-Skarmeta[1], J. Piernas[1] and M. Delgado[2]

[1] Dept. Informática y Sistemas. Universidad de Murcia, Spain
[2] Dept. Ciencias de la Computación e I.A. Universidad de Granada, Spain

Abstract. In this paper we present an efficient method for estimating the significant points of a gray level image by means of a fuzzy clustering algorithm. This method can be used to reduce the resolution of the image so it can be transmited and later reconstructed with the greatest reliability. We will show how using less than $0,01$ of the original information it is possible to reconstruct an image with a considerable level of detail.

1 Introduction

Clustering of numerical data forms the basis of many classification and system modeling techniques. The purpose of clustering is to distill natural groupings of data from a large data set, producing a concise representation of the original information. Fuzzy clustering in particular has shown excellents results, because it has a no restrictive interpretation of the membership of the data to the different clusters. Within the fuzzy clustering algorithms, the Fuzzy C-means [1] has been widely studied and applied in different environments [3] [4].

In our case we are interested in the reduction of a gray level image replacing it by its more significant points, so that in a later moment we can reconstruct the original image with the greatest precision, or in other words with the less information loss. The number of points must be as small as possible in relation with the points of the original image, but trying to preserve the more information as possible of the original image, in order that the later reconstruction allows its regeneration it with great reliability.

In order to obtain these significant points, we use the Mountain Clustering Algorithm. This is a fuzzy clustering algorithm developed by Yager and Filev [7] which can be used for estimating by means of a simple and effective algorithm, the number and location of fuzzy cluster centers . Their method is based in griding the data space and computing a score value for each grid point based on its distance to the actual data points; a grid point with many data points nearby will have a high potential value. The grid point with the highest potential value is chosen as the first cluster center. Once a centroid is detected the potential of all grid points is reduced according to their distance from the cluster center. This process is repeated until the potential of all grid points fall below a threshold.

* This work has been partially supported by CICYT project TIC95-1019

This method is specially suitable for our problem, because we work with retinal or polar images. Images which griding try to simulate the eyes vision using a radial griding over a circular image, having in this way a great resolution near the central point of focus and with a smaller resolution as we move away from the focus. This is an important difference with other applications of the Mountain Clustering Algorithm [2]. In this context a natural alternative is to reduce the number of points in each radial axis or radii, searching for the most representative points corresponding to the original data set, producing in this way a concise representation of the original information. As we have indicated, this is the main objective of clustering in general and fuzzy clustering in particular. Furthermore, as the representation we have adopted in the image treatment is similar in comparison with the Mountain Clustering technique, this latter is an excellent candidate as a clustering algorithm and an alternative to other proposals in the literature [5].

Regardless of other possible applications of this technique as image compression or image storage, in our case it is part of a vision component of an autonomous system which main objective is to capture images and send them to a remote controller where the whole image is reconstructed to be processed, or just a partial zone where some object of interest could be located based on some predictions is reconstructed.

In the next section we present the basic ideas behind the Mountain Clustering Algorithm. Then in section 3 we show how we apply this technique in our image reduction/reconstruction method using the centroids obtained by the Mountain Clustering algorithm. In section 4 we discuss the experimental results obtained with our method, and finally in section 5 we indicate some future trends.

2 Mountain Clustering Algorithm Overview

In the following we shall briefly explain the basis ideas of the Mountain Clustering method (MC). For simplicity, we shall focus on two dimensional space, but the generalisation of the result is straightforward.

The MC can be seen to be a three step process. In the first step we discretize the object space and in doing so generate the potential cluster centers. The second step uses the observed data, the objects to be clustered, to construct the mountain function. The third step generates the cluster centers by a iterative destruction of the mountain function.

First Step

Assume the data consist of a set of q points (x_k, y_k) in the $\Re^2$ space. We restrict ourselves to the rectangular subspace $X \times Y$ of $\Re^2$ containing the data points. The first step in the MC is to form a discretization of $X \times Y$ space by griding X and Y with r_1 and r_2, respectively, equidistant lines (although this is not obligatory). The intersection of these grid lines, called nodes, form out set of potential cluster centers. We shall denote this set as N, and an element in N as N_{ij} and with (X_i, Y_j) indicating the node obtained by the intersection of the grid lines passing through the lines at X_i and at Y_j.

As we shall subsequently see the purpose of this discretization is to turn the continuos optimisation problem of finding the centers into a finite one.

Second Step

In this step we shall construct the mountain function M, which is defined on the space N of potential cluster centers:

$$M : N \longrightarrow \Re$$

The mountain function M is constructed from the observed data by adding an amount to each node in N proportional to that nodes distances from the data point. More formally for each point N_{ij}, (X_i, Y_j), in N

$$M(N_{ij}) = \sum_{k=1}^{q} e^{-\alpha \times d(N_{ij}, O_k)}$$

where O_k is the kth data point (x_k, y_k), α is a constant and $d(N_{ij}, O_k)$ is a measure of distance between N_{ij} and O_k typically, but not necessarily, measured as

$$d(N_{ij}, O_k) = (X_i - x_k)^2 + (Y_j - y_k)^2$$

Obviously the closer a data point is to a node the more it contributes to the score at that node. It is evident from the construction of the mountain function that its value are approximations of the density of the data points in the neighbourhood of each node. The higher the mountain function value at a node the larger is its potential to be a cluster center.

Third Step

The third step in the MC is to use the mountain function to generate the cluster center. Let the node N_1^* be the grid point with maximal total score, the peak of the mountain function. We shall denote its score by $M_1^* = Max_{ij}[M(N_{ij})]$. If there are more than one maxima we select randomly one of them. We designate this node as the first cluster center and indicate its coordinates by $N_1^* = (x_1^*, y_1^*)$. We next must look for the next cluster center, so we must eliminate the effect of the cluster center just identified because usually this peak is surrounded by a number of grid points that also have high scores. This means that we must revise the mountain function for all the other nodes. The process of revision can be seen as a destruction of the mountain function, and is realised by subtracting from the total score of each node a value that is inversely proportional to the distance of the node to the just identified cluster center, as well as being proportional to the score at this just identified cluster center. More specifically, we form a revised mountain function $\hat{M}_2$ also defined on N such that

$$\hat{M}_2(N_{ij}) = \hat{M}_1(N_{ij}) - M_1^*(N_{ij}) * e^{-\beta \times d(N_1^*, N_{ij})}$$

where $\hat{M}_1$ is the original mountain function M, β is a positive constant, N_1^* and M_1^* are the location of and score at the just identified cluster center and $d(N_1^*, N_{ij})$ is a distance measure.

We now use the revised mountain function $\hat{M}_2$, to find the next cluster center by finding the location N_2^*, and score M_2^*, of its maximal value. N_2^* becomes our second cluster center. We then revise our mountain function to obtain $\hat{M}_3$ as

$$\hat{M}_3\left(N_{ij}\right) = \hat{M}_3\left(N_{ij}\right) - M_3^*\left(N_{ij}\right) * e^{-\beta \times d(N_2^*, N_{ij})}$$

and so on.

This process, that can be seen as a destruction of the mountain function, ends when the score of the last found cluster center is less than a constant δ. This means that there are only very few points around the last cluster center and it can be omitted.

The main advantage of this method is that it does not require a predefined number of clusters. It determines the first m cluster centers that satisfy the stopping rule, starting from the most important ones which are characterised with maximal value of the mountain functions at nodes $N_1^*, N_2^*, ..., N_m^*$ with coordinates $(x_1^*, y_1^*), (x_2^*, y_2^*), ..., (x_m^*, y_m^*)$.

3 Image Reduction/Reconstruction

3.1 Mountain Function Alternatives

The first question to address in the applicability of the Mountain Clustering Algorithm is the determination of the possible centroid candidate points. As we have indicate in the introduction we are working with a retinal or polar vision where we transform a rectangular image obtained from a camara, in a circular one where the new image points are distributed in an equidistant way around concentric circumferences with different radii from a focal point.

Although in this image treatment there are areas in the original image that are not considered and which information is lost as can be seen in figure 1, it has shown good behaviour in machine vision though it let simplificate the original image and concentrate the vision in the objects around the focal point in a similar way as human beings do.

In our case we are going to trace r radii and s concentric circumferences, having in this way $r * s + 1$ possible centroid candidates including the focal point or centre of the image. Along each radial axis or radii, we will obtain C centroids using the Mountain Clustering Algorithm, thus finally the original image will be replaced by a collection of $C * r + 1$ centroids that will concentrate the more important information representing the image and that will permit us later its reconstruction.

Once we have established the centroid candidates, the next step is to define the mountain function to be aplied to each of them. In this point a critical decision appears that will determine the success of our method. What we mean by a centroid?. There are different alternatives:

- A centroid will represent a weighted average of the different gray levels of the points around it.

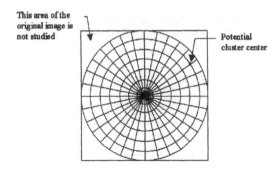

Fig. 1. Griding applied to the Image

- A centroid candidate will be more candidate as its gray level is equal or similar to the gray levels of the points that surround it.
- A centroid candidate will be more candidate as its gray level is different to the gray levels of the points that surround it.

The first alternative leads to consider as centroids only the points that are in light areas of the images, reducing the possibilities of the points in darker areas. The mountain function value for each candidate is calculated by means of the gray level sum of all the points around the candidate, weighting each gray level by a distance exponential function.

The second alternative indicate us that the centroids are those candidate points whose gray level is much similar to those points around it. The mountain function value for each candidate point is calculate as the inverse of the neighbourhood points' gray level sum, weighting each gray level by means of a distance exponential function to the candidate.

The third alternative places the centroids in those zones where a uniform color does not exist. The mountain function value is calculate in a similar way as in the second alternative, but without consider the inverse of the sum.

Within the different alternatives, the third one has shown the best results. The reasons are clear: from an image we are interested in the points that represent the most significant information, in order to use them in the later reconstruction. This reconstruction will be better as we can recognize the objects that are present in the image. Thus, the points that contain the most representative information will be those that are in the border between objects, small areas that are characterized by significant changes in the gray level of the points. In this way, the third alternative for the mountain function is the more adequate in order to represent this situation.

In our case, the mountain function is defined using the expression:

$$M(N_{ij}) = \sum_k |(Gray(N_{ij}) - Gray(O_k))| * e^{-\alpha \times d(N_{ij}, O_k)}$$

where N_{ij} is a centroid candidate, $Gray(\cdot)$ is a function that returns the gray level of the original image for each point, α is a constant, $d(\cdot)$ is the euclidian distance and O_k is a point in the neighbourhood of N_{ij}. We must remark that in our method, during the mountain function evaluation in each point, we are not going to consider all the possible image points as it is done in the original Mountain Clustering Algorithm. Instead we only take into acccount the original image points that are close to each centroid candidate, within certain context using ideas from [4] [6].

Once we have obtained the c possible centroid candidate in each radii, and once we have calculated the mountain function value in each of these points, what remains is to apply the Mountain Clustering Algorithm as indicated in section 2, in order to find the C centroids over each of the radii.

3.2 Image Reconstruction from the Centroids

Once we have found the C centroids over each of the radii, we need a method to reconstruct the original image starting from this reduced number of points. We must remark that we have obtained excellent results for real images of $450 * 350$ pixels, using 18 centroids for each of the 50 radius. In other words, with our method we can reconstruct the image by means of only $900 (= 18 * 50)$ centroids representing just the $0,0057$ of the $157.500 (= 450 * 350)$ points of the real image.

In order to reconstruct the original image we must proceed in the following way. We can consider the gray level of each point as the return value of a two variables function, x and y which represents the coordinates of the point. The graphical representation of this function will be a "colour mountain" where the hills represent the light zones and the valleys the dark ones. The centroids represent the points of this colour montain that will let us reconstruct it with minor information loss.

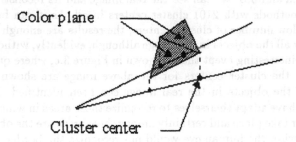

Fig. 2. Colour Plane

As we can see in figure 2, the centroids of each radius are paired with the centroids of the next radius, so a quadrilateral is formed by the two centroids of one radii and the two centroids of the other one. This quadrilateral will be divided in two components with three points each and two of them in common. The objective of this operation is to characterize by means of the three points just indicated the "colour plane", which equation lets us infer the corresponding colour to any point within the triangle formed by the three points. Figure 2 show these ideas.

In this way and using all the quadrilaterals that the centroids generate, we can reconstruct the original image.

4   Experimental Results

Experiments have been realized with real images and synthetic ones and several interesting conclusions have been obtained.

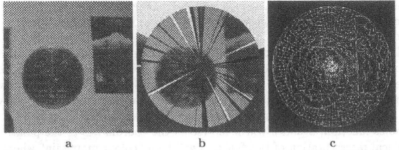

Figure 3: Original Image / Reconstructed Image / Clusters Center and Quadrilaterals

In figures 3.a and 3.b[3] we can see the real image and its reconstruction after applying our methods with 2101 cluster centers in all. As it can be noted and in spite of the low number of cluster centers, the results are enough satisfactory noticing clearly all the objects in the image although, evidently, without the same resolution. An interesting event that is shown in Figure 3.c, where quadrilaterals constituted by the cluster centers for the above image are shown up, is that the borders of the objects in the real image can been identified. That is, the cluster centers have adapt theirselves to recognize those areas in which important changes of color take place and certainly among these areas are the object borders because, otherwise, the human eye would not recognize the border either.

It is important to emphasize that in spite of the image darkness, we have obtained very similar results to the above ones with different iluminations. Hence we can conclude that our method is not much sensitive to image ligthing, being this a very interesting property.

Figures 4.a 4.b and 4.c show the application of our method on a synthetic image. In this case, we can note some anomalies in the reconstructed image that distort it. These anomalies arise because two adjacent radii cross very different zones, so cluster centers do not uniformly distribute on each radius, but tends

---

[3] The reconstructed image has a little noise due to the reproduction process

248

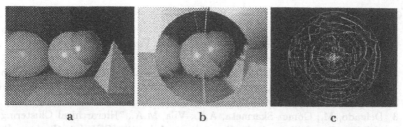

a   b   c

Figure 4: Original Image / Reconstructed Image / Clusters Center and Quadrilaterals

to concentrate on certain radii parts. Moreover as the number of cluster centers over both radius is the same and they match theirselves correlatively (the first with the first, the second with the second, and so on), it is posible that some quadrilaterals constituted by the cluster centers get longer longitudinally following radius direction, giving rise to the appearance of that "peaks" in the reconstructed image.

There are several posibilities that can help us to alleviate these anomalies. One solution is to increase the number of radii. Another one is to draw a new radius between two existing radii when the cluster center behaviour is enough different over both. One additional solution is to restrict cluster centers mobility on a radii. In order to achieve this, we can divide the radius in several parts (like having concentric rings) and fixing the number of cluster centers for each part. In this way, we achieve an uniforme distribution of cluster centers along each radius.

5 Conclusion and Future Trends

In this paper we have shown how the use of a fuzzy clustering technique like the Mountain Clustering Algorithm allows the reduction of the information needed to be maintained/transmited, relative to a gray level image preserving the more information about it, so later it can be reconstructed with greater reliability. The examples presented have shown that the method obtain good results either in the case of real images with adequate or slight ilumination, or in the case of synthetic images.

Although the results obtained indicate the good performance of the method proposed, there are some points that need further improving. Specifically the mountain function evaluation is a computational expensive process. In order to improve its evaluation it could be interesting to consider just a random collection of points in the centroids' neighbourhood. Another important aspect to reconsider is the way the centroid candidates are elected. Instead of assigning them to fixed positions over the radius, we could give them more movility, so they could move to more significant zones in relation to the final reconstruction of the image, although with an increase in the complexity of the algorithm.

References

1. Bezdek, J.C., "Pattern Recognition with Fuzzy Objective Function Algorithms", Plenun, New York, 1981.
2. Chiu, L.S., "A Cluster Estimation Method with Extension to Fuzzy Model Identification", *Proceedings of the 3rd IEEE Int. Conference on Fuzzy Systems*, Orlando, USA, pp. 1240-1245, 1994.
3. Delgado, M.; Gómez-Skarmeta, A.F.; Vila, M.A., "Hierarchical Clustering to Validate Fuzzy Clustering", *Proceedings of the 4th IEEE Int. Conf. on Fuzzy Systems*, pp. 1807-1812, 1995.
4. Delgado, M.; Gómez-Skarmeta, A.F.; Martín, F., "Generating Fuzzy Rules using Clustering Based Approach", *Proc. Third European Congress on Fuzzy and Intelligent Technologies and Soft. Computing*, Aachen, Germany, 810-814, 1995.
5. Djirar, S.; Banks, S.P.; Williams, K., "Dynamical Maps Applied to Image Compression", *Proc. Sixth Int. Conference Information Procesing and Management of Uncertainty in Knowledge-Based Systems IPMU'96*, Granada, Spain, pp. 641-645, 1996.
6. Karayiannis, N.B., "MECA: Maximun Entropy Clustering Algorithm", *Proceedings of the 3rd IEEE Int. Conference on Fuzzy Systems*, Orlando, USA, pp. 630-635, 1994.
7. Yager, R.R.; Filev, D.P., "Generation of fuzzy rules by mountain Clustering", *Tech. Report* MII-1318, 1993.

Predicates and Superimposed Coding: An Application to Image Recognition

Bernd-Jürgen Falkowski

FH Stralsund
FB Wirtschaft
Grosse Parower Str. 145
D-18435 Deutschland

Abstract. First of all we show that the evaluation of predicates leads to a classical problem of retrieval on secondary keys. After a brief sketch of the signature file technique we collect some known probabilistic results. This allows us to treat two distinct mathematical models simultaneously. We then consider the so-called false drop probability in the context of key-based image recognition. It is shown that known optimality results do not necessarily remain valid in this situation. This suggests shifting the emphasis from optimizing the false drop probability to optimizing the search time (both aims do not necessarily coincide). Thus we are led to consider a 2-level scheme which is simple compared to other 2-level schemes which have been used before. This scheme can be employed because of a suboptimal choice of parameters. Finally theoretical predictions derived from a somewhat crude abstract model are validated by experimental work carried out on a Smalltalk prototype. Encouraging results are obtained.

1 Introduction

Ever since Minsky and Papert's seminal study [14] the fast evaluation of predicates has been considered an important problem. It turns out that there is a close connection between this and the efficient answering of inclusive queries, which may be viewed as a special case of retrieval on secondary keys. Such queries may briefly be described as follows: One is given a collection of records R_i which are identified by record-id's. Each record has associated with it a set of keys $\mathcal{K}_i := \{K_{i_1}, K_{i_2}, \ldots, K_{i_r}\}$. One is then interested in finding those records R_i whose key sets $\mathcal{K}_i$ contain a given set of query keys $\mathcal{K}^Q := \{K_1^Q, K_2^Q, \ldots, K_s^Q\}$, where $s \leq r$.

The problem arises for example in the context of full text retrieval. In this situation the records might be documents (or other pieces of text) whilst the key sets would be given as the sets of distinct words appearing in the respective documents. Looking for all documents containing a given set of words would be possible by defining $\mathcal{K}^Q$ as this set.

Of course, instead of words describing a document we could equally well consider words describing an image and would then enter the realm of image recognition. Indeed, we might even consider the row (or column) vectors of a bit matrix representing the image as descriptors to arrive at the problem formulated in abstract terms above. Text retrieval and image recognition may be treated in the same way from an abstract point of view, although there are some significant differences in practice as we shall see below. Note, however, that in this paper we restrict ourselves to key-based image recognition as opposed to content-based image recognition. Thus our results in this direction may be seen as complementing those obtained e.g. in the QBIC project, cf. [8].

Although a considerable number of papers have been written on the general subject, cf. e.g. [12] for an extensive treatment, there is in essence only the method of inverted files applied to solve the problem in a commercial environment. In view of its practical importance this seems somewhat surprising and even more so, since the method of inverted files possesses some serious drawbacks. Roughly speaking, it requires that all possible key values, together with a list of pointers to those records in which they occur, be kept as an index. Large overheads (typically 50%-300%) and deterioration of performance for large query sets $\mathcal{K}^Q$ are the consequences.

This is probably the reason why signature file techniques using superimposed coding have recently been investigated in this context, cf. e.g. [7], [6], [2], [13], [1]. The underlying mathematical model for very small query sets $\mathcal{K}^Q$ (corresponding e.g. to full text retrieval as described above) is by now reasonably well understood, cf. e.g. [15], [17], [5], and optimal parameter sets can be obtained by minimizing the so-called false drop probability. For our purposes this method seems particularly attractive since the connection to certain perceptrons can be explicitly established.

For image recognition, however, usually comparatively large $\mathcal{K}^Q$ are required, cf. e.g. [18]. It seems that hitherto it has been tacitly assumed that optimality conditions remain essentially unchanged in this situation, cf. e.g. [18], p.347. In this paper we show that such an assumption is not justified in general. On the one hand this is regrettable since the available technical results can only be used to derive upper bounds for the false drop probability. On the other hand we show that for large $\mathcal{K}^Q$ suboptimal parameter sets may be used whilst still keeping the false drop probability negligibly small. This then suggests a shift of emphasis from optimizing the false drop probability to optimizing the search time for the signature file (both aims not necessarily coinciding here!). Thus we are led to construct and analyze a simple 2-level signature file scheme, where in contrast to the scheme described in [16] we are able to use the same signature length for both levels. An implementation of this scheme as a Smalltalk V prototype produces encouraging experimental results.

The paper is organized as follows:

In section 2 we describe the connection between the evaluation of special predicates (masks) and retrieval on secondary keys. Via the positive normal form theorem the method can, in principle, be employed to compute general predicates as well.

In section 3 we briefly review the signature file technique and and comment on its relation to certain perceptrons. We also address the problem of hash collisions.

In section 4 we give sketch proofs of known probabilistic results which allows us to deal with two slightly different models (corresponding to text retrieval and image recognition) simultaneously on the abstract level.

Having thus prepared the ground we employ classical analysis methods to derive a known approximation for the false drop probability, cf. e.g. [15], and an important restriction under which it is valid in section 5 . Moreover we show that this false drop probability always possesses a unique minimum (implicitly given as solution of a certain equation), which for small $\mathcal{K}^Q$ was known. This result appears to be new and gives a hint that optimal parameter sets might not remain so for larger $\mathcal{K}^Q$. In Example 4 we demonstrate that this is indeed true.

In section 6 we deal with image recognition. In the first part of this section we derive bounds on the minimum of the approximate false drop probability. This is followed by the investigation of an abstract model which leads to the construction and analysis of a simple 2-level signature file scheme. In the second part of this section we report on experimental work in order to validate the theoretical predictions for our 2-level scheme.

The concluding remarks in section 7 contain a (necessarily subjective) valuation of the results presented.

2 Predicates and Retrieval on Secondary Keys

In [14] the evaluation of predicates plays an important role in connection with perceptrons. In this section we show how this problem is closely related to a special case of retrieval on secondary keys.

2.1 The Positive Normal Form Theorem

Let R be an arbitrary but finite set of cardinality n. Then according to [14], a predicate is a function $\psi : 2^R \rightarrow \{0, 1\}$. Given the elements $x_1, x_2, \ldots, x_n$ of R, a subset A may be identified with a point in $\{0, 1\}^n$ by setting $A \leftrightarrow (\chi_A(x_1), \chi_A(x_2), \ldots, \chi_A(x_n))$, where χ_A is the characteristic function of A.

Hence a predicate ψ may be viewed as a function $\psi' : \{0, 1\}^n \rightarrow \{0, 1\}$ defined by $\psi'(\chi_A(x_1), \chi_A(x_2), \ldots, \chi_A(x_n)) = \psi(A)$. Taking this point of view one easily obtains the following theorem, cf. also [14], p.34.

Theorem 1. *Every predicate*

$$\psi' : \{0, 1\}^n \rightarrow \{0, 1\}$$

may be written as

$$\psi'(y_1, y_2, \ldots, y_n) = \sum_{i=1}^{s} \alpha_i \prod_{j=1}^{m_i} y_{ij}$$

for some integers α_i, m_i, s, *where* $y_{ij} \in \{y_1, y_2, \ldots, y_n\}$.

Proof: Use the disjunctive normal form theorem of propositional calculus.

□

Note that if $B := \{x_{i1}, x_{i2}, \ldots, x_{im_i}\} \sqsubseteq R$, then the function $\psi'_B(y_1, y_2, \ldots, y_n) := \prod_{j=1}^{m_i} y_{ij}$ defines a special predicate ψ_B given by

$$\psi_B(A) = \psi'_B(\chi_A(x_1), \chi_A(x_2), \ldots, \chi_A(x_n)) = \prod_{j=1}^{m_i} \chi_A(x_{ij}).$$

Thus $\psi_B(A) = 1 \Leftrightarrow B \sqsubseteq A$. Such a predicate is called a mask in [14]. Using this terminology we see that the positive normal form theorem (Theorem 1) says: Every predicate may (in fact uniquely) be written as a linear combination of masks. Hence it becomes clear that the fast evaluation of masks presents an important problem.

Considering only sets A of fixed cardinality $|A| = r$ the problem of evaluating masks on these sets is seen to be equivalent to the following classical problem concerning a special case of retrieval on secondary keys.

Suppose that one is given a collection of records R_i which are identified by record-id's. Each record has associated with it a set of keys $\mathcal{K}_i := \{K_{i_1}, K_{i_2}, \ldots, K_{i_r}\}$. One is then interested in finding those records R_i whose key sets $\mathcal{K}_i$ contain a given set of query keys $\mathcal{K}^Q := \{K_1^Q, K_2^Q, \ldots, K_s^Q\}$, where $s \leq r$.

Note here that the set B occurring in the description of masks corresponds to $\mathcal{K}^Q$ and the sets A correspond to the $\mathcal{K}_i$. Since for the applications we have in mind this notation seems more appropriate, we are going to adhere to it in the sequel.

Of course, the connection between information retrieval (in a wide sense) and perceptrons/predicates has been known for quite some time, cf. e.g. [14], p. 188. However, using the probabilistic technique of superimposed coding and signature files in the retrieval context brings to light some quite explicit correspondences.

3   The Signature File Technique

In this section we briefly describe the signature file scheme and the associated search procedure in order to fix the terminology. We also comment on the connection to perceptrons and analyze the problem of hash collisions which seems to have been neglected to some extent in the past.

3.1   Generating Signatures

Suppose that we are given a collection of records R_i which are identified by their record-id's and characterized by a set of keys $\mathcal{K}_i := \{K_{i_1}, K_{i_2}, \ldots, K_{i_r}\}$, (a record might, e.g., be a block of text in some documents whilst the keys might be the distinct 'non-common' words occurring in the text). We compress the information contained in $\mathcal{K}_i$ as follows:

a) Let $\mathcal{K}$ be the set of possible keys and $BCW(b, k)$ the set of binary code words of length b and weight k. Then we define a mapping $f : \mathcal{K} \to \text{BCW}(b, k)$. The value for f is obtained by setting precisely k bits to 1 in the binary code word (bcw). The bit positions are chosen pseudo-randomly and depend on a numerical encoding of the key in question. For details see e.g. [15].

b) For any set of keys $\{K_1, K_2, ..., K_n\}$ one can then generate a 'signature' s by setting $s := \bigvee_{i=1}^{n} f(K_i)$, where '$\bigvee$' denotes the componentwise 'OR'-operation, (this 'superimposing' explains the term superimposed coding!). The signatures together with the record-id's are finally collected together to form the so-called signature file. Note at this point that we are going to neglect important implementation details, like storage of the signature file, to a large extent in this paper. For these the interested reader may, e.g. consult [3], [6], [13], [18].

Example 1. Suppose that we have a record R_i with associated $\mathcal{K}_i :=$ $\{K_{i_1}, K_{i_2}, K_{i_3}\}$. Let $b = 8$ and f be defined on $\mathcal{K}_i$ by $f(K_{i_1}) :=$ $(0, 1, 0, 0, 0, 1, 0, 0); f(K_{i_2}) := (0, 0, 0, 0, 1, 1, 0, 0); f(K_{i_3}) := (1, 0, 0, 0, 1, 0, 0, 0)$. Then the associated signature s_i is given by $s_i := (1, 1, 0, 0, 1, 1, 0, 0)$.

3.2 The Search Procedure

The search for a record whose key set contains a given set of query keys $\mathcal{K}^Q :=$ $\{K_1^Q, K_2^Q, ..., K_s^Q\}$ now proceeds as follows:

a) Form the signature s^Q for $\mathcal{K}^Q$ as described above.

b) Find those record-id's (and hence records) whose signatures s_i satisfy

$$s_i \wedge s^Q = s^Q \tag{1}$$

Here '$\wedge$' denotes the componentwise 'AND'-operation.
Instead of giving a detailed proof we demonstrate the procedure by referring to Example 1.

Example 2. Suppose $\mathcal{K}^Q := \{K_{i_2}, K_{i_3}\}$, then s^Q is given by $(1, 0, 0, 0, 1, 1, 0, 0)$ and hence clearly we have $s_i \wedge s^Q = s^Q$.
Note that we could formulate equation 1 equivalently as follows: Inspect all those bit positions of s_i where s^Q has a 1. If all of them contain 1s then s_i qualifies. That is to say that the bit positions of s^Q which contain 0s are 'don't care'-positions.

Unfortunately equation 1 gives a necessary but in general not sufficient condition for s_i to qualify.

Example 3. Suppose $\mathcal{K}^Q := \{K\}$, where $f(K) := (0, 1, 0, 0, 1, 0, 0, 0)$. Then $s^Q :=$ $(0, 1, 0, 0, 1, 0, 0, 0)$ and $s_i \wedge s^Q = s^Q$, but clearly $\mathcal{K}^Q$ is not contained in $\mathcal{K}_i$. Hence R_i would be a 'false drop'.

Clearly, controlling the number of false drops in a signature file search constitutes an important problem which we shall address in section 3.

3.3 Comments

Recall for the moment that in terms of masks equation (1) may be interpreted as

$$\psi_B(A) = 1 \Rightarrow$$
$$s^B \wedge s^A = s^B.$$

We define predicates φ_i by $\varphi_i(A) := i^{th}$ component of s^A, where s^A is the signature associated with the set A. We set $\varphi := (\varphi_1, \varphi_2, \ldots, \varphi_b)$ and denote by $w(s^B)$ the weight of s^B.

Then it is interesting to observe that equation (1) may in these terms be rewritten as

$$s^B \bullet \varphi(A) > w(s^B) - 1,$$

where $\bullet$ stands for the scalar product. Hence the connection to perceptrons, and in particular Gamba Perceptrons, cf. [14], p. 228 is made explicit.

3.4 Hash Collisions

The function f defined above hashes keys (hopefully!) almost uniformly onto $\binom{b}{k}$ bcw's. Of course, it is most desirable that there should be few collisions. In this context the following theorem provides information for a judicious choice of the parameters b and k.

Theorem 2. *Suppose that the total number of keys contained in our collection of records is p. Suppose further that $n := \binom{b}{k}$ and $p \ll n$. Then the probability pr, that no collision occurs during hashing, is approximately given by*

$$\log pr = \frac{p}{n}(1/2 - p)$$

Proof: The probability of no collision is given by

$$pr = \frac{n(n-1)\cdots(n-p+1)}{n^p} = \frac{n!}{(n-p)!n^p}$$

We approximate $n!$ and $(n-p)!$ using Stirling's formula, cf. [9], p. 159, to obtain

$$pr \approx \frac{\sqrt{2\pi n}(n/e)^n}{\sqrt{2\pi(n-p)}(\frac{n-p}{e})^{n-p}n^p}$$
$$= e^{-p}(\frac{n}{n-p})^{n-p+1/2}.$$

Hence

$$\log pr = -(n - p + 1/2)\log(1 - p/n) - p$$
$$\approx \frac{p}{n}(n - p + 1/2) - p = \frac{p}{n}(1/2 - p).$$

$\square$

Comments

(i) The exact formula for pr is obtained in analogy to the solution of the classical birthday problem, cf. e.g. [10], p. 36.

(ii) $p \ll n$ may be assumed since otherwise there would certainly be too many collisions.

(iii) As a consequence of Theorem 2 the following restriction on b and k should hold to obtain a pr close to 1:

$$\frac{p}{n}(p - 1/2) \ll 1 \tag{2}$$

4 Probabilistic Results

In this section we derive some basic probabilistic results (sketch proofs) concerning the false drop probability (FDP). To this end the principle of inclusion and exclusion is applied. Our methods allow us to treat two distinct models simultaneously. We also give conditions under which the two models coincide.

4.1 The Principle of Inclusion and Exclusion (PIE)

Since the PIE is used extensively in the calculations of this section we restate it here.

Theorem 3 PIE. *Given a set X with n elements and m subsets $S_1, S_2, \ldots, S_m$ of X. Then the number of elements of X which do not belong to $\bigcup_{i=1}^{m} S_i$ is given by*

$$n - \sum_{1 \leq k_1 \leq m} |S_{k_1}| + \sum_{1 \leq k_1 < k_2 \leq m} |S_{k_1} \cap S_{k_2}| - \sum_{1 \leq k_1 < k_2 < k_3 \leq m} |S_{k_1} \cap S_{k_2} \cap S_{k_3}|$$
$$+ \cdots + (-1)^m |S_1 \cap S_2 \cap \ldots \cap S_m|$$

where $|.|$ denotes the cardinality operator.

Proof: Use induction on m.

□

4.2 The False Drop Probability

In order to obtain the false drop probability we first of all have to calculate the probability that a signature contains 1s in t prespecified bit positions $I(t) = (i_1, i_2, \ldots, i_t)$. So let $\mathcal{S}_1 :=$ set of all r-tuples of bcw's of length b and weight k, where all bcw's are to be distinct; and let $\mathcal{S}_2 :=$ set of all r-tuples of bcw's of length b and weight k. It will be seen below that the choices of $\mathcal{S}_1$ respectively $\mathcal{S}_2$ correspond to the text retrieval respectively image recognition models. Then we have

$$|\mathcal{S}| = \begin{cases} \left(\binom{b}{k}\right) & \text{if} \quad \mathcal{S} = \mathcal{S}_1 \\[2mm] \binom{b}{k}^r & \text{if} \quad \mathcal{S} = \mathcal{S}_2 \end{cases}$$

and thus we set $|\mathcal{S}| := g(b, k, r)$. With these conventions we obtain

Theorem 4. *The probability $p(I(t))$ that the signature generated from r keys (or bcw's) contains 1s in t prespecified bit positions given by $I(t) := (i_1, i_2, \ldots, i_t)$ is given by*

$$p(I(t)) = \frac{1}{g(b, k, r)} \left[\sum_{i=0}^{t} (-1)^i \binom{t}{i} g(b - i, k, r) \right]$$

Proof: Consider the sets $S_1, S_2, \ldots, S_t$ defined by $S_k := \{\underline{x} \in \mathcal{S} \mid s_{i_k} = 0\}$, where $\underline{x} = (x_1, x_2, \ldots, x_r)$ and the signature $\underline{s} = (s_1, s_2, \ldots, s_b) := \bigvee_{i=1}^{r} x_i$. Then use the PIE.

$\square$

Theorem 4 gives us the probability that a record qualifies in the search process accidentally provided that we know the weight of the query signature. Unfortunately, however, this is not known in advance in general. Hence we proceed to calculate the probability $p(t)$ that a signature generated from r keys (or bcw's) has weight t. The result is contained in the following theorem (where we employ the same notation as in Theorem 4)

Theorem 5. *The probability $p(t)$ that a signature generated from r keys (or bcw's) has weight t is given by*

$$p(t) = \frac{\binom{b}{t}}{g(b, k, r)} \left[\sum_{i=0}^{t} (-1)^i \binom{t}{i} g(t - i, k, r) \right]$$

Proof: Let $I(t) := (i_1, i_2, \ldots, i_t)$ as in Theorem 4 and consider the sets $S_1', S_2', \ldots, S_t'$ defined by $S_k' := \{\underline{x} \in \mathcal{S} \mid s_{i_k} = 0 \text{ and } s_i \neq 0 \text{ for } i \notin \{i_1, i_2, \ldots, i_t\}\}$. Again using the PIE first compute $p'(I(t))$, namely the probability that the signature contains 1s in t prespecified bit positions and nowhere else. The result follows by observing that there are $\binom{b}{t}$ possibilities for specifying t bit positions.

$\square$

From Theorem 5 we can now calculate the expected weight $\overline{w}$ of the query signature. In fact we obtain

Theorem 6. *The expected weight $\overline{w}$ of a signature generated from r keys is given by*

$$\overline{w} = b \left[1 - \frac{g(b - 1, k, r)}{g(b, k, r)} \right]$$

Proof: Use Theorem 5 and (7) and (23) of [11, section 1.2.6].

□

These results may finally be put together to obtain the false drop probability where we take as the weight of the query signature its expected value.

Theorem 7. *The false drop probability FDP, that is to say the probability that a record with r associated keys will qualify accidentally if a search is carried out in which s keys are specified, is given by*

$$FDP = \frac{1}{g(b,k,r)} \sum_{i=0}^{\overline{w}} (-1)^i \binom{\overline{w}}{i} g(b-i,k,r) \ where \ \overline{w} = b[1 - \frac{g(b-1,k,s)}{g(b,k,s)}]$$

Proof: Theorem 4, Theorem 5, Theorem 6.

□

Of course, from a practical point of view Theorem 7 is not all that useful since the formula, although exact, is not easy to handle in calculations. We shall, however, come back to this at a later stage.

4.3 Coincidence of the Two Models

In section 4.2 we were able to derive results for two different models simultaneously. The first of those corresponds, for example, to full text retrieval where documents are partitioned into blocks containing an equal number of distinct non-common words (common words are e.g. articles, pronouns, etc., that is to say words which are presumably of no interest if full text retrieval is carried out), cf. e.g. [5]. The second model corresponds e.g. to image recognition since the descriptive keys for images need not necessarily be distinct (think of identical bit vectors occurring in a bit matrix representing an image). Hence it is perhaps remarkable that under certain circumstances the two models coincide as far as the *FDP* is concerned. In fact we obtain

Theorem 8. *Let*

$$g_1(b,k,r) = \binom{\binom{b}{k}}{r}, \quad g_2(b,k,r) = \binom{b}{k}^r$$

Then if $(r-1)^2 \ll \binom{b-\overline{w}}{k}$ *we have*

$$\frac{g_1(b-i,k,r)}{g_1(b,k,r)} \approx \frac{g_2(b-i,k,r)}{g_2(b,k,r)} \qquad for \qquad 0 \le i \le \overline{w}$$

and moreover

$$\frac{g_1(b-1,k,s)}{g_1(b,k,s)} \approx \frac{g_2(b-1,k,s)}{g_2(b,k,s)} \qquad for \qquad s \le r$$

Hence the corresponding FDPs coincide.

Proof: Let us, for brevity of writing, set

$$\alpha := \binom{b-i}{k}, \beta := \binom{b}{k}$$

then we have

$$\frac{g_1(b-i,k,r)}{g_1(b,k,r)} = \prod_{l=0}^{r-1} \frac{\alpha-l}{\beta-l} = (\frac{\alpha}{\beta})^r \prod_{l=0}^{r-1} \frac{1-\frac{l}{\alpha}}{1-\frac{l}{\beta}}.$$

Now note that

$$\prod_{l=0}^{r-1} \frac{1-\frac{l}{\alpha}}{1-\frac{l}{\beta}} \geq [1-\frac{r-1}{\alpha}]^{r-1} \geq 1 - \frac{(r-1)^2}{\alpha} \qquad \text{if} \qquad \frac{r-1}{\alpha} \leq 1$$

and

$$\prod_{l=0}^{r-1} \frac{1-\frac{l}{\alpha}}{1-\frac{l}{\beta}} \leq \frac{1}{1-\frac{(r-1)}{\beta}} \leq \frac{1}{1-\frac{(r-1)^2}{\beta}} \qquad \text{if} \qquad \frac{r-1}{\beta} \leq 1.$$

It follows from these inequalities that

$$\frac{g_1(b-i,k,r)}{g_1(b,k,r)} \approx (\frac{\alpha}{\beta})^r = \frac{g_2(b-i,k,r)}{g_2(b,k,r)} \qquad \text{if}$$

$(r-1)^2 \ll \alpha (\Rightarrow (r-1)^2 \ll \beta)$. The theorem is an immediate consequence of these approximations.

□

At this point some comments appear appropriate.

4.4 Comments

(i) The results in this section are not new (apart from possibly Theorem 8). It seemed, however, worthwhile to sketch them again in order to get simplified proofs based on the PIE according to a suggestion in [12]. Explicit (albeit rather complicated) proofs may be found in [15].

(ii) The difference between the two models considered may be seen in analogy to the difference between sampling with and without replacement. It is of course well-known that with a large population these models are almost identical and our approximation in Theorem 8 may be interpreted along these lines.

5 An Approximation to the FDP and its Minimum

From now on we are going to concentrate on the model corresponding to image recognition. In this section we shall derive an approximation for the FDP and a resulting restriction on the parameters. Moreover we shall prove that this approximation to the FDP always possesses a unique minimum.

5.1 The Approximation to the FDP

Taking $g(b, k, r) = \left(\frac{b}{k}\right)^r$ in Theorem 7 the false drop probability FDP is given by

$$FDP = \sum_{i=0}^{\overline{w}}(-1)^i C_i^{\overline{w}}[C_k^{b-i}/C_k^b]$$

where $\overline{w} = b[1 - (1 - k/b)^s]$ and C_y^x denotes the binomial coefficient (elsewhere written as $\binom{x}{y}$). For this FDP we obtain the following approximation.

Theorem 9. *Let the FDP be as above. Then the approximation*

$$FDP \approx [1 - (1 - k/b)^r]^{\overline{w}}$$

where $\overline{w} = b[1 - (1 - k/b)^s]$ holds provided that

$$\frac{rk\overline{w}(\overline{w} - 2)}{4(b - k)(2b - \overline{w} + 2)} \ll 1 \qquad (3)$$

Proof: We shall approximate C_k^{b-i}/C_k^b in the expression for the FDP given above.

$$C_k^{b-i}/C_k^b = \prod_{l=0}^{k-1} \frac{b - i - l}{b - l}$$

$$= \prod_{l=0}^{i-1} \frac{b - k - l}{b - l}$$

$$= [1 - k/b]^i \prod_{l=1}^{i-1}[1 - \frac{lk}{(b - k)(b - l)}]$$

Noting that $k \ll b$ for practical applications we set $\eta := \frac{k}{b-k}$ and expand $\prod_{l=1}^{i-1}[1 - \frac{lk}{(b-k)(b-l)}]$ in powers of η to obtain, cf. [15],

$$\prod_{l=1}^{i-1}[1 - \frac{lk}{(b - k)(b - l)}] = 1 - a\eta + O(\eta^2).$$

Here

$$a = \sum_{l=1}^{i-1} \frac{l}{b - l} < \frac{1}{b - i + 1} \sum_{l=1}^{i-1} l = \frac{i(i - 1)}{2(b - i + 1)}$$

Hence $[C_k^{b-i}/C_k^b]^r \approx [1 - k/b]^{ri}$ provided that $\frac{rki(i-1)}{2(b-k)(b-i+1)}$ is small compared to 1.

We also observe that in the expression for the FDP each summand contains a factor $C_i^{\overline{w}}$ which is maximal near $i = \overline{w}/2$. Hence we shall obtain a good approximation for the FDP as

$$FDP = \sum_{i=0}^{\overline{w}}(-1)^i C_i^{\overline{w}}[1 - k/b]^{ri} = [1 - (1 - k/b)^r]^{\overline{w}}$$

provided that $\frac{rk\mu(\mu-1)}{2(b-k)(b-\mu+1)}$ is small compared to 1, where $\mu = \overline{w}/2$. Substituting for μ we finally obtain as restriction under which our approximation is valid $\frac{rk\overline{w}(\overline{w}-2]}{4(b-k)(2b-\overline{w}+2)}$ is small compared to 1.

$\square$

We hasten to point out that inequality (3) severely restricts the size of $\overline{w}$ which is unfortunate since we are interested in applications where $\overline{w}$ cannot be assumed to be too small. Nevertheless optimal parameter sets have been computed by even further approximating the expression for the FDP given in Theorem 9, cf. e.g. [15], p. 1630, and used in situations where $\overline{w}$ is by no means small, cf. e.g. [18], p. 356.

Thus we proceed to derive conditions for optimal parameter sets, without further approximatimg the FDP, in the next section.

5.2   The Minimum of the Approximate FDP (AFDP)

For technical convenience we shall minimize

$$\log AFDP = \overline{w}\log[1 - (1 - k/b)^r]$$

where $\overline{w} = b[1 - (1 - k/b)^s]$ according to Theorem 9. If we set $u := [1 - k/b]^r$ and $\gamma := s/r$ then this amounts to finding the minimum of the function F given by $F(u) := [1 - u^\gamma]\log[1 - u]$. Clearly F is defined and continuous for $0 < u < 1$. It may be extended to a continuous function on $[0, 1]$ by setting $F(0) = F(1) = 0$. Thus clearly it will attain a minimal value on the compact set $[0, 1]$. The hard part is to show that this minimum is in fact unique.

First of all note that the derivative of F (denoted by F') is given by

$$F'(u) = -\gamma u^{\gamma-1}\log(1 - u) - \frac{1 - u^\gamma}{1 - u}$$

and thus $F'(u) = 0$ iff

$$(1 - u)\log(1 - u) = -\frac{1 - u^\gamma}{\gamma}u^{1-\gamma}.$$

Hence our problem is reduced to finding the points of intersection of the two curves G_1, G_2 given by

$$G_1(u) = (1 - u)\log(1 - u) \qquad \text{for} \qquad 0 < u < 1,$$
$$G_1(0) = G_1(1) = 0,$$
$$G_2(u) = -\frac{1 - u^\gamma}{\gamma}u^{1-\gamma} \qquad \text{for} \qquad 0 < u < 1,$$
$$G_2(0) = G_2(1) = 0.$$

The following lemma concerning properties of G_1 and G_2 is easily proved by elementary analysis.

Lemma 10. *The following statements about G_1 and G_2 are true:*

(i) $G_1'(u) = -\log(1-u) - 1$ *for* $0 < u < 1$, $G_1'(0) = -1, G_1'(1) = \infty$.

(ii) $G_2'(u) = 1 + \frac{1-\gamma}{\gamma}(1 - u^{-\gamma})$ *for* $0 < u < 1$, $G_2'(0) = -\infty, G_2'(1) = 1$.

(iii) G_1 *has a unique minimum for* $u = u_{min_1} = 1 - e^{-1} \approx 0.632$, *where* $G_1(u_{min_1}) = -e^{-1} \approx 0.368$.

(iv) G_2 *has a unique minimum for* $u = u_{min_2} = (1-\gamma)^{\frac{1}{\gamma}}$, *where* $G_2(u_{min_2}) = -(1-\gamma)^{\frac{1-\gamma}{\gamma}}$.

(v) $G_1''(u) = \frac{1}{1-u} > 0$ *for* $0 < u < 1$ *and hence* G_1 *is convex.*

(vi) $G_2''(u) = (1-\gamma)u^{-\gamma-1} > 0$ *for* $0 < u < 1$ *and hence* G_2 *is convex.*

The facts of Lemma 10 are summarized in Figure 1.

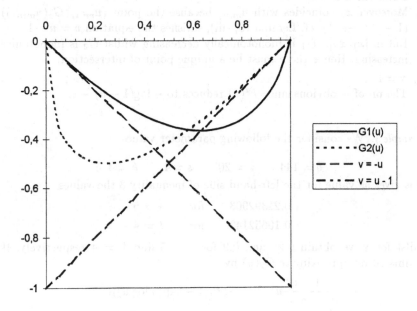

Fig. 1. Functions appearing in Lemma 10

With the aid of the diagram shown in Figure 1 it is now possible to prove the following theorem.

Theorem 11. *The log of the AFDP (and hence the AFDP) has a unique minimum given by*

a) *the unique solution of the equation*

$$(1 - u)\log(1 - u) = -\frac{1 - u^\gamma}{\gamma}u^{1-\gamma}$$

in $(0,1)$ *provided that* $0 < \gamma < 1$,

b) $u = 1 - e^{-1}$ *if* $\gamma = 1$.

Proof:

a) $0 < \gamma < 1$

Consider the above picture and note that because of (v) and (vi) in Lemma 10 (convexity) and because of (i) and (ii) in Lemma 10 G_1 and G_2 can only intersect to the right of the straight line $v = -u$ and to the left of the straight line $v = u - 1$. For the same reasons the intersection point(s) can only lie between $u = x_1$ and $u = x_2$ (these coordinates being obtained by intersecting $v = -u$ with $G_2(u)$ respectively $v = u - 1$ with $G_1(u)$).

Next note that x_1 must lie to the right of u_{min_2} since the point $(u_{min_2}, G_2(u_{min_2})) = ((1 - \gamma)^{1/\gamma}, -(1 - \gamma)^{\frac{1-\gamma}{\gamma}})$, cf. Lemma 10, (iv), lies below the line $v = -u$.

Moreover x_2 coincides with u_{min_1} because the point $(u_{min_1}, G_1(u_{min_1})) = (1 - e^{-1}, -e^{-1})$, cf. Lemma 10, (iii), satisfies the equation $v = u - 1$.

But in $[x_1, x_2]$ G_1 is monotonically decreasing whilst G_2 is monotonically increasing. Hence there must be a <u>unique</u> point of intersection.

b) $\gamma = 1$

The proof is obvious since $F'(u)$ reduces to $-\log(1 - u) - 1$.

$\square$

Example 4. We consider the following parameter values

$$b = 144 \qquad r = 20 \qquad s = 4 \qquad k = 4,5$$

This gives as value for the left-hand side of inequality 3 the values

$$0.21392903 \qquad \text{for} \qquad k = 5$$
$$0.10652142 \qquad \text{for} \qquad k = 4$$

whilst for γ we obtain 0.25 and 0.2 for $k = 5$ and $k = 4$ respectively. If in Lemma 10 we approximate $G_2(u)$ by

$$-\frac{1 - u^\gamma}{\gamma}u^{1-\gamma} = -1/\gamma(1 - \exp(\gamma \log u))u^{1-\gamma}$$

$$\approx u \log u \qquad \text{for small} \qquad \gamma$$

then we obtain as approximate solution for the minimum of the AFDP $u_{min} = 1/2$ which has been computed before, cf. e.g. [15], [17].

However, $k = 5$ leads to a value for u which is given by $u = 0.493226228$, whilst $k = 4$ leads to a value for u which is given by $u = 0.569260266$. The corresponding values for the AFDP are given by $2.4929 * 10^{-6}$ and $2.4368 * 10^{-6}$ respectively.

Hence the value of the AFDP for $u = 1/2$ is no longer optimal in this situation!

Comments

(i)   The difference in the $AFDP$ in Example 4 for $k = 5$ respectively $k = 4$ is, of course, only very small and hence the FDP may still be optimal for $u = 1/2$.

(ii) Example 4 does, however, show that the FDP does not necessarily have a minimal value for $u = 1/2$ if s is large.

(iii) The restriction in inequality 3 is a severe one and does not allow us to choose s as large as would be desirable for the purposes of image recognition.

(iv) In spite of comment (iii) above we may still exploit our results since the values for the $AFDP$ which can be obtained observing the restriction in inequality 3 constitute an upper bound for the FDP since, ceteris paribus, increasing s decreases the FDP as is easily seen from the exact formula given in section 5.1.

(v)   Example 4 also clearly shows that the FDP is going to be extremely small for large values of s.

6   Image Recognition

In this section we are first of all going to derive bounds for the u-value corresponding to the minimum of the AFDP as given in Theorem 8. In order to gain further insight into the situation we consider an abstract model. This motivates the construction and analysis of a simple 2-level signature file scheme. Finally we present experimental results in order to validate the theory. In the whole section we mainly consider cases where s is not too small this being characteristic for image recognition, cf. e.g. [18].

6.1   Bounds for the Minimum of the AFDP

In Theorem 10 the minimum value for u is (apart from the case $\gamma = 1$) only given implicitly. Here we derive upper and lower bounds for it.

Theorem 12. *The $AFDP$ as in Theorem 9 has a minimum at $u = u_{min}$, where*

$$\frac{1}{1 + e} < u_{min} < 1 - e^{-1}$$

Proof: A glance at Figure 1 immediately shows that an upper bound for u_{min} is given by $1 - e^{-1}$. A lower bound, however, is considerably more difficult to obtain.

Looking at Figure 1 again and connecting the points $(u_{min_2}, G_2(u_{min_2}))$ and $(1, 0)$ by a straight line l, we observe that, because of the convexity of G_2, the minimum must lie to the right of the intersection point of l with the line $v = -u$.

Hence we shall take the u-coordinate of this intersection point as a lower bound $u_{lower}(\gamma)$. A little calculation shows that

$$u_{lower}(\gamma) = \frac{\delta(\gamma)^{1-\gamma}}{1 - \delta(\gamma) + \delta(\gamma)^{1-\gamma}} = \frac{1}{\delta(\gamma)^{\gamma-1} + \gamma}$$

where $\delta(\gamma) := (1 - \gamma)^{\frac{1}{\gamma}}$. Now

$$\frac{d}{d\gamma}[\delta(\gamma)^{\gamma-1}] = \frac{\delta(\gamma)^{\gamma-1}}{\gamma}[1 + \frac{\log(1-\gamma)}{\gamma}]$$

and $1 + \frac{\log(1-\gamma)}{\gamma} < 0$ for $0 < \gamma < 1$. Hence $\delta(\gamma)^{\gamma-1}$ is decreasing in γ for $0 < \gamma < 1$ and a lower bound for $u_{lower}(\gamma)$ is consequently obtained as

$$\lim_{\gamma \to \infty} \frac{1}{1 + \delta(\gamma)^{\gamma-1}} = \frac{1}{1 + e}$$

$\square$

Example 5. Let us consider the example given in [18], p. 356, where $b = 600$, $k = 10$, and $r = 40$ (i.e. each image is described by 40 key values). Note here that the key values may be verbal descriptors of the image as in [18] as well as bit vectors representing pixels of the image. Moreover we assume that a query contains 5 words (i.e. $s = 5$) as in [18]. In order to see whether the AFDP gives a valid approximation in this situation we compute the left-hand side of the restriction in inequality 3 to obtain 0.329392436. Hence it is slightly doubtful whether the AFDP may be considered as a valid approximation to the FDP even in this situation where s is still small compared to r. If, on the other hand, we take $k = 6$ (the other parameters remaining unchanged) then inequality 3 certainly holds but the value for u (0.6689971758) is already well outside the optimal region described in Theorem 11 above. In this case inequality 2 is still satisfied but for even smaller k this restriction certainly becomes important.

We conclude from the above example that in a realistic setting it will be difficult to obtain an optimal choice of parameters since the restriction in inequality 3 is likely to imply that the approximation $AFDP$ is possibly no longer valid. However, due to the extremely small FDP for larger s-values a suboptimal choice of parameters may well be good enough for practical purposes and we are indeed going to give experimental results supporting this statement below.

6.2 An Abstract Model

In order to gain more insight into the situation we calculate the expected number of bit references (ENB) needed for an unsuccessful search. We assume that we are given N images where each one of these is described by r key-values, each of which is given by a bit vector of length r'. For our (somewhat crude) model let us suppose that in order to verify false drops rr' bits have to be inspected. We further assume that signatures have length b and that all Nb bits of the signature file have to be inspected. Whilst both of these assumptions can certainly be improved in practice (using e.g. bit slice organization for the storage of the signature file, cf. [15]) they still suffice to show that false drops may be neglected for s not too small. To be more precise, we obtain

Lemma 13. *Under the assumptions stated above the expected number of bit references (ENB) needed for an unsuccessful search is given by*

$$ENB = \underbrace{N * FDP * r * r'}_{I} + \underbrace{Nb}_{II}$$

Proof: Simply observe that I results from the verification of false drops whilst II stems from the signature file search.

$\square$

Example 6. Consider the parameter set $b = 600$, $k = 6$, $r = 40$, $r' = 50$, $s = 5$ (as in Example 5) with $N = 10000$. Then summand II in Example 5 is given by $6 * 10^6$, whilst summand I (with AFDP instead of FDP) is approximately given by $2 * 10^7 * 7.6 * 10^{-15} = 15.2 * 10^{-8}$.

Hence it becomes clear that the false drops may be completely neglected provided that s is not too small. This is the case even if we choose our parameter s in a suboptimal fashion as in Example 6 above. Note that we have considered a particularly simple storage scheme for the signature file in Lemma 13. As hinted at before several better storage schemes are known but one usually has to pay a price in the form of update difficulties, more intricate programming, etc. for these.

The (straightforward) result of Lemma 13 does, however, suggest a shift of emphasis. Instead of concentrating on minimizing the number of false drops it seems more useful to try to speed up the search of the signature file even if this is achieved at the cost of a suboptimal parameter set with respect to false drops. Thus we are led to explore a simple 2-level signature file scheme which becomes possible by adopting a suboptimal value for k.

6.3 A Simple 2-Level Signature File Scheme

We propose to compress the information contained in the signature file even further as follows: We 'OR' together r_1 signatures in the original signature file SF and deposit the resulting signatures in a new signature file SF_1.

The search procedure is then carried out in SF_1 first in order to eliminate non-qualifying sections of SF. Of course, if too many signatures in SF are 'OR'-ed together the resulting signatures in SF_1 may contain too many 1s to be of real use and indeed, in the extreme case of almost all 1s occurring in the signatures of SF_1, may be harmful, since additional search time for SF_1 is needed. A brief analysis for the approximate false drop probability $AFDP_1$ in SF_1 runs as follows (cf. Theorem 9): We assume that each signature in SF has $\overline{w_s}$ bits randomly set to 1. Then we obtain the following theorem.

Theorem 14.

$$\log AFDP_1 \approx \overline{w} \log[1 - (1 - \overline{w_s}/b)^{r_1}]$$

where $\overline{w}$ is as in Theorem 9. This approximation is valid provided that

$$\frac{r_1 \overline{w_s} \overline{w}(\overline{w} - 2)}{4(b - \overline{w_s})(2b - \overline{w} + 2)} \ll 1 \tag{4}$$

holds.

Proof: Immediate from Theorem 9.

□

Let us now abbreviate the $AFDP$ (for the signature file SF) and the $AFDP_1$ (for the signature file SF_1) by p respectively p_1 and let us further assume that r_1 divides N (in order to avoid purely technical complications). Then setting $N_1 := N/r_1$ we may calculate the expected number of bit references ENB_1 needed for an unsuccessful search within the 2-level scheme. We obtain the following lemma.

Lemma 15. *The expected number of bit references ENB_1 needed for an unsuccessful search within the 2-level scheme is given by*

$$ENB_1 = \underbrace{N_1 * b}_{I'} + \underbrace{N_1 * p_1 * r_1 * b}_{II'} + \underbrace{N_1 * p_1 * r_1 * p * r * r'}_{III'}$$

where we have made assumptions similar to the ones described before Lemma 13.

Proof: Summand I' results from the search of SF_1, II' stems from the restricted search of SF, and III' accounts for the verification of false drops. Of course, the result is only valid if inequality 2, inequality 3, and inequality 4 hold.

□

It is interesting to observe that if p and p_1 are sufficiently small, which is most likely for image recognition, then $ENB/ENB_1 \approx r_1$ and hence the 2-level scheme will improve the number of bits to be inspected by approximately a factor r_1. This improvement is obtained at the cost of an extra overhead of approximately (size of SF)/ r_1. Of course, this is a result which we would have expected.

Example 7. Consider the following set of parameters: $b = 600$, $k = 6$, $r = 40$, $s = 2$, $N = 10000$, $r_1 = 5$. (This continues Example 6 except that we have chosen $s = 2$ in order to take care of the restriction in inequality 4). Then certainly inequality 2 andinequality 3 are satisfied as was shown in Example 6. Now for the left-hand side of inequality 4 we obtain 0.292651811 which might just be considered satisfactorily small. This then leads to an $AFDP_1$ given by 0.17950863 which is still unpleasantly large. However, seeing that this was obtained for $s = 2$ we might expect (without being able to prove it since our approximations are no longer valid) that for larger s there will result a very much better false drop probability.

Thus we are led to conduct some experimental work in order to confirm our expectations.

6.4 Experimental Results

In order to validate the predictions of our somewhat crude abstract model described in sections 6.2 and 6.3 we used a full-text retrieval prototype implemented in Smalltalk V for Windows, for details see [4]. We wished to avail ourselves of this prototype without too many programming alterations and hence simulated the keys as random strings of 0s and 1s. We considered 10000 such 'images' ($N = 10000$) where each 'image' contained 40 keys ($r = 40$) of length 50 ($r' = 50$). The signature length was set to 600 ($b = 600$) to create a situation comparable to the one considered in [18], p. 358. The signatures in SF (10000) were compressed further by 'OR'-ing together five at a time ($r_1 = 5$) resulting in a second level signature file SF_1 of size 2000. Twenty unsuccessful searches were conducted and elapsed times and false drops were measured and averaged over these twenty searches in order to eliminate undesirable side effects due to e.g. automatic garbage collection as far as possible (some inaccuracies due to, e.g., cache effects, certainly remained!). The experiments were conducted on a standard Pentium-PC (60 MHz-clock, 8 MB main memory) giving the results summarized in Table 1.

Note that for $k = 10$ we would expect approximately 10 false drops on the average for $s = 1$ according to the theory. Hence the value obtained experimentally, namely 11.1, seems quite reasonable considering the inherent inaccuracies. Note also that even for the case $s = 1$ our 2-level scheme performed only insignificantly worse than the single-level scheme. If, on the other hand, we deduct the time taken to generate the query signature from the times T_1 and T_2 (for queries of large files this will surely be insignificant!) then we obtain for the 2-level scheme in the case $k = 6$ even for $s = 5$ an improvement of search time by a factor 3.45 compared to the single-level scheme. This goes up to a factor close to 4 for larger s (neglecting obvious measurement inaccuracies). These improvements seem to validate the predictions of our somewhat crude model rather nicely. Note also that if we consider our 'images' as bit matrices consisting of 40 (number of keys) * 50 (length of keys) = 2000 bits then we used an overhead of 30% (length of signature = 600) for the single-level scheme whilst for the 2-level scheme an additional overhead of 6% was needed, which seems quite acceptable.

7 Concluding Remarks

A close look at the problem of evaluating predicates showed that there is an intimate connection to a special version of retrieval on secondary keys. This could be made quite explicit by the use of probabilistic techniques which lead to an analysis of signature file methods for image recognition.

Hitherto the false drop probability seems to have been considered a quantity of paramount importance for signature file techniques. Whilst this is certainly true for queries containing a very small number of keys we have shown that this is not so in the case of large query weight (this being relevant in the context of image recognition). Moreover it turned out that the minimum value of the false drop probability in this case did not necessarily occur for signatures having approximately one half of their entries equal to 1. Indeed signatures containing slightly fewer ones appeared to be preferable. Noting that for signatures containing significantly fewer 1s large query weights still produced very few false drops we were led to consider a simple 2-level signature file scheme which indeed provided significant improvements in this case. Hence for image recognition it seems useful to study parameter sets which are suboptimal with respect to the false drop probability.

We also wish to point out again that we haven't paid much attention to storage details (apart from treating the 2-level scheme). Of course, the signature files considered may still be stored in a very much more sophisticated manner to effect further improvements, cf. e.g. [3], [15], [18],[16]. If, however, a situation is envisaged where frequent updates of the signature file(s) have to take place or where programming effort is of major concern, then our simple 2-level scheme may profitably be employed.

At any rate, the results presented above indicate that for image recognition (especially in the context of large binary objects) signature files may hold a great deal of promise and could perhaps soon be exploited commercially.

The meaning of the symbols appearing in this table is as follows:

k : weight of bcw's

s : number of keys used for the query

T_1 : elapsed average time for searching SF_1 and SF (2-level scheme)

T_2 : elapsed average time for searching SF (single-level scheme)

AFD : average number of false drops in SF

AFD_1 : average number of false drops in SF_1

ST : average time taken to generate the query signature

$k = 10$

s	T_1	T_2	AFD	AFD_1	ST
1	548.5	454.0	11.1	1379.65	73.5
5	354.5	568.0	0.0	314.15	161.0
10	418.0	662.5	0.0	49.55	292.0
20	644.5	918.0	0.0	2.35	536.5
30	885.0	1169.5	0.0	0.1	777.0
40	1142.0	1420.0	0.0	0.05	1043.0

$k = 8$

s	T_1	T_2	AFD	AFD_1	ST
1	474.5	478.5	12.5	1103.7	67.0
5	311.0	566.0	0.0	106.45	163.5
10	382.5	672.5	0.0	6.2	280.5
20	638.0	903.5	0.0	0.0	524.0
30	862.5	1148.0	0.0	0.0	747.5
40	1097.0	1390.5	0.0	0.0	986.0

$k = 6$

s	T_1	T_2	AFD	AFD_1	ST
1	419.5	469.5	19.4	831.4	71.0
5	270.5	558.0	0.0	23.7	153.0
10	390.0	677.5	0.0	0.2	245.5
20	601.5	898.5	0.0	0.0	502.0
30	841.0	1126.0	0.0	0.0	723.5
40	1078.0	1367.0	0.0	0.0	955.5

Tab. 1. Time and False Drop Averages (Times are given in milliseconds)

References

[1] Chang, W.W., and Schek, H.J. A Signature Access Method for the Starburst Database System. In Proceedings of the 15th International Conference on VLDB (Amsterdam, Aug. 1989), pp. 145-153

[2] Ciaccia, P., and Zezula, P. Estimating Accesses in Partitioned Signature File Organizations. ACM Trans. Inf. Syst., 11, 2, (April 1993), pp. 133-142

[3] Deppisch, U. S-tree: A dynamic balanced signature index for office retrieval. In Proceedings of the ACM Conference on Research and Development in Information Retrieval Pisa, (Sept. 8-10, 1986), pp. 77-87

[4] Falkowski, B.-J. Probabilistic Perceptrons. Neural Networks, 8, 4, (1995), pp. 513-523

[5] Faloutsos, C., and Christodoulakis, S. Signature files: An access method for documents and its analytical performance evaluation. ACM Trans. Off. Inf. Syst., 2, 4, (1984), pp. 267-268

[6] Faloutsos, C., and Christodoulakis, S. Description and performance analysis of signature file methods for office filing. ACM Trans. Off. Inf, Syst., 5, 3, (1987), pp. 237-257

[7] Faloutsos, C. Signature-based text retrieval methods: A survey. In IEEE Computer Society Technical Committee on Data Engineering. Special issue on document retrieval, 13, 1, (March 1990), pp. 25-32

[8] Faloutsos, C., Barber, R., Flickner, M., Hafner, J., Niblack, W., and Petkovic, D. Efficient and Effective Querying by Image Content. J. of Intelligent Information Systems, 3, (1994), pp. 231-262

[9] Forster, O. Analysis 1. Vieweg & Sohn, Braunschweig, Wiesbaden, (1980)

[10] Gray, J.R. Probability. Oliver & Boyd, Edinburgh and London, (1967)

[11] Knuth, D.E. The Art of Computer Programming, Vol. I, Fundamental Algorithms, Reading, MA: Addison-Wesley, (1973)

[12] Knuth, D.E. The Art of Computer Programming, Vol. III Sorting and Searching, Reading MA: Addison-Wesley, (1973)

[13] Lee, D.L., and Leng, C. Partitioned signature files: Design issues and performance evaluation. ACM Trans. Inf. Syst. 7, 2, (April 1989), pp. 158-180

[14] Minsky, M.L., and Papert, S.A. Perceptrons. The MIT Press, (1988)

[15] Roberts, C.S. Partial match retrieval via the method of superimposed codes. Proc. IEEE 67, 12 (Dec. 1979), pp. 1624-1642

[16] Sacks-Davis, R., Kent, A., and Ramamohanarao, K. Multikey Access Methods Based on Superimposed Coding Techniques. ACM Transactions on Database Systems, 12, 4, (December 1987), pp. 655-696

[17] Stiassny, S. Mathematical analysis of various superimposed coding methods. American Documentation 11, 2, (Feb. 1960), pp. 155-169

[18] Zezula, P., Rabbiti, F., and Tiberio, P. Dynamic Partitioning of Signature Files. ACM Trans. Inf. Syst. 9, 4, (Oct. 1991), pp. 336-369

Locally Adaptive Fuzzy Image Enhancement

H. R. Tizhoosh, G. Krell, B. Michaelis

Otto-von-Guericke-University Magdeburg
Institute for Measurement Technology and Electronics
D-39016 Magdeburg, P.O. Box 4120, Germany
tizhoosh/krell/michaelis@ipe.et.uni-magdeburg.de

Abstract. In recent years, some researchers have applied the concept of fuzziness to develop new enhancement algorithms. The global fuzzy image enhancement methods, however, fail occasionally to achieve satisfactory results. In this work, we introduce a locally adaptive version of two existing fuzzy image enhancement algorithms to overcome this problem. Keywords: fuzzy image enhancement, minimization of fuzziness, histogram hyperbolization

1 Introduction

Since one can consider an image as an array of fuzzy singletons [5], it is possible to apply fuzzy set theory [8] to develop new enhancement techniques for image processing. The use of fuzzy techniques in the image processing is appropriate due to the uncertainties within the processing and the subjective nature of many tasks, e.g. judgment of image quality. In comparison to the classical look-up operations, fuzzy methods offer a flexible knowledge-based approach to image enhancement. In this work we briefly describe two existing fuzzy enhancement algorithms. The selection of these methods is because of the fact that they are simple for implementation and not expensive in computing. Furthermore, we introduce a fast and reliable locally adaptive version of these techniques that can be easily extended for other fuzzy enhancement methods.

2 Image enhancement: minimization of fuzziness

The idea of minimization of fuzziness as a possibility to increase the image contrast is described in [3][4]. In this approach, the gray-levels g_{mn} are fuzzified by following membership function:

$$\mu_{mn} = G(g_{mn}) = \left[1 + \frac{g_{max} - g_{mn}}{F_d} \right]^{-F_e} , \tag{1}$$

where the parameters F_e and F_d are calculated with respect to the corss-over point of the membership function. Using the intensification operator [9] we can define a transformation T_1 to modify the membership values:

$$T_1(\mu_{mn}) = \mu'_{mn} = \begin{cases} T_1'(\mu_{mn}) = 2 \cdot [\mu_{mn}]^2 & 0 \le \mu_{mn} \le 0.5 \\ T_1''(\mu_{mn}) = 1 - 2 \cdot [1 - \mu_{mn}]^2 & 0.5 \le \mu_{mn} \le 1 \end{cases} \quad (2)$$

In general, the transformation T_r is defined as successive applications of T_1 by the following recursive relation:

$$T_r(\mu_{mn}) = T_1\{T_{r-1}(\mu_{mn})\} \qquad r = 1, 2, \cdots \qquad (3)$$

The application of the intensification operator reduces the fuzziness of a fuzzy set and makes it crisper (Fig. 1). After modification of the membership values we can generate new gray-levels using inverse membership function as follows:

$$g'_{mn} = G^{-1}(\mu'_{mn}) = g_{max} - F_d\left((\mu'_{mn})^{\frac{-1}{F_e}} - 1\right) \quad . \qquad (4)$$

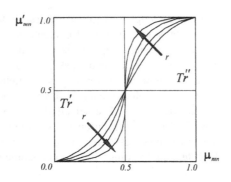

Fig.1. successive application of intensification operator. The greater r, the crisper the membership function becomes [3].

The minimization of fuzziness is also a suitable method for image thresholding. The parameter r in equation (3) allows the user to use an appropriate level of enhancement for domain-specific situations [6].

3 Image enhancement: fuzzy hyperbolization

The idea of histogram hyperbolization and fuzzy hyperbolization is described in [1] and [7], respectively. After selection of a suitable membership function and a parameter β, called the fuzzifier, the membership values are calculated for each gray-level. Here, we use a very simple membership function:

$$\mu(g_{mn}) = \frac{g_{mn} - g_{min}}{g_{max} - g_{min}} \qquad (5)$$

where g_{min} and g_{max} are the minimum and maximum gray-level in the image. A membership modification is carried out with the fuzzifier β:

$$\mu'(g_{mn}) = \left[\mu(g_{mn})\right]^{\beta}. \tag{6}$$

The new gray-levels can be generated regarding to the desired maximum gray level L as follows:

$$g'_{mn} = \left(\frac{L-1}{e^{-1}-1}\right) \bullet \left[e^{-\mu(g_{mn})^{\beta}} - 1\right] \tag{7}$$

By selection of a specific membership function the fuzzy hyperbolization can be used for segmentation and edge detection [7].

4 Locally adaptive implementation

The classical approach to locally adaptive image enhancement, e.g. for histogram equalization, defines an n x m neighborhood and moves the center of this area from pixel to pixel [2]. At each location, the histogram of the subimage is calculated to obtain a histogram equalization function. This function is finally used to map the level of the pixel centered in the neighborhood [2]. For calculation of membership values in (1) and (5) we need only minimum and maximum gray-level g_{min} and g_{max}. Therefore, we can find the minimum and maximum for some subimages and interpolate these values to obtain corresponding values for each pixel (Fig. 2).

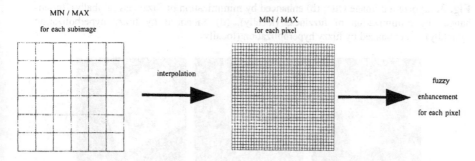

Fig. 2. Computing of parameter of membership functions through interpolation of results in subimages.

These will not be the exact values really existing for each pixel, but since we are using the concept of fuzziness we do not need precise data. This kind of local enhancement is very fast compared with classical approaches. The disadvantage of computing with minimum and maximum gray-levels, however, is that outliers could falsify the result of membership generation. This can be prevented either by selection of sufficiently great subimages or, more efficient, by histogram calculation in each subimage.

Now, we have for all pixels of the image a corresponding minimum and maximum. The enhancement can be applied to each pixel. Fig. 3 and fig.4 show some results of global and locally adaptive enhancement. The intensification operator is applied only once to the membership values. For fuzzy hyperbolization the fuzzifier is set to 0.75 that means a slight brightening of the result image. For both methods in local version we have used 30 × 30 windows to find minimum and maximum intensities.

Fig. 3. (a) original image Otto, (b) enhanced by minimization of fuzziness (globally), (c) enhanced by minimization of fuzziness (locally), (d) enhanced by fuzzy hyperbolization (globally), (e) enhanced by fuzzy hyperbolization (locally).

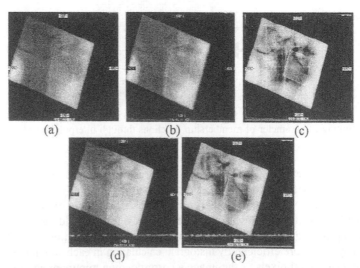

Fig. 4. (a) original megavoltage image , (b) enhanced by minimization of fuzziness (globally), (c) enhanced by minimization of fuzziness (locally), (d) enhanced by fuzzy hyperbolization (globally), (e) enhanced by fuzzy hyperbolization (locally).

5 Conclusion

Fuzzy approaches to image enhancement provide us with a powerful mathematical framework to develop improvement techniques. For some cases, the global fuzzy techniques fail to achieve satisfactory results. In our investigation, we have developed a fast and reliable locally adaptive method which allow us to apply the concept of fuzziness to contrast enhancement problems in the image processing.

References

1. Banks, S., Signal Processing, Image Processing and Pattern Recognition, Prentice Hall International, 1990.
2. Gonzalez, R. C., Woods, R. E., Digital Image Processing, Addison-Wesley Publishing Group, second edition, 1992.
3. King, R.A., Pal, S. K., Image Enhancement Using Smoothing with Fuzzy Sets, IEEE Transactions on Systems, Man and Cybernetics, vol. SMC-11, no. 7, pp. 494-501, July 1981.
4. King, R. A., Pal, S. K., On edge detection of X-ray images using fuzzy sets, IEEE Transactions on Pattern Analysis and Machine Intelligence, Vol. PAMI-5, No. 1, pp. 69-77, 1983.
5. Pal, S. K., Fuzziness, Image Information and Scene Analysis. In: Yager, R., Zadeh, L. A.: An Introduction to Fuzzy Logic Applications in Intelligent Systems, pp. 147 - 183, Kluwer Academic Publishers, 1992.
6. Ross, T. J., Fuzzy Logic with Engineering Applications, McGraw-Hill , USA, 1995.
7. Tizhoosh, H. R., Fochem, M., Image Enhancement with Fuzzy Histogram Hyperbolization, proceedings of EUFIT'95, vol. 3, pp. 1695 - 1698, Aachen, Germany, 1995.
8. Zadeh, L. A., Fuzzy Sets, Information and Control 8, p. 338 - 353, 1965.
9. Zadeh, L. A., A Fuzzy-Set-Theoretic Interpretation of Linguistic Hedges, J. Cybern., vol. 2, no. 2, pp. 4-34, 1972.

Fuzzy Modeling of Conceptual Spaces

Silvescu Adrian

Faculty of Mathematics, Computer Science Departament,
Bucharest University

Abstract. It is introduced a new concept: H-sets which generalises the notion of fuzzy sets and is based on the concept of hemilattice. Then, we extend the attribute-value conceptual spaces over hemilattices, which gives us a kind of H-set, and prove some of their properties .

1 Introduction

Our goal is to define a structure for a space of real concepts R which should provide us a way to test some order relations such as subsumption and to operate generalisation and unification. Given two real concepts **a** and **b**, we denote **a** $\leq$ **b** and we read "**b** subsumes **a**" or "**a** is subsumed by **b**" if **a** is a more particular concept than **b** or, equivalently, **b** is a more general concept than **a**. By generalisation of two real concepts we denote the most particular concept more general than both of them; by unification we denote the most general concept more particular than the given concepts.

2 Hemilattices and Hemilatticeal Sets

In this section we will introduce a new concept which generalises the fuzzy sets. This is the notion of a set over a hemilattice. First we will present some preliminary definitions.

Definition 1. We say that $(H, \vee, \wedge)$ is a hemilattice iff

- $(H, \leq)$ is a partially ordered set
- $\forall a,b \in H \Rightarrow a \geq a \wedge b$ and $b \geq a \wedge b$
- $\forall a,b \in H \Rightarrow a \leq a \vee b$ and $b \leq a \vee b$.

Remark. Every lattice is a hemilattice.

Proposition 2. *If $(H, \vee, \wedge, 0, 1, \leq)$ is a hemilattice with 0 first element and 1 last element then 0 is idempotent relatively to $\wedge$ and 1 is idempotent relatively to $\vee$.*

Proof. Let $h \in H$, then $0 \wedge h \leq 0 \Rightarrow 0 \wedge h = 0$. Let $h \in H$, then $1 \vee h \geq 1 \Rightarrow 1 \vee h = 1$. $\qquad\qquad\square$

Definition 3. Let $(H, \vee, \wedge)$ be a hemilattice, we say that it has the:

- c-property iff $\vee$ and $\wedge$ are commutative operations;
- a-property iff $\vee$ and $\wedge$ are associative operations;
- d-property iff $\vee$ and $\wedge$ are distributive operations each over the other;
- m-property iff $\forall x, y, z \in H$, $x \leq z \Rightarrow x \vee (y \wedge z) = (x \vee y) \wedge z$;
- 01-property iff 0 is neutral element for $\vee$ and 1 is neutral element for $\wedge$.

Definition 4. Let $(H, \vee, \wedge, 0, 1, \leq)$ be a hemilattice with 0 as first element and 1 as last element and let X be an ordinary set. We define a hemilatticeal set A as a function $f_A : X \to H$.

Notation: We denote the set of H-sets of X as

$$\mathcal{F}_X^H = \{f_A | f_A : X \to H\}$$

We denote the set of characterisic functions of ordinary sets as

$$\Xi_X = \{\chi_A | \chi_A : X \to \{0, 1\}\}$$

Remark. Obviously $\Xi_X \subseteq \mathcal{F}_X^H$, which means that the ordinary sets are a particular case of H-sets of X.

Definition 5. We introduce two operations on hemilatticeal sets:

- Intersection: $(f_A \cap f_B)(x) = f_A(x) \wedge f_B(x)$ $\forall x \in X$;
- Union: $(f_A \cup f_B)(x) = f_A(x) \vee f_B(x)$ $\forall x \in X$;

Remark. The sets $\emptyset$ and X could be described as follows:

- $\emptyset$: $f_\emptyset(x) = 0, \forall x \in X$;
- X: $f_X(x) = 1, \forall x \in X$.

Definition 6. The set of hemilatticeal sets can be ordered as follows:

$$f_A \leq f_B \Leftrightarrow \forall x \in X \ f_A(x) \leq f_B(x)$$

Proposition 7. If $(H, \vee, \wedge, 0, 1, \leq)$ is a hemilattice with 0 as first element and 1 as last element then $(\mathcal{F}_X^H, \cup, \cap, f_\emptyset, f_X, \leq)$ is a hemilattice with $f_\emptyset$ as first element and f_X as last element.

Proof. Obviously $(\mathcal{F}, \subseteq)$ is partially ordered.
Let $f_A, f_B \in \mathcal{F}_X^H$; let $x \in X$ then
$(f_A \cap f_B)(x) = f_A(x) \wedge f_B(x) \leq f_A(x), \forall x \in X \Rightarrow f_A \cap f_B \leq f_A$ and
$(f_A \cap f_B)(x) = f_A(x) \wedge f_B(x) \leq f_B(x), \forall x \in X \Rightarrow f_A \cap f_B \leq f_B$.
Let $f_A, f_B \in \mathcal{F}_X^H$; let $x \in X$; then
$(f_A \cup f_B)(x) = f_A(x) \vee f_B(x) \geq f_A(x), \forall x \in X \Rightarrow f_A \cup f_B \geq f_A$ and
$(f_A \cup f_B)(x) = f_A(x) \vee f_B(x) \geq f_B(x), \forall x \in X \Rightarrow f_A \cup f_B \geq f_B$.
Let $f_A \in \mathcal{F}_X^H$; let $x \in X$ then $f_\emptyset(x) = 0 \leq f_A(x) \Rightarrow f_\emptyset \leq f_A$.
Let $f_A \in \mathcal{F}_X^H$; let $x \in X$; then $f_X(x) = 1 \geq f_A(x) \Rightarrow f_X \geq f_A$ $\qquad \square$

Proposition 8. *If $(H, \vee, \wedge, 0, 1, \leq)$ is a lattice with 0 as first element and 1 as last element then $(\mathcal{F}_X^H, \cup, \cap, f_\emptyset, f_X, \leq)$ is a lattice with $f_\emptyset$ as first element and f_X as last element.*

Proof. Let $f_C \in \mathcal{F}_X^H$ so that $f_C \leq f_A$, $f_C \leq f_B$ and $f_C \geq f_A \cap f_B$ then

$$f_C \leq f_A \;\Rightarrow\; \forall x \in X \;\; f_C(x) \leq f_A(x)$$

$$f_C \leq f_B \;\Rightarrow\; \forall x \in X \;\; f_C(x) \leq f_B(x)$$

$$f_C \geq f_A \cap f_B \;\Rightarrow\; \forall x \in X \;\; f_C(x) \geq f_A(x) \wedge f_B(x)$$

From the precendind three statements and from the fact that H is a lattice it follows that $\forall x \in X \;\; f_C(x) = f_A(x) \wedge f_B(x)$ which implies that $f_C = f_A \cap f_B$. Let $f_C \in \mathcal{F}_X^H$ so that $f_C \geq f_A$, $f_C \geq f_B$ and $f_C \leq f_A \cup f_B$ then

$$f_C \geq f_A \;\Rightarrow\; \forall x \in X \;\; f_C(x) \geq f_A(x)$$

$$f_C \geq f_B \;\Rightarrow\; \forall x \in X \;\; f_C(x) \geq f_B(x)$$

$$f_C \leq f_A \cup f_B \;\Rightarrow\; \forall x \in X \;\; f_C(x) \leq f_A(x) \vee f_B(x)$$

From the precendind three statements and from the fact that H is a lattice it follows that $\forall x \in X \;\; f_C(x) = f_A(x) \vee f_B(x)$ which implies that $f_C = f_A \cup f_B$ □

Remark. If $(H, \vee, \wedge, 0, 1, \leq)$ is a (hemi)lattice with 0 as first element and 1 as last element has any of a, c, d, m, 01-properties then $(\mathcal{F}_X^H, \cup, \cap, f_\emptyset, f_X, \leq)$ is a (hemi)lattice with a, c, d, m, 01-property, respectively. Proof is omited because it is obvious.

Examples of Hemilattices:

0. The *singular hemilattice* $(\{\#\}, \mathbf{max}, \mathbf{min}, 0, 1, \leq)$ where $\mathbf{max}$, $\mathbf{min}$, 0, 1, and $\leq$ have the usual meaning is obviously a hemilattice, because it is a lattice.

1. The *boolean hemilattice* $(\{0, 1\}, \mathbf{max}, \mathbf{min}, 0, 1, \leq)$ where $\mathbf{max}$, $\mathbf{min}$, 0, 1, and $\leq$ have the usual meaning is obviously a hemilattice, because it is a lattice.

2. The *probability hemilattice:* $([0, 1], \oplus, \cdot, 0, 1, \leq)$; $\cdot$ is the usual multiplication, $\leq$ is the usual order relation and $\oplus$ is defined by

$$a \oplus b = a + b - ab$$

Proposition 9. $\oplus$ *is well defined.*

Proof. $\oplus$ is well defined iff $0 \leq a \oplus b \leq 1$.
$0 \leq a \oplus b \Leftrightarrow 0 \leq a + b - ab \Leftrightarrow 0 \leq a + b(1 - a)$ which is true, because $a \geq 0$, $b \geq 0$ and $a \leq 1 \Rightarrow (1 - a) \geq 0$.
$a \oplus b \leq 1 \Leftrightarrow a + b - ab \leq 1 \Leftrightarrow 0 \leq 1 - a - b + ab \Leftrightarrow 0 \leq (1 - a)(1 - b)$ which is true, because $a \leq 1 \Rightarrow (1 - a) \geq 0$ and $b \leq 1 \Rightarrow (1 - b) \geq 0$. □

Proposition 10. $([0,1], \oplus, \cdot, 0, 1, \leq)$ *is a hemilattice.*

Proof. $([0,1], \oplus, \cdot, 0, 1, \leq)$ is a hemilattice iff $([0,1], \leq)$ is partially ordered, $ab \leq a$, $ab \leq b$, $a \oplus b \geq a$ and $a \oplus b \geq b$.

$([0,1], \leq)$ is obviously partially ordered because it is totally ordered.

$b \leq 1$ and $a \geq 0 \Rightarrow ab \leq a$;

$a \leq 1$ and $b \geq 0 \Rightarrow ab \leq b$;

$a \oplus b \geq a \Leftrightarrow a + b - ab \geq a \Leftrightarrow b - ab \geq 0 \Leftrightarrow b(1-a) \geq 0$, which is true;

$a \oplus b \geq b \Leftrightarrow a + b - ab \geq b \Leftrightarrow a - ab \geq 0 \Leftrightarrow a(1-b) \geq 0$, which is true. $\square$

Remark. The probability hemilattice is not a lattice, because $0.5 \cdot 0.5 = 0.25 \neq 0.5$; $0.5 \oplus 0.5 = 0.75 \neq 0.5$

3. The *fuzzy hemilattice* $([0,1], \mathbf{max}, \mathbf{min}, 0, 1, \leq)$ is obviously a hemilattice, because it is a lattice.

4. The *trust intervals hemilattice* $(\mathcal{I}_{[0,1]}, \overline{\cup}, \cap, \subseteq)$ where $\mathcal{I}_{[0,1]} = \{[a,b] | 0 \leq a \leq b \leq 1\}$, $\mathbf{max}$, $\mathbf{min}$ and $\cap$ have the usual meaning and $\overline{\cup}$ is defined by $[a_1, b_1] \overline{\cup} [a_2, b_2] = [\mathbf{min}\{a_1, a_2\}, \mathbf{max}\{b_1, b_2\}]$ is obviously a lattice, thus it is a hemilattice.

3 Conceptual Spaces over Hemilattices

In this section we will introduce the notion of conceptual spaces over hemilattices by naturally generalising the bare conceptual spaces, which are a particular instance of conceptual spaces over hemilattices, i. e. they are conceptual spaces over the boolean (hemi)lattice. The main result of this section is Theorem 8 stating that the conceptual space over a hemilattice is a hemilattice that preserves a, c and 01-properties and does not preserve d and m-properties of the initial hemilattice.

Definition 1. An attribute-value conceptual space is a structure

$$\mathcal{C} = (A, \{V_a\}_{a \in A})$$

where A is a set of attributes and for every attribute $a \in A$, $V_a \neq \emptyset$ is a set of atomic values that attribute a can take. An element $m \in \mathcal{C}$ is a set

$$m = \{(a,v) | a \in A, v \in V_a\}$$

Example : A={size,colour} V_{size}={small,medium,big} V_{colour}={red,green, yellow,brown} the concept of apple would be represented as apple ={(size,medium), (colour,red),(colour,yellow),(colour,green)} and the concept of red-apple as red-apple={(size,medium),(colour,red)}.

Definition 2. Let $(H, \vee, \wedge, 0, 1, \leq)$ be a hemilattice with first and last element. We define the H-conceptual space over the hemilattice H as

$$\mathcal{C}^H = (A, \{V_a\}_{a \in A}, H)$$

An element of the H-conceptual space $\mathcal{C}^H$ is a set

$$m = \{(a, v, h) | a \in A, v \in V_a, h \in H\}$$

such that $\forall (a, v, h), (a', v', h') \in m, a = a', v = v' \Rightarrow h = h'$.

Definition 3. We introduce $\mathrm{pr}_a : \mathcal{C}^H \to \mathcal{F}_{V_a}^H$,

$$\mathrm{pr}_a(m)(v) = h \Leftrightarrow (a, v, h) \in m$$

called the projection on attribute a.

Definition 4. We introduce $\mathrm{pr}_{a,v} : \mathcal{C}^H \to H$,

$$\mathrm{pr}_{a,v}(m) = h \Leftrightarrow (a, v, h) \in m$$

called the projection on attribute-value a, v.

Remark. $\forall m \in \mathcal{C}^H \; m = \bigcup_{a \in A} \bigcup_{v \in V_a} \{a\} \times \{v\} \times \mathrm{pr}_{a,v}(m) = \bigcup_{a \in A} \times (V_a, pr_a(V_a))$.

Remark. We observe that

$$\mathcal{C}^H \simeq \mathcal{F}^H_{\bigcup_{a \in A} \{a\} \times V_a}$$

and the following structure: $(\mathcal{C}^H, \cup, \cap, f_\emptyset, f_{\bigcup_{a \in A} \{a\} \times V_a}, \subseteq)$ is a (hemi)lattice with first and last element if H is and will preserve the eventual properties of H.

Notation: We denote:

- the void concept: $\emptyset = f_\emptyset$;
- the most general concept: $\square = f_{\bigcup_{a \in A} \{a\} \times V_a}$

Definition 5. We define $\Psi : \mathcal{C}^H \to \mathcal{C}^H$ as follows

$$\Psi(\alpha) = \begin{cases} \alpha \text{ if } \forall a \in A \; pr_a(\alpha) \neq f_\emptyset^a \\ f_\emptyset \text{ if } \exists a \in A \; pr_a(\alpha) = f_\emptyset^a \end{cases}$$

Notation: We will denote $\mathcal{C}_\emptyset^H = Im\Psi(\mathcal{C}^H) \subseteq \mathcal{C}^H$.

Definition 6. We introduce two operations on $\mathcal{C}_\emptyset^H$:

- Unification: $\alpha \mathcal{U} \beta = \Psi(\alpha \cap \beta) \quad \forall \alpha, \beta \in \mathcal{C}_\emptyset^H$
- Generalization: $\alpha \mathcal{G} \beta = \alpha \cup \beta \quad \forall \alpha, \beta \in \mathcal{C}_\emptyset^H$

Proposition 7. *If H is a hemilattice with 01-property then :*

- $\Box \mathcal{U} \alpha = \alpha \mathcal{U} \Box = \alpha \quad \forall \alpha \in C^H_\theta$
- $\oslash \mathcal{U} \alpha = \alpha \mathcal{U} \oslash = \oslash \quad \forall \alpha \in C^H_\theta$
- $\Box \mathcal{G} \alpha = \alpha \mathcal{G} \Box = \Box \quad \forall \alpha \in C^H_\theta$
- $\oslash \mathcal{G} \alpha = \alpha \mathcal{G} \oslash = \alpha \quad \forall \alpha \in C^H_\theta$

Proof.

$$\Box \mathcal{U} \alpha = \Psi(\Box \cap \alpha) = \Psi(\alpha) = \alpha$$

$$\oslash \mathcal{U} \alpha = \Psi(\alpha \cap \oslash) = \Psi(\oslash) = \oslash$$

$$\Box \mathcal{G} \alpha = \alpha \cup \Box = \Box$$

$$\oslash \mathcal{G} \alpha = \alpha \cup \oslash = \alpha$$

$\Box$

Theorem 8. *If $(H, \vee, \wedge, 0, 1, \leq)$ is a (hemi)latice with 0 first element and 1 last element then*

$$(C^H_\theta, \mathcal{G}, \mathcal{U}, \oslash, \Box, \leq)$$

is a (hemi)lattice with $\oslash$ first and $\Box$ last element.

Proof. Obviously $\oslash$ and $\Box$ are first respectively last element. Obviously $\mathcal{G}$ has the requested properties if H is a hemilattice and is a supremum if H is lattice. All we have to prove now is about $\mathcal{U}$.
Let $\alpha, \beta \in C^H_\theta$ then $\alpha \mathcal{U} \beta = \Psi(\alpha \cap \beta) =$

$$\begin{cases} \alpha \cap \beta \ if \ \ \forall a \in A \ pr_a(\alpha \cap \beta) \neq f^a_\theta \\ \oslash \quad otherwise \end{cases}$$

which is obviously enclosed in α and β.
For the case when H is lattice let $\alpha, \beta, \gamma \in C^H_\theta$ so that $\gamma \leq \alpha$, $\gamma \leq \beta$ and $\gamma \geq \alpha \mathcal{U} \beta$ then from the first two results that $\gamma \subseteq \alpha \cap \beta$. we have two cases:
a) $\forall \ a \in A; \ pr_a(\alpha \cap \beta) \neq f^a_\theta$ then $\alpha \mathcal{U} \beta = \alpha \cap \beta$ and results $\gamma \supseteq \alpha \cap \beta$ q.e.d.
b) $\exists \ a \in A; \ pr_a(\alpha \cap \beta) = f^a_\theta$ then $\alpha \mathcal{U} \beta = f_\theta = \oslash$ but $\gamma \subseteq \alpha \cap \beta \Rightarrow \exists \ a \in A \ pr_a(\gamma) = f^a_\theta$ but $\gamma \in C^a_\theta \Rightarrow \gamma = f_\theta = \oslash$ q.e.d. $\Box$

Proposition 9. *If H is as above and has c, a and 01-properties then C^H_θ has c, a and 01-properties.*

Proof. The a and c-properties of generalisation results from the the same properties of $\cup$. The 01-property results from 7. All we have to prove now is the c and a-property of unification.
c-property: $\alpha \mathcal{U} \beta = \Psi(\alpha \cap \beta) = \Psi(\beta \cap \alpha) = \Psi(\beta \mathcal{U} \alpha)$.
a-property: $(\alpha \mathcal{U}(\beta \mathcal{U} \gamma)) = (\alpha \mathcal{U} \Psi(\beta \cap \gamma)) = \Psi(\alpha \cap \Psi(\beta \cap \gamma)) =$

$$\begin{cases} \Psi(\alpha \cap (\beta \cap \gamma)) \ if \ \ \forall a \in A \ pr_a(\beta \cap \gamma) \neq f^a_\theta \\ \oslash \quad\quad\quad\quad otherwise \end{cases}$$

$$\begin{cases} \alpha \cap \beta \cap \gamma \; if \;\; \forall a \in A \; pr_a(\beta \cap \gamma) \neq f_\emptyset^a \;\; and \;\; pr_a(\alpha \cap \beta \cap \gamma) \neq f_\emptyset^a \\ \oslash \qquad otherwise \end{cases}$$

But $pr_a(\alpha \cap \beta \cap \gamma) \neq f_\emptyset^a \Rightarrow pr_a(\beta \cap \gamma) \neq f_\emptyset^a$

$$\begin{cases} \alpha \cap \beta \cap \gamma \; if \;\; \forall a \in A \; pr_a(\alpha \cap \beta \cap \gamma) \neq f_\emptyset^a \\ \oslash \qquad otherwise \end{cases}$$

In an analoguous way we get the right hand formula. □

Remark. If H is as above and has d and m-properties then $C_\emptyset^H$ does not have d and m-properties.

Counterexample

Let H be the boolean hemilatice. In the following counterexamples we will not emphasize the 3-uples with 0 on the third position and we will replace a 3-uples of the form (a,v,1) with (a,v) obtaining a conceptual space isomorph with the boolean one but wich facilitate the calculus.

1. Distributivity $C^H = (\{1,2\},\{\{a,b,c,d\},\{a,b,c,d\}\},H)$
a) $\mathcal{G}$ over $\mathcal{U}$

$$\alpha = \{(1,c),(2,d)\}, \beta = \{(1,a),(2,a)\}, \gamma = \{(1,a),(2,b)\}$$

$$\alpha\mathcal{G}(\beta\mathcal{U}\gamma) = (1,c),(2,d)$$

$$(\alpha\mathcal{G}\beta)\mathcal{U}(\alpha\mathcal{G}\gamma) = (1,c),(1,a),(2,d) \neq (1,c),(2,d)$$

b) $\mathcal{U}$ over $\mathcal{G}$

$$\alpha = \{(1,c),(2,d)\}, \beta = \{(1,a),(2,d)\}, \gamma = \{(1,c),(2,b)\}$$

$$\alpha\mathcal{U}(\beta\mathcal{G}\gamma) = (1,c),(2,d)$$

$$(\alpha\mathcal{U}\beta)\mathcal{G}(\alpha\mathcal{U}\gamma) = \emptyset \neq (1,c),(2,d)$$

2. Modularity $C^H = (\{1,2\},\{\{a,b,c\},\{a,b,c\}\},H)$

$$\alpha = \{(1,b),(2,c)\}, \beta = \{(1,a),(2,b)\}, \gamma = \{(1,b),(2,b),(2,c)\}$$

$$\alpha \leq \gamma$$

$$\alpha\mathcal{G}(\beta\mathcal{U}\gamma) = (1,b),(2,c)$$

$$(\alpha\mathcal{G}\beta)\mathcal{U}\gamma = (1,b),(2,b),(2,c)$$

Definition 10. Let $\mathcal{R}$ be a space of real concepts. We say that C_H is an attribute-value decomposition of $\mathcal{R}$ if there is an injective moprphism $\Phi \; \mathcal{R} \rightarrow C_H$ which perserves the order and operations.

4 Further Extensions

The H-conceptual spaces described above could be extended by allowing the attributes to take composed values, meaning values that could be themself concepts. With this improvement the new concepts could be represented like trees, dags, or graphs. But this is not a problem because we can represent any graph as the list of his edges.

Being more explicit if we have a graph with weighted edges (as a general concept wold look like) and the edge between nodes n_i and n_j has the weight $w_{i,j}$ then we would have an attribute in the concept named $n_i n_j$ with the value $w_{i,j}$ so we would be able to use Theorem 8.

5 Concluding Remarks

By introducing this new concepts of hemilattice and conceptual spaces over hemilattices we are now able to treat problems that uses conceptual spaces such as the learnig algorithms ID3,C4.5 or Tensor product varible binding in a new fuzzyfied way and to improve their resemblance with real spaces of concepts.

References

[1] Birkoff G., Lattice Theory , 1973.
[2] Gazdar Gerald, Mellish Chris, Natural Language Processing in LISP , 1989.
[3] Mellish Chris, Term Encodable description spaces, Logic Programing:New Frontiers Ed. D.Borough , 1992.
[4] Smolensky Peter, Tensor Product Variable Binding and reprezentation of symbolic structures in Connectionist Systems, Artificial Intelligence, 46 , 1990.
[5] Vaida Dragos, Algebric Models in Syntax,Semantics and Paralelism Graduate Course, University of Bucharest, 1994.
[6] Yager Ronald,Dimitar Filev, Essential of Fuzzy Modelling and Control , 1992.
[7] Zadeh Ltofi, Fuzzy sets, Information and Control, 8 , 1965.

Construction of a Fuzzy Model with Few Examples

Farida Benmakrouha

Département d'Informatique
INSA de Rennes
FRANCE
benma@irisa.fr

Abstract. This article suggests an automatic initialization of a fuzzy rule basis from available numerical data, as well as two methods of extension allowing the incremental updating of the rule basis.

1 Introduction

Fuzzy systems are universal approximators [1] used for identification and control. They are particularly useful when classical models are ineffective i.e. when processes are nonlinear or too complex because of the number of parameters involved. Unfortunately, the parameter identification - consequents and membership functions - is difficult when data are insufficient. In fact, in industrial processes, specially real-time processes, off-line experiments are limited. In on-line experiments, some rules are never learned. This lack of data is pointed out by Sugeno and Yasukawa in [2] and by M Lutaud-Brunet in [3]. In [4], Kóczy and Kovács evoke another problem of fuzzy modeling, namely a large combinational which increase computational time. They suggest use of sparse rule bases where membership functions don't cover the whole input space and propose the interpolation method to cover all cases. In this paper, we are interested in automatic extraction of knowledge from numerical data in fuzzy systems and in incremental construction of rule basis. We present two methods for automatic initialization of consequents which make it possible to learn all the rules and so greatly improve the performance of fuzzy systems. After this presentation, we emphasize the impact of this initialization on output precision by using two examples previously studied by many authors. At last, we discuss stability of the methods with different sampling.

2 Initialization algorithm

2.1 Fuzzy Model (FIS)

The object of this study is the initialization of a fuzzy rule basis from available numerical data. These data, supposed *a priori* scarce, are used by extension. With each additional piece of information, it is possible to update the rule basis. These are of the form :

$$\textbf{If } x_1 \text{ is } A_{1i} \textbf{ and } x_2 \text{ is } A_{2i} ... \textbf{and } x_p \text{ is } A_{pi}$$
$$\textbf{then } y \text{ is } C_i \tag{1}$$

Furthermore, we take following non-restrictive hypotheses :

- A_{ji} are fuzzy subsets making a *strong fuzzy partition* of input space. We have

$$(\forall x) \sum \mu_A(x) = 1.$$

- C_i are crisp values representing the modal points of output fuzzy sets. The fuzzy systems considered are therefore a simplified form of Tagaki-Sugeno model [5].

The inferred output is :

$$y(x) = \frac{\sum_{i=1}^{N} \alpha_i(x) \times C_i}{\sum_{j=1}^{N} \alpha_j(x)} \tag{2}$$

where $\alpha_i(x)$ is the truth value of the rule i. We suppose that membership functions are set *a priori*. Learning then consists in modifying consequents. We suppose that there exists a learning set $\mathcal{L} = \{(\mathbf{x}_i, d_i)\}$, where $\mathbf{x}_i$ is an input vector and d_i, the corresponding output. The gradient method consists in modifying C_i at each presentation of examples from the error $(y(\mathbf{x}_i) - d_i)$:

$$\Delta C_i = -\epsilon(y(x) - d_i) \times \frac{\alpha_i(x)}{\sum_{i=1}^{N} \alpha_i(x)} \tag{3}$$

2.2 Initialization algorithms

The purpose of these methods is to initialize by approximative values consequents $C = (C_1, C_2, ..., C_N)$ where N is number of rules. The strong fuzzy partitions on input space define a grid G formed by modal points of fuzzy subsets. In these points, noted $(u^i)_{i=1}^{N}$, one and only one rule applies and we have : $SIF(u^i) = C_i$. An initialization algorithm proposed [6] consists in searching among learning points those closer to grid points, to infer a value of C. The algorithm is :

> **for** each rule i **do**
>   search the couple $(\mathbf{x}^*, d^*)$ of $\mathcal{L}$ such as :
> $\alpha_i(\mathbf{x}^*) = Max_{\mathbf{x} \in \mathcal{L}} \alpha_i(\mathbf{x})$
> $C_i = d^*$
> **done**

But the efficiency of this method depends on the distribution of learning points. With a bad repartition, several consequents C_i can be indefinite and output value given by (2) is too imprecise.

The extension methods that we suggest give an initial value to indefinite consequents, their value deduced by interpolation of neighbouring points. We will show that these methods are interesting when we have a reduced number of examples for they extend locally a limited knowledge.

More formally, if G is the grid defined above, S a subset of G, the problem is to find a function $g' : G \to R$ with $g'|_S = g$ where $g : S \to R$, $S \sqsubseteq G$, $g(k) = C(k)$ initialized by the above algorithm. We note $V_i =$ the neighbouring points $\in G$ of the rule number $i, i \in G$.

Two methods are implemented to determine this function g'.

2.3 Linear method

We define a series of sets $S_i \sqsubseteq G$ and functions $g_i : S_i \to R$:

$$S_0 = S \tag{4}$$

$$g_0 = g \tag{5}$$

$$S_{i+1} = S_i \cup (\cup_{k \in S_i} V_k) \tag{6}$$

$$g_{i+1}(k) = g_i(k), k \in S_i \tag{7}$$

$$g_{i+1}(k) = average_{j \in V_k \cap S_i} g_i(j), k \notin S_i \tag{8}$$

The series S_i is increasing, bounded and stationary.

It stops when $S_i = G$. We take $g' = g_i$ and we have $g'|_S = g$. We note $Lin(g)$ this function g'.

We have if g, g_1 and g_2 are interpolated functions on S :

$$\sup_{x \in G} |Lin(g_1)(x) - Lin(g_2)(x)|$$
$$\leq \sup_{x \in S} |g_1(x) - g_2(x)| \tag{9}$$

and $\forall x$:

$$\inf_{s \in S} g(s) \leq Lin(g)(x) \leq \sup_{s \in S} g(s) \tag{10}$$

2.4 Harmonious method

This method is proposed by Le Gruyer et Archer in [7]. It is interesting because it has some properties such as stability. This point will be discussed further. We note $VS_i = i$, if $i \in S$ and $VS_i = V_i$ otherwise.

We define a series of functions $g_i : G \to R$:

$$g_0(k) = g(k), k \in S \qquad (11)$$

$$g_0(k) = 0, k \notin S \qquad (12)$$

$$g_{i+1}(k) = \tfrac{1}{2} \sup_{j \in VS_k} g_i(j) + \tfrac{1}{2} \inf_{j \in VS_k} g_i(j),$$
$$k \in G \qquad (13)$$

The authors show that the series converges when i tends towards ∞. We stop when $\sup_{k \in G} |g_{i+1}(k) - g_i(k)| < \epsilon_1$, for an ϵ_1 and $g' = g_i$. We note $Harm(g)$ this function g'. It is shown that if g, g_1 and g_2 are interpolated functions on S

$$\sup_{x \in G} |Harm(g_1)(x) - Harm(g_2)(x)|$$
$$= \sup_{x \in S} |g_1(x) - g_2(x)| \qquad (14)$$

and

$$Lip(Harm)(g) = Lip(g) \qquad (15)$$

where Lip(g) is the Lipschitz constant

$$Lip(g) = \sup_{i \neq j \in S} \frac{|g(i) - g(j)|}{d(i,j)} \qquad (16)$$

where d(i,j) is the minimum number of immediate neighbours between i and j
If S_1 and $S_2 \subseteq S$, we also have:

$$\sup_{x \in G} |Harm(g/S_1)(x) - Harm(g/S_2)(x)|$$
$$\leq 4 * Lip(g) * \delta(S_1, S_2) \qquad (17)$$

where

- $\delta(S_1, S_2) = \sup(d1, d2)$
- $d1 = \sup_{i \in S_1} \inf_{j \in S_2} d(i,j)$
- $d2 = \sup_{j \in S_2} \inf_{i \in S_1} d(i,j)$

We also have $\forall x$:

$$\inf_{s \in S} g(s) \leq Harm(g)(x) \leq \sup_{s \in S} g(s) \qquad (18)$$

3 Experimentations

We performed two series of measures.

3.1 First Series

We discuss our methods using the well-known example, proposed by Box et Jenkins [9] giving CO_2 concentration according to gas flow rate in a boiler.
Data file is constitued by 296 couples $(u(t), y(t))$ where $u(t)$ is the gas flow rate and $y(t)$ the CO_2 concentration.
The chosen model is $y(t) = f(y(t-1), u(t-3))$. We compared the results of a fuzzy system (1) with two different initializations, one without interpolation, the other with interpolation, using the two methods. Initialization was done with a variable number n of examples and 25 membership functions. The mean square error, performance index of system, is calculated on the totality of measures (292 points).

The error given below is the mean square error before learning.

n	without interpolation	method 1	method 2
1	1595.81	10.370	10.370
5	1212.81	9.13	10.208
30	2.54	0.97	0.98
80	1.27	0.62	0.64
100	0.46	0.46	0.46

Table 1. First Series .

3.2 Second series

We took the regression function proposed in [8] :

$$g(x_1, x_2) = 1.9(1.35 + e^{x_1} \sin(13(x_1 - .6)^2)$$
$$e^{-x_2} \sin(7x_2)) + 4. \tag{19}$$

Our model uses 7 membership functions by input.

We trained on n points with a random choice and predicted on 296 points.

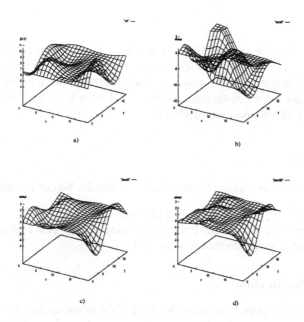

Fig. 1. Error FIS(x,y) - g(x,y).

We visualised the function g(x,y) on Fig a. On Fig b,c,d, we show the error
obtained FIS(x,y) - g(x,y) for :

- n=5, without interpolation (fig b)
- n=5 with interpolation (fig c)
- n=50 with interpolation and at the end of 10 iterations (fig d)

We note in Fig c that the error is important at the edge of the grid, due to a
small number of points.

The mean square error is before learning :

n	without interpolation	with interpolation
5	27.07	0.63
50	3.76	0.45
100	0.18	0.18

Table 2. Second Series.

Remarks :

- Interpolation gives good results particularly when the number of indefinite consequents is important, which happens statistically more often when we have few examples. For the case n = 100, interpolation may not be the best method since most consequents have a value after initialization.
- The two methods give similar results.

4 Stability

In this section, we are interested in stability of results before learning.

- in time, by comparing results with different sampling.
- in space, by comparing immediate neighboring values. This comparison means only if approximate function is continous.

4.1   Stability in time

Let E_1 and E_2 be different sampling leading, after initialization, to the determination of two sets S_1 and S_2, of two functions g_1 and g_2, then after interpolation of two sets of consequents C and C'. A and A' are the membership functions. Let be :

- $y(x)$, output of the fuzzy system = (A,C)
- $y'(x)$, output of the fuzzy system = (A,C')

We have $\forall x$:

$$
\begin{aligned}
|y(x) - y'(x)| &= \\
| \sum_{i=1}^{N} \alpha_i(x) \times C_i - \sum_{i=1}^{N} \alpha_i(x) &\times C_i'| \\
&\leq \sum_{i=1}^{N} \alpha_i(x) \times |C_i - C_i'| \\
&\leq \sup_{i=1}^{N} |C_i - C_i'| \\
\leq \sup(\sup_{k_1 \in S_1} g_1(k_1), \sup_{k_2 \in S_2} &\, g_2(k_2)) \\
= \sup_{k_1 \in S_1, k_2 \in S_2} |g_1(k_1) &- g_2(k_2)|.
\end{aligned} \tag{20}
$$

In the two methods, interpolation does not emphasize gaps due to different sampling. But, this overestimation, although interesting, does not take into account sampling distribution. According to (17), only the second method garantees stability, when sampling is slightly modified. The first method may introduce important gaps, as we show in the example below.

Let f(x,y) = 2.4 x -2.4 y +6 a function to approximate by a fuzzy system (2), $0 \leq x,y \leq 10$ We take 5 membership functions by input, by splitting interval [0,10] into equal parts.

Let

- $S_1 = \{ (0,2.5),(10,0) \}$
- $S_2 = \{ (0,2.5),(7.5,0) \}$
- $g_1(0,2.5) = g_2(0,2.5) = 0$
- $g_1(10,0) = 30$
- $g_2(7.5,0) = 24$

Let be $y/_{Lin}$ (resp $y/_{Harm}$) value inferred with linear method (resp harmonious method). At point x = (5,0) :

- $y/_{Lin}(x) = 10$, $y'/_{Lin}(x) = 24$
- $y/_{Harm}(x) = 18$, $y'/_{Harm}(x) = 18$

Sampling sets S_1, S_2 are very close, they only differ from one another in one point. However, $y/_{Lin}(x)$ is very different than $y'/_{Lin}(x)$, that is not the case in harmonious method.

4.2   Stability in space

Let x , x' be neighbors values, we compare y(x) and y (x'). We suppose $x = (x_1, x_2, x_i, ..., x_p)$, and $x' = (x'_1, x_2, x_i, ..., x_p)$ where p is the number of inputs. i.e. x and x' are immediate neighbors and differ only on one component. We may suppose, without lost of generality, that it is the first component.
With a fuzzy strong partition, only 2^p rules apply. We also suppose that x and x' are close and are in the same fuzzy interval (first interval for example).

We have $\forall i \neq 1,2$:

- $\mu_{1i}(x_1) = \mu_{1i}(x'_1) = 0$

where μ_{ik} is the membership degree. Finally, we have :

- $\mu_{11}(x_1) = 1 - \mu_{21}(x_1)$
- $\mu_{11}(x'_1) = 1 - \mu_{21}(x'_1)$

and then :

$$|y(x) - y(x')| \leq$$
$$|\textstyle\sum_{i=1}^{N} \alpha_i(x) \times C_i - \sum_{i=1}^{N} \alpha_i(x') \times C_i| \leq$$
$$\sup_{i,j} |C_i - C_j| \qquad (21)$$

where C_i and C_j are neighbouring in the grid G. In this case too, the second method is stable according to (15).
If S_1 and g_1 are those defined above, the values inferred at the points x = (5,0), x'= (7.5,0) are :

- $y/_{Lin}(x) = 10$, $y/_{Lin}(x') = 30$
- $y/_{Harm}(x) = 18$, $y/_{Harm}(x') = 24$

x and x' are immediate neighbors, and $y/_{Lin}(x)$, $y/_{Lin}(x')$ are very different, while $y/_{Harm}(x)$, $y/_{Harm}(x')$ are "sufficiently" close.

5 Conclusion

In this article, we have presented an automatic initialization of a fuzzy rule basis. We have shown that these methods, when we have few examples, greatly improve performance of fuzzy systems before learning. We think that these methods, combined with efficient learning can give good results.

References

1. Castro J.L., "Fuzzy logic controllers are universal approximators", *IEEE Trans. on SMC*, Vol. 25 n° 4, April 1995.
2. Sugeno M.,Yasukawa T. "A Fuzzy-Logic-Based Approch to Qualitative Modeling" , *IEEE Trans. on Fuzzy Systems*, Vol 1 n° 1, February 1993.
3. Lutaud M. "Identification et Contrôle de processus par Réseaux Neuro-Flous", Thèse de Doctorat de l'Université d'Evry Val d'Essonne
4. Kóczy L.T., Kovács S., "Linearity and the cnf property in linear fuzzy rule interpolation", Proceedings of the Third IEEE International Conference on Fuzzy Systems, Orlando, pp. 870-875 (1994)
5. Takagi T., Sugeno M., "Fuzzy identification of systems and its applications to modeling and control", *IEEE Trans. on SMC*, SMC-15(1), Janv 1985.
6. Glorennec PY., "Quelques aspects analytiques des systèmes d'inférence floue", *Journal Européen des Systèmes Automatisés*, Vol. 30, n° 2-3, 1996.
7. Le Gruyer E., Archer JC., "Stability and convergence of extension schemes to continuous functions in general metric spaces", *SIAM J. Math. Anal.*, Vol. 27, n° 1, 1996.
8. Hwang J.N., Lay S.R., Maechler M., Martin D., Schimert J. , "Regression modeling in backpropagation and projection pursuit learning", *IEEE Trans. Neural Networks* , Vol. 5 n° 3, May 1994
9. G. Box, G. Jenkins, "Time series analysis: forecasting and control" (Holden-Day, San Francisco, 1976).

Rare Fault Detection
by Possibilistic Reasoning

Marc Thomas, Andreas Kanstein, and Karl Goser

Microelectronics Department,
University of Dortmund, 44221 Dortmund, Germany.
E-mail: thomas@luzi.e-technik.uni-dortmund.de

Abstract. Kernel based neural networks with probabilistic reasoning
are suitable for many practical applications. But influence of data set
sizes let the probabilistic approach fail in case of small data amounts.
Possibilistic reasoning avoids this drawback because it is independent of
class size.
The fundamentals of possibilistic reasoning are derived from a probabil-
ity/possibility consistency principle that gives regard to relations. It is
demonstrated that the concept of possibilistic reasoning is advantageous
for the problem of rare fault detection, which is a property desired for
semiconductor manufacturing quality control.

1   Introduction

Possibilistic reasoning allows the implementation of a possibility based classifier
analogous to a classifier based on Bayes' theorem. The new concepts main advantage is found in problems of rare fault classification. The possibilistic reasoning
approach is derived analogue to the probabilistic one. Possibilistic reasoning is
based on a probability/possibility consistency principle different from the definition of Dubois and Prade [1].

Applications under investigation are dealing with fault detection in semiconductor fabrication such as defect density analysis, production line analysis,
and tests of power semiconductor devices. In these cases faults are often limited
to about a few percent [2]. Furthermore, problems are growing with separating
different types of faults.

The possibility based classifier is implemented in a kernel based neural network similar to radial basis function networks that implement a suboptimal
Bayesian classifier. Kernel based neural networks are a favoured technique for
classifying tasks. They provide quick learning, an interpretable structure and
robustness. These features result from their equivalence to certain fuzzy systems and from kernel adaptation by competitive learning of data clusters. That
accounts for the growing interest in this neural network type.

2   Motivation

The usual way of implementing a classifier in a kernel based neural network is a
radial basis function network (RBFN). The kernel neurons are trained by com-

petitive learning to represent clusters of data. The composition of the activations of kernel neurons approximates a posteriori probabilities of classes. In this way the network implements a suboptimal Bayes classifier. This concept has been proven very powerful in many applications.

The main disadvantage of Bayesian classifiers is their inability to classify rare faults. If the size of the classes differ strongly, the interesting class representing faults is likely to be covered by probabilities of classes with a large number of data that represent the normal case. The Bayesian classifier might also fail due to badly estimated probability distributions if the number of samples is very low. Figure 1 displays both cases in drastic but illustrative examples. Misclassification occurs by choosing a disadvantageous criterion, because rare fault detection is made impossible by a dominating probability of the non-fault case. Low a priori probability can cause this effect, shown in Fig. 1(a). Misclassification also occurs by badly estimated probability distributions. Consider two adjacent uniform distributions, one small and one large. Estimating the probability distributions by very few examples can result in a dominating situation as shown in Fig. 1(b).

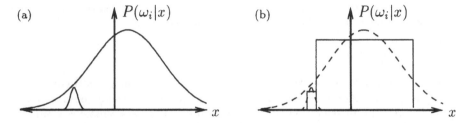

Fig. 1. Misclassification of rare faults in Bayesian classifiers due to dominating probability distributions. (a) Overlapping of distributions due to low a priori probability of rare faults. (b) Bad estimation of adjacent uniform distributions of different size due to very few samples of data.

These problems are very likely to appear in semiconductors manufacturing. Because of processing complexity the number of different fault classes is very large while the amount of data of single classes is low. The interest in the analysis of data that is collected in these processes is growing. The new concept of possibilistic reasoning implemented in a kernel based neural network is a promising approach because of the features listed above. In the following, the derivation of a decision rule from the definition of possibility distributions and a consistency principle is presented.

3   Possibilistic Approach

The set M of possible elements x includes all events that can ever appear and can be defined by probability values.

$$M := \{x \in X \mid P(x) > 0\} \quad .$$

The membership function of element x is given by:

$$\mu_M(x) = \begin{cases} 1 \text{ for } P(x) > 0 \\ 0 \text{ for } P(x) = 0 \end{cases} .$$

Because the set M is not known it has to be estimated using given data. Considering the uncertainty in this process, continuous possibility values in [0;1] are introduced. The similarity to fuzzy sets is obvious.

Definition 1. The mapping $\Pi: X \to [0;1]$, $X \subseteq \mathbb{R}^d$ is a possibility distribution.

A distinction between continuous and discrete universes, as necessary for probabilities, is not needed for these distributions. In the following continuous possibility distributions are denoted as $\pi(x)$.

To use possibilities as probabilities are in Bayes' theorem a connection between both has to be defined. The known consistency principles of Zadeh [3] or Dubois and Prade [1] are based on absolute values and do not regard relations as preferable for classification. The presented consistency keep this in mind.

Definition 2. A possibility and a probability distribution are *relational consistent*, iff

$$\forall x_1, x_2 \in \mathbb{R}^d : \quad \pi(x_1) > \pi(x_2) \Rightarrow p(x_1) \geq p(x_2) .$$

Equivalent to this definition is $\forall x_1, x_2 \in \mathbb{R}^d : p(x_1) \geq p(x_2) \Rightarrow \pi(x_1) > \pi(x_2)$. The properties of probabilities, the common concept of both and the relational consistency lead to the well known t-norm for intersection and s-norm for union of sets.

By using probabilities a distinction between the normal and the conditional case is necessary. To facilitate a decision as in Bayes classification, conditional possibility is required. Because t-norm and s-norm operators for intersection and set union of possibilities are not based on the property of dependent or independent events. Then the following relation can be used: $\Pi(A \cap B) = \tau(\Pi(A|B), \Pi(B))$; ($\tau$ is t-norm). Then $\Pi(A|B)$ can be calculated by $\Pi(B|A)$, $\Pi(A)$ and $\Pi(B)$ for a bijective operator or special cases.

Possibilistic Classification

The decision criterion is derived by minimizing the error possibility given by $S_{i \neq r} \Pi(\omega_i)$ with $\{\omega_i\}$ all decisions and ω_r the resulting decision. Using the possibilistic approach analogous to Bayes decision theory, this criterion results in the following decision rule:

$$\text{Choose } \omega_r \text{ with } \max_{\omega_i \in \Omega} \tau \left(\underset{j}{S} \tau(K(\mathbf{x}, \mathbf{m}_{ij}, \mathbf{s}_{ij}), \Pi(\omega_{ij}|\omega_i)), \Pi(\omega_i) \right) .$$

$\Pi(\omega_{ij}|\omega_i)$ and $\Pi(\omega_i)$ are the a priori possibilities and are free parameters of the classifier. τ is a t-norm and S denotes an s-norm. The trained clusters ω_{ij} given by centers $\mathbf{m}_{ij}$ and widths $\mathbf{s}_{ij}$ are specified by the kernel function $K()$

which computes the possibility $\pi(\mathbf{x}|\omega_{ij})$ of a vector $\mathbf{x}$ belonging to the cluster ω_{ij} of class ω_i. In contrast to probabilistic density estimation the kernel function need only be monotonous in distance. The used clustering algorithm is an on-line version of k-means with kernel function as distance measure. The process of classification, especially the clustering part is described in detail in [4].

4 Example

To demonstrate the target of possibilistic classification, an artificial dataset is used. The data are given by two overlapping gaussian distributions with one of them having a five times larger standard deviation and ten times greater amount of samples.

Fig. 2. Two overlapping gaussian distributions A and B

An RBF network, recognizing two clusters, gives a global error of 9.1% in hold-one-out. Best result by possibilistic reasoning is a value of 15.9%, keeping the a priori possibilities $\Pi(\omega_i)$ fixed to one. But examination of RBF-classification shows that always the bigger class A is determined. This is in final no classification for all inputs at all. Using possibilistic decision making results in an error of 16.9% in class A and 10% in B. This allows usable classification despite of a larger global error.

5 Conclusion

The presented approach for classification based on possibilistic reasoning allows to detect rare classes while keeping the facility of an implementation in a kernel based neural network. As the classical radial basis function approach stays as a special case of this network, the operation range of that neural network method is extended. The approach demonstrated here facilitates to handle the named task in semiconductor environment.

Acknowledgment: This work is supported by the "Deutsche Forschungsge-meinschaft (DFG)", grant number Go-379/11 and a grant in the Collaborative Research Center on Computational Intelligence (SFB 531).

References

1. D. Dubois and H. Prade, *Fuzzy sets and systems: Theory and Applications*, vol. 144 of *Mathematics in science and engineering*. Boston: Academic Press, 1980.
2. Ö. Hallberg, "Facts and fiction about the reliability of electronics," in *Proceedings of ESREF'95*, (Bordeaux, France), pp. 39–46, 1995.
3. L. A. Zadeh, "Fuzzy sets as a basis for a theory of possibility," in *Fuzzy Sets and Systems 1*, pp. 3–28, North-Holland, 1978.
4. A. Kanstein and K. Goser, "Dynamic learning of radial basis functions for fuzzy clustering," in *Proceedings of IWANN'95*, (Málaga, Spain), pp. 513–518, 1995.
5. R. Duda and P. Hart, *Pattern Classification and Scene Analysis*. New York: J. Wiley, 1973.
6. J. Moody and C. Darken, "Learning with localized receptive fields," in *Proc. 1988 Connectionist Summer School*, (San Mateo), pp. 133–143, Morgan Kaufmann, 1988.
7. E. Parzen, "On estimation of a probability density function and mode," *Ann. Math. Stat.*, vol. 33, pp. 1065–1076, 1962.

Fuzzy Zeroes and
Indistinguishability of Real Numbers

Bernard De Baets[1,*], Milan Mareš[2] and Radko Mesiar[3]

Department of Applied Mathematics and Computer Science[1]
University of Gent. Krijgslaan 281 (S9). B-9000 Gent, Belgium

ÚTIA, Academy of Sciences of the Czech Republic[2]
P.O. Box 18, 182 08 Praha 8, Czech Republic

Department of Mathematics, Slovak Technical University[3]
Radlinského 11, 81368 Bratislava, Slovakia

Abstract. In this paper, the well-known indistinguishability problem of real numbers is addressed. It is explained that T-equivalences form a suitable mathematical model for dealing with this problem. Firstly, it is shown that, for a continuous Archimedean t-norm T with additive generator f, from any pseudo-metric a T-equivalence can be constructed , by applying the pseudo-inverse of f. Secondly, a particular pseudo-metric d_g on $\mathbb{R}$ is constructed from a scale or generator g. It is investigated how this pseudo-metric can be transformed into a T-equivalence on $\mathbb{R}$. The answer lies in the study of the T-idempotents of the T-addition of fuzzy numbers. It is explained that by suitably modelling 'fuzzy zero', the pseudo-metric d_g allows to propagate this 'indistinguishability from 0' across the real line, thus obtaining a description of indistinguishability of real numbers in general.

Keywords: T-equivalence; fuzzy number; fuzzy zero; generator; T-idempotent; shape; t-norm.

1 Indistinguishability of Real Numbers

In real world problems, the exact value of real numbers is of secondary importance: rather the magnitude of data plays a principal role. The concept of magnitude is formulated by a context dependent identification process: two real numbers a and b are identified provided the absolute value of their difference is smaller than a given positive real number ϵ. Formally, we can write $a \approx_\epsilon b \Leftrightarrow |a - b| < \epsilon$. The relation $\approx_\epsilon$ clearly is reflexive and symmetric, but not transitive. This observation goes back to Poincaré's paradox [13,14]:

$$A = B \quad , \quad B = C \quad , \quad A \neq C,$$

---

* Post-Doctoral Fellow of the Fund for Scientific Research – Flanders (Belgium).

where the expression $A = B$ is interpreted as "A and B are indistinguishable". In [5], Höhle addresses fundamental problems related to non-transitivity, and remarks that replacing the relation $\approx_\epsilon$ by its transitive closure offers no solution (it identifies all real numbers). He furthermore discusses a mathematical model compatible with Poincaré's paradox: the concept of a fuzzy equality (also called likeness relation [15]), i.e. a reflexive, symmetric and W-transitive binary fuzzy relation (with W the Lukasiewicz t-norm [16]). Notice that the concept of a fuzzy equality is dual to that of a pseudo-metric: for a given fuzzy equality E, the mapping $d = 1 - E$ is a pseudo-metric.

Two real numbers a and b are called *indistinguishable* w.r.t. a fuzzy equality E if and only if $E(a, b) > 0$ [5]. The fuzzy equality on $\mathbb{R}$ hidden in the concept of 'magnitude' is given by

$$E_\epsilon(a, b) = \max(1 - \frac{1}{\epsilon}|a - b|, 0).$$

Obviously, a and b are indistinguishable w.r.t. E_ϵ if and only if $a \approx_\epsilon b$; hence, fuzzy equalities seem to be appropriate for dealing mathematically with the intuitive concept of 'magnitude' [5].

An important generalization of fuzzy equalities are T-equivalences [2], with T an arbitrary t-norm [16]. The aim of this paper is to investigate how such a T-equivalence on the real line can be constructed. The idea is to describe the indistinguishability of real numbers from 0, leading to the concept of a fuzzy zero (more formally, a T-idempotent), and to propagate this description, by means of a simple pseudo-metric based on the concept of a scale or generator, across the whole real line, thus describing indistinguishability of real numbers in general.

2   T-equivalences and Pseudo-metrics

Let us start this section by recalling some basic concepts. A T-equivalence E [2], with T a t-norm, on a universe X is a reflexive, symmetric and T-transitive binary fuzzy relation on X, i.e. for any (x, y, z) in X^3 it holds that

$$T(E(x, y), E(y, z)) \leq E(x, z).$$

Note that T-equivalences are also called indistinguishability operators [17] or equality relations [7]. The *equivalence class* of $x \in X$ w.r.t. E is the fuzzy set $[x]_E$ in X defined by $[x]_E(y) = E(x, y)$. A T-equivalence E on $\mathbb{R}$ is called *admissible* [6] if and only if $[x]_E$ is a *fuzzy number*, for any $x \in \mathbb{R}$, where a fuzzy number A is defined as an $\mathbb{R} \to [0, 1]$ mapping such that there exists $a \in \mathbb{R}$ with $A(a) = 1$ and $A|_{]-\infty, a]}$ is increasing and $A|_{[a, +\infty[}$ is decreasing.

We will now recall our recent results concerning the construction of pseudo-metrics from T-equivalences and vice versa [3,4]. For the notion of an additive generator f and its pseudo-inverse $f^{(-1)}$, we refer to [8,16]. If the cardinality of the universe X is smaller than 3, then for any t-norm T, any T-equivalence E on X and any additive generator f, it obviously holds that the mapping $d = f \circ E$ is a pseudo-metric on X.

Theorem 1. [3] *Consider a universe X with $\#X > 2$, a t-norm T^* with additive generator f and a t-norm T. Then T^* is weaker than T if and only if for any T-equivalence E on X, the $X^2 \longrightarrow [0, \infty]$ mapping $d = f \circ E$ is a pseudo-metric on X.*

In the converse problem, namely the construction of T-equivalences from pseudo-metrics, continuous additive generators play an important role.

Theorem 2. [3] *Consider a pseudo-metric d on a universe X and a continuous Archimedean t-norm T^* with additive generator f, then the binary fuzzy relation $E = f^{(-1)} \circ d$ in X is a T^*-equivalence on X.*

As corollaries of this theorem, we rediscover the following results of Jacas and Recasens [6] (see also [1]). Consider a *scale s*, i.e. a monotonic $\mathbb{R} \to \mathbb{R}$ mapping. Then clearly the $\mathbb{R}^2 \to \mathbb{R}^+$ mapping d_s, defined by $d_s(x, y) = |s(x) - s(y)|$, is a pseudo-metric on $\mathbb{R}$. Applying Theorem 2, it then easily follows that the binary fuzzy relation E_1, defined by

$$E_1(x, y) = \max(1 - |s(x) - s(y)|, 0)$$

is a W-equivalence on $\mathbb{R}$. Similarly, the binary fuzzy relation E_2, defined by

$$E_2(x, y) = \min \left(\frac{s(x)}{s(y)}, \frac{s(y)}{s(x)} \right),$$

with s a positive scale, is a P-equivalence on $\mathbb{R}$ (with P the algebraic product).

Theorem 2 is rather general, as it deals with arbitrary pseudo-metrics. However, our primary interest goes to pseudo-metrics of the type d_s. More formally, we are interested in finding necessary and sufficient conditions under which the binary fuzzy relation $E_{g,\varphi} = \varphi \circ d_g$, with g a *generator* [1], i.e. an increasing scale that runs through the origin, and φ a *shape (function)* [1], i.e. a decreasing $\mathbb{R}^+ \to [0, 1]$ mapping such that $\varphi(0) = 1$, is a T-equivalence on $\mathbb{R}$; admissibility clearly is always fulfilled. Note that a shape φ can be identified with a positive fuzzy number, i.e. $\varphi(x) = 0$ for any $x < 0$. We have discovered that the above problem is closely related to the study of 'fuzzy zeroes'.

3   Fuzzy Zeroes

It is well known that for the addition of real numbers there exists only one idempotent, i.e. only one real number x for which $x + x = x$, namely $x = 0$. The corresponding problem for fuzzy numbers has been addressed by Marková [11]. Consider a t-norm T, then a fuzzy number I is called a T-idempotent if and only if $I \oplus_T I = I$, where $I \oplus_T I$ is, as usual, defined by

$$I \oplus_T I(y) = \sup_{x \in \mathbb{R}} T(I(x), I(y - x)).$$

Less formally, T-idempotents could be called 'fuzzy zeroes' (see also [9,10]). In view of determining suitable shapes for the above-mentioned problem, we restrict our discussion of T-idempotents to positive fuzzy numbers.

For the minimum operator M, the problem of finding M-idempotents is easily solved [1]. Indeed, the only positive M-idempotents are the shapes φ_c, $c \in [0, 1]$, defined by

$$\varphi_c(x) = \begin{cases} 1 & , \text{ if } x = 0 \\ c & , \text{ otherwise} \end{cases}$$

Notice that the shape φ_1 is a T-idempotent for any t-norm T. For an arbitrary continuous t-norm, the problem is a lot harder, and has been solved only recently by Marková-Stupňanová in [12].

Theorem 3. [12] *Consider a continuous t-norm T with ordinal sum representation $(\langle \alpha_k, \beta_k, h_k \rangle; k \in K)$, i.e. $(]\alpha_k, \beta_k[; k \in K)$ is a disjoint system of open subintervals of $[0, 1]$, h_k is a continuous strictly decreasing $[\alpha_k, \beta_k] \to [0, \infty]$ mapping with $h_k(\beta_k) = 0$, $k \in K$, and for any $(x, y) \in [0, 1]^2$:*

$$T(x, y) = \begin{cases} h_k^{-1}(\min(h_k(\alpha_k), h_k(x) + h_k(y))) & , \text{ if } (x, y) \in [\alpha_k, \beta_k]^2, k \in K \\ \min(x, y) & , \text{ elsewhere} \end{cases}$$

A continuous shape φ different from φ_1 is a T-idempotent if and only if there exists $k_0 \in K$ such that $\beta_{k_0} = 1$, $\mathrm{rng}\, \varphi \subseteq [\alpha_{k_0}, \beta_{k_0}]$ and $h_{k_0} \circ \varphi$ is subadditive on $\mathbb{R}^+$.

For a continuous Archimedean t-norm in particular, this leads to the following theorem.

Theorem 4. [12] *Consider a continuous Archimedean t-norm T with additive generator f and a continuous shape φ such that $\lim_{x \to +\infty} \varphi(x) = 0$. Then φ is a T-idempotent if and only if $f \circ \varphi$ is subadditive on $\mathbb{R}^+$.*

Consequently, for a continuous Archimedean t-norm T with additive generator f, the shape $\varphi = f^{(-1)}$ is a T-idempotent.

4 T-equivalences Generated by Fuzzy Zeroes

We now return to our basic problem, i.e. the determination of necessary and sufficient conditions that have to be imposed on the generator g and the shape φ for $E_{g,\varphi}$ to be a T-equivalence. Note that for the shapes φ_0 and φ_1, the fuzzy relations E_{g,φ_0} and E_{g,φ_1} $(= \mathbb{R}^2)$ are crisp equivalence relations, and hence also T-equivalences for any t-norm T.

Theorem 5. [1] *Consider a t-norm T, a generator g and a shape φ. Let $H = \{|g(u) - g(v)| \mid (u, v) \in \mathbb{R}^2\}$. If*

$$(\forall x \in H)(\varphi \oplus_T \varphi(x) = \varphi(x)),$$

then the fuzzy relation $E_{g,\varphi}$ is a T-equivalence on $\mathbb{R}$.

This theorem implies that if φ is a T-idempotent, then the fuzzy relation $E_{g,\varphi}$ is a T-equivalence on $\mathbb{R}$. In other words, choosing an appropriate fuzzy zero, any pseudo-metric d_g allows to describe T-indistinguishability of real numbers in general. This again generalizes the results of Jacas and Recasens [6].

The converse of Theorem 5 can also be proven, provided that we are working with continuous generators.

Theorem 6. [1] *Consider a t-norm T, a continuous generator g and a shape φ. Let $H = \{|g(u) - g(v)| \mid (u,v) \in \mathbb{R}^2\}$. If the fuzzy relation $E_{g,\varphi}$ is a T-equivalence on $\mathbb{R}$, then*

$$(\forall x \in H)(\varphi \doteq_T \varphi(x) = \varphi(x)).$$

A simplified version of the above results is obtained for a continuous generator g satisfying at least one of the conditions $\lim_{r \to -\infty} g(x) = -\infty$ or $\lim_{r \to +\infty} g(x) = +\infty$. In that case, the fuzzy relation $E_{g,\varphi}$ is a T-equivalence on $\mathbb{R}$ if and only if φ is a T-idempotent. In this particular situation, the generator has no influence on the fact whether $E_{g,\varphi}$ is a T-equivalence or not.

Acknowledgement

The research reported on in this paper was partially supported by Grants GAČR 402/96/0414, VEGA 1495/94 and 95/5305/471.

References

1. B. De Baets, M. Mareš and R. Mesiar, *T-partitions of the real line generated by idempotent shapes*, Fuzzy Sets and Systems (submitted).
2. B. De Baets and R. Mesiar, *T-partitions*, Fuzzy Sets and Systems (to appear).
3. B. De Baets and R. Mesiar, *Pseudo-metrics and T-equivalences*, J. Fuzzy Mathematics (to appear).
4. B. De Baets and R. Mesiar, *Metrics and T-equalities*, J. Math. Anal. Appl. (submitted).
5. U. Höhle, *Fuzzy equalities and indistinguishability*, Proceedings of the First European Congress on Fuzzy and Intelligent Technologies (Aachen, Germany, September 1993) (H.-J. Zimmermann, ed.), vol. 1, 1993. Aachen, pp. 358–363.
6. J. Jacas and J. Recasens, *Fuzzy numbers and equality relations*, Proceedings of the Second IEEE International Conference on Fuzzy Systems (San Francisco, California, March 1993), 1993, pp. 1298–1301.
7. F. Klawonn and R. Kruse, *From fuzzy sets to indistinguishability and back*, Proceedings of the First ICSC International Symposium on Fuzzy Logic (Zürich, Switzerland, May 1995) (N. Steele, ed.), ICSC Academic Press, 1995, pp. A57–A59.
8. E.-P. Klement, R. Mesiar and E. Pap, *Triangular norms*, (in preparation).
9. M. Mareš, *Computation over fuzzy quantities*, CRC-Press, Boca Raton, 1994.
10. M. Mareš, *Fuzzy zero, algebraic equivalence: yes or no?*, Kybernetika **4** (1996), 343–351.
11. A. Marková, *Idempotents of the T-addition of fuzzy numbers*, Tatra Mountains Math. Publ. (to appear).
12. A. Marková-Stupňanová, *Idempotents of the addition of fuzzy intervals based on a continuous t-norm*, Fuzzy Sets and Systems (submitted).
13. H. Poincaré, *La science et l'hypothèse*, Flammarion, Paris, 1902.
14. H. Poincaré, *La valeur de la science*, Flammarion, Paris, 1904.
15. E. Ruspini, *A new approach to clustering*, Information and Control **15** (1969), 22–32.
16. B. Schweizer and A. Sklar, *Probabilistic metric spaces*, North-Holland, New York, 1983.
17. E. Trillas and L. Valverde, *An inquiry into indistinguishability operators*, Aspects of vagueness (H. Skala, S. Termini and E. Trillas, eds.), Reidel, Dordrecht, 1984, pp. 231–256.

Detection of Sleep with New Preprocessing Methods for EEG Analysing

A. Berger, D.P.F. Möller, M. Reuter

Institute of Computer Science; Technical University Clausthal
Julius-Albert-Str. 4; 38678 Clausthal-Zellerfeld (Germany)
email: Angelika.Berger@tu-clausthal.de

Abstract. A selforganized map was designed to learn and detect sleep stages will be described. Initial the input data were preprocessed with Difference Power Spectra (DPS (german: DLS)). The associative fields of the Kohonen map are directly transformed by frequency spectra. Interference phenomena are probably indicating the influence of biosignals with a source in the reticular system of a human brain.

1. Theory

1.1 Sleep and sleep detection

Many functional processes of sleep are not well explored and are combined with so far unknown questions. Known by today is, that the normal sleep is devided into 4 to 5 sleep periods, whereby every sleeping period is finished after 60 to 90 minutes with a phase of rapid eyes movement (REM).

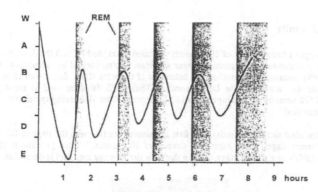

Fig. 1: Sleeping periods and sleep stages of one night [1]

Sleep differs between 5 sleeping stages. Each of these sleeping stages are an important part of the sleepling period. Sleep begins with becoming tired (stage A), getting deeper in the low (stage B), middle (stage C) and the deep-sleep-stage (stage D). With the increasing depth of sleep the charakteristica of sleep electroencephalography is changing. The frequency spectrum of awakefullness is dominated by -waves (8-12 Hz). These characteristical spectrum decreases to dominant ∂-waves (0.5-4 Hz) of the deep-sleep-stage. The electrical activity of the brain, detected as EEG and the features of their frequency spectra are the important points to detect and analyse the different sleep stages.

1.2 Structure of sleep electroencephalography signals

With the fast fourier transformation (FFT) a algorithm exists to compute the frequency spectrum of the brain signals in an acceptable time. During the detection of the electricity brain signals noise and technical artefacts are overlapping these signals. The result is, that the use of this overlapped signals destroy the ability to develop neural classificators for the identification of sleep stages, as the correlations between the linear indepent sleep feature vectors and the noisy contributions will force the net to evaluate non-characteristical classification concepts. The resulting classifier will be a non-workable version of a sleep-detector. Hence a sufficient preprocessing-procedure has to be choosen to evaluate a real problem oriented representation of the given EEG-datas.

Fig. 2: Preprocessing of sleep EEG

1.3 Detection of sleep stages using selforganized maps

When a real problem oriented representation of the EEG-datas is evaluated, it should be possible to use selforganized Kohonen nets to evaluate associative fields for signals with same feature-relationship coded in the EEG similar to the Brodmannschen fields of the neocortex.

This classifiers should be able to separate the different brain activities depending to the stages of awakefullness, by mapping the sensory input to topological arears of higher intensity. Therefore the change of the awakefullness will be detectible in a deterministical way by overwatching the change of the areas of higher intersity of the Kohonen map and a supervising of the sleep of the probants is possible.

2. Experimental results

To evaluate sleep stages a time window of 1024 points was used. Next the FFT and the Auto Power Spectrum of the time signals is computed. The resulting power spectrum is preprocessed by calculating the Difference Power Spectum (DPS). Because of the stochastical behaviour of the brain signal data a well choosen grid must be found to compute the clusters for the DPS algorithm. These DPS datas are used as input patterns in a Kohonen map of 20*20 neurons with an input layer of 512 neurons. After 10 learning cycles the learning phase was succesfully determined.

The used data set described sleeping records of the first 22 minutes, starting with the stage of full awakefullness, followed by the different stages of the beginning sleep after 10 minutes. Within 10 minutes the neural net identified the sleep EEG's as a similar stage. During this time the -wave spectrum is dominant.

Fig. 3: Associative fields till the 10th minute. Afterwards the test person falls asleep

At the end of this phase the test persons suddenly fell into deeper sleep stages announced by the neural net by changing the arears of higher intensities. A „jumping" between small associative fields on the Kohonen map can be detected whereby this jumping converge slowly to an field at the lower right side of the map. At this time the characteristic wave spectrum of sleep EEG changed from -wave dominated spectrum to the deep sleep-stage describing a ∂-wave dominated spectrum.

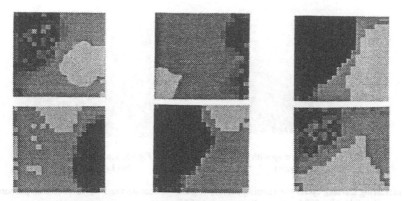

**Fig. 4. Associative field on the Kohonen map between the 10th and the 22th minute
(from upper left to the lower right corner)**

Scientific studies [5] are indicating, that parts of the reticular system of human brain control the awakeness (vigilance) and the sleep depht stage. In Fig. 1 it has been shown, that the depht of sleep is also depending on the numbers of sleeping periods the test person has passed before. The frequency spectrum of sleep EEG is also modified during some sleeping periods during the night. Our results suspect that there is a relation between the explained interference phenomenon and the deep sleep control, which can be detected by brain electrical potentials.

After the test person has fallen asleep into deeper sleep stages the characteristical frequency spectrum of sleep EEG changed to the ∂-waves. During this time of increasing sleep depth the electrical potentials of the brain are in an instabil stage. Some other experiments compared the preprocessed DPS as sensory input with the non preprocessed autopower spectra. A neural net was conditioned with this autopower spectra. After a learning phase of 10 cycles no separations could be seen on the net. Even after we repeated these experiments with longer learning cycles no better testing results could be noticed.

Fig. 5a: Kohonen map with autopower
spectra after 5000 learning cycles

Fig. 5b: Kohonen map with DLS
after 10 learning cycles

By using the Difference Power Spectrum all maschine- und environment noise were eliminated, whereby the DPS representation can be seen as a numeric filter, which points out the relevant feature-specific information for sleep-detection. The DPS represented the desired presentation of sleep EEG as sufficiant sensory input for neural nets [3]. With this preprocessing step the neural net become able to point out the desired separation-features for classifying characteristical sleep stages. Results from other investigations showed that the DPS-representantiation seemed to be a comfortable and important preprocessing-step conditioning neural nets for an employment in the sleep like narcotic stage detection during operations too.

We supposed, that the conditioning behaviour of the neural net with DPS input would be increase with higher learning cycles. To our astonishment, we noticed the contrary learning success.

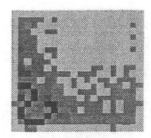

Fig 6a: Kohonen map with
10 learning cycles
　　　　　　　　　　　　　　　　Fig 6b: Kohonen map with
　　　　　　　　　　　　　　　　1000 Lerning cycles

With increasing learning cycles the learning success is going to become more worse. A similar phenomenon has been seen while using 200 learning patterns instead of 50.

During adaptive learning a selforganized map modifies exhibitable the weights of neurons which are in the neighborhood of an so called „winner neuron". This winner neuron has the maximal similarity to the given sensory input. The exhibit environment of the winner neurons will be reduced sucsessive. This depends on the number of given learning cycles. If the sensory input has minimal correlations, activation fields are build, which respond to similar sensory input with similar reactions in the activation niveau.

If the intra sensory relations are minimal no reactions like in Fig. 6 described would be seen. Our experimental results can only be interpreted, that, although we used DPS as a preprocessing step, correlated information is hidden in sleep EEG signals. This information overlapps like an interfere phenomenon onto the conditioning behaviour and modifies this in an negativ way, as a DPS eliminates clusters of present constant noise only. This overlapping informations have to be small dots in the frequency spectrum, otherwise they would be eliminated.

At the end we had to learn, when more learning cycles used the map separated the feature-pace in an uneffective way by extracting more non-interesting linear independent features of the EEG and mapping them in linear independent areas.

3. Discussion

From all our experiments it can be shown that the different sleep-stages are represented by clear topological structures on the Kohonen map. By supervising the time behaviour of these areas of higher intensity, sleep behaviour can be classified as shown in Fig. 7. On the right hand inside of each line you can see the autopowerspectrum and resulting power spectrum, which is preprocessed by calculating the Difference Power Spectrum (DPS). The selforganised was feeded by this DPS. On the left hand side you the see the responding neural activity on the Kohonen map.

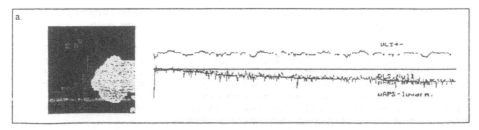

Fig. 7a: Difference Power Spectra (DLS) and selforganized maps (SOM) during the stages of deeper sleep

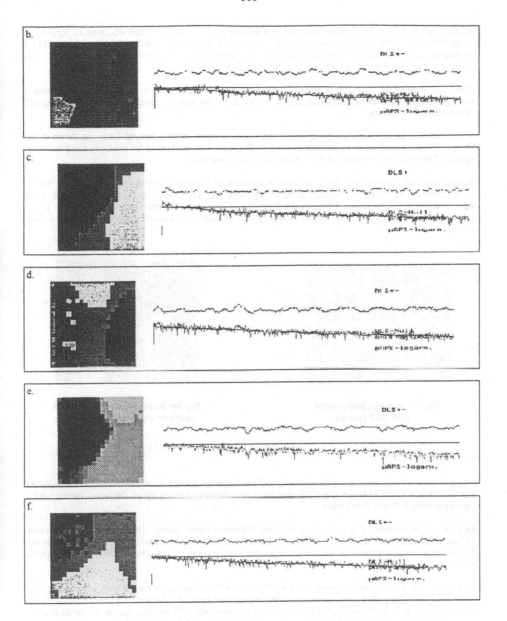

Fig. 7b-f: Difference Power Spectra (DLS) and selforganized maps (SOM) during the stages of deeper sleep

The Fig. 7a-f clearly point out, that the stages of instabil entire brain activity can also be seen in the topological maps of the neural net. In the deep-sleep stages D and E a regulation process arrises which stabilises the different EEG stages. This convergation process can also be seen on the Kohonen map, where stages in maps with similar neural activity are converging.

Interpreting the results of the Kohonen map it has to be discussed how the function transformation behaviour of the frequency spectra to the maps and its neural activity is done.

From a mathematical point of view we can define a selforganised map as an operator $\emptyset_{som}$ also. In dependence to the DPS (Fig. 7) this operator creates a function f_{DPS} of the fourier transformation which is represented by a niveau of activity in form of a matrix. Therefore this operator $\emptyset_{som}$ is a function of the frequency- and amplitude spectra and represents the correlated time signal in a suffcient way.

There are indications, that the amplitude spectra is a neglect input pattern of $\emptyset_{som}$. To verify this a selforganised map was feeded with so called DAP spectra during the learning process. DAP spectra are strongly standardized DPS spectra.

The Kohonen map represent different niveaus of activity in coloured coded associated fields. As shown in fig. 8a no fields of association were detected. So these DAP spectra were exchanged by positively elongated DAP spectra. The former non represented fields of association are now visible.

So the amplitude spectra leads to an increase of intensity of the neural activity, but not to a topological modification on the process of adaptation of the selforganised map. This leads us to the fact, that the only input pattern that controls the operator $\emptyset_{som}$ is the frequency spectra. A validation of the classified states of the process by the selforganised map can be compared with characteristic frequency spectra of different sleep phases.

 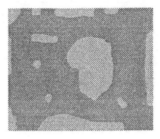

Fig. 8a: Kohonen map, feeded
with DAP spectra

Fig. 8b: Kohonen map, feeded with
positively elongated DAP spectra

4. Summary

The selforganised map represent a transformation of the frequency spectra on neural activities. This means, that a validation of the Kohonen map leads to a direct classification of characteristic frequency spectra which are representing different sleep phases.

As a step of preprocessing it is necessary to calculate Difference Power Spectra (DPS) of the EEG to garanty the neural detection and classification of the sleep phases. As one result the learning expense of the neural net decrease to minimal time periods. Therefore it is possible to create individual topological sleep-logs of patients in a short time.

As a framework the complete frequency spectra including the higher harmonic spectra were investigated. We suppose that disturbing phenomenons of interference, which were detected in longtime learning phases with the DPS, depend on the electric potential of the brain which are contolled by the reticular system. While the electric activity of the brain is blocked in deeper sleep stages, sleeping control of the reticular system can be detected as an electrical activity all the time.
Our present research put the localisation of this spectra and its later neural analysis in the focus.

310

5. Literature

1. Jovanovic, U.J.: „Schlaf und Traum - Physiologische und psychologische Grundlagen, Störungen und Ihre Behandlungen"; Gustav Fischer Verlag; Stuttgart 1974
2. Werner, H.: „Time Trajectories in Sleep Stages"; Proceedings EUFIT, September 1996, pp. 2107-2111; Aachen.
3. Reuter, M.; Berger, A.; Elzer, P.: „A proposed method for Representing the Real-Time Behaviour of Technial Processes in Neural Net"; Proceedings IEEE Conference of Man Maschine and Cybernetics, Vancover 1995
4. Reuter, M.: „Frequency Difference Spectra and Their Use as a New Preprocessing Step for Acoustic Classifiers/Identifiers", EUFIT 93 Proceedings, September 1993, pp 436-442, Aachen.
5. Bennett, T.L.: „Brain and Behaviour"; Brooks/Cole Publishing Company; Monterey California 1977

Application of Evolutionary Algorithms to the Problem of New Clustering of Psychological Categories Using Real Clinical Data Sets

Th. Villmann[†], B. Villmann[‡] and C. Albani[‡]

[†]Institut für Informatik
Universität Leipzig
D–04109 Leipzig, Augustusplatz 10/11, Germany
(corresponding author)
email: villmann@informatik.uni-leipzig.de

[‡]Klinik und Poliklinik für Psychotherapie und psychosomatische Medizin
Universität Leipzig,
D–04107 Leipzig, Karl–Tauchnitz–Str. 25, Germany

Abstract

One of the mostly used method for acquisition of structures of interpersonal relationships in the area of psychodynamic psychotherapy research is the method of the 'Core Conflictual Relationship Theme' which allows a standardization of phrases in so–called standard categories. We re-clustered these categories on the basis of a set of real clinical data by application of evolutionary algorithms which leads to an improvement of the clustering in comparison to earlier resulted cluster distribution. For the evolutionary algorithms we used a special $(\mu * \lambda)$–spring–off strategy balancing between the well known (μ, λ)– and $(\mu + \lambda)$–strategy. Furthermore, we developed a special migration scheme for handling the dynamic of subpopulations.

1 Introduction

Modern psychology uses all the standard methods of mathematical statistics to extract relevant features, structural informations and other data from several therapeutical approaches. One of the mostly used method for acquisition of structures of interpersonal relationships in the area of psycho–dynamic psychotherapy research is the method of the 'Core Conflictual Relationship Theme' (CCRT) developed by LUBORSKY , [Lub77, Sol93]. The method investigates short stories about relationships, so–called *relationship-episodes*, which are often reported by the patients in their therapeutical sessions [DTR+93]. In each of these episodes were the components *wish of the subject, response of the object* and *response of the subject* encoded which were used to perform the CCRT.

BARBER ET AL. [BCCL90] determined a system of $s_{max} = 34$ so-called *standard categories* S_j to classify the wishes of patients in the episodes. Examples for such standard categories are: '... I want to be accepted ...', '... I want to be understood ...', '... I want to be successful ...', etc., i.e. the standard categories describes an aspect in verbal manner. Of course, these categories are often correlated in meaning because of a similarly describing psychological topic, i.e., it is obviously to see that there exists overlapping categories in the above sense. Therefore, they are collected in $c_{max} = 8$ *clusters* C_k to reduce these correlations (BARBER ET AL., [BCCL90]). Then in further work instead of the standard categories the clusters are used for the specification of the episode, whereby the clusters should have different meaning contents (in the sense of disjunction). The scheme of mapping the standard categories onto the respective clusters is also predefined in the above mentioned work [BCCL90] which, in general, leads to an improvement of the reliability of the CCRT-method [LBS89, LCC90].[1] The number and the interpretation of the clusters as well as the assignment of the standard categories are resulted from the experience of several number of psychotherapists using conventional statistic methods.

However, as mentioned in [AVV+96] the clusters also are still correlated again. Furthermore, several considerations have shown that *the used scheme of assignment leads still to low reliability* rates for the CCRT–method and misunderstanding in investigations based on this scheme. For further and more detailed critical remarks we refer to [AVV+96].

Yet, the correlation between the standard categories as well as between the clusters are difficult to capture, because of the non–measurable structure.

Hence, the problem is now, how one can reform the clusters of standard categories in such a manner that the correlations will be reduced in a faithful way using the underlying meaning, additional therapeutical knowledge etc.. Furthermore we have to pay attention to the assumption that the set **S** of standard categories will not be changed during the reclustering and in addition that the number of clusters is pretend.[2]

In the present article we applied evolutionary algorithms to solve this clustering problem. For a better handling we transformed it in a partitioning one. The new clusters are compared with the original one and , second, with also new clusters obtained using the factor decomposition method.

2 Clustering by evolutionary algorithms according to a $(\mu * \lambda)$–strategy

For solving the above described problem at first several psychotherapists judge (rate) a large number of therapy interviews of several patients. The raters determined the most relevant standard category S_{j*} of all wishes in the episodes

---

[1] Analog categories and clusters exist for the response of object and subjects, respectively. However, the here reported results are examplary and transformable to the other components.

[2] The last condition is necessary because of the compatibility to other approaches in psychodynamic psychotherapy research and methods which extend the CCRT–method.

and, *in addition,* a second one S_{j+}, which have to be different from the first one but it is also well describing the considered episode (wish). All these pairs $p_i = \left(S^i_{j*}, S^i_{j+} \right)$, $i = 1 \ldots N$, form a database P which implicitly contains

κ–coefficient	meaning
$\kappa < 0.1$	no agreement
$0.1 \leq \kappa < 0.4$	weak agreement
$0.4 \leq \kappa < 0.6$	clear agreement
$0.6 \leq \kappa < 0.8$	strong agreement
$0.8 \leq \kappa$	nearly complete agreement.

Table 1: Different values for the weighted concordance coefficient κ and the respective meaning for the agreement of the appearance of the considered observables

information of the correlations between the standard categories. In our case the database P contains $N = 5383$ pairs p_i. If the clusters are determined in a most faithful way according to an arbitrary clustering algorithm, both the most and the second relevant standard category should belong to the same cluster after this procedure. Thereby, for clustering one has to take into account the hidden correlation information in the database P.

To extract this features we used two methods: first the factor decomposition method (FDM) and second evolutionary algorithms (EAs)[3] to obtain new cluster distributions. For compatibility we chose the number of new clusters also to be $c_{\max} = 8$ as pretend in (LUBORSKY , [LBS89] and [BCCL90]). The resulted mappings of standard categories onto the clusters are then compared via the weighted concordance coefficient κ of a two–dimensional 34×34 contingence table. In general, the weighted concordance coefficient κ measures the simultaneous appearance of the considered observables [Fle81, Gjo88, KF81, Coh68]. In our application the observables are the respective clusters $C_{k(j*,i)}$ and $C_{k(j+,i)}$ onto which the both standard categories S^i_{j*} and S^i_{j+} of a pair $p_i \in P$ are mapped, respectively. Then, the κ–coefficient measures the agreement of $C_{k(j*,i)}$ and $C_{k(j+,i)}$ for all pairs $p_i \in P$. The computation of κ uses the emergence of so–called structural zeros (because of the fact that $S^i_{j*} \neq S^i_{j+}$ is always assumed) . The resulted κ–coefficients may be interpreted as depicted in Tab. 1 [Sac92].

---

[3]Thereby EAs comprise genetic algorithms according to HOLLAND [Hol75], evolutionary programming according to FOGEL [Fog95] and evolutionary strategies according to [Sch81] and [MGSK88].

At first we computed the concordance coefficient κ for the original clusters of the standard categories as defined in [BCCL90] and we found the value $\kappa = 0.33$ which corresponds to an only weak agreement (see Tab. 1), i.e. the chosen assignment of standard categories to the clusters is not optimal. In the next steps we tried to improve the clustering of the standard categories using the above suggested approaches.

2.1 Application of the Factor decomposition method

For a later comparison with the results of EAs we applied the FDM to extract new clusters as a beginning consideration. For the FDM first the correlation matrix of the standard categories according to the used database P was determined and then stepwise the relevant principle components were evaluated. The final clustering was then derived via the *VARIMAX–method* (CLAUSS&EBENER [CE78]).

This approach generates a cluster distribution of the standard categories which yields for the weighted κ–coefficient the value $\kappa = 0.44$. This κ–value may be interpreted as a clear agreement according to Tab. 1 . Moreover, it is a significant improvement against the κ–value obtained from the original clustering.

2.2 Evolutionary algorithms

However, using the above FDM it is difficult to integrate specific psychotherapeutic knowledge into the decorrelation scheme or other conditions for clustering, for instance the balance with respect to the number of standard categories belonging to the several clusters. In contradiction, EAs give the possibility to formulate explicitly the conditions which one want to optimize. Therefore, we tried a first ansatz to classify the standard categories using EAs. For a better handling we reformulate the problem as follows: the task is to find a *partition* of the set **S** of all s_{max} standard categories under certain conditions, which may be specified by the fitness function F, i.e., we now have to solve a partitioning problem.

The concept of a partition may be introduced in the following way [HHV95]: A *partitioning* of a nonempty set U related to a nonempty set V is an unique and surjective mapping

$$\Phi : U \rightarrow V . \tag{2.1}$$

Then a *partition* Ψ_Φ of U related to the partitioning Φ is given by

$$\Psi_\Phi = \left\{ \Phi^{-1}(v) \mid v \in \text{cod}(\Phi) \right\} , \tag{2.2}$$

whereby $\text{cod}(\Phi)$ is the range of Φ.

For solving the partitioning problem a string (or individual) in a population of an EA describes a certain partition. Thereby each component of an individual represent one standard category and the value of it determines the cluster onto which the category will be mapped. The mutation of a individual is defined as a random change of the mapping for a random selected individual component. The mutation rate also was here an object of optimization itself, i.e. the mutation rate was a additional individual component which was suggested in [Mic96, Rec73]. The crossover was the usual one point crossover. The individuals where selected for the above genetic operations randomly with respect to their actual fitness value.

2.2.1 The $(\mu * \lambda)$–strategy for selection

For the off–spring generation we used a mixture of the (μ, λ)– and the $(\mu + \lambda)$–strategy (in the notation of SCHWEFEL, [Sch81]) which balances the advantages of both strategies [Mic96]. This approach was introduced to solve hard (very large) partitioning problems in VLSI–design and denoted by $(\mu * \lambda)$–strategy [HHV95, HHV96].

In this approach again μ individuals produce the λ preliminary off–springs. However, in the selection step the μ_t best individuals of the old generation and the λ new ones are allowed for comparison with respect to their fitness to generate the final off–spring generation of μ individuals. Thereby μ_t depends on time:

$$\mu_t = \text{int}\left[(\mu - \mu_\tau) \cdot \gamma(t)\right] + \mu_\tau \qquad (2.3)$$

whereby int$[x]$ is standing for the integer value of x. The function $\gamma(t)$ is of decreasing sigmoid type with $0 \leq \gamma(t) \leq 1$. Then we have for the initial value $\mu_0 = \mu$ and $\lim_{t \to \infty} \mu_t = \mu_\tau$. μ_τ codes a survival probability for the parent individuals in the limit $\lim_{t \to \infty} \mu_t$. In this way we get a soft change from the (μ, λ)– to the $(\mu + \lambda)$–strategy, what we call a $(\mu * \lambda)$–strategy, whereby best individuals always will be preserved.

2.2.2 The migration scheme for subpopulations

The above explained modifying and selection operators are integrated in a *migration scheme* of topological ordered subpopulations Π_i. In this model, originally introduced by TOTH ET AL. [TL93] and here applied in an extended approach, the subpopulation are arranged on a topological structure which is often chosen to be a regular lattice, for instance a ring, a quadratic lattice or a cube.[4] A visiting (migration) from individuals between neighbored (closed) subpopulations is allowed, i.e. some individuals are visiting neighbored populations for a short time with respect to the topological order and according to the range of neighborhood function.

During the evaluation of each subpopulation the neighborhood function h is applied which determines the number of visiting individuals from each other subpopulation. At the begin t_0 of the evolution process the range h of the neighborhood comprise nearly the complete set of subpopulations and it decreases exponentially during the time according to

---

[4]In general, other arrangements are are also admissible. Then the lattice can be defined by connection matrix which descibes the neighborhood relations.

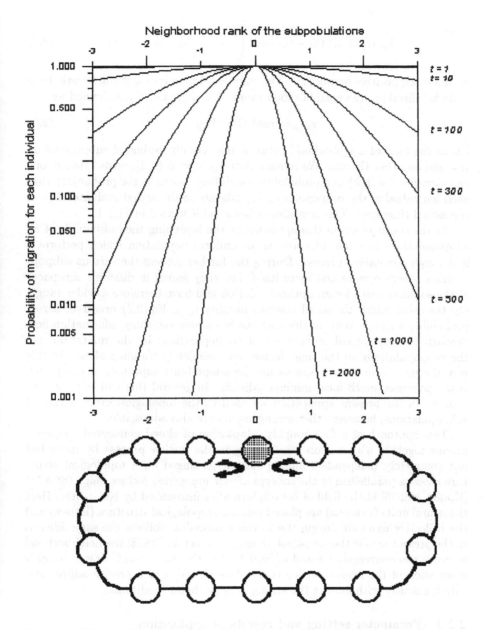

Figure 1: Illustration of the migration scheme. The subpopulations (here depicted as circles) are arranged on ring as topological structure. The actually considered subpopulation i^* is shaded. Individuals from the other subpopulations may migrate (visit for one time step) the actual subpopulation according to a probability which is determined by the neighborhood rank of its subpopulation (shown above). Thereby the range of neighborhood defined by the neighborhood function $h_{i^*}(t,k)$ is shrinking during the evolution process as also depicted in the above diagram.

$$h_{i^*}(t,k) = (1.0 - \epsilon_h) \cdot \exp\left(-\frac{r_{i^*,k} \cdot (t - t_0)^2}{\sigma_h^2}\right) + \epsilon_h \qquad (2.4)$$

with a small positive number ϵ_h. The value σ_h determines a characteristic time scale for shrinking the neighborhood range. The function $r_{i,k}$ is defined as

$$r_{i^*,k} = \text{rank}\,(\Pi_{i^*}, \Pi_k)\ , \qquad (2.5)$$

i.e. as the rank of neighborhood between the actually evaluated subpopulation Π_{i^*} and an other Π_k one. We remark that $h_{i^*}(t,k) \in (0,1]$ yields. Hence, one can interpret $h_{i^*}(t,k)$ as a probability which here describes the probability that each individual of the subpopulation Π_k migrate to the actual evaluated Π_{i^*} in the actual time step. The migration scheme is illustrated in Fig. 1.

As the consequence of this approach at the beginning the individuals of all subpopulations one can interpret as an uniform population which perform a first rough adaptation process. During the further process the various subpopulations separate more and more itself, i.e. they search in different subspaces of the solution space for an optimal solution and have therefore a wider range. On the other hand, the small positive number ϵ_h in Eq.(2.4) preserves a rest probability for migration. In this way one has a non–vanishing information flow through the topological ordered set of subpopulations by the migration, i.e. the subpopulations all the time do not act completely standing alone. In this way the migration scheme speeds up the adaptability especially if the search space possesses much local minima [Mic96]. In general this will improve the results. In the present application we used as the topological order a ring of subpopulations, however other arrangements are also admissible.

This approach of a first roughly adaptation of closed connected subpopulations together with a more fine tuning in the further process by more but not completely independent subpopulations arranged on a topological structure shows a parallelism to the concept of *self–organizing feature maps* (SOFM) [Koh84, Koh95] in the field of neuroinformatics introduced by KOHONEN. Here the neural units (neurons) are placed also on a topological structure (lattice) and the collective dynamic during the learning procedure follows the same idea as in the present article the subpopulations. As shown in [RS88] the neighborhood improve the convergence speed of SOFM. On the other hand, first a roughly adaptation of the neuron lattice takes place changing to a precise adjustment simultaneously with loosing the strong neighborhood conditions.

2.2.3 Parameter setting and results of application

The fitness–function $F(s)$ of an individual s in this first study was a simple one. It was be composed using only two additive terms without using explicit therapeutic expert knowledge. The first term counts for all pairs $p_i = \left(S_{j^*}^i, S_{j+}^i\right)$ of the database P whether $S_{j^*}^i$ and S_{j+}^i belong to different clusters according the partition of the set S represented by the individual s, i.e. it rates the misclassifications. Of course, therapeutical experiences influence the database P and, hence, also the fitness function. In this way an implicit knowledge processing of the several therapeutists was implemented.

The second term of $F(s)$ takes the balance of the number of standard categories in the several clusters C_k into account, which is measured by the variance σ of the number of standard categories belonging to the clusters:

$$F(s) = \alpha \cdot \frac{\text{number of misclassifications}}{N} + \beta \cdot \tanh(\gamma \cdot \sigma) \ . \qquad (2.6)$$

The variables α and γ play the role of free selectable parameters, whereby α is related to the quantity β by the normalization condition $\alpha + \beta = 1$. For instance, the choice of $\gamma = 0$ leads as the result of application of an EA to such a partition of **S** that all standard categories are collected in only one cluster. We used in our computations the empirically determined values $\alpha = 0.5$ and $\gamma = 0.5$ and as initial values for the number of all individuals $\mu_{\text{all}} = 400$. The individuals were even distributed onto 10 subpopulations arranged on a ring as topological order, i.e. to each subpopulation belong $\mu_0 = 40$ individuals whereby is set to $\mu_\tau = 1$. We trained the ensemble during $t_{\max} = 5000$ time steps. The characteristic time scale for decreasing the neighborhood between the subpopulation was defined as $\sigma_h = \frac{t_{\max}}{3}$ in (2.4).

The application of this configuration yields a final partition the weighted concordance coefficient of which was $\kappa = 0.49$. This indicates again a clear agreement according to Tab. 1 . However, the new κ–value is the higher than the value obtained from the FDM and, therefore, of course much better than this one received from the original cluster mapping.

3 Conclusions

We have shown in this presentation that evolutionary algorithms are a proper method also for applications in psychosomatic medicine. We demonstrated this approach for a problem of finding clusters in a set of standard categories the correlations between it are implicitly given by a coupled appearance in a database of real clinical data. The solution obtained by evolutionary algorithms was compared with the result computed by a standard method of cluster analyzing, the factor decomposition method, and with the original clustering introduced by BARBER ET AL.. Our solutions received from both methods, the factor decomposition method and evolutionary algorithms suggest a reformulation of these original clusters. The best result we obtained from the evolutionary algorithm approach.

For the evolutionary algorithms we used $(\mu * \lambda)$–strategy as a balance between the classical $(\mu + \lambda)-$ and (μ, λ)–strategy together with a migration scheme. Thereby the subpopulation where arranged on a topological order, which here was chosen as a ring. This approach shows a parallelism to the concept of self-organizing feature maps in the field of neural computation.

However, the fitness function F up to now does not contains any explicit psychotherapeutic knowledge which we want include in our further work. Furthermore, a testing phase for the capability of the new clusters in practical work of therapeutists have to be carried out.

ACKNOWLEDGEMENT: THE AUTHORS WOULD LIKE THANK D. POKORNY, G. BLASER (BOTH UNIVERSITY ULM) AND R. HAUPT (UNIVERSITY LEIPZIG) FOR HELPFUL DISCUSSIONS.

References

[AVV+96] C. Albani, B. Villmann, Th. Villmann, M. Geyer, D. Pokorny, G. Blaser, and H. Kächele. Kritik und erste Reformulierung der kategorialen Strukturen der Methode des Zentralen Beziehungs–Konflikt–Themas (ZBKT). *Psychotherapie, Psychosomathik und medizinische Psychologie - Springer–Verlag*, page submitted, 1996.

[BCCL90] J. Barber, P. Crits-Christoph, and L. Luborsky. A guide to the CCRT Standard Categories and their classification. In L. Luborsky and P. Crits-Chrostoph, editors, *Understanding Transference*, pages 37–50. Basic Books New York, 1990.

[CE78] G. Clauß and H. Ebener. *Grundlagen der Statistik*. Verlag Volk und Wissen Berlin, 1978.

[Coh68] J. Cohen. Weighted kappa. *Psychological Bulletin*, 70:213–220, 1968.

[DTR+93] R. W. Dahlbender, L. Torres, S. Reichert, S. Stübner, G. Frevert, and H. Kächele. Die Praxis des Beziehungsepsioden–Interviews. *Zeitschrift für Psychosomatische Medizin und Psychoanalyse*, 56:490–495, 1993.

[Fle81] J. L. Fleiss. *Statistical Methods for Rates and Proportions*. Wiley, New York, 2nd edition, 1981.

[Fog95] D. B. Fogel. *Evolutionary Computation: Towards a New Philosophy of Machine Intelligence*. IEEE Press, Piscataway, NJ, 1995.

[Gjo88] Th. Gjorup. The Kappa coefficient and the prevalence of a diagnosis. *Methods of Information in Medicine*, 27:184–186, 1988.

[HHV95] K. Hering, R. Haupt, and Th. Villmann. An Improved Mixture of Experts Approach for Model Partitioning in VLSI–Design Using Genetic Algorithms. Technical Report 14, University of Leipzig / Inst. of Informatics, Germany, 1995.

[HHV96] K. Hering, R. Haupt, and Th. Villmann. Hierarchical Strategy of Model Partitioning for VLSI–Design Using an Improved Mixture of Experts Approach. In *Proc. Of the Conference on Parallel and Distribute Simulation (PADS)*, pages 106–113. IEEE Computer Society Press, Los Alamitos, 1996.

[Hol75] J.H. Holland. *Adaptation in Natural and Artificial Systems*. University of Michigan Press, 1975.

[KF81]   M. S. Kramer and A. R. Feinstein. The biostatistics of concordance. *Clinical Pharmacology and Therapeutics*, 29:111–123, 1981.

[Koh84]   Teuvo Kohonen. *Self-Organization and Associative Memory*. Springer, Berlin, Heidelberg, 1984. 3rd ed. 1989.

[Koh95]   Teuvo Kohonen. *Self-Organizing Maps*. Springer, Berlin, Heidelberg, 1995.

[LBS89]   L. Luborsky, J. Baber, and P. Schaffer. The assessment of the CCRT: comparison of tailor–made with standard category rating scales on a specimen case. In *Progress in Assessing Psychodynamic Functioning: A Comparison of Four Methods on a Single Case*. Society of Psychotherapy Research, Toronto, 1989.

[LCC90]   L. Luborsky and P. Crits-Christoph. *Understanding Transference*. Basic Books, New York, 1990.

[Lub77]   L. Luborsky. The core conflictual relationship scheme. In N. Freedman and S. Grand, editors, *Communicative Structure and Psychic Structures*. Plenum Press New York, 1977.

[MGSK88]   H. Mühlenbein, M. Gorges-Schleuter, and O. Krämer. Evolution Algorithm in Combinatorial Optimization. *Parallel Computing*, (7):65–88, 1988.

[Mic96]   Z. Michalewicz. *Genetic Algorithms + Data Structures = Evolution Programs*. Springer–Verlag Berlin Heidelberg New York, third, revised and extended edition, 1996.

[Rec73]   I. Rechenberg. *Evolutionsstrategie - Optimierung technischer Systeme nach Prinzipien der biologischen Information*. Fromman Verlag Freiburg (Germany), 1973.

[RS88]   H. Ritter and K. Schulten. Convergence properties of Kohonen's topology preserving maps: fluctuations, stability, and dimension selection. *Biol. Cyb.*, 60(1):59–71, 1988.

[Sac92]   L. Sachs. *Angewandte Statistik*. Springer Verlag, 7-th edition, 1992.

[Sch81]   H.-P. Schwefel. *Numerical Optimization of Computer Models*. Whiley and Sons, 1981.

[Sol93]   S. Soldz. Understanding transference: The CCRT method. *Psychotherapy Research*, 3:69–73, 1993.

[TL93]   Gábor J. Tóth and András Lőrincz. Genetic algorithm with migration on topology conserving maps. In Stan Gielen and Bert Kappen, editors, *Proc. ICANN'93, Int. Conf. on Artificial Neural Networks*, pages 605–608, London, UK, 1993. Springer.

Forecasting Sales Using Neural Networks

Frank M. Thiesing and Oliver Vornberger

Department of Mathematics and Computer Science
University of Osnabrück
D-49069 Osnabrück, Germany
E-Mail: frank@informatik.uni-osnabrueck.de

Abstract. In this paper, neural networks trained with the back-propagation algorithm are applied to predict the future values of time series that consist of the weekly demand on items in a supermarket. The influencing indicators of prices, advertising campaigns and holidays are taken into consideration. The design and implementation of a neural network forecasting system is described that has been installed as a prototype in the headquarters of a German supermarket company to support the management in the process of determining the expected sale figures. The performance of the networks is evaluated by comparing them to two prediction techniques used in the supermarket now. The comparison shows that neural nets outperform the conventional techniques with regard to the prediction quality.

1 Introduction

A central problem in science is predicting the future of temporal sequences. Examples range from forecasting the weather to anticipating currency exchange rates. The desire to know the future is often the driving force behind the search for laws in science and economics.

In recent years many sophisticated statistical methods have been developed and applied to forecasting problems [1], however, there are two major drawbacks to these methods. First for each problem an individual statistical model has to be chosen that makes some assumptions about underlying trends. Second the power of deterministic data analysis can be exploited for single time series with some hidden regularity (though strange and hard to see but existent), however, this approach fails for multidimensional time series with mutual non-linear dependencies.

As an answer to the weakness of statistical methods in forecasting multidimensional time series an alternative approach gains increasing attraction: neural networks [2]. The practicability of using neural networks for economic forecasting has already been demonstrated in a variety of applications, such as stock market and currency exchange rate prediction, market analysis and forecasting time series of political economy [3, 4, 5, 6].

The approaches are based on the idea of training a feed-forward multi-layer network by a supervised training algorithm in order to generalize the mapping between the input and output data and to discover the implicit rules governing the movement of the time series and predict its continuation in the future. Most of the proposals deal with one or only few time series.

There are two main streams in the manner of presenting the data to the nets. In *explanatory* forecasting the values of several different but interesting economic indicators at time t are used as the components of the input vector during the training and the value at time $t + 1$ of the time series to be predicted as the corresponding desired output. This approach is based on the assumption that the development of various phenomena in the economy are essential to study the behavior of the time series in discussion.

In *time series* prediction forecasting is realized by treating n successive past values of the time series in discussion ending up at time t as an input vector and the value at time $t + 1$ as the corresponding desired output. This technique processes the times series in a sliding window of width n. The underlying assumption of this procedure is that any information that is necessary to predict the future behavior is hidden in the time series only.

In this paper, neural networks trained with the *back-propagation* algorithm [7] are applied to predict the future values of 20 time series that consist of the weekly demand on items in a German supermarket. An appropriate network architecture will be presented for a mixture of both explanatory and time series forecasting. Unlike many other neural prediction approaches described in the literature, we compare the forecasting quality of the neural network to two prediction techniques currently used in the supermarket. This comparison shows that our approach produces good results.

2 Time Series Considered

The times series used in this paper consist of the sales information of 20 items in a product group of a supermarket. The information about the number of items sold and the sales revenue are on a weekly basis starting in September 1994. There are important influences on the sales that should be taken into consideration: advertising campaigns sometimes combined with temporary price reductions; holidays shorten the opening hours; the season has an effect on the sales of the considered items.

We take the sales information, prices and advertising campaigns from the cash registers and the marketing team of the supermarket. The holidays are calculated. For the season information we use the time series of the turnover sum in DM of all items of this product group as an indicator. Its behavior over a term of 19 months is shown in figure 1.

We use feed-forward multilayer perceptron (MLP) networks with one hidden layer together with the back-propagation training method. In order to predict the future sales the past information of n recent weeks is given in the input layer. The only result in the output layer is the sale for the next week.

Due to the purchasing system used in the supermarket there is a gap of one week between the newest sale value and the forecasted week. In addition the pricing information, advertising campaigns and holidays are already known for

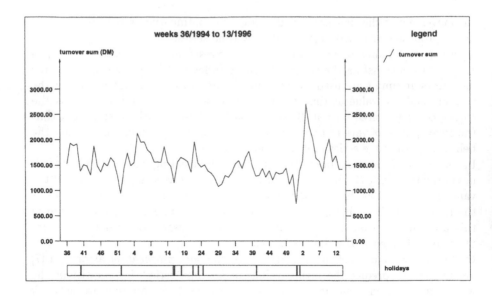

Fig. 1. turnover sum in DM of the product group September 1994 to March 1996

the future, when the forecast is calculated. This information is also given to the input layer as shown in figure 2.

3   Preprocessing the Input Data

An efficient preprocessing of the data is necessary to input it into the net. In general it is better to transform the raw time series data into indicators that represent the underlying information more explicitly. Due to the sigmoidal activation function of the back-propagation algorithm the sales information must be scaled to $]0, 1[$. The scaling is necessary to support the back-propagation learning algorithm [8]. We tested several scalings (z_t) for the sale and the turnover time series $x = (x_t)$:

$$z_t = \frac{x_t - \min(x)}{\max(x) - \min(x)} \cdot 0.8 + 0.1 \qquad \text{resp.}$$

$$z_t = \frac{x_t - \mu}{c \cdot \sigma} + 0.5$$

where *min* and *max* are the minimum and maximum values of time series x and μ and σ are the average and the standard deviation. c is a factor to control the interval of the values.

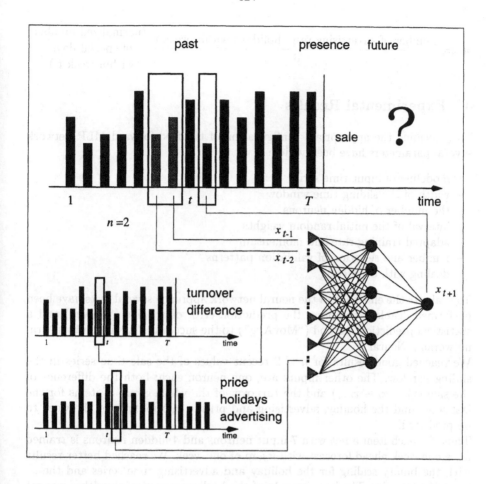

Fig. 2. input and output of the MLP

For the prices the most effecting indicator is the price change. So the prices are modeled as follows:

$$pri_t := \left\{ \begin{array}{rl} 0.9 & : \text{ price increases} \\ 0.0 & : \text{ price keeps equal} \\ -0.9 & : \text{ price decreases} \end{array} \right\} \begin{array}{l} \text{within} \\ \text{week } t \end{array}$$

For both the time series of holidays and advertising campaigns we tested binary coding and linear aggregation to make them weekly. Their indicators are:

$$y_t := \left\{ \begin{array}{ll} 0.9 & : \text{ if there is a holiday resp. advertising within week } t \\ 0.0 & : \text{ otherwise} \end{array} \right\} \text{ resp.}$$

$$y_t := \frac{\text{number of advertising resp. holidays within week } t}{6} \quad \begin{array}{l}(\text{normalized number} \\ \text{of special days} \\ \text{within week } t \text{ })\end{array}$$

4 Experimental Results

To determine the appropriate configuration of the feed-forward MLP network several parameters have been varied:

- modeling of input time series
- width of the sliding time window
- the number of hidden neurons
- interval of the initial random weights
- adapted training rate and momentum
- number and selection of validation patterns
- dealing with overfitting

To evaluate the efficiency of the neural network approach several tests have been performed. Table 1 compares the prediction error of a naive ("Naive") and a statistical prediction method ("MovAvg") to the successive prediction by neural networks ("Neural").

We reached good results for $n = 2$ recent values of the sale time series in the sliding window. The other inputs are, one neuron each: both the difference of the sale ($x'_t = x_t - x_{t-1}$) and the turnover of the whole group of items for the last week and the holiday, advertising and pricing information for the week to be predicted.

Thus, for each item a net with 7 input neurons and 4 hidden neurons is trained for a one week ahead forecast with a gap of one week. We reached better results with the binary scaling for the holiday and advertising time series and the σ-μ-scaling for sales. The learning rate of the back-propagation algorithm was set to 0.3 with a momentum of 0.1. The initial weights were chosen from [-0.5, 0.5] by chance. The training was validated by 12 patterns and the stopped at the minimum error.

4.1 Naive Prediction

The naive prediction method uses the last known value of the time series of sales as the forecast value for the future. In our terms: $\hat{x}_{t+1} := x_{t-1}$. This forecasting method is often used by the supermarket's personnel.

4.2 Statistical Prediction

The statistical method is currently being used by the supermarket's headquarters to forecast sales and to guide personnel responsible for purchasing. It calculates the moving average of a maximum of nine recent weeks, after these sale values have been filtered from exceptions and smoothed.

Table 1. prediction error: RMSE/Mean

Item	Mean	Neural	MovAvg	Naive
036252	7.79	1.037	1.217	1.181
078924	8.97	1.084	1.409	1.551
180689	9.92	0.973	1.243	1.231
215718	12.73	0.468	0.601	0.612
215732	12.97	0.325	0.402	0.395
215749	8.63	0.485	0.471	0.535
228558	17.40	0.481	0.709	0.923
229104	63.88	0.431	0.591	0.649
289573	8.29	0.992	1.154	1.325
304962	15.35	0.971	1.267	1.523
341110	6.79	0.672	0.815	0.755
341127	5.79	0.953	1.114	1.293
362238	10.88	1.221	1.573	2.138
372206	9.69	3.119	3.253	4.513
392785	19.79	0.513	0.624	0.534
399883	11.47	0.407	0.530	0.444
468978	23.99	0.263	0.318	0.387
468985	17.19	0.942	1.135	1.420
567411	3.95	0.586	0.569	0.623
852234	7.75	0.891	1.125	1.211
Average	—	0.841	1.006	1.162

4.3 Comparison of Prediction Techniques

To measure the error the root mean squared error (RMSE) is divided by the mean value ("Mean") of the time series. The results are calculated for the successive prediction of the 22 weeks 44/1995 to 13/1996. In this weeks there are influences of many campaigns and Christmas holidays.

Based on the information in table 1 the naive approach is outperformed by the two other methods. For 18 of the 20 items the prediction by the neural network is better than the statistical prediction method.

A close inspection of the times series favored by the statistical approach shows that these are very noisy without any implicit rules that could be learned by the neural network. Especially item 567411 has an average weekly sale of less than 4 items.

Figure 3 shows the predicted values for item 468978 calculated by the statistical and neural approach. The price, advertising and holiday information is included in this figure.

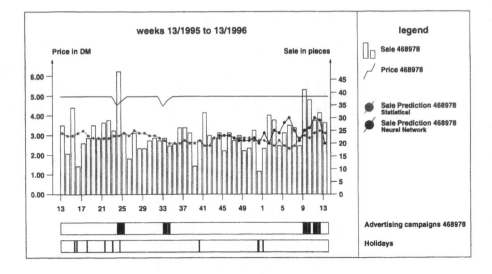

Fig. 3. sale prediction for an item by statistical and neural approach

5 Conclusions and Future Research

For a special group of items in a German supermarket neural nets have been trained to forecast future demands on the basis of the past data augmented with further influences like price changing, advertising campaigns and holiday season information. The experimental results show that neural nets outperform the naive and statistical approaches that are currently being used in the supermarket.

In contrast to many other neural prediction approaches our procedure preprocesses the data of all kinds of time series in the same manner and uses the same network architecture for the prediction of all 20 time series of sales. The parameter optimization is based on all of the time series instead of on one special item.

The program runs as a prototype and handles only a small subset of the supermarket's inventory. Future work will concentrate on the integration of our forecasting tool into the whole enterprise data flow process. Since a huge number of varying products have to be managed a selection process has to be installed that discriminates between steady time series suitable for conventional methods and chaotic candidates which will be processed by neural nets.

The prototype is part of an automatic forecasting system that is able to take the raw data, do the necessary preprocessing, train the nets and produce an appropriate forecast. The next steps will be the development of additional adaptive transformation techniques and methods to test the significance of inputs which can be used to reduce the complexity of the nets.

328

In addition other training algorithms for neural networks will be compaired. This will be done by the *Stuttgart Neural Network Simulator SNNS*[9].

References

1. Weigend A.S., Gershenfeld, N.A., *Time Series Prediction: Forecasting the Future and Understanding the Past*, Addison-Wesley, 1994.
2. Rojas, R., *Neural Nets*, Springer, 1996.
3. Schöneburg, E., "Stock Price Prediction Using Neural Networks: An Empirical Test," *Neurocomputing*, 2, 1, 1991.
4. Refenes A.N., Azema-Barac M., Chen L., Karoussos, S.A., "Currency Exchange Rate Prediction and Neural Network Design Strategies," *Neural Computing & Applications*, 1(1) pp. 46–58, 1993.
5. Chakraborty, K., Mehrotra, K., Mohan, C.K., Ranka, S., "Forecasting the Behaviour of Multivariate Time Series Using Neural Networks," *Neural Networks*, Vol. 5, pp. 961–970, 1992.
6. Freisleben, B., Ripper, K., "Economic Forecasting Using Neural Networks," *Proceedings of the 1995 IEEE International Conference on Neural Networks*, Vol. 2, pp. 833–838, Perth, WA., 1995.
7. Vemuri, V.R., Rogers, R.D., *Artificial Neural Networks - Forecasting Time Series*, IEEE Computer Society Press, 1994.
8. Rehkugler, H., Zimmermann, H.G., *Neuronale Netze in der Ökonomie (in German; Neural Networks in Economics)*, Verlag Vahlen, München 1994.
9. *SNNS Stuttgart Neural Network Simulator User Manual, Version 4.1*, University of Stuttgart, Report No. 6/1995.

Word-Concept Clusters in a Legal Document Collection

T.D. Gedeon [1], R.A. Bustos [1], B.J. Briedis [1],
G. Greenleaf [2] and A. Mowbray [3]

[1] Department of Information Engineering
School of Computer Science Engineering
The University of New South Wales
Sydney NSW 2052, Australia

School of Law
[2] The University of New South Wales
[3] University of Technology, Sydney

http://www.cse.unsw.edu.au/~tom

Abstract. For very large document collections or high volume streams of documents such as information resources on the web, finding relevant documents is a major information filtering problem. Traditional full text retrieval methods can not locate documents which use specialised synonyms or related concepts to the formal query. This is particularly a problem in legal document collections, since lawyers use normal words with specialised meanings which vary subtly between legal sub-domains. We use a neural network approach to learn synonyms and related clusters of words defining similar concepts from a sample document set. We demonstrate that our clusters of words are qualitatively useful, in the legal domain in particular, and can thus be used for high throughput information filtering to find documents likely to contain concepts relevant to a user's information need.

1 Background

The Australasian Legal Information Institute (AustLII), was established by the University of New South Wales and the University of Technology, Sydney. Funding for 1995 was provided to Greenleaf Mowbray and Gedeon by the Department of Employment, Education and Training, and supplemented by the two Universities. Further funding has been received from the Law Foundation of NSW for 1996, and from the Australian Research Council for 1996-1998 to Gedeon Greenleaf and Mowbray. The work reported in this paper was supported by the latter grant.

The high volume use of the legal materials available via the internet on AustLII provides an invaluable research opportunity in information filtering, retrieval and index generation, particularly for neural networks which require large numbers of instances for training. AustLII the World Wide Web site (http://www.AustLII.edu.au)

came up on the web at the beginning of July 1995, and by August of 1995, AustLII was averaging 4,000 hits per work day. By the end of August 1996, the AustLII site was averaging 38,000 hits per work day.

2 Introduction

The problem domain is the provision of sophisticated access to legal information via AustLII, which allows the modelling of the complex interconnections possible between sources of information, which does not require expensive expert intervention to maintain, and is adaptive to user needs.

Hypertext meets the first two criteria, our aim is to use neural network and other AI learning techniques to discover useful connections [3] based on the document collections themselves, and to maintain and enhance the hypertext structure [4] based on observation of user interaction with the AustLII internet resource.

Users face a difficult task when formulating queries for boolean retrieval: words must be selected that will retrieve the documents wanted, but fail to retrieve unwanted documents. Blair [1] has suggested that this is an unreasonable expectation of users and that retrieval performance of boolean retrieval systems is seriously limited as a result. In situations where high recall is desired (as for most legal tasks) we can add words to the query that will have the least negative effects on precision.

3 Neural network experiment

We create a network consisting of an input and output node for each word, connected by hidden units. Training patterns are generated using each document in the collection. One input only is activated for each input vector, corresponding to a word that occurs in the document under consideration.

The corresponding output vector consists of the word frequencies of all of the words occurring in the document. A pattern is generated in this way for each word in each document. The network is trained on these patterns, and the back propagation algorithm is used to generalise an output vector of word activation terms most similar to the training examples for the given input.

The word activation values can be ranked in descending order to discover the most important related words.

3.1 Preliminary work

This paper continues a preliminary experiment using INSPEC (computer science: neural networks) abstracts [5].

The aim is to use on-line legal data from AustLII, and also address the issue of sub-dividing large documents for retrieval [9].

3.2 Network topology

One hundred words were selected from the collection using a cumulative [2] inverse document term weight [8] method, and an input and output unit created for each. Varying numbers of hidden units were tested over 700 epochs, and on the basis of performance a network with 10 hidden units was constructed.

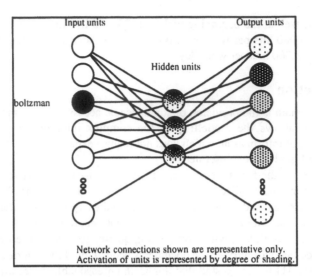

Fig. 1. Network schematic

3.3 Pattern Generation

Every occurrence of an indexed word in the document collection generates a training pattern. Input vectors can be described as input categories, since only a single word unit is activated for the pattern. This unit is activated with a magnitude of 1. The corresponding target output vector for each category is the document word frequency profile of the document containing the input word.

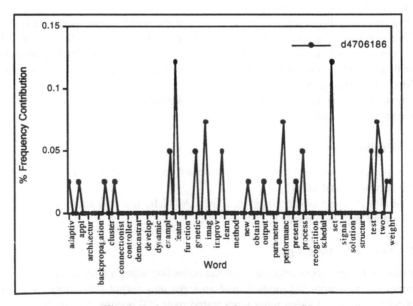

Fig. 2. A document word frequency profile

Document profiles were calculated by normalising the word frequency of each indexed word. The 2,191 abstracts generated 10,220 patterns.

The 621 legal documents (web pages on AustLII) generated 34,128 patterns, from which 22,760 were used for training and 11,368 for validation. The documents were Residential Tenancies Tribunal cases dealing with rental bonds. This data set was chosen because the cases are short and small in number, and are thus similar to our previous work using computer science abstracts [5].

3.4 Inherent error

This task required that the network generalise from the training set which includes multiple target output vectors for each input category. As a result, the training set could be considered to contain an amount of 'inherent' error, since the network would be unable to find a set of weights that exactly satisfies all of the mappings from the input categories to the document profiles.

To quantify this error so that the learning performance of the network can be evaluated, the minimum inherent error was estimated by calculating the total sum of squares of the difference between the maximum and minimum values for each term of the pattern vectors.

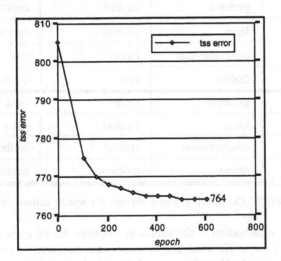

Fig. 3. Total Sum Squares error during training

For each category i with target vector $V_i[t_1, t_2, ..., t_n]$, with terms $V_i t_j$ the minimum inherent error is:

$$\sum_{j=1}^{N} \left(max\left(V_x t_j \right) - min\left(V_y t_j \right) \right)^2 \tag{1}$$

where x, y range over the vectors for each category (input term), and N is the number of terms. Clearly, this method will underestimate the inherent error except in the degenerate case where all the values are clustered together with just one outlier. However, the calculated value of 256.7 for our data set is an indication of the minimum error possible on this data set. Considering the minimum inherent error, the final TSS of 764 for 10,220 patterns appears acceptable. Networks with varying numbers of hidden units produced similar TSS error profiles.

3.5 Clusters Generated

Clusters are generated by activating a category and ranking the output word units by activation. The most highly activated word units were selected for each input category.

word	technique	Neural Network	Total Co-occ.	Ave. Co-occ.
connectionist		solution	model	model
		problem	fuzzy	fuzzy
		nonlinear	author	author
		filter	learn	learn
fuzzy		perform	control	controller
		learn	method	function
		computational	model	rule
		feature	rule	control
nonlinear		example	model	model
		input	method	filter
		connectionist	control	method
		neuron	problem	neuron

Table 1. INSPEC Comp.Sci. example clusters for words: action, consent, sign

The most striking observation is the similarity between the co-occurence measure cluster contents for a specific word, and even more the similarity of cluster contents for different source words. Note that none of the neural network derived clusters for each of the source words has any overlap.

word	technique	Neural Network	Total Co-occ.	Ave. Co-occ.
action		sum	tenant	tenant
		item	landlord	landlord
		pay	premise	premise
		provide	rent	act

consent	paid	tenant	compensation
	circumstance	landlord	claim
	follow	premise	clean
	item	agreement	tenant
sign	agreement	tenant	said
	pay	landlord	show
	set	premise	rent
	july	agreement	premise

Table 2. AustLII Legal example clusters for words: action, consent, sign

The network clusters are reasonably good, with the words found having some plausible conceptual relationships in the context of legal documents. The total co-occurence statistical measure is essentially useless in this data set, with most input words words producing very similar clusters. The average co-occurence clusters showed less of this effect, however, the words in the clusters are qualitatively less satisfactory than the network produced clusters.

We examined the document vectors we produced from the documents, below is a representation of the vectors using a 4 point scale.

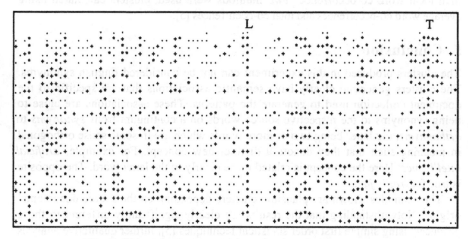

Fig. 4. Compressed representation of a few document vectors

Some words clearly occur in most documents and hence appear as a solid column in Figure 4. In particular, two words are identified. The letter L shows the location in the document representation of landlord, and T shows tenant. This explains the failure of the total co-occurence measure. It is worth noting that the neural technique has shown itself to be robust in this situation, and still produces useful clusters of words.

Issues remaining to be resolved are:

- whether the two statistical techniques can be improved to provide more reasonable comparison for the the neural network;
- improvement of the word selection technique, in that clearly the inclusion of landlord and tenant in the 100 words used to train the neural networks was not ideal in this case; and
- the contribution of the method used for scaling word importance in the production of the document vectors.

4 Results

The examples shown in the previous section are clearly not English synonyms. The backpropagation procedure which produced these words based on the nonlinear nature of artificial neurons is sensitive to the statistical distribution of the collection frequency data the network has been trained on.

The final decision on their usefulness remains to be tested in practice to determine whether the collections of words has value in denoting concepts. Note that Kumar and Lindley [6] have shown that even trigram information traces are suitable for hypertext information retrieval. We retain more information than they do. We believe that our clusters are adaptive to, and reflect higher order statistical dynamic information about the words in the specific (sub-)collection.

To test this assertion, we have performed a simple first order statistical analysis of the clusters, to determine if the network was producing clusters of greater complexity than local word co-occurrence. Two methods were used, clusters calculated using average word co-occurrences and total co-occurrences [5].

5 Conclusion

The clusters produced in this experiment can not be considered English synonyms. The clusters generated clearly have semantic associations that are specific to the document collection used to generate the patterns. These associations are close to being synonyms to the respective source words on a continuum from *synonyms* to *orthognyms*, the latter being two or more words which are in some sense orthogonal in meaning such that they denote a distinct new concept. For example, *artificial intelligence*. Here the concept is related more clearly to the latter word. The opposite is true for *traffic jam*.

The examples shown are plausible clusters depending on the source documents' origins. Notwithstanding that the results are clearly significantly different to those produced using simple first order statistical techniques [5], further qualitative analysis using a comprehensive domain thesaurus is required to understand more fully the semantic value in these clusters.

It is apparent that words occurring with high frequency in the collection are more often included in clusters, than those occurring only infrequently in the collection. A possible solution to this problem would be to use an alternative method when generating training patterns. The relatedness function proposed by Wilks [8] takes into account word frequency and could be modified to apply to single documents rather than

336

the whole collection. It may also be possible to speed the network learning and improve generalisation by scaling the network training patterns. The danger in this approach is that the already noisy relations inherent in the data may be obscured.

The technique described here has possible practical application to off-line processing of retrieval collections, and with further development, automated generation of synonyms that are domain specific [2]. Thesauri are useful to augment users queries, however the high costs of maintenance means that they can rarely be truly domain specific. Query enhancement strategies to improve information retrieval will become more practical when such thesauri are more readily available. The techniques described here apply back propagation neural networks to this problem in a way that has not been reported elsewhere.

References

1. Blair D.C. *Language and Representation in Information Retrieval*, Amsterdam, Elsevier, 1990.
2. Bustos, R.A. and Gedeon, T.D. "Learning Synonyms and Related Concepts in Document Collections," in Alspector, J., Goodman, R. and Brown, T.X. *Applications of Neural Networks to Telecommunications 2*, pp. 202-209, Lawrence Erlbaum, 1995.
3. Gedeon, T.D., Johnson, L. and Mital, V. "Neural Networks for Information Retrieval," in Mital, V. and Johnson, L. *Advanced Information Systems for Lawyers*, pp. 268-277, Chapman & Hall, 1992.
4. Gedeon, T.D. and Mital, V. "Information Retrieval in Law using a Neural Network Integrated with Hypertext," *Proceedings International Joint Conference on Neural Networks*, pp. 1819-1824, Singapore, 1991.
5. Gedeon, T.D. and Bustos, R.A. "Word-Concept Clusters in Document Collections," *Proceedings Australian Document Computing Conference*, pp. 21-24, Melbourne, 1996.
6. Kumar, V.R. and Lindley, C.A. "Improving Decision Support Through Hypermedia", *Proceedings, 3rd ACM Golden-West International Conference on Intelligent Systems*, Kluwer Academic Publishers, Las Vegas, 1994.
7. Salton, G. *The SMART Retrieval System - Experiment in Automatic Document Processing*, Englewood Cliffs, Prentice-Hall, 1971.
8. Wilks, Y., Guthrie, L., Guthrie, J. and Cowrie, J. "Combining Weak Methods in Large-Scale Text Processing" in Jacobs, P.S. *Text-Based Intelligent Systems: Current Research and Practice in Information Extraction and Retrieval*, Lawrence Erlbaum Associates, Hillsdale, New Jersey, at pp. 35, 1992.
9. Zobel, J., Moffat, A., Wilkinson, R. and Sacks-Davis, R. "Efficient Retrieval of Partial Documents," *Information Processing and Management*, vol. 31, no. 3, pp. 361-377, 1995.

A Retrospective View and Outlook
on Evolutionary Algorithms

Lawrence J. Fogel

Natural Selection, Inc.
3333 North Torrey Pines Ct., Suite 200
La Jolla, CA 92037
lfogel@natural-selection.com

Abstract

Evolutionary algorithms have been studied for over 35 years. This paper provides a brief summary of the similarities and differences of various methods in evolutionary computation, as well as some ideas for future avenues of research.

1 Introduction

Recent years have seen a rapid increase in the use of simulated evolution to address difficult problems in machine learning. Efforts within the field of *evolutionary computation* have generally followed three main lines of investigation: (1) *genetic algorithms*, (2) *evolution strategies*, or (3) *evolutionary programming*. These techniques are broadly similar. Each relies on a population of contending trial solutions which are subjected to random alterations and compete to be retained as parents of successive progeny. The differences between the methods concern the level in the hierarchy of evolution being modeled: the chromosome, the individual, or the species.

Genetic algorithms model evolution as a succession of changing gene frequencies, with contending solutions being analogous to chromosomes. The space of possible solutions is searched by applying transformations to the trial solutions as observed in the chromosomes of living organisms: crossover, inversion, point mutation, and so forth (Holland 1975). Solutions are typically made to propagate into future generations with a likelihood that is proportional to their fitness relative to all other existing solutions. In this manner, it is hoped that the population will recombine building blocks of independently discovered solutions with above-average fitness and thereby generate new, improved solutions over successive generations (Goldberg 1989).

In contrast, evolution strategies (Rechenberg 1965; Schwefel 1995) and evolutionary programming (Fogel 1962; Fogel et al. 1966) (also sometimes collectively referred to as *evolutionary algorithms*, as opposed to genetic algorithms; Mühlenbein 1992; Fogel 1993a; Goldberg 1994) model evolution as a process of adaptive behavior of individuals or species, respectively, rather than of adaptive genetics. Populations of trial solutions are evolved, but the solutions are modified such that there is a continuous range of possible new behaviors while at the same time maintaining a strong behavioral link between parents and offspring. For example, if the solutions are represented as real-valued vectors, a typical mutation operation is to add a multi-variate zero mean Gaussian random variable to each parent. Selection operates, either deterministically or probabilistically, respectively, to eliminate the worst solutions in the population rather than promote copies of those with above-average fitness.

2 Retrospective Differences

Although the above methods of simulating evolution are broadly similar, important differences exist in the philosophy that each adopts to problem solving. A keystone of genetic algorithms is a bottom-up perspective. The traditional view has been that it is worthwhile to code problems into strings of bits (simulating genes along a chromosome), then create new solutions by crossing over existing ones. This method was founded on the belief that crossover will work to combine building blocks of "good genes" from individual bit strings and thereby construct superior solutions. But restrictions on the representation and modes of genetic variation may or may not be useful, depending on the specific circumstance.

When such bottom-up approaches have been applied to real-world optimization problems, they have often prematurely converged to suboptimal solutions. This has led to a need for additional rules to compensate for the remaining error, for example, the use of steepest descent. Under such circumstance it is appropriate to question the validity of the original hypothesis that optimal solutions can be created by the repeated assembly of suboptimal subsections of complete solutions.

The genetic algorithm replicates with greater frequency those genes (i.e. building blocks) associated with above-average performance (an answer to the credit assignment problem). But it is improper to attribute value to individual components, genes, or any other individual parts in a collective whole. When viewed in isolation, these components have no value. Worth is only realized in their purposeful interaction. By analogy, if asked "What is the value of a particular playing card in your hand?" your response should be "That's not a proper question." The individual card has no worth. It is useful only when it is part of a hand of other cards, and the utility of the entire hand depends on the other cards in the hand, the cards in the other players' hands, the game being played, and the ability of the other players. It is inappropriate to estimate the worth of individual genes because they generally affect more than one phenotypic trait. Credit assignment can be a useful practice in certain cases, but it generally leads to suboptimal designs in complex systems.

In contrast, the methods of evolution strategies and evolutionary programming have not relied on the recombination of building blocks. Instead they have recognized that an optimal design is likely to comprise components that strongly interact in some purposeful manner. Natural evolution designs just such seamless systems without separate optimization of the components. Selection acts from the top down. To survive, each creature must meet the challenge of its environment: find sufficient food, avoid predation, find adequate shelter, and so forth. The measure of its success is the suitability of its complete expressed behavior, not the "goodness" of its individual components or subcomponents. Following this perspective, entire collections of parameters to be evolved are simultaneously varied, often using continuous random perturbations such as Gaussian or Cauchy distributions (and operators with similar properties in discrete representations).

Although genetic algorithms, evolution strategies, and evolutionary programming have all been used with success in addressing difficult engineering challenges, it is perhaps not surprising to find that evolution strategies and evolutionary programming have outperformed genetic algorithms on a variety of optimization problems. These include the traveling salesman problem (Fogel, 1993b), automatic control (Fogel, 1994), and pattern recognition (Rizki et al., 1993). Initial comparisons by Fogel and Atmar (1990) indicated that purely random mutations could be more profitably applied to general function optimization experiments using linear systems than could crossover. More recently, Fogel and Stayon (1994) directly compared results from evolutionary programming with published results using genetic algorithms conducted by Schraudolph and Belew (1992). The results favored evolutionary programming across a wide range of functions, including a test suite of functions that have been used by the genetic algorithm community for tuning their procedures. Bäck (1996) (and Bäck and Schwefel, 1993) have compared evolution strategies to genetic algorithms on more complex functions and found evolution strategies to be superior.

The "no free lunch" theorems of Wolpert and Macready (1997), as well as other related theoretical results offered in Fogel and Ghozeil (1997), demonstrate that there can be no one superior approach to addressing all problems. But the empirical evidence to date suggests that real-world problems pose challenges that may not be particularly well suited for piecemeal analysis, and it seems prudent to pay greater attention to evolution strategies and evolutionary programming for tackling these significant challenges.

3 Perspective on Future Efforts

At this point it would appear wise to devise a framework for addressing issues that remain in evolutionary computation. These concern both basic and applied research.

With regard to basic research, there are many fundamental facets of evolutionary algorithms where our knowledge is embarrassingly poor. For example, there are no generally useful guidelines for trading off population size for the number of function evaluations, the type of variation or selection, and so forth. Some preliminary efforts have been made in Jog et al. (1989), Bäck (1996), and Gehlhaar and Fogel (1996), but these are only a first step. Moreover, the theory regarding the rates of convergence of evolutionary algorithms on different functions is very limited. Bäck (1996) and Fogel (1995) provide some overviews, and Beyer (1995) offered some new results on multirecombination, but the required assumptions to perform the analysis (i.e., strongly convex functions) limits its utility.

One aspect of basic research where there has been considerable effort is the area of self-adaptation. Work dates back to Reed et al. (1967) and Rechenberg (1973), with more substantive recent efforts in Schwefel (1981), Fogel et al. (1991), Angeline (1996), and many others. The traditional interest has been in evolving parameters for continuous mutation operators, but some work has now been directed to self-adaptation in discrete spaces (e.g. Angeline et al., 1996; Chellapilla and Fogel, 1997). These investigations deserve greater attention.

With regard to applied research, the potential areas for application are diverse. These include: (1) optimal forecasting in light of an arbitrary payoff matrix, (2) designing optimal neural networks and/or fuzzy systems, (3) identifying chaotic signals in noise, (4) data mining, (5) image processing, (6) biochemistry and medical diagnosis, and (7) modeling economic systems and ecosystems. Applied research is directed by the problem at hand and a review of current proceedings in evolutionary algorithms indicates several specific possibilities (e.g., Fogel et al., 1996; Voigt et al., 1996; Angeline et al., 1997).

In order to best pursue the open problems in evolutionary algorithms, these problems must be identified, categorized, and prioritized. This presumes an overall purpose of the general community, which must be quantified. Effort is required to properly define each concern in terms of its relative importance, the degrees of achievement that may be anticipated, the likelihood in this regard, the associated time and cost required, and their associated degree of criticality. This can be expressed as a Valuated State Space and normalizing function that can be used to measure the current overall degree of success and that of prospective situations that may grow out of any investment strategy.

The next step is therefore to weight the concerns in relative importance, define scales of achievement for each of these items, then identify the current state of the art. The difference between the value associated with the current level and the desired level multiplied by the influence coefficient then forms a preliminary basis for ranking opportunities for investment.

The last step is to estimate the time and cost needed to overcome these noted deficiencies. Here again, the ranking may change for, in some cases, a small investment might have a large payoff in measurable terms.

4 Conclusion

As is so often the case, nature separately solves the same problem in similar but different ways. In recent decades, different evolutionary computation techniques were devised at different times, independently, and in different settings. Only in recent years have these techniques been viewed as different avenues with respect to a broad subject of evolutionary computation. No doubt, progress will be made at an increasing rate, for a wide diversity of applications is in view and the computational equipment far exceeds what would have been dreamed of when the field originated more than 35 years ago.

Although individual projects can contribute in their own right, an overview of the problems that deserve to be addressed in consort may serve to stimulate more appropriate investment of time, effort, and of course, funding by various sponsoring agencies. This may cross-fertilize the projects and at the same time build a more cohesive community devoted to evolutionary computation.

5 References

P.J. Angeline (1996) "The effects of noise on self-adaptive evolutionary optimization," *Evolutionary Programming V: Proceedings of the Fifth Annual Conference on Evolutionary Programming*, L.J. Fogel, P.J. Angeline, and T. Bäck (eds.), MIT Press, Cambridge, MA, pp. 433-439.

P.J. Angeline, D.B. Fogel, and L.J. Fogel (1996) "A comparison of self-adaptation methods for finite state machines in dynamic environments," *Evolutionary Programming V: Proceedings of the Fifth Annual Conference on Evolutionary Programming*, L.J. Fogel, P.J. Angeline, and T. Bäck (eds.), MIT Press, Cambridge, MA, pp. 441-449.

P.J. Angeline, R.G. Reynolds, J.R. McDonnell, and R.C. Eberhart (eds.) (1997) *Evolutionary Programming VI: Proceedings of the Sixth Annual Conference on Evolutionary Programming*, Springer, Berlin.

T. Bäck (1996) *Evolutionary Algorithms in Theory and Practice*, Oxford, NY.

T. Bäck and H.-P. Schwefel (1993) "An overview of evolutionary algorithms for parameter optimization," *Evol. Comp.*, Vol. 1:1, pp. 1-24.

H.-G. Beyer (1995) "Toward a theory of evolution strategies: on the benefit of sex — the $(\mu/\mu,\lambda)$-theory," *Evol. Comp.*, Vol. 3:1, pp. 81-111.

K. Chellapilla and D.B. Fogel (1997) "Exploring Self-Adaptive Methods to Improve the Efficiency of Generating Approximate Solutions to Traveling Salesman Problems Using Evolutionary Programming," *Evolutionary Programming VI: Proceedings of the Sixth Annual Conference on Evolutionary Programming*, P.J. Angeline, R.G. Reynolds, J.R. McDonnell, and R.C. Eberhart (eds.), Springer, Berlin, in press.

D.B. Fogel (1993a) "On the philosophical differences between evolutionary algorithms and genetic algorithms," *Proceedings of the Second Annual Conference on Evolutionary Programming*, D.B. Fogel and W. Atmar (eds.), Evolutionary Programming Society, La Jolla, CA, pp. 23-29.

D.B. Fogel (1993b) "Applying evolutionary programming to selected traveling salesman problems," *Cybernetics and Systems*, Vol. 24, pp. 27-36.

D.B. Fogel (1994) "Applying evolutionary programming to selected control problems," *Comp. Math. Applic.*, Vol 27:11, pp. 89-104.

D.B. Fogel (1995) *Evolutionary Computation: Toward a New Philosophy of Machine Intelligence*, IEEE Press, NY.

D.B. Fogel and J.W. Atmar (1990) "Comparing genetic operators with Gaussian mutations in simulated evolutionary processing using linear systems," *Biological Cybernetics*, Vol. 63, pp. 111-114.

D.B. Fogel, L.J. Fogel and J.W. Atmar (1991) "Meta-evolutionary programming," *Proc. of the Asilomar Conf. on Signals, Systems and Computers*, R.R. Chen (ed.), Maple Press, San Jose, CA, pp. 540-545.

D.B. Fogel and A. Ghozeil (1997) "A note on representations and operators," *IEEE Trans. Evolutionary Computation*, Vol. 1:2, in press.

D.B. Fogel and L.C. Stayton (1994) "On the effectiveness of crossover in simulated evolutionary optimization," *BioSystems*, Vol 32:3, pp. 171-182.

L.J. Fogel (1962) "Autonomous automata," *Industrial Research*, Vol. 4, pp. 14-19.

L.J. Fogel, A.J. Owens and M.J. Walsh (1966) *Artificial Intelligence through Simulated Evolution*, John Wiley, NY.

L.J. Fogel, P.J. Angeline, and T. Bäck (eds.) (1996) *Evolutionary Programming V: Proceedings of the Fifth Annual Conference on Evolutionary Programming*, MIT Press, Cambridge, MA.

D.K. Gehlhaar and D.B. Fogel (1996) "Tuning evolutionary programming for conformationally flexible molecular docking," *Evolutionary Programming V: Proceedings of the Fifth Annual Conference on Evolutionary Programming*, MIT Press, Cambridge, MA, pp. 419-429.

D.E. Goldberg (1989) *Genetic Algorithms in Search, Optimization, and Machine Learning*, Addison-Wesley, Reading, MA.

D.E. Goldberg (1994) "Genetic and evolutionary algorithms come of age," *Communications of the ACM*, Vol. 37, pp. 113-119.

J. H. Holland (1975) *Adaptation in Natural and Artificial Systems*, Univ. Mich. Press, Ann Arbor.

P. Jog, J.Y. Suh, and D. Van Gucht (1989) "The effects of population size, heuristic crossover and local improvement on a genetic algorithm for the traveling salesman problem," *Proceedings of the Third Intern. Conf. on Genetic Algorithms*, J.D. Schaffer (ed.), Morgan Kaufmann, San Mateo, CA, pp. 110-115.

H. Mühlenbein (1992) "Evolution in time and space — the parallel genetic algorithm," Foundations of Genetic Algorithms, G.J.E. Rawlins (ed.), Morgan Kaufmann, San Mateo, CA, pp. 316-337.

I. Rechenberg (1965) "Cybernetic solution path of an experimental problem," Royal Aircraft Establishment, Library Translation No. 1122, August.

J. Reed, R. Toombs, and N.A. Barricelli (1967) "Simulation of biological evolution and machine learning. I. Selection of self-reproducing numeric patterns by data processing machines, effects of hereditary control, mutation type and crossing," *J. Theoret. Biol.*, Vol. 17, pp. 319-342.

M.M. Rizki, L.A. Tamburino, and M.A. Zmuda (1993) "Evolving multi-resolution feature detectors," *Proceedings of the Second Ann. Conf. on Evolutionary Programming*, D.B. Fogel and W. Atmar (eds.), Evolutionary Programming Society, La Jolla, CA, pp. 108-118.

N.N. Schraudolph and R.K. Belew (1992) "Dynamic parameter encoding for genetic algorithms," *Machine Learning*, Vol. 9:1, pp. 9-22.

H.-P. Schwefel (1981) Numerical Optimization of Computer Models, John Wiley, Chichester, U.K.

H.-P. Schwefel (1995) *Evolution and Optimum Seeking*, John Wiley, NY.

H.-M. Voigt, W. Ebeling, I. Rechenberg, and H.-P. Schwefel (eds.) (1996) *Parallel Problem Solving from Nature 4*, Springer, Berlin.

D. Wolpert and W.G. Macready (1997) "No free lunch theorems for optimization," *IEEE Trans. Evolutionary Computation*, Vol. 1:1, in press.

A Self Tuning Fuzzy Controller

Sireesh Kumar Pandey

Technical Institute Of Cybernetics, Technical University Of Wroclaw, Poland

Abstract. Real industrial processes can never be modelled perfectly as simple as the linear first and second order systems. They have such marked characteristics as high-order, dead-time, non-linearity etc., and may be affected by noise, load disturbance and other ambient conditions that cause parameter variation and sudden model structural change. The existing theories can no longer provide systematic and robust tuning laws for these complex situations. The operator intuitively regulates the executor to control the process by watching the error and the change rate of the error between the system's output and the set-point value. Usually fuzzy control rules are constructed by summarising the manual control experiences of an operator who has been controlling the industrial process skilfully and successfully.

In the presence of substantial parameter changes, however, or major external disturbances, PID-systems usually are faced with a trade-off between fast reaction with significant overshoot or smooth but slow reactions, or they even run into problems in stabilising the system at all. In this paper, fuzzy control adaptive system monitors its own performance and adjusts its control mechanism to improve performance for slowly time-varying processes. The whole controlling process is automatically adjusted on-line in response to the varying control situation with certain updating scheme. In this manner, an adaptive fuzzy controller can able to handle the complex situations and variety of non-linearities even when subject to random disturbances.

Keywords : *Self-tuning fuzzy controller; scaling factors; tuning rule; membership function; adaptive control.*

1 Introduction

In the nearly twenty-five years, fuzzy set theory has become a valuable, if somewhat controversial tool for modelling uncertainty. Fuzzy control is a special form of knowledge-based control system. In designing a fuzzy control system, the precise mathematical model of target process is not needed. Only the relevant experiences and heuristic concerning the process are utilised to form a set of fuzzy control rules. These rules are linguistic in nature and often use the simple cause-effect relationship to link a fuzzy partitionary of certain state-space of the plant with a fuzzy partitioning of the control action. The linguistic description is constructed subjectively on the basis of the priori knowledge about the process. Thus the source for deriving the linguistic rules is the expert's direct knowledge of the process. It is this knowledge that is expressed in the form of logical rules. The direct approach to fuzzy modelling, based solely upon the use of expert's description of the functioning of the system, has some inherent limitations. Hence, sometimes it is difficult to acquire good control performance for the process from which there has been little experience. Therefore, there is need to tune the rules and the scaling factors for good performance after the establishment of control rules.

The essential properties of the resulting fuzzy controllers are their high flexibility which enables the application to varying problems and the transparence of the control action and the control parameters, respectively. A fuzzy controller, due to its structure, has a greater degree of freedom in achieving self-tuning. A fuzzy controller composed of control rules of conditional linguistic statements on the relationship between input and output variables has the enticing advantages to emulate the behaviour of a human operator and to deal with model uncertainty. Hence, this can be said that the fuzzy control system is a real-time expert system, implementing a part of human operator's or process engineer's expertise which does not lend itself to being easily expressed in PID-parameters or differential equations but rather in situation/action rules. For example

IF room temperature is *low* and environmental temperature is *slightly increasing* **THEN** energy supply is *medium positive*.

The performance of the closed-loop system is improved by modifying the fuzzy rules which sum up people's common sense and experience. The fuzzy controller exhibits superior applicability and considerable robust effects to the conventional PID controller [6, 11, 38, 40, 42, 47]. Takagi and Sugeno [45, 46] proposed a fuzzy identification algorithm for modelling a human operator's control action. A typical example of the model-based tuning laws is the famous Ziegler-Nichols tuning formula [25]. Shao proposed a self-organising controller used in real time. The control policy of the controller was capable of developing and improving automatically [43].

While a controller is at work, uncertainties such as load disturbance, parameter perturbation etc. always exist in the system. Conventional tuning techniques for PID controller usually produces an unsatisfactory control performance. To deal with this problem, the autotuning is then introduced [1, 9, 10, 19, 24, 26, 30, 34, 36]. In many cases, the designer does not fully understand the plant parameters and the basic physical processes in the plant. Adaptive controller is then introduced to cope with these problems [2, 18, 41]. But due to their complexity, the applicability of adaptive controllers is sometimes limited because adaptive algorithms do not cope with the process in real-time.

In this paper, an adaptive fuzzy control system observe some characteristics such as rise-time, overshoot, undershoot and steady state error of output response of the process having varying parameters with respect to time as indices to tune the parameters of a PI controller in real time. My aim was to choose the simplest possible controller that meets the specifications. Therefore, in this paper, a PI controller has been taken for the purpose of research and computer simulation works. The presented method in this paper does not need a precise description of the process i.e., in the form of a dynamic model. The method speeds up the convergence of the process output to a given set-point and slows down the oscillatory nature of the process. The fuzzy controller keeps an eye on error and appropriate rational action takes to bring the process back to course which is generated by a suitable defuzzyfying process.

2 The basics of fuzzy logic

A fuzzy set is defined by a "membership function" that can assume an infinite number of values, any real number in the closed interval [0,1]. A fuzzy set X i.e., "universe of discourse" can be expressed flexibly in terms of the membership function μ_c :

$$\mu_c : X \to [0,1] \qquad (1)$$

A fuzzy set, like an ordinary set, can assume a variety of forms such as a continuous set or a discrete set. The value of $\mu_c(x)$ is at least 0 and not exceed 1. The concepts of complements, intersection, multiplication and unions are defined for fuzzy sets as in the case of crisp sets and their membership function satisfying the following relations:

$$\mu_{A'}(x) = 1 - \mu_A(x) \qquad (2)$$

$$\mu_{A \cap B} = \min(\mu_A(x), \mu_B(x)) \qquad (3)$$

$$\mu_{A \cup B} = \max(\mu_A(x), \mu_B(x)) \qquad (4)$$

$$\mu_{AB}(x) = \mu_A(x) \cdot \mu_B(B) \qquad (5)$$

for all $x \in X$.

For example, let us define two fuzzy sets A and B in the following manner :

$$A = \frac{0.5}{x_1} + \frac{0.2}{x_2} + \frac{0.7}{x_3} + \frac{1}{x_4},$$

$$B = \frac{0.3}{x_1} + \frac{0.6}{x_2} + \frac{0.4}{x_3} + \frac{0.5}{x_4}$$

We obtain,

$$A + B = \frac{0.5}{x_1} + \frac{0.6}{x_2} + \frac{0.7}{x_3} + \frac{1}{x_4},$$

$$A \cap B = \frac{0.3}{x_1} + \frac{0.2}{x_2} + \frac{0.4}{x_3} + \frac{0.5}{x_4},$$

$$A \cdot B = \frac{0.15}{x_1} + \frac{0.12}{x_2} + \frac{0.28}{x_3} + \frac{0.5}{x_4},$$

$$A' = \frac{0.5}{x_1} + \frac{0.8}{x_2} + \frac{0.3}{x_3}.$$

Consider the fuzzy rule having one antecedent and one conclusion, i.e.

IF A is A_k **THEN** B is B_k , where k = 1 ... n.

where $A_k \in X$ is the antecedent universe of discourse and $B_k \in Y$ the conclusion universe of discourse and have a linguistic variable. By a linguistic variable we mean a variable whose values are words or sentences in a natural or artificial language [12]. For example, Water is a linguistic variable if its values are linguistic rather than numerical i.e., very cold, cold, not very cold, warm, not very hot, hot, very hot etc. The relation between the antecedent A_k and conclusion B_k can be described using fuzzy sets in terms of their membership functions. The function of the rules is to map the inputs of the rule base to the output of the rule base through a fuzzy logic inference. The shape and description precision of membership functions depends on the type of process. For example, the term "hot" for the water can mean different things depending on whether the water is to be used for bathing purpose or in a heating plant. Therefore, it is very necessary to express membership function in a variety of forms and describe the precision required for the control objective. The main objective of fuzzy logic is to

1. keep rise-time short,
2. keep minimum possible overshoot,
3. have short settling time, etc.

It is possible to design a P-, PD-, PI-, PID-like fuzzy based controller. Here, these type of fuzzy based controllers are presented whereas the process state variables representing the contents of the rule-antecedent and rule-consequent are taken as follows :

- steady state error, denoted by e,
- change-of-error denoted by Δe,
- sum-of-errors denoted by δe,
- control output signal denoted by u,
- change-of-control output signal denoted by Δu,

where, $\Delta e(k) = e(k) - e(k-1)$,

$\Delta u(k) = u(k) - u(k-1)$ and k is the sampling time.

The P-like fuzzy based controller consists of rules of the form :

    IF e(k) is \<linguistic value\>

    THEN u(k) is $u(k-1) + k_p(e(k) - e(k-1))$.

The PD-like fuzzy based controller consists of rules of the form :

    IF e(k) is \<linguistic value\> and $\delta e(k)$ is \<linguistic value\>

    THEN u(k) is $u(k-1) + k_p(e(k) - e(k-1)) + k_d / T \cdot (e(k) - 2e(k-1) - e(k-2))$.

The PI-like fuzzy based controller consists of rules of the form :

    IF e(k) is \<linguistic value\> and $\Delta e(k)$ is \<linguistic value\>

    THEN u(k) is $u(k-1) + k_p(e(k) - e(k-1)) + k_i \cdot e(k)$.

The PID-like fuzzy based controller consists of rules of the form :

    IF e(k) is \<linguistic value\> and $\Delta e(k)$ is \<linguistic value\> and $\delta e(k)$ is
    \<linguistic value\>

    THEN u(k) is $u(k-1) + k_p(e(k) - e(k-1)) + K_i \cdot e(k) +$

    $+ k_d / T(e(k) - 2e(k-1) - e(k-2))$,

where, k_p, k_i and k_d are the proportional, integral and derivative gain coefficients.

3 The structure of a fuzzy based controller

Generally, a fuzzy system consists of four main parts : a fuzzification interface, a knowledge base, decision-making logic and a defuzzification interface. The results of fuzzy inference is given as a fuzzy variable or variables; defuzzification is necessary before it or they can be used in control operation to any process. In ordinary fuzzy control, the area bisecting point, peak point and center of gravity point in relation to the membership functions are used and defuzzification of an inference result must be a single processing operation. However, the main problem in designing of a fuzzy based system is in choosing these above said design parameters i.e., fuzzification, defuzzification methods of the universe of discourse, choice of the shape of membership functions, rule base etc.

3.1 The fuzzification part of the fuzzy based controller

The fuzzification part consists of a predefined set of linguistic values. Its role is to convert numerical inputs to the fuzzy system into fuzzy inputs to the rule base. In other words, we can say that the fuzzification part maps the physical values of the current process state variables into a normalised universe of discourse in order to make it compatible with the fuzzy set representation of the process state variable in the rule-antecedent.

3.2 The knowledge base

The knowledge base consists of two sections : a database and a rule base. The database provides the fuzzy set definitions of the linguistic values for the linguistic variables of the fuzzy system. In this way, it provides the necessary information for the proper functioning of the fuzzification part, the rule base and the defuzzification part of the fuzzy system. The rule base represents a mapping of the system from a set of fuzzy inputs to a set of fuzzy outputs. The rules are fuzzy conditional statements and represent in a structured way the control policy of an experienced process operator in the form of a set of production rules.

3.2.1 Choice of membership functions

The most popular choices for the shape of the membership function include triangular, trapezoidal and bell-shaped functions which are shown in Fig. 1. It is observed that the parametric, functional description of the triangular shaped membership function is the most economic one. This explains the pre-dominant use of this type of membership functions. The shape of the membership does not play a significant role, but trapezoidal function are responsible for a slower rise time. The triangular and trapezoidal membership functions have the following mathematical form :

$$\Lambda(\upsilon; \alpha, \beta, \lambda) = \begin{cases} 0 & \mu < \alpha \\ \dfrac{\mu - \alpha}{\beta - \alpha} & \alpha \le \mu \le \beta \\ \dfrac{\alpha - \mu}{\beta - \alpha} & \beta \le \mu \le \gamma \\ 0 & \mu > \gamma \end{cases} \tag{6}$$

(triangular function)

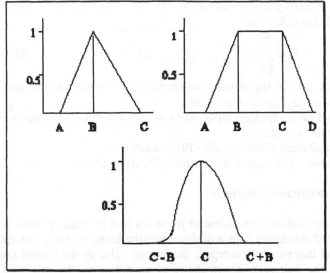

Fig. 1. The membership functions

Generally, it is recommended that the triangular membership functions describing the linguistic variables in the rule base should be symmetrical.

$$\Pi(\mu;\alpha,\beta,\delta)=\begin{cases} 0 & \mu<\alpha \\ \dfrac{\mu-\alpha}{\beta-\alpha} & \alpha\le\mu<\beta \\ 1 & \beta\le\mu\le\gamma \\ \dfrac{\gamma-\mu}{\delta-\gamma} & \gamma<\mu\le\delta \\ 0 & \mu>\delta \end{cases} \qquad (7)$$

(trapezoidal function)

3.2.2 Choice of process state and control output variables

The choice of process state and control output variables depends on the choice of designing a P-, PI-, PD or PID-like fuzzy based controller. To a great extent, it is also depends on the type of process to be controlled. In this paper, the process state variables representing the contents of the "if-part of a rule" are taken as:

- approximation error, denoted by Q,
- the maximum numerical value of response of the controlled process in a close-loop system, denoted by Output_{max} .
- the number of cross-points between set-point value i.e., y_{sp} and response of the controlled process in a close-loop system i.e., y(t), termed by Occurrence.

where, Occurrence = Number of cross-points for ($y_{sp} - y(t) \ge \pm 0.1\%$);

Output$_{max}$ = max{y(t), y$_{sp}$} and t is the sampling time. Also the approximation-error (criterion) Q is defined as:

$$Q = \sqrt{\int_0^T [(y_{set\ point} - y_{process}(\tau)]^2 d\tau}$$ (8)

where, $y_{process}(\tau)$ is the process output variable having varying parameters with respect to time due to some noises etc.

The control output (process input) variables representing the contents of the "then-part of a rule" are taken as:

- gain coefficient of the controller PI, denoted by k_p,
- integration coefficient of the controller PI, denoted by k_i.

3.3 The decision-making logic

The decision-making logic is one of the main part of fuzzy systems and it has the capability of simulating human decision making based on fuzzy concepts. The main function of this part is to compute the overall value of the control output variable based on the individual contributions of each rule in the rule base. Each such individual contribution represents the values of the control output variables as computed by a single rule. The current linguistic value of the process state variable is matched to each rule-antecedent and in this way, the decision-making logic establishes a degree of match for each rule defined in rule base. The value of the control output variable is determined on the basis of this degree of match. Hence, the overall fuzzy value of the control output variable is the set of all control output values of the matched rules.

3.4 The defuzzification part of the fuzzy based controller

The defuzzification part converts the fuzzy output of the rule-base into a non-fuzzy (numerical) value. The most common defuzzification procedure is the center of gravity method :

$$u = \frac{\sum_{i=1}^{1} u_i \mu_c(u_i)}{\sum_{i=1}^{1} \mu_c(u_i)}$$ (9)

This method is computationally rather complex and therefore, results in quite slow inference cycles. Hence, the height method is oftenly used for the defuzzification purpose as this method is computationally simple and very quick in comparison to other ones :

$$u^* = \frac{\sum_{k=1}^{m} c^{(k)} . f_k}{\sum_{k=1}^{m} f_k}$$ (10)

where f_k is the height.

The middle of maxima and first of maxima methods are also oftenly used as a defuzzification procedure and are relatively fast methods. The center of sums and middle of maxima methods are not generally used for a defuzzification procedure because the computational complexity of theses methods depend on the shape of the membership functions.

4 Simulation results

Let us the process y(t) has time-varying coefficients with respect to time due to some noises, load disturbances etc. The step response of the process y(t) is generated and the criterion Q (equ. 8) is calculated for the time T = 100 units of time. In the same duration of time, the other process state variables i.e., Output$_{max}$ and Occurrence are also determined. These process state variables are used to define the fuzzy logic and after defuzzification, the obtained fuzzy value (control output variable) is used to actuate the process by regulating the parameters of controller which is connected to the fuzzy part. The diagram for the whole process is presented in Fig. 2. The two controllers PI have their contribution in the process diagram shown in Fig. 2. The one controller has the fixed control coefficients k_p and k_i in a close-loop system (obtained by minimising the criterion Q for any initial process) and the other one has the varying control coefficients k_p and k_i which are the output of the fuzzy logic. The fuzzy part of the system is enabled only when $\Delta Q = Q(t)-Q(t-1) \neq 0$, where t is the present value of time. There are no changes in the values of control output variables (fuzzy output) until $\Delta Q = 0$.

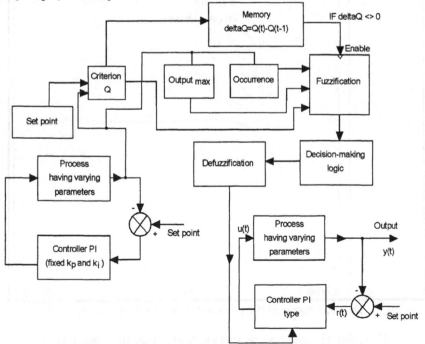

Fig. 2. The process diagram with fuzzy logic

Let us consider the model which is generally used in the field of energetic specially for the boilers. The mathematical model for the process can be taken as follows:

$$G_o(s) = \frac{k}{(Ts+1)^n} = \frac{1}{(s+1)^3} \qquad (11)$$

In this case, a PI (Proportional - Integration) controller is used as a follow-up of the fuzzy logic. This PI controller takes r(t) and the output of the fuzzy part, as its inputs, and produces the final control signal u(t) as its output. The PI relationship between u(t) and r(t) can be given as :

$$u(t) = k_p\, r(t) + k_i \int r(t)dt, \qquad (12)$$

where, k_p, k_i are two adjustable coefficients and the input signal is 1(t). To achieve a desired response, we have to tune the parameters of the controller PI and in this manner, we also minimise the criterion Q. The optimal control coefficients (i.e. criterion Q is minimum i.e., Q = 1.4935, no overshoot and short settling-time) for the model given in the equ. (11) are :

$$k_p = 0.63, \qquad k_i = 1.95. \qquad (13)$$

The step response of the process (equ. (11)) in a close-loop system for the control coefficients given in equ. (13) is shown in Fig. 3.

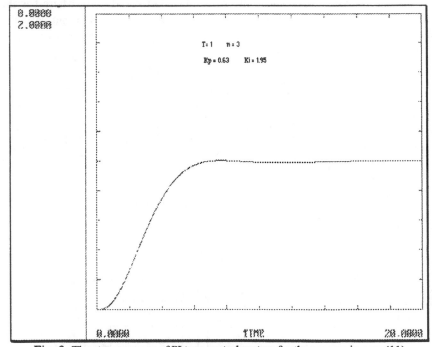

Fig. 3. The step response of PI type control system for the process in equ. (11)

Now, due to some reasons (noises, load disturbance, bad quality of coal, moistured coal etc.), there are sudden parameter variations in the process, i.e. the time constant (T) and order (n) of the transfer function given in the equ. (11) jump to a new value. Therefore it is possible that the our process is no more optimal (in the sense of minimising the criterion Q, settling-time and overshoots) or an unstable process in a close-loop system for the control coefficients given in equ. (13). Therefore, we have to adjust the control coefficients in the controller PI connected to the fuzzy part for the new (changed) parameters of the model in order to stabilise and optimalize (the criterion Q must be minimum) the process in a close-loop system . The relation Q = f(n,T) is shown in Fig. 4 where "n" the order of transfer function (equ. (11)) changes in the interval 2≤n ≤8 and "T" the time constant of transfer function (equ. (11)) changes in the interval 0.1≤T ≤1.9.

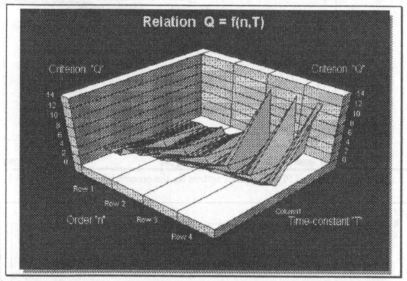

Fig. 4. The relation between criterion Q and time constant and order of transfer function

The Figs. 5, 6, 7 and 8 show the effects of parameters i.e., the time constant and order of transfer function on the process in a close-loop system for the control coefficients given in equation (13) which were quite good for the process having T = 1 and n = 3. As we can see from the Figs. 5, 6, 7 and 8 that these control coefficients are no longer suitable for the changed process. The process with new (changed) parameters is either no more optimal i.e., have overshoots, longer settling-time and longer rise time or in some cases becomes unstable.

Membership functions : In a great extent, the control algorithms in fuzzy logic are based on shape and type of the membership functions. The decision on the shapes of the membership functions are largely ad hoc, and based on the insights in the natures of the underlying non-fuzzy variables and the control problems at hand. On the basis of the Figs. 4, 5, 6, 7 and 8 and on the way of other simulation data's and experiences, the membership functions for the criterion Q, Output$_{max}$ and Occurrence has been determined and are shown in Figs. 9, 10 and 11 respectively. As well as the membership functions for the adjustable control coefficients k_p and k_i are also shown in Figs. 12 and 13 respectively.

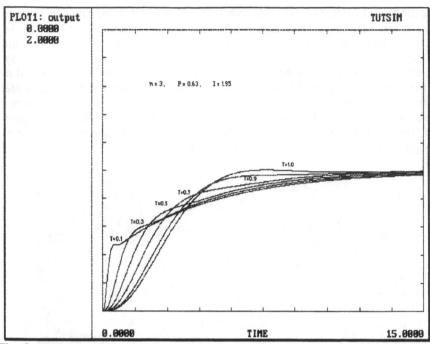

Fig. 5. The step response of PI type control system for the process having varying T∈[0.1, 1]

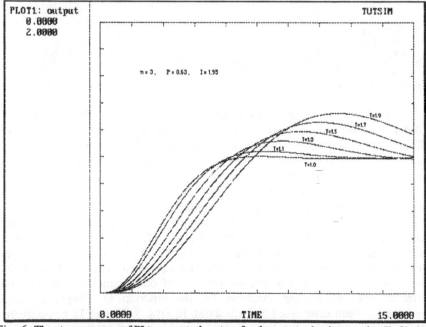

Fig. 6. The step response of PI type control system for the process having varying T∈[1, 1.9]

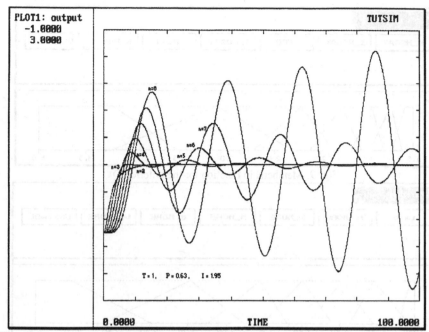

Fig. 7. The step response of PI type control system for the process having varying n ∈ [2, 8]

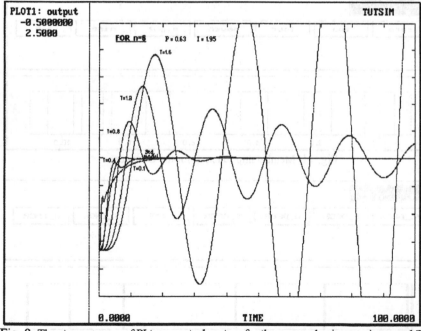

Fig. 8. The step response of PI type control system for the process having varying n and T

355

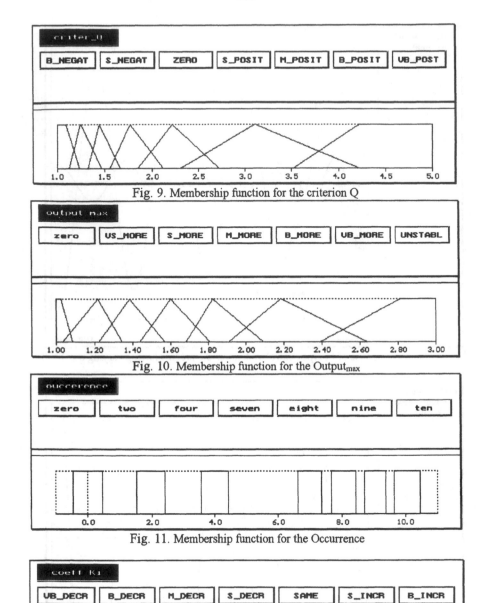

Fig. 9. Membership function for the criterion Q

Fig. 10. Membership function for the Output$_{max}$

Fig. 11. Membership function for the Occurrence

Fig. 12. Membership function for adjustable parameter k_i for the fuzzy based controller

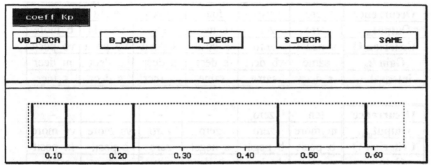

Fig. 13. Membership function for adjustable parameter k_p for the fuzzy based controller

Fuzzification and defuzzification : In the fuzzification part, the relation $Q = f(n,T)$ (Fig. 4) has been used to define the fuzzy decision making logic.

Criterion Q = {B_NEGAT, S_NEGAT, ZERO, S_POSIT, M_POSIT, B_POSIT, VB_POSIT}

where, B_NEGAT - big negative, S_NEGAT - small negative, ZERO - no change in the process, S_POSIT - small positive, M_POSIT - medium positive, B_POSIT - big positive, VB_POSIT - very big positive values of criterion Q,

Occurrence = {ZERO, TWO, FOUR, SEVEN, EIGHT, NINE, TEN}

Output$_{max}$ = {ZERO, VS_MORE, S_MORE, M_MORE, B_MORE, VB_MORE, UNSTABL}

where, ZERO - no change in the process, VS_MORE = very small more, S_MORE - small more, M_MORE - medium more, B_MORE - big more, VB_MORE - very big more, UNSTABL : unstable process.

The parameter k_p and k_i are the adjustable parameters of the controller PI, connected with the fuzzy part, depending upon the parameters of the process i.e., T and n.

k_p = {VB_DECR, B_DECR, M_DECR, S_DECR, SAME}

k_i = { VB_DECR, B_DECR, M_DECR, S_DECR, SAME, S_INCR, B_INCR}

where, VB_DECR - very big decrease, B_DECR - big decrease, M_DECR - medium decrease, S_DECR - small decrease, SAME - no change, S_INCR - small increase, B_INCR - big increase for the values of control coefficients.

The table I shows *if-then* rules which are defined for the process given in equation (11).

It is very essential to reverse the operation of the fuzzifying process so as to be understandable for the controller. The defuzzification has been done with the help of Height method given in equation (10). The defuzzified control signal is simply provided to the controller PI to adjust it's parameters k_p and k_i in order to stabilise the process.

Examples :

1). Let us consider that the parameters (time constant T and order n) of the process has been changed from $T = 1$ and $n = 3$ to $T = 1.4$ and $n = 5$ due to some noises.

$$G_{changed}(s) = \frac{k}{(Ts+1)^n} = \frac{1}{(1.4s+1)^5} \qquad (14)$$

Occurrence	zero	two	four	seven	eight	nine
Output$_{max}$	zero	zero	vs_more	s_more	unstabl	b_more
Criterion Q	s_negat	zero	s_posit	s_posit	vb_posit	vb_posit
Gain k$_p$	same	vb_decr	s_decr	s_decr	b_decr	m_decr
integral k$_i$	s_decr	same	same	s_incr	s_decr	s_incr

Occurrence	ten	zero	-	-	-	-
Output$_{max}$	m_more	zero	zero	zero	vs_more	vs_more
Criterion Q	b_posit	b_negat	s_negat	zero	s_posit	m_posit
Gain k$_p$	m_decr	b_decr	vb_decr	same	s_decr	s_decr
integral k$_i$	same	b_decr	vb_decr	same	s_incr	s_incr

Occurrence	-	-	-	-	-	-
Output$_{max}$	s_more	s_more	m_more	m_more	s_more	unstabl
Criterion Q	m_posit	b_posit	b_posit	m_posit	s_posit	vb_posit
Gain k$_p$	s_decr	m_decr	b_decr	b_decr	s_decr	b_decr
integral k$_i$	b_incr	b_incr	s_decr	s_decr	same	same

Occurrence	-	-	-	-	-	-
Output$_{max}$	b_more	b_more	vs_more	b_more	vb_more	unstabl
Criterion Q	b_posit	m_posit	zero	vb_posit	vb_posit	b_posit
Gain k$_p$	m_decr	b_decr	same	m_decr	m_decr	s_decr
integral k$_i$	s_incr	same	same	b_incr	b_incr	s_decr

Table I. "if-then" rules for the process in the equation (11)

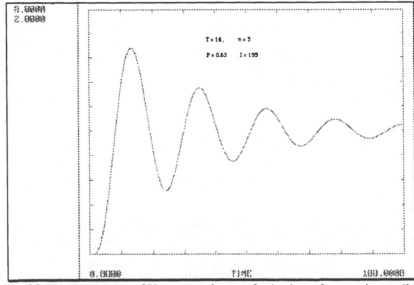

Fig. 14. The step response of PI type control system for the changed process in equ. (14)

The step response of this changed process has the oscillatory character in a close-loop system for the control coefficient given in equ. (13) and is shown in Fig. 14.

$$Q = 3.162, \text{Output}_{max} = 1.6780, \text{Occurrence} = -. \qquad (15)$$

On the basis of these process state variables (input of fuzzy part) given in equ. (15), the fuzzy logic proposes the following control coefficients for the changed process :

$$k_p = 0.119, k_i = 1.4395. \qquad (16)$$

The step response of the changed process in a close-loop system for the control coefficient given in equ. (16) shown in Fig. 15.

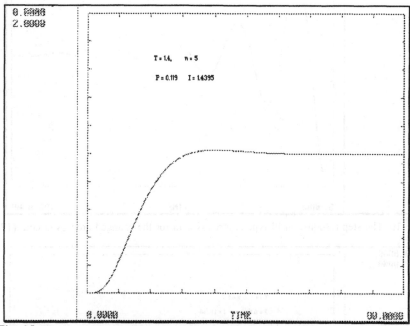

Fig. 15. The step response of PI type control system for the changed process in equ. (14) having control coefficients proposed by fuzzy logic.

2). Let us consider that the parameter (order n) of the process has been changed from $n = 3$ to $n = 6$ due to some noises and load disturbances..

$$G_{changed}(s) = \frac{k}{(Ts+1)^n} = \frac{1}{(s+1)^6} \qquad (17)$$

The step response of this changed process has the overshoots and an oscillatory behaviour in a close-loop system for the control coefficient given in equ. (13) and is shown in Fig. 16.

$$Q = 2.692, \text{Output}_{max} = 1.6020, \text{Occurrence} = 10. \qquad (18)$$

On the basis of these process state variables (input of fuzzy part) i.e. the criterion Q, Output_{max} and Occurrence given in equ. (18), the fuzzy logic proposes the following control coefficients for the changed process :

$$k_p = 0.1686, k_i = 1.8348. \qquad (19)$$

The step response of the changed process in a close-loop system for the control coefficient given in equ. (19) shown in Fig. 17.

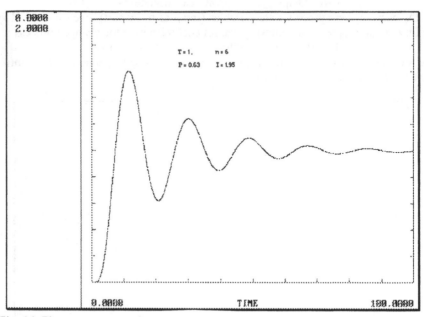

Fig. 16. The step response of PI type control system for the changed process in equ. (17)

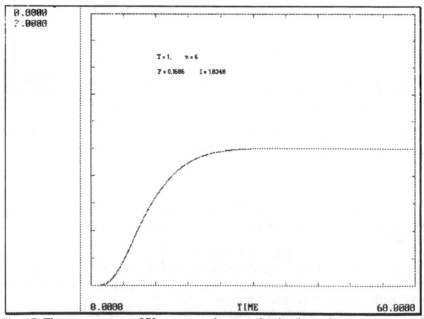

Fig. 17. The step response of PI type control system for the changed process in equ. (17) having control coefficients proposed by fuzzy logic

Hence, on the basis of above given examples, we can say that the adaptive fuzzy control system can successfully cope with the processes having time-varying coefficients.

5 Conclusion

Simulations analysis presented in this paper indicate that the adaptive fuzzy technique has a performance comparatively superior to that of a PI (PID) controller when noise was present and the process has time-varying coefficients. The presented method successfully handle the problem of parameters variations in the process which are caused by the different types of noises. The step responses, for the changed processes, proposed by the fuzzy decision-making logic have no overshoot and very short settling-time. However, it is very important to choose the tuning ranges (ranges for control coefficients) properly. The choice of a tuning range directly affects the step response of the process. The choice of rule-base of a fuzzy decision-making logic has also great effect on the performance of the system.

An expert knowledge is very necessary to build the fuzzy based systems. Thus this can be said that fuzzy control is based on a set of fuzzy rules which sum up people's common sense and experience. The rules are linguistic statements expressed mathematically through the concepts of fuzzy sets and correspond the action of a human operator would take when controlling a given process.

References

1. A.E.B. Ruano, P.J. Fleming, D.I. Jones : Connectionist approach to PID autotuning. IEE Proceedings -D 139, 279 - 285 (1992).
2. A.M. Annaswamy, K.S. Narendra : Stable adaptive systems. Prentice-Hall, Englewood Cliffs, New Jersey (1989).
3. A.O. Esogbue, M. Theologidu, K. Guo : On the application of fuzzy set theory to the optimal flood control problem arising in water resources systems, Fuzzy Sets and Systems, vol. 48,155 - 172 (1992).
4. B.A.M. Wakileh, K.F. Gill : Robot control using self-organising PID controller, IEE Proceedings-D, 138(3), 303 - 311 (1991).
5. B.P. Grahem, R. B. Newell : Fuzzy adaptive control of a first order process. Fuzzy Sets and Systems, vol. 3, 47 - 65 (1989).
6. B.W. Hogg, Q.H. Wu : On line evaluation of auto-tuning optimal PID controller on micromachine system. International Journal of Control 53, 751 - 769 (1991).
7. C. Altrock, B. Krause, H.J. Zimmermann : Advanced fuzzy logic control a model car in extreme situation, Fuzzy Sets and Systems, vol. 48, 41 - 52 (1992).
8. C. Altrock, H.O. Arend, B. Krause, C. Steffens, E. Behrens-Römmler : Adaptive fuzzy control applied to home heating system, Fuzzy Sets and Systems, vol. 61, no. 1, 29 - 35 (1994).
9. C. Zervos, P.R. Belanger, G.A. Dumont : On PID controller tuning using orthonormal series identification. Automatica 24, 165 - 175 (1988).
10. C.H. Chou, H.C. Lu : A heuristic self-tuning fuzzy controller. Fuzzy Sets and Systems 61, 249 - 264 (1994).

11. C.H. Chou, H.C. Lu : Real-time fuzzy controller design for hydraulic servo system. International Journal of Computers in Industry 22, 129 - 142 (1993).

12. D. Driankow, H. Hellendoorn, M. Reinfrank : An Introduction to Fuzzy Control. Springer Verlag (1993).

13. E. H. Maudani, J. J. King : The application of fuzzy control system to industrial processes. Automatica, tom 13, 235 - 242 (1977).

14. E. Mishkin, L. Braun : Adaptive control system. McGraw-Hill electrical and electronic engineering series.

15. E.H. Maudani : An application of fuzzy algorithms for control of simple dynamic plant. Proc. Ins. Elec. Eng., tom 121, 1585 - 1588 (1974).

16. E5AF/E5EF Fuzzy Temperature Controller, Operation Manual firmy OMRON (1992).

17. F. Steimann, K.P. Adlassig : Clinical monitoring with fuzzy automata, Fuzzy Sets and Systems, vol. 61, no. 1, 37 - 42 (1994).

18. F.C. Teng, H.R. Sirisena : Self-tuning PID controllers for dead time processes. IEEE Transaction on Industrial Electronics IE-35, 119 - 125 (1988).

19. G.C. Goodwin, K.S. Sin : Adaptive filtering prediction and control. Prentice-Hall, Englewood Cliffs, New Jersey (1984).

20. G.E. Rotstein, D.R. Levin : Simple PI and PID tuning for open-loop unstable systems, Industrial Engineering Chemistry Research, vol. 30, no. 8, 1864 - 1869 (1991).

21. H. Schüdel : Utilization of fuzzy techniques in intelligent sensors, Fuzzy Sets and Systems, vol. 63, no. 3, 271 - 292 (1994).

22. H. Ying : Analytical structure of a two-input two-output fuzzy controller and its relation to PI and multilevel relay controllers, Fuzzy Sets and Systems, vol. 63, no. 1, 21 - 33 (1994).

23. J. J. Buckley : Universal fuzzy controllers. Automatica, vol. 28,1245-1248 (1992).

24. J. Litt : An expert system to perform on-line controller tuning. IEEE Control Systems Magazine 11, 18 - 23 (1991).

25. J.G. Ziegler, N.B. Nichols : Optimum setting for automatic controllers. Transaction AMSE 65, 433 - 444 (1943).

26. J.H. Kim, K.K. Choi : Self-tuning discrete PID controller. IEEE Transaction on Industrial Electronics IE-34, 298 - 300 (1987).

27. J.H. Kim, K.K. Choi : Self-tuning discrete PID controller, IEEE Transactions on Industrial Electronics, IE-34, 298 - 300 (1987).

28. J.J. Saade : Towards intelligent radar systems, Fuzzy Sets and Systems, vol. 63, no. 2, 141 - 157 (1994).

29. K. Hirota : Industrial Applications of Fuzzy Technology, Springer-Verlag, (1993).

30. K. J. Åström : Towards intelligent PID control. Automatica 28, 1 - 9 (1992).

31. L.J. Huang, M. Tomizuka : A Self-Raced Fuzzy Tracking Controller for Two-Dimensional Motion Control, IEEE Trans. on Systems, Man and Cybernetics, vol. 20, no. 5,1115 - 1124 (1990).

32. L.M. Jia, X.D. Zhang : Distributed intelligent railway traffic control based on fuzzy decision making, Fuzzy Sets and Systems, vol. 62, no. 3, 255 - 265 (1994).

33. M. Maeda, S. Murakami : A self-tuning fuzzy controller. Fuzzy Sets and Systems 1992, vol. 51, 29 - 40 (1992).

34. P. Dorato : A historical review of robust control. IEEE Control Systems Magazine 7, 44 - 47 (1987).
35. P. Guillemin : Universal motor control with fuzzy logic, Fuzzy Sets and Systems, vol. 63, 339 - 348 (1994).
36. P.J. Gawthrop, P.E. Nomikos : Automatic tuning of commercial PID controllers for single-loop and multiloop applications. IEEE Control Systems Magazine 10, 34 - 42 (1990).
37. R. Palm : Fuzzy Controller for a Sensor Guided Robot Manipulator, Fuzzy Sets and Systems, vol. 20, no. 1, 133 - 149 (1989).
38. R.J. Mulholland, K.L. Tang : Comparing fuzzy logic with classical controller designs. IEEE Transaction on Systems, Man, and Cybernetics 17, 1085 - 1087 (1987).
39. R.M. Tong : A control engineering review of fuzzy systems, Automatica, t. 13, 559 - 569 (1977).
40. S. Daley, K.F. Gill : A study of fuzzy logic controller robustness using the parameter plant. International Journal of Computers in Industry 7, 511 - 522 (1986).
41. S. Daley, K.F. Gill : Comparison of fuzzy logic controller with a P+D control law. Transaction of the ASME Journal of Dynamic Systems, Measurement and Control 111, 128 - 137 (1989).
42. S. Sastry, M. Bodson : Adaptive control : Stability convergence, and robustness. Prentice-Hall, Englewood Cliffs, New Jersey (1989).
43. S. Shao : Fuzzy self-organizing controller and its application for dynamic processes. Fuzzy Sets and Systems 26, 151 - 164 (1988).
44. S.Z. He, S.T.F.L. Xu, P.Z. Wang : Fuzzy self-tuning of PID controllers. Fuzzy Sets and Systems, vol. 56, 37 - 46 (1993).
45. T. Takagi, M. Sugeno : Derivation of fuzzy control rules from human operator's control actions. Proc. IFAC Symp. on Fuzzy Information, Knowledge Representation and Decision Analysis, Marseilles, France, 55 - 60 (1983).
46. T. Takagi, M. Sugeno : Fuzzy identification of systems and its application to modelling and control. IEEE Trans. Sys. Man. Cyb. 15, 116 - 132 (1985).
47. W. Pedrycz : Fuzzy control and fuzzy systems. Research Studies Press LTD., Taunton, Somerset, England (1989).
48. W.H. Bare, R.J. Mulholland, S.S. Sofer : Design of a self-tuning rule based controller for a gasoline refinery catalytic reformer, IEEE Transactions on Autom. Control, vol. 32, no. 2, 156 - 164, (1990).
49. W.J. Kicert, H.R. Nauta Lemke : Application of a fuzzy controller in a warm water plant, Automatica, tom 12, 301 - 308, (1976).
50. Y. Yamashita, S. Matsumoto, M. Suzuki : Start-up of Catalityc Reactor by Fuzzy Controller, J. Chemical Engineering of Japan, vol. 21, 277 - 281 (1988)

A New Type of Fuzzy Logic System for Adaptive Modelling and Control

J. Zhang, A. Knoll and K. V. Le

Faculty of Technology, University of Bielefeld,
33501 Bielefeld, Germany

Abstract. We present a new type of fuzzy controller constructed with the B-spline model and its applications in modelling and control. Unlike the other normalised parameterised set functions for defining fuzzy sets, B-spline basis functions do not necessarily span from membership value 0 to 1, but possess the property "partition of unity". These B-spline basis functions are automatically determined after the input space is partitioned. By using "product" as fuzzy conjunction, "centroid" as defuzzification, "fuzzy singletons" for modelling output variables and adding *marginal linguistic terms*, fuzzy controllers can be constructed which have advantages like smoothness, automatic design and intuitive interpretation of controller parameters. Furthermore, both theoretical analysis and experimental results show the rapid convergence for tasks of data approximation and unsupervised learning with this type of fuzzy controller.

1 Introduction

In most fuzzy systems, linguistic terms are defined with *fuzzy numbers*, i.e. normalised, closed, convex fuzzy sets. In approximate reasoning, usually only the qualitative information is referred to, thus the final result is not very sensitive to the shape and the height of the fuzzy sets. However, if a fuzzy logic system is applied in modelling or control problems, the shape as well as the position of fuzzy sets are implicitly or explicitly synthesised both in the inference procedure and in the defuzzification. Therefore, the specification of the fuzzy sets for both the IF- and THEN-part is worth being discussed in more detail.

IF-part. All fuzzy controllers employ true fuzzy sets for modelling linguistic terms for each input. The input space is partioned into overlapping cells, which reflects the vague modelling of linguistic concepts on one side and enables the continuous transition of output values on the other side.
The IF-part of a rule is generally modelled as

$$(x_1 \text{ is } A_{i_1}^1) \text{ and } (x_2 \text{ is } A_{i_2}^2) \text{ and } \ldots (x_n \text{ is } A_{i_n}^n),$$

where x_j is the j-th input $(j = 1, \ldots, n)$ and $A_{i_j}^j$ is the i-th linguistic term defined on x_j. The "and"-operation is implemented with a so-called $t-$norm, which is represented by "min" or "product" in most applications.
While discrete representation of fuzzy sets avoids the on-line function evaluation on a fuzzy hardware chip, parameterised representation is adopted

in fuzzy controllers running on a general-purpose, non-fuzzy computer architecture. In up-to-date control applications, mainly triangle and trapezoid set functions are used. Recently, Fuzzy Basis Functions based on Gaussian functions are also proposed for function approximation, [7]. In [6] several exotic functions like "Cauchy", "*sinc*", "Laplace", "Logistic", "Hyperbolic Tangent" are introduced and their abilities of function approximation are compared[1]. However, all the above set functions need additional special parameters apart from the partition positions (called *knots* in the following) on the universe of discourse of each input. Since the knots are the only intrinsic parameters resulting from the partition of the input space, the selection and tuning of these additional parameters are neither natural nor intuitive.

THEN-part. The classical fuzzy controller of Mamdani type is based on the idea of directly using symbolic rules for control tasks. A rule has the form

> IF $\quad (x_1$ is $A_{i_1}^1)$ and $(x_2$ is $A_{i_2}^2)$ and ... and $(x_n$ is $A_{i_n}^n)$
> THEN y is B_k,

where B_k is a fuzzy set with the same properties as that used in the "IF-part", $k = 1, \ldots, t$, and t is the total number of linguistic terms for modelling the output y. The aggregation of output values of all the firing rules are realised either by the "max"-operator [5] or simple addition [4], where the second method is a small variation of the first one and even more simple to compute.

Another important type of fuzzy controllers is based on the TSK (Tagaki-Sugeno-Kang) model. A rule using a TSK model of order 1 looks like:

> IF $\quad (x_1$ is $A_{i_1}^1)$ and $(x_2$ is $A_{i_2}^2)$ and ... and $(x_n$ is $A_{i_n}^n)$
> THEN $y = a_0^i + a_1^i x_1 + \cdots + a_n^i x_n$,

where $a_0^i, a_1^i, \ldots, a_n^i$ are the coefficients of a simplified local linear model. These parameters can be identified by optimising a least squares performance index using the data acquired by observing a skilled human operator's control action. The recent work with TSK model shows that it is a suitable function approximator. However, some authors [1] pointed out that the TSK model is a multi-local-model black-box. Obviously, the knowledge acquisition with this model is indirect and not intuitive.

We propose an approach which can systematically build the fuzzy sets for linguistic terms of the IF-part while the fuzzy sets of the THEN-part can be adapted through learning. The model of linguistic terms in our approach is based on B-spline basis functions, a special set of piecewise polynomial curves.

---

[1] The experimental results in [6] show that the non-convex functions *sinx* works generally better for a quick and accurate function approximation. Nevertheless, this function possesses more than one peak and thus cannot be assigned an appropriate linguistic meaning.

2 B-Spline Basis Functions as Fuzzy Sets

2.1 Definition

In our previous work [11,9] we compared the basis functions of periodical *Non-Uniform B-Splines* (NUBS) with a fuzzy controller. In this paper, we also follow the usage of this type of NUBS basis functions (*B-functions* for short).

Assume x is a general input variable of a control system which is defined on the universe of discourse $[x_0, x_m]$. Given a sequence of ordered parameters (knots): $(x_0, x_1, x_2, \ldots, x_m)$, the i–th normalised B-spline basis function (B-function) $X_{i,k}$ of order k is defined as:

$$X_{i,k}(x) = \begin{cases} \begin{cases} 1 & \text{for } x_i \leq x < x_{i+1} \\ 0 & \text{otherwise} \end{cases} & \text{if } k = 1 \\ \frac{x - x_i}{x_{i+k-1} - x_i} X_{i,k-1}(x) + \frac{x_{i+k} - x}{x_{i+k} - x_{i+1}} X_{i+1,k-1}(x) & \text{if } k > 1 \end{cases}$$

with $i = 0, 1, \ldots, m - k$. The B-functions of order 1, 2, 3, 4 are illustrated in Fig. 1.

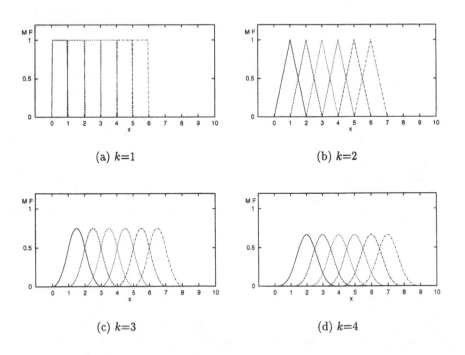

(a) $k=1$ (b) $k=2$

(c) $k=3$ (d) $k=4$

Fig. 1. Fuzzy sets defined by B-spline basis functions of different orders (examples of uniform cases).

The important properties of B-functions are:

Partition of unity:	$\sum_{i=0}^{m} X_{i,k}(x) = 1.$
Positivity:	$X_{i,k}(x) \geq 0.$
Local support:	$X_{i,k}(x) = 0$ for $x \notin [x_i, x_{i+k}].$
C^{k-2} *continuity:*	If the knots $\{x_i\}$ are pairwise different from each other, then $X_{i,k}(x) \in C^{k-2}$, i.e. $X_{i,k}(x)$ is $(k-2)$ times continuously differentiable.

2.2 Real and Virtual Linguistic Terms

It is assumed that linguistic terms are to be defined over $[x_0, x_m]$, the universe of an input variable x of a fuzzy controller. They are referred to as *real linguistic terms*. In order to maintain the "partition of unity" for all $x \in [x_0, x_m]$, some more B-functions should be added at both ends of $[x_0, x_m]$. They are called *marginal B-functions*, defining *virtual linguistic terms*. *Real* and *virtual* linguistic terms are denoted as A_i in Fig. 2:

- In case of order 2, no marginal B-function is needed, Fig. 2(a).
- In case of order 3 or 4, two marginal B-functions are needed, one for the left end and another for the right end, Fig. 2(b), (c).
- Generally, $((k+1)\ div\ 2)$ marginal B-functions are needed.

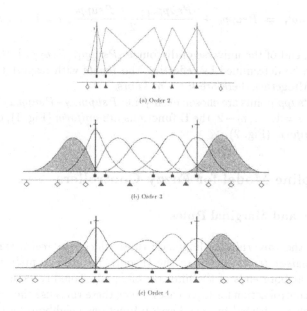

(a) Order 2

(b) Order 3

(c) Order 4

Fig. 2. Non-uniform B-functions of different orders defined for *real* and *virtual* linguistic terms by the same knot vector (*Peak supprot points:* □; *Iknots:* △; *Eknots:* ◇; *virtual linguistic terms:* shaded).

2.3 Peak Support Points and Knots

In fuzzy set theory, the *support* of a fuzzy set A within a universal set X is the crisp set that contains all the elements of X that have non-zero membership grades in A. If a B-function of order k ($k > 1$) is used for modelling a fuzzy set, it possesses only one peak which has the largest membership grade. The support point of this peak, denoted as the *Psupp*-point (*peak support point*), can be defined as $Psupp(A) = \{x|A(x) = maximum\}$, where A is defined by a B-function.

A B-function representing A_i is defined by the *knots*, the boundary points of the *support* of A_i. The complete *knots* consist of two parts, the *interior knots* (noted as *Iknots*) which lie within the universe of discourse and *Extended knots* (*Eknots*) which are generated at both ends of the universe for defining the marginal linguistic terms. Generally, $m - (k \bmod 2)$ interior knots are needed, where

- m is the number of the real linguistic terms, and
- k is the order of the B-functions ($k \leq m$).

If k is even, the interior knots coincide with the *Psupp*-points. If k is odd, the $m - 1$ interior knots can be determined by

$$Iknot_i = Psupp_i + \frac{Psupp_{i+1} - Psupp_i}{2}, \quad i = 1, \ldots, m - 1. \qquad (1)$$

At each end of the universe of discourse $[Psupp_1, Psupp_m]$, $((k + 1) \ div \ 2)$ *Eknots* can be determined by reflecting the *Iknots* with respect to $Psupp_1$ and $Psupp_m$. Altogether there are $k + m$ knots.

If the *Psupp*-points are chosen evenly, i.e. $Psupp_{i+2} - Psupp_{i+1} = Psupp_{i+1} - Psupp_i$ for $i = 1, \ldots, m - 2$, the B-functions are *uniform* (Fig. 1), otherwise they are *non-uniform* (Fig. 2).

3 B-Spline Model for Fuzzy Controllers

3.1 Core and Marginal Rules

We define the *core rules* as linguistic rules which use *real linguistic terms*. If *virtual linguistic terms* appear in the IF-part, in order to maintain the output continuity at both ends of the universe of x, additional rules are needed to describe the control action for these cases. Since these rules use the *virtual linguistic terms* which are defined by membership functions neighbouring the ends of the universe of each variable, they are called *marginal rules*. The output value of each *marginal rule* is selected just as the output value of the "nearest" *core rule*, i.e. the rule using the directly adjacent linguistic terms in its IF-part (Fig. 3).

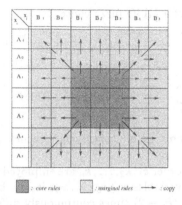

| : core rules | : marginal rules | ⟶ : copy |

Fig. 3. The outputs of the marginal rules are copied from that of the neighbouring core rules.

3.2 A B-Spline Interpolator

Since a MIMO (Multiple-Input-Multiple-Output) rule base is normally divided into several MISO (Multiple-Input-Single-Output) rule bases, we consider only the MISO case. Under the following conditions:

- periodical B-spline basis functions as membership functions for inputs;
- fuzzy singletons as membership functions for outputs;
- "product" as fuzzy conjunctions;
- "centroid" as defuzzification method;
- addition of "virtual linguistic terms" at both ends of each input variable and
- extension of the rule base for the "virtual linguistic terms" by copying the output values of the "nearest" neighbourhood;

the computation of the output of such a fuzzy controller is equivalent to that of a *general B-spline hypersurface*. Generally, we consider a MISO system with n inputs $x_1, x_2, \ldots, x_n$. A $Rule(i_1, i_2, \ldots, i_n)$ with the n conjunctive terms in the IF-part is given in the following form:

> IF $\quad (x_1$ is $X^1_{i_1,k_1})$ and $(x_2$ is $X^2_{i_2,k_2})$ and ... and $(x_n$ is $X^n_{i_n,k_n})$
> THEN y is $Y_{i_1 i_2 \ldots i_n}$,

where

- x_j: the j-th input $(j = 1, \ldots, n)$,
- k_j: the order of the B-spline basis functions used for x_j,
- $X^j_{i_j,k_j}$: the i-th linguistic term of x_j defined by B-spline basis functions,
- $i_j = 1, \ldots, m_j$, representing how fine the j-th input is fuzzy partitioned,
- $Y_{i_1 i_2 \ldots i_n}$: the control vertex (deBoor points) of $Rule(i_1, i_2, \ldots, i_n)$.

Then, the output y of a MISO fuzzy controller is:

$$y = \frac{\sum_{i_1=1}^{m_1} \cdots \sum_{i_n=1}^{m_n} (Y_{i_1,\dots,i_n} \prod_{j=1}^{n} X_{i_j,k_j}^{j}(x_j))}{\sum_{i_1=1}^{m_1} \cdots \sum_{i_n=1}^{m_n} \prod_{j=1}^{n} X_{i_j,k_j}^{j}(x_j)} \tag{2}$$

$$= \sum_{i_1=1}^{m_1} \cdots \sum_{i_n=1}^{m_n} (Y_{i_1,\dots,i_n} \prod_{j=1}^{n} X_{i_j,k_j}^{j}(x_j)) \tag{3}$$

This is called a *general NUBS hypersurface*, which possesses the following properties:

- If the B-functions of order $k_1, k_2, \dots, k_n$ are employed to specify the linguistic terms of the input variables $x_1, x_2, \dots, x_n$, it can be guaranteed that the output variable y is $(k_j - 2)$ times continuously differentiable with respect to the input variable $x_j, j = 1, \dots, n$ (see Fig. 4 for an example).
- If the input space is partitioned fine enough and at the correct positions, the interpolation with the B-spline hypersurface can reach a given precision.

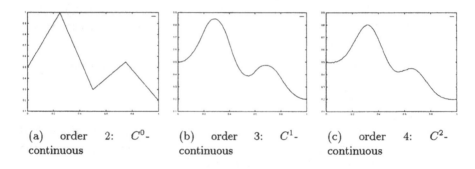

(a) order 2: C^0-continuous

(b) order 3: C^1-continuous

(c) order 4: C^2-continuous

Fig. 4. Trajectory of the output y with respect to the input x. The core rule set of 5 rules: $CRS = \{\text{IF } x \text{ is } A_i \text{ THEN } y \text{ is } Y_i, i = 1, \dots, 5\}$, where the universe of discourse of x is defined on $[0, 1]$ and Y_1 to Y_5 are fuzzy singletons with the following values: 0.5, 1.0, 0.3, 0.55, 0.2.

We will show later in section 4 that the output of the fuzzy controller can be flexibly adapted to anticipated values by adjusting the positions of the fuzzy singletons (control vertices) of the *core rules* after the order of the B-functions and the linguistic terms used in the IF-part are chosen.

3.3 SISO Systems

To illustrate the corresponding procedures of fuzzification, inference and de-fuzzification with our B-spline fuzzy controller, we first consider a system with

single input and single output (SISO), Fig. 5. For clarity, we suppose the input
is covered with four B-functions of order 2.

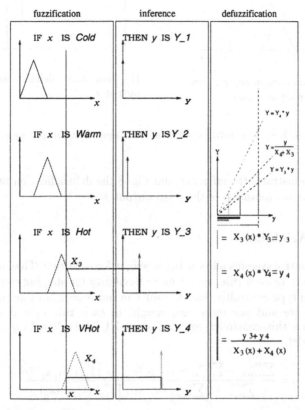

Fig. 5. A SISO system with B-functions of order 2 ($X_i(x)$: firing strength of rule i; y_i:
the contribution of rule i to the output).

3.4 MISO Systems

We further illustrate the principle of the fuzzy controller by an example with two
input variables (x and y) and an output z. The control vertices of the output
are Z_1, Z_2, Z_3, Z_4. Fig. 6(a) and 6(b) depict the linguistic terms of the input and
output.

The rule base consists of four rules:
Rule
1) IF x is X_1 and y is Y_1 THEN z is Z_1
2) IF x is X_1 and y is Y_2 THEN z is Z_2
3) IF x is X_2 and y is Y_1 THEN z is Z_3
4) IF x is X_2 and y is Y_2 THEN z is Z_4

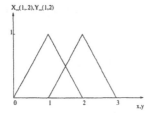
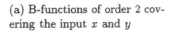

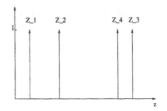

(a) B-functions of order 2 covering the input x and y

(b) Singeltons defined on the output z

Fig. 6. Linguistic terms used in a Two-Input-One-Output system

Fig. 7 demonstrates the inference and Fig. 8 the defuzzification procedure by illustrating the computation of the crisp output z.

3.5 Rule Weighting

Some fuzzy control systems allow using a weight for each rule. That provides one more parameter to each rule, surely more flexibility for shaping control surface in a Mamdani type controller, but it results in more work for fine-tuning.

In fact, if we add one more rule weight to each rule in a B-spline fuzzy controller, then this controller corresponds a NURBS (*Non-Uniform Rational B-Spline*) model:

$$y = \frac{\sum_{i_1=1}^{m_1} \cdots \sum_{i_n=1}^{m_n} w_{i_1,\dots,i_n} Y_{i_1,\dots,i_n} \prod_{j=1}^{n} X_{i_j,k_j}^{j}(x_j)}{\sum_{i_1=1}^{m_1} \cdots \sum_{i_n=1}^{m_n} w_{i_1,\dots,i_n} \prod_{j=1}^{n} X_{i_j,k_j}^{j}(x_j)}$$

Experience of using B-splines in CAD shows that by using sufficient B-spline basis functions, NUBS of non-rational form may approximate any shape to a given precision, [2]. NURBS curves and surfaces are mainly used for exactly modelling special analytical functions like a circle, square, etc. The control vertex of each rule in a B-Spline fuzzy controller plays the role of the rule weight as well as the control action. Therefore we adopt the NUBS of non-rational form for constructing fuzzy controllers.

3.6 Acceleration of Rule Evaluation

The index coding of the B-functions makes the evaluation of fuzzy rules highly efficient. For an input $x \in (x_i, x_{i+1})$, it is known that exact k (the order of B-functions) linguistic terms will be activated, i.e. k B-functions $X_{i,k}(x)$, $X_{i-1,k}(x)$, $\dots$, $X_{i-k+1,k}(x) > 0$. All the other linguistic terms are unactivated. In the whole rule base with n inputs, exact k^n rules are firing for any input vector in the universe of discourse.

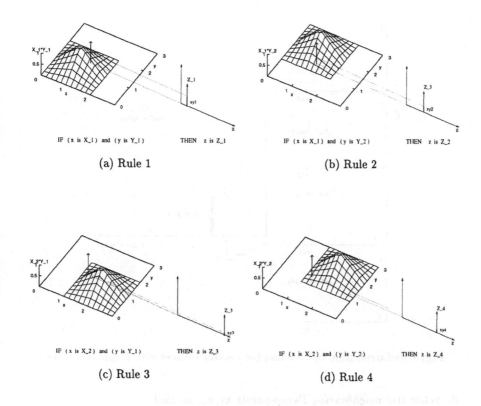

IF (x is X_1) and (y is Y_1)   THEN z is Z_1

(a) Rule 1

IF (x is X_1) and (y is Y_2)   THEN z is Z_2

(b) Rule 2

IF (x is X_2) and (y is Y_1)   THEN z is Z_3

(c) Rule 3

IF (x is X_2) and (y is Y_2)   THEN z is Z_4

(d) Rule 4

Fig. 7. Inference with a B-spline fuzzy controller with two inputs

4 Adaptation and Learning

4.1 Adaptation of Peak Support Points

We developed an algorithm for adapting the *Psupp*-points. This algorithm is a modified algorithm for self-organising neural networks. On the left and right neighbour of a *Psupp*-point $\mathbf{x}$, two new *Psupp*-points $\mathbf{x}_l$ and $\mathbf{x}_r$ are selected and the output values of the controller, noted as y_l and y_r, for these two points are computed. If the desired training data y^d is greater or smaller with a threshold $\theta \geq 0$ than both y_l and y_r, a modification of $\mathbf{x}$ is necessary.

Assume $\mathbf{x}_1, \mathbf{x}_2, \ldots, \mathbf{x}_m$ are the Psupp-points of the input variables and $\mathbf{X}_{1,k}, \mathbf{X}_{2,k}, \ldots, \mathbf{X}_{m,k}$ are their corresponding linguistic terms (B-functions of order k).

1. *Apply a new training input-output pair $(\mathbf{x}, y^d)$.*

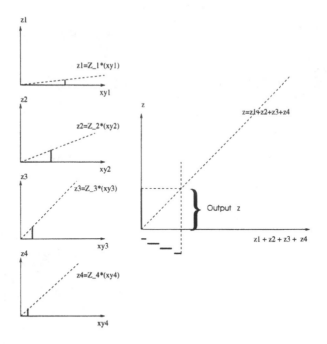

Fig. 8. Defuzzification by blending the control vertices with B-function values

2. *Select two neighbouring Psupp-points* x_l, x_r, *so that*
   $x_i \leq x_l \leq x \leq x_r \leq x_j$, *for* $i = 1, \ldots, l - 1$ *and* $j = r + 1, \ldots, m$
   Compute y_l, y_r. *Assume that* $y_l \leq y_r$
3. *If* $y^d \leq y_l - \theta$:
   Modify x_l: $x_l = x_l + (x - x_l) \cdot X_{l,k}(x)$
   If $y_r + \theta \leq y^d$:
   Modify x_r: $x_r = x_r + (x - x_r) \cdot X_{r,k}(x)$
4. *Optimise the control vertices of the output variables (see section 4.2).*
5. *Continue with 1.*

As an example we show the approximation of a function $\sin(2\pi x^2)$. The function has a minimum at $x = 0.86$ and a maximum at $x = 0.5$. At the beginning five *Psupp*-points are evenly distributed within the interval $[0, 1]$. All the five control vertices are set to 0. The training data are randomly generated from the interval $[0, 1]$. In the following implementation, we select θ as 0.2. The dashed curve represents the desired values, the solid curve the output of the controller. The points depicted as "◊" are the control vertices, whose abscissas are the *Psupp*-points.

4.2 Supervised Learning of Control Vertices

Assume $\{(X, y_d)\}$ is a set of training data, where

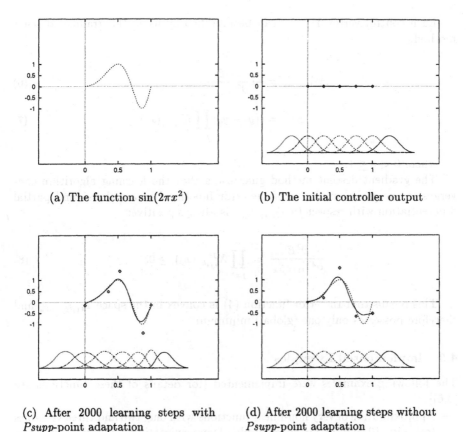

(a) The function $\sin(2\pi x^2)$

(b) The initial controller output

(c) After 2000 learning steps with *Psupp*-point adaptation

(d) After 2000 learning steps without *Psupp*-point adaptation

Fig. 9. Approximation of the function $y = \sin(2\pi x^2)$. The initial uniform B-functions are adapted to non-uniform ones.

- $\mathbf{X} = (x_1, x_2, \ldots, x_s)$: the input data vector,
- y_d : the desired output for $\mathbf{X}$.

The squared error is computed as:

$$E = \frac{1}{2}(y_r - y_d)^2, \tag{4}$$

where y_r is the current real output value during training.

The parameters to be found are $Y_{i_1, i_2, \ldots, i_n}$, which make the error in (4) as small as possible, i.e.

$$E = \frac{1}{2}(y_r - y_d)^2 \equiv \text{MIN.} \tag{5}$$

Each control vertex $Y_{i_1,...,i_n}$ can be modified by using the gradient descent method:

$$\Delta Y_{i_1,...,i_n} = -\epsilon \frac{\partial E}{\partial Y_{i_1,...,i_n}} \tag{6}$$

$$= \epsilon(y_r - y_d) \prod_{j=1}^{n} X_{i_j,k_j}^j(x_j) \tag{7}$$

where $0 < \epsilon \leq 1$.

The gradient descent method guarantees that the learning algorithm converges to the global minimum of the error function because the second partial differentiation with respect to $Y_{i_1,i_2,...,i_n}$ is always positive:

$$\frac{\partial^2 E}{\partial^2 Y_{i_1,...,i_n}} = \prod_{j=1}^{n} X_{i_j,k_j}^j(x_j) \geq 0. \tag{8}$$

This means that the error function (4) is convex in the space $Y_{i_1,i_2,...,i_n}$ and therefore possesses only one (global) minimum.

4.3 Implemented Examples

The following examples were implemented (for details of these functions see [3,6]):

- Approximation of very non-linear functions with one, two and three variables (see Fig. 10 and 11 for an example of approximating a function with two variables).
- Parameter identification and time series prediction.
- Supervised learning of "Truck Backer-Upper" (see Fig. 12 for the learned control surfaces and Fig. 13 for two test examples) and "Inverse Kinematics".

The results show that B-spline fuzzy systems can achieve the same and sometimes better modelling effect as done by the ANFIS (Adaptive-Network-Based Fuzzy Inference System) [3] and "Additive Systems" [6] methods, but B-spline fuzzy systems converge faster in all these cases.

4.4 Unsupervised Learning

In unsupervised learning, it is usually possible to define an "evaluation function" if the desired data of the output are unknown. Such an evaluation function should describe how "good" the current system state $((x_1, x_2, ..., x_n), y)$ is. For each input vector, an output is generated. With this output, the system transits to another state. The new state is compared with the old one; an adaptation is performed if necessary.

Assume the evaluation function, denoted by $F(\cdot)$, possesses a bigger value for a better state, i.e. for two states A and B, if A is better than B, then

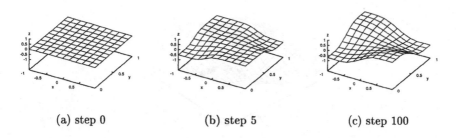

(a) step 0   (b) step 5   (c) step 100

Fig. 10. The control surfaces during the optimisation for approximating the function $z = sin(2\pi x) \cdot cos(\pi y)$, with $-1 \leq x \leq 1$ and $0 \leq y \leq 1$. The membership functions defining the real and virtual linguistic terms of x and y are 7 B-functions of order 3 for both x and y.

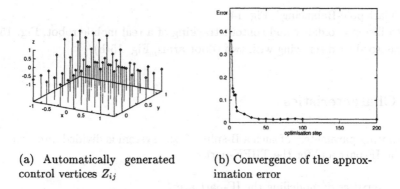

(a) Automatically generated control vertices Z_{ij}

(b) Convergence of the approximation error

Fig. 11. Learning results after 100 optimisation steps of the problem shown in Fig. 10.

$F(A) \geq F(B)$. The adaptation of the control vertices can be performed with a similar representation as in supervised learning. Assume that the desired state is A_d. The change of control vertices can be written as:

$$\Delta Y_{i_1,...,i_n} = S \cdot \epsilon \cdot |F(B) - F(A_d)| \cdot \prod_{j=1}^{n} X_{i_j,k_j}(x_j). \qquad (9)$$

where

$$S = sign(F(A) - F(B)) * sign(F(B) - F(A_d)) * sign(y) \qquad (10)$$

represents the direction to modify the control vertex.

By appropriately defining certain cost functions, we have successfully applied the B-spline fuzzy systems to the following control problem:

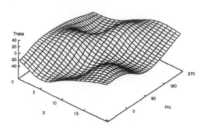

Fig. 12. Solution of the "truck backer-upper" problem: output θ after learning (θ was initialised as zero for all inputs). Both inputs are covered with 5 B-functions of order 3. Altogether 14 training trajectories are prepared.

- "Cart-pole-Balancing", Fig. 14;
- Collision-avoidance and contour-tracking of a real mobile robot, Fig. 15(a);
- Sensor-based screwing with two robot arms, Fig. 15(b).

5 Characteristics

Computing parameters of such a B-spline fuzzy system is divided into two steps: for the IF-part and for the THEN-part.

Characteristics of modelling the IF-part are:
- Considering the granularity of the input space and the extrema distribution of the control space if known, the fuzzy sets can be generated using the recursive computation of B-spline basis functions. This approach provides an natural, automatic approach to generate the information granularity proposed by Zadeh [8].
- These fuzzy sets can be further adapted during the generation of the whole system.

Characteristics for generating the THEN-part are:

- Fuzzy singletons can be initialised with the values acquired from expert knowledge. These approximately determined parameters will be fine-tuned with learning algorithms.
- For supervised learning, the differentials of the Square-Error with respect to control vertices are convex functions. Therefore, rapid convergence for supervised learning is guaranteed, and a reinforced learning process can also converge quite well if the change of cost function is piecewise approximately linear with the change of control vertices.

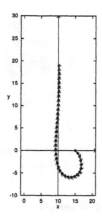 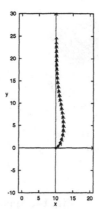

(a) Starting position: $x = 15, \phi = -30$.

(b) Starting position: $x = 10, \phi = 30$.

Fig. 13. Two motion trajectories produced by the trained fuzzy controller of the "truck backer-upper" problem.

- The control space changes locally while the control vertices are modified. Based on this feature, the control vertices can be optimised gradually, area-by-area. This is especially important for a high-dimensional control space with a large amount of parameters.

6 Discussions

The curse of dimensionality. Like all other types of fuzzy controllers, B-spline type controllers cannot avoid this problem, i.e. the number of rules increases exponentially with the number of inputs. However, as shown in section 3.6, the evaluation time of the whole rule base can be reduced form m^n to k^n, where m is the number of linguistic terms for input, k the order of B-functions and k is usually selected as 2 or 3 for most applications. The adaptation of knots also contributes to the efficient utilisation of linguistic terms. Additionally, the learning process of a MISO system does not suffer from the problem of local minima even for a high-dimensional control space thanks to the local influence of control vertices.

Conversion back to classic type. For some applications, e.g. data mining or qualitative analysis, we may find the large number of fuzzy singletons too numerical and want to approximately transform them into a small number of linguistic terms. Such a transformation can best be realised by fuzzy c-means clustering: given the number of linguistic terms we want, the fuzzy singletons can be grouped naturally into fuzzy sets, which represents a fuzzy partition of the output variable, Fig. 16.

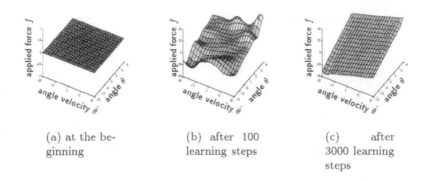

(a) at the beginning

(b) after 100 learning steps

(c) after 3000 learning steps

Fig. 14. Applied force f to the cart-pole system

In this way, our approach can optimally combine the numerical interpolation with linguistic interpretation, it is thus very promising for a wide range of applications in adaptive modelling and control.

References

1. H. Bersini and G. Bontempi. Now comes the time to defuzzify neuro-fuzzy models. In *Proceedings of the FLINS Workshop on Intelligent Systems and Soft Computing for Nuclear Science and Industry*, pages 130–139, 1996.
2. T. Dokken, V. Skytt, and A. M. Ytrehus. *The role of NURBS in geometric modelling and CAD/CAM*. In "Advanced Geometric Modelling for Engineering Applications", edited by Krause, F.-L.; Jansen, H. Elsevier, 1990.
3. J.-S. R. Jang. ANFIS: Adaptive-network-based fuzzy inference system. *IEEE Transactions on System, Man and Cybernetics*, 23(3):665–685, 1993.
4. B. Kosko and J. A. Dickerson. *Function Approximation with Additive Fuzzy Systems*, chapter 12, pages 313–347. In "Theoretical Aspects of Fuzzy Control", edited by H. T. Nyuyen, M. Sugeno and R. R. Yager, John Wiley & Sons, 1995.
5. E. H. Mamdani. Twenty years of fuzzy control: Experiences gained and lessons learned. *IEEE International Conference on Fuzzy Systems*, pages 339–344, 1993.
6. S. Mitaim and B. Kosko. What is the best shape of a fuzzy set in function approximation. In *IEEE International Conference on Fuzzy Systems*, 1996.
7. L. Wang. *Adaptive Fuzzy Systems and Control*. Prentice Hall, 1994.
8. L. A. Zadeh. Fuzzy logic = computing with words. *IEEE Trans. on Fuzzy Systems*, 4(2):103–111, 1996.
9. J. Zhang and A. Knoll. Constructing fuzzy controllers with B-spline models. In *IEEE International Conference on Fuzzy Systems*, 1996.
10. J. Zhang, K. V. Lee, and A. Knoll. Unsupervised learning of control spaces based on b-spline models. Submitted to IEEE International Conference on Fuzzy Systems, 1997.
11. J. Zhang, Y. v. Collani, and A. Knoll. On-line learning of b-spline fuzzy controller to acquire sensor-based assembly skills. In *Proceedings of the IEEE International Conference on Robotics and Automation*, 1997.

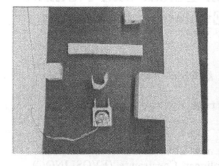

(a) Motion control of a mobile robot. The robot learns to avoid obstacles and to track an unknow contour. For details see [10].

(b) Two cooperating manipulators for fixture-less assembly. Compensation motions are determined by a self-learning B-spline controller with force/moment sensor readings, [11].

Fig. 15. Applications of unsupervised learning in robot systems.

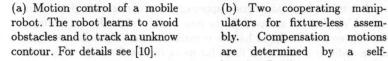

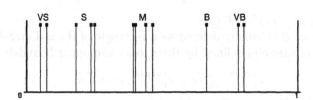

Fig. 16. All fuzzy singletons can be clustered into linguistic terms.

Scalar Optimization with Linear and Nonlinear Constraints Using Evolution Strategies

To Thanh Binh & Ulrich Korn

Institute of Automation, University of Magdeburg, Germany

Abstract. This paper introduces a new EVOlution Strategy for scalar optimization with LInear and NOnlinear Constraints (EVOSLINOC) which is robust to obtain a good approximation of a feasible global minimum. EVOSLINOC is based on the new concept of $\mathcal{C}$-, $\mathcal{F}$- and $\mathcal{N}$-fitness allowing systematically to handle constraints and (in)feasible individuals. In addition a number of ideas for using (in)feasible niche individuals which enable to explore new feasible areas and to make the population quickly to evolve towards a feasible global minimum is proposed. The performance of the EVOSLINOC can be successfully evaluated on many benchmark optimization problems [11, 10].

1 Introduction

1.1 Optimization problem

The general numerical scalar optimization problems with linear and nonlinear constraints can be formally stated as:

$$\min_{\mathbf{x}} \ f(\mathbf{x})$$

where $\mathbf{x} = (x_1, \ldots, x_n)^T \in \mathcal{F} \subseteq \mathcal{S}$.

The *search space* $\mathcal{S}$ is usually defined as a rectangle of the n-dimensional space $\mathcal{R}^n$ (domains of variables defined by their lower and upper bounds):

$$x_i^{(lower)} \leq x_i \leq x_i^{(upper)}, \forall i = \overline{1, n},$$

whereas the *feasible region* $\mathcal{F} \subseteq \mathcal{S}$ is defined by a set of m additional linear and/or nonlinear constraints ($m \geq 0$):

$$g_j(\mathbf{x}) \leq 0, \forall j = \overline{1, m}.$$

Constraints in terms of equations, e.g. $h(\mathbf{x}) = 0$, are not considered here because they can be replaced by a pair of inequalities:

$$-h(\mathbf{x}) - \epsilon \leq 0,$$
$$h(\mathbf{x}) - \epsilon \leq 0,$$

where an additional parameter ϵ is used to define the precision of the system. Therefore they seem to be special cases of constraints in terms of inequalities.

1.2 Motivation

Most of the current research on applications of evolutionary algorithms (EAs) to solve this optimization problem is based on the concept of penalty functions, i. e, the constrained optimization is converted into the unconstrained optimization of an auxiliary scalar function which is created from the given objective function and constraints [5, 6, 9, 10]. The basic problem of all penalty approaches is *how to design an auxiliary scalar function*. Therefore, they can be very well suitable for some optimization problems but they are disappointing for other ones.

Based on eliminating infeasible individuals from the population the main idea of traditional evolution strategies (ESs) seems to be unsuitable for solving this problem. ESs cannot start until a feasible starting population is generated. The finding feasible individuals is itself a difficult problem especially in cases the ratio between the feasible and search region is too small.

To overcome these drawbacks a new ES for handling linear and nonlinear constraints should satisfy the following requirements:

- It is not necessary to create any auxiliary function and to give a feasible starting point before optimization.
- The handling of the objective function and constraints is performed separately.
- The generating infeasible individuals is allowed.

This paper is organized as follows. The next section introduces a new representation of individuals that is suitable for the given optimization task. The handling feasible and infeasible individuals is discussed in section 3. The first experimental results of the EVOSLINOC on some test cases are presented in section 5.

2 Representation of an individual

For handling linear and nonlinear constraints it is necessary to introduce a suitable representation of an individual. As in traditional ESs, each individual consists of a vector of objective variables $\mathbf{x} = (x_1, x_2, \cdots, x_n)^T$ (a point in the search space), a strategy parameter vector $\mathbf{s} = (s_1, s_2, \cdots, s_n)^T$ (a vector of standard deviations). To evaluate the fitness of an individual the two following measures have to be taken into account:

- an objective function value $f(\mathbf{x})$ (so-called $\mathcal{F}$-fitness in the objective function space),
- a degree of violation of constraints or degree of (in)feasibility (it is called $\mathcal{C}$-fitness in the constraint space).

A problem how to evaluate the $\mathcal{C}$-fitness of an individual will be discussed here.

Let $c_i(\mathbf{x}) = \max\{g_i(\mathbf{x}), 0\}$, $\forall i = \overline{1, m}$, the $\mathcal{C}$-fitness of an individual can be charaterized by a vector:

$$\mathcal{C}(\mathbf{x}) = (c_1(\mathbf{x}), c_2(\mathbf{x}), \cdots, c_m(\mathbf{x})), \tag{1}$$

(a point in the constraint space $\mathcal{R}_c^m$). Clearly, $\mathcal{C} \equiv \mathbf{0}$ for feasible individuals and $\mathcal{C} > \mathbf{0}$ for infeasible ones. In this sense the original point of the constraint space corresponds to the feasible region of the search space. An advantage to using this $\mathcal{C}$-fitness is that it enables precisely to represent an individual in the constraint space. However it has the following disadvantages:

- When the number of constraints is big, a large amount of memory is necessary to save $\mathcal{C}(\mathbf{x})$ for every individual.
- It takes much time to make a decision which infeasible individuals are better than other ones or should be selected for the next generation due to the fact that the comparision between infeasible individuals essential is the comparision between vectors $\mathcal{C}(\mathbf{x})$.

To avoid these problems the following measure of violation based on the *distance* between a point $(c_1(\mathbf{x}), c_2(\mathbf{x}), \cdots, c_m(\mathbf{x}))$ and the original point in the constraint space, is used [12]:

$$\mathcal{C}(\mathbf{x}) = \left(\sum_{i=1}^{m} [c_i(\mathbf{x})]^p \right)^{\frac{1}{p}}, \quad (p > 0). \tag{2}$$

The second measure of violation has experimentally shown to be as good as the first one and acceptable. Therefore it should be used to design the EVOSLINOC. By this way, the population can be divided into classes; individuals of the same class have the same value of $\mathcal{C}$ (i. e. the same distance to the feasible region) and they are said to be individuals of the $\mathcal{C}$-class. Using this concept, the 0-class includes all feasible individuals; individuals of the higher classes are *farther* from the feasible region than ones of the lower classes.

Thus, an individual can be represented as follows:

$$\mathcal{I}nd \overset{\text{def}}{=} (\mathbf{x}, \mathbf{s}, f(\mathbf{x}), \mathcal{C}(\mathbf{x})).$$

3 Handling feasible and infeasible individuals

Different to traditional ESs, both feasible and infeasible individuals can live in the population simultaneously. The EVOSLINOC allows mutation and reproduction operators to generate both feasible and infeasible offspring. Therefore it is necessary to check whether offspring are better than their parent (by mutation and reproduction) or to select better individuals for the next generation (by selection).

Up to now, there is no general method for comparision between infeasible and feasible individuals or between two infeasible individuals together [10]. For the EVOSLINOC the following selection criteria are recommended:

Criterion 1. An individual of the C_1-class ($C_1 > 0$) is said to be better than the other of the C_2-class iff $C_1 < C_2$.

In other words, infeasible individuals of higher classes are worse than the other of lower classes. Using this criterion the new ES will drive the population towards the feasible region before trying to search for a feasible global minimum.

Criterion 2. Among individuals of the same class, better individuals have lower $\mathcal{F}$-fitness values.

During the search process for the feasible region the population at some stage of the evolution process may contain some feasible and infeasible individuals. If the criterion 1 is extended for use with 0-class (that means feasible individuals are always better than infeasible ones), the population would be soon feasible because all feasible individuals will be kept in the population unless a feasible offspring with a lower $\mathcal{F}$-fitness is generated. In many optimization problems the feasible region is non-convex or the ratio between the feasible and search region (denoted by ρ) is too small (for example, a feasible region defined by constraints in terms of functional equations) so that feasible offspring cannot be generated even from feasible individuals after many generations. In these cases the feasible population very slowly converges to a feasible global minimum. The following criteria help the evolution strategy avoid such situation and is based upon the intuitive conviction as below:

For many optimization problems, feasible individuals are *not always* better than all infeasible ones. For example, in *Fig. 1* the infeasible individual marked by 'b' lies nearer to the feasible global minimum than the feasible one marked by 'a'. For the same strategy parameter vector s it is hopeful that the infeasible individual 'b' generates feasible offspring which are better than offspring of the feasible individual 'a'. Here the infeasible individual 'b' is *not better* than other feasible individuals, but it is at least *acceptable* for the next generation. Such infeasible individuals should be kept in the population and can be included into a special class of infeasible individuals in a neigbourhood of the feasible region.

Criterion 3 (Extra class). Infeasible individuals up to the C_{extra}-class (i. e. individuals of C-classes so that $0 < C \le C_{extra}$) are said to be in the same class (called the *extra class*).

385

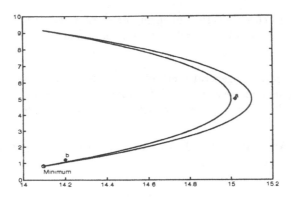

Fig. 1. An infeasible individual (marked 'b') of an extra class

The problem of how to choose the best value of C_{extra} is far from trivial. With a lower value of C_{extra} it is more difficult to locate offspring in the extra class. Otherwise the choosing a higher value of C_{extra} leads to generate more infeasible individuals (in higher classes). Our first experiments showed that the value $C_{extra}=0.1$ was preferable.

The following criterion allows feasible individuals to generate infeasible offspring:

Criterion 4. An infeasible offspring of a feasible individual is said to be viable iff it belongs to the so-called extension $C_{extension}$-class defined by:

$$C_{extension} = \max\{C_{extra}, C_{pop}\},$$

where C_{pop} is the highest class in the population (corresponding to the infeasible individual with the biggest distance to the original point of the constraint space).

4 Niche individuals

Traditional ESs [2, 7] have shown that so-called *niche* (feasible) individuals play an important role to overcome local minima by multi modal optimization problems. Therefore it is also significant to use niche (in)feasible individuals for the new evolution strategy. A niche feasible individual can be characterized by the two following properties [2, 7]:

– Their $\mathcal{F}$-fitness is as small as possible.
– They are as far as possible from the current best feasible individual.

Mathematically, a niche feasible individual should have a small value of the so-called niche fitness ($\mathcal{N}_f$-fitness) [2, 7]:

$$\mathcal{N}_f(\mathbf{x}) = \frac{f(\mathbf{x}) - f(\mathbf{x}_{fbest})}{\|\mathbf{x} - \mathbf{x}_{fbest}\|^\beta}, \quad (\mathbf{x} \neq \mathbf{x}_{fbest}) \tag{3}$$

where:

- $\mathbf{x}_{fbest}$ and $f(\mathbf{x}_{fbest})$ are an objective variable vector and the $\mathcal{F}$-fitness of the best feasible individual, respectively.
- $\|.\|$ denotes a norm in the n-dimensional parameter space.
- β is a scalar value ($\beta = 1, 2, \cdots$).

The left problem is to choose niche infeasible individuals from the current population. For optimization problems for which feasible individuals are hard to find, the population of some first generations contains infeasible individuals only. In this situation the $\mathcal{C}$-fitness is used to help the evolution strategy in shifting the population towards feasibility. To avoid problems for which the infeasible population converges to (possible) local minima of the $\mathcal{C}$-fitness, the following criterion is recommended to choose niche infeasible individuals:

Criterion 5. If there is no feasible individual in a population, niche infeasible individuals should have an as small value of $\mathcal{C}$-fitness as possible and be as far as possible from the best current infeasible individual.

Similar to *Eq. 3* the niche fitness of infeasible individuals ($\mathcal{N}_i$-fitness) can be evaluated by:

$$\mathcal{N}_i(\mathbf{x}) = \frac{\mathcal{C}(\mathbf{x}) - \mathcal{C}(\mathbf{x}_{ibest})}{\|\mathbf{x} - \mathbf{x}_{ibest}\|^\beta}, \quad (\mathbf{x} \neq \mathbf{x}_{ibest}) \tag{4}$$

where $\mathbf{x}_{ibest}$ and $\mathcal{C}(\mathbf{x}_{ibest})$ are an objective variable vector and the $\mathcal{C}$-fitness of the best infeasible individual, respectively. Notice that traditional ESs [2] have also shown that the use of niche individuals may help the ES in finding many global minima with the same objective function value located at different points of the parameter space. For constrained optimization problems for which the feasible region is disconnected the $\mathcal{C}$-fitness (see *Eq. 2*) has the (global) smallest value 0 at different subsets of the feasible region. Therefore it is meaningful to use niche infeasible individuals to explore other subsets of the feasible region.

When there exist some feasible individuals in a population, the ES tries to generate infeasible individuals in a neighbourhood of the feasible region (in a $\mathcal{C}_{extension}$-class). The best infeasible individuals are not only infeasible ones with lower $\mathcal{C}$-fitness, but they are other ones lying nearer to global feasible minima (niche infeasible individuals). Because global feasible minima are unknown, it can be expected that they are still far from the best current feasible individual. Therefore:

Criterion 6. If at least one feasible individual exists in the population, niche infeasible individuals should have an as small value of C-fitness as possible and be as far as possible from the best current feasible individual.

The niche fitness of infeasible individuals ($\mathcal{N}_{if}$-fitness) can be evaluated by:

$$\mathcal{N}_{if}(\mathbf{x}) = \frac{C(\mathbf{x})}{\|\mathbf{x} - \mathbf{x}_{fbest}\|^\beta}. \tag{5}$$

Our initial experiments showed that this criterion was useful for many optimization problems.

5 Some test cases

In this section, some test cases have been carefully selected to illustrate the efficiency of the EVOSLINOC. Most of them are described as benchmark optimization problems and used to compare or demonstrate the current research methods for the optimization with linear and nonlinear constraints [8, 10]. For all test cases the following important parameters of the ES were used:

- Population size $= 100$
- Number of niche feasible individuals $= 5$
- Number of niche infeasible individuals $= 5$
- Number of parents for mutation and reproduction $= 10$
- Number of offspring per mutation $= 5$.

5.1 Test Case 1

This optimization problem (taken from [6], Test Case 6 of [10]) is

$$minimize \ f(\mathbf{x}) = (x_1 - 10)^3 + (x_2 - 20)^3$$

subject to nonlinear constraints:

$$(x_1 - 5)^2 + (x_2 - 5)^2 \geq 100$$
$$(x_1 - 6)^2 + (x_2 - 5)^2 \leq 82.81$$

and bounds: $13 \leq x_1 \leq 100$ and $0 \leq x_2 \leq 100$. In this case, the feasible region is nonconvex and $\rho = 0.0066^1$. The known global solution is $\mathbf{x}^* = (14.095, 0.84296)^T$ and $f(\mathbf{x}^*) = -6961.81381$. From an infeasible starting point $\mathbf{x}_0 = (20.1, 5.84)^T$ lying far from the feasible region the evolution of the population is shown in *Fig. 2*. For all runs the search for the feasible region ends

---

[1] From now, ρ was determined experimentally by generating one milion random points in the search space S and checking if they belong to $\mathcal{F}$

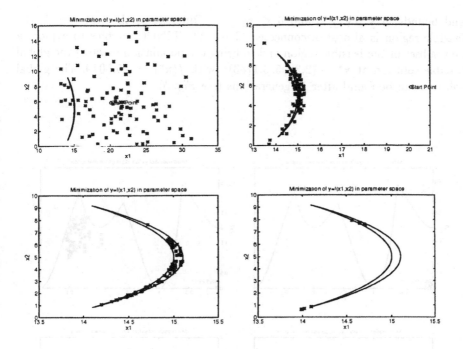

Fig. 2. Evolution of the population for Test Case 1

after less than 10 generations; the population moves then into the feasible global minimum and concentrates on it in less than 400 generations. It is interesting to note that the ES is more robust and stable by finding the global feasible minimum than other tools [11, 2, 7].

5.2 Test Case 2

The optimization problem (taken from [6, 11]) is

$$minimize \ f(\mathbf{x}) = -x_1 - x_2,$$

subject to nonlinear constraints:

$$2x_1^4 - 8x_1^3 + 8x_1^2 + 2 - x_2 \geq 0$$
$$4x_1^4 - 32x_1^3 + 88x_1^2 - 96x_1 + 36 - x_2 \geq 0$$

and bounds: $0 \leq x_1 \leq 3$ and $0 \leq x_2 \leq 4$. Different to the test case 1, the feasible region is almost disconnected. The EVOSLINOC enables to explore a new subset of the feasible region and to bring the population towards the global feasible solution at $\mathbf{x}^* = (2.3295, 3.1783)^T$ with $f(\mathbf{x}^*) = -5.5079$. The global solution can be found after 10 generations (see *Fig. 3*).

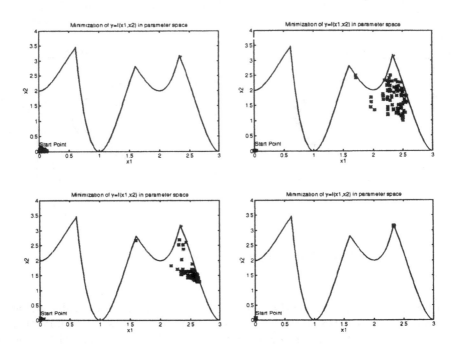

Fig. 3. Evolution of the population for Test Case 2

5.3   Test Case 3

This optimization problem has been used to illustrate *the death penalty method* for an evolution strategy [1] and to compare some evolutionary algorithms ([11], Test Case 7 of [10]). Minimize a function:

$$f(\mathbf{x}) = x_1^2 + x_2^2 + x_1 x_2 - 14x_1 - 16x_2 + (x_3 - 10)^2 + 4(x_4 - 5)^2 + (x_5 - 3)^2 +$$
$$2(x_6 - 1)^2 + 5x_7^2 + 7(x_8 - 11)^2 + 2(x_9 - 10)^2 + (x_{10} - 7)^2 + 45;$$

subject to the constraints:

$$105 - 4x_1 - 5x_2 + 3x_7 - 9x_8 \geq 0$$
$$-10x_1 + 8x_2 + 17x_7 - 2x_8 \geq 0$$
$$8x_1 - 2x_2 - 5x_9 + 2x_{10} + 12 \geq 0$$
$$3x_1 - 6x_2 - 12(x_9 - 8)^2 + 7x_{10} \geq 0$$
$$-3(x_1 - 2)^2 - 4(x_2 - 3)^2 - 2x_3^2 + 7x_4 + 120 \geq 0$$
$$-x_1^2 - 2(x_2 - 2)^2 + 2x_1x_2 - 14x_5 + 6x_6 \geq 0$$
$$-5x_1^2 - 8x_2 - (x_3 - 6)^2 + 2x_4 + 40 \geq 0$$
$$-.5(x_1 - 8)^2 - 2(x_2 - 4)^2 - 3x_5^2 + x_6 + 30 \geq 0$$

and bounds: $-10 \leq \mathbf{x} \leq 10$. This function is quadratic and has the global minimum at:

$$\mathbf{x}^* = (2.172, 2.364, 8.774, 5.096, .9907, 1.4306, 1.322, 9.829, 8.280, 8.376)^T,$$

where $f(\mathbf{x}^*) = 24.3062$.

Up to now, the best solution which had the value of 25.653 was found in [1], but it is necessary to initialize a population by feasible individuals [11]. The EVOSLINOC reached the value less than 25.00 for all test runs and the best solution among them had the value of 24.61.

5.4 Test Case 4

The Test Case is selected to compare the result by using the EVOSLINOC with results of other methods ([5], [4], Test Case 4 of [10]). Minimize a function:

$$f(\mathbf{x}) = 5.3578547x_3^2 + 0.8356891x_1x_5 + 37.293239x_1 - 40792.141;$$

subject to three double nonlinear inequalities:

$$0 \leq 85.334407 + 0.0056858x_2x_5 + 0.00026x_1x_4 - 0.0022053x_3x_5 \leq 92$$
$$90 \leq 80.51249 + 0.0071317x_2x_5 + 0.0029955x_1x_2 + 0.0021813x_3^2 \leq 110$$
$$20 \leq 9.300961 + 0.0047026x_3x_5 + 0.0012547x_1x_3 + 0.0019085x_3x_4 \leq 25$$

and bounds: $78 \leq x_1 \leq 102$, $33 \leq x_2 \leq 45$, $27 \leq x_i \leq 45$, $\forall i = 3, 4, 5$. EVOSLINOC provided the best value: $f_{min} = -31025$ at

$$\mathbf{x}^* = (78.0001, 33.0072, 27.0760, 44.9872, 44.9591)^T$$

that is better than in [5] ($f_{min} = -30005.7$) and in [4] ($f_{min} = -30665.5$).

5.5 Test Case 5

It is the Test Case 9 in [10] to illustrate *the method of superiority of feasible points* and can be described as below:

$$\min_{\mathbf{x}} f(\mathbf{x}) = (x_1 - 10)^2 + 5(x_2 - 12)^2 + x_3^4 + 3(x_4 - 11)^2 +$$
$$10x_5^6 + 7x_6^2 + x_7^4 - 4x_6 x_7 - 10x_6 - 8x_7;$$

subject to the constraints:

$$127 - 2x_1^2 - 3x_2^4 - x_3 - 4x_4^2 - 5x_5 \geq 0$$
$$282 - 7x_1 - 3x_2 - 10x_3^2 - x_4 + x_5 \geq 0$$
$$196 - 23x_1 - x_2^2 - 6x_6^2 + 8x_7 \geq 0$$
$$-4x_1^2 - x_2^2 + 3x_1 x_2 - 2x_3^2 - 5x_6 + 11x_7 \geq 0$$

and bounds: $-10 \leq x_i \leq 10$, $\forall i = \overline{1,7}$. The function has its global minimum at

$$\mathbf{x}^* = (2.330499, 1.951372 - 0.47754, 4.3657, -0.62448, 1.0381, 1.5942)^T,$$

where $f(\mathbf{x}^*)$=680.6300573. The EVOSLINOC approached very closely the global optimum after less than 400 generations for all runs.

6 Conclusion

In this paper the EVOSLINOC allowing to handle linear and nonlinear constraints was proposed. Experiments on many benchmark optimization problems indicated that the EVOSLINOC was robust and gave good performances. Interesting is that the paper provides some new ideas for effectively handling (in)feasible individuals: niche infeasible individuals can be *better* than some feasible ones. The use of niche infeasible individuals enables the ES to avoid an unnecessary concentration of the population at local minima and quickly to shift the population towards the feasible global minima. The EVOSLINOC was implemented in a MATLAB–based[2] environment [3]. It is still under development, and there are some possible modification and extensions in the future version.

References

1. T. Bäck, F. Hoffmeister, and H.-P. Schwefel. A survey of evolution strategies. In R.K. Belew and L.B. Booker (Eds.), Proceedings of the 4th International Conference on Genetic Algorithms, Morgan Kaufmann, pages 2–9, 1991.

---

[2] MATLAB is the Trademark of the MathWorks, Inc.

2. T. Binh. Eine Entwurfsstrategie für Mehrgrößensysteme zur Polgebietsvorgabe. PhD thesis, Institute of Automation, University of Magdeburg, Germany, 1994.
3. T. Binh, U. Korn, and J. Kliche. *Evolution Strategy Toolbox for use with MATLAB*. Technical report, Institute of Automation, University of Magdeburg, Germany, Mar. 1996.
4. D. Himmelblau. Applied Nonlinear Programming. McGraw-Hill, 1992.
5. A. Homaifar, S. Lai, and X. Qi. Constrained optimization via genetic algorithms. Simulation 62, 4:242–254, 1994.
6. J. Joines and C. Houck. On the use of non-stationary penalty functions to solve nonlinear constrained optimization problems with gas. In Z. Michalewicz, J. D. Schaffer, H.-P. Schwefel, D. B. Fogel, and H. Kitano (Eds.), Proceedings of the First IEEE International Conference on Evolutionary Computation, pages 579–584, 1994.
7. J. Kahlert. Vektorielle Optimierung mit Evolutionsstrategien und Anwendung in der Regelungstechnik. Forschungsbericht VDI Reihe 8 Nr.234, 1991.
8. Z. Michalewicz. Genetic algorithms, numerical optimization and constraints. In L. J. Eshelman (Ed.), Proceedings of the 6th International Conference on Genetic Algorithms,Morgan Kaufmann, pages 151–158, 1995.
9. Z. Michalewicz. Heuristic methods for evolutionary computation techniques. Journal of Heuristics 1(2), pages 177–206, 1995.
10. Z. Michalewicz, D. Dasgupta, R. Riche, and M. Schoenauer. Evolutionary algorithms for constrained engineering problems. In Y. Davidor, H.-P. Schwefel, and R. Manner (Eds.), Proceedings of the 6rd Conference on Parallel Problems Solving from Nature, Springer Verlag, 1996.
11. Z. Michalewicz and C. Nazhiyath. Genocop III: A co–evolutionary algorithm for numerical optimization problems with nonlinear constraints. In D. B. Fogel (Ed.), Proceedings of the 2th IEEE International Conference on Evolutionary Computation, pages 647–651, 1995.
12. J. Richardson, M. Palmer, G. Liepins, and M. Hilliard. Some guidelines for genetic algorithms with penalty functions. Proceedings of the 3rd International Conference on Genetic Algorithms, Los Altos, CA, Morgan Kaufmann Publishers, pages 191–197, 1989.

Continuous Control of a Synchronous Generator's Active and Reactive Power Connected to Network by Using Fuzzy Neural Networks Algorithms

Muğdeşem TANRIÖVEN   Celal KOCATEPE   Mehmet UZUNOĞLU

Yıldız Technical University, Electric-Electronic Faculty
Electrical Eng. Dept. 80750 Yıldız-İstanbul

Abstract : The fuzzy logic controller based on the fuzzy set theory provides a useful tool for converting the linguistic control rules from the expert knowledge into automatic control rules [1]. By using fuzzy automatic rules from the heuristic or mathematical strategies, complex processes can be controlled effectively in many situations. But the most important and difficult point is how to obtain the proper control rules for a given system. In addition to that Artificial Neural Networks (ANN) is successfully used in many areas such as fault detection, control and signal processing in our daily technology. Artificial Neural Networks have nonlinear structure and this is an effective feature that it approaches to the results of learning phase. Then, it gives results in test phase in short time (the degree about 10^{-3} second). It is a very preferable according to the other approaching methods.

In this study, generalized fuzzy logic controller was resembled by writing its software in Q - basic programming language. Active and reactive power control of a generator connected to infinite bus system was fulfilled effectively by the fuzzy logic controller which is supported by artificial neural networks. As well as this, the software of the system's simulation results were given.

1 Introduction: Power system operation considered so far was under conditions of steady load. However, both reactive and active power demands are never steady and they continually change with rising or falling trend. Steam input to turbo generators (or water input to hydro-generators) must, therefore, be continuously regulated to match the reactive power demand with reactive power generation, otherwise the voltages at various system busses may go beyond the prescribed limits [2]. In modern large interconnected systems, manual regulation is not feasible and therefore automatic generation and voltage regulation equipment is installed on each generator.

2 Active and Reactive Power Control of a Synchronous Generator Connected to Networks

Where the word network's means is that it has a very small internal impedance, which can be ignored, with a constant voltage and is infinite compare with considered machine [3]. Because of its connection of such a network, both terminal voltage (Vt) and parameters of synchronous machine (Ra,Xs) are constant. Thus for the reactive and active power adjustment of a synchronous generator, we can change both the excitation current and the torque is given to machine shaft , respectively.

2.1 Reactive Power Adjustment of a Synchronous Generator

Excitation current control of synchronous generator is operated parallel to network (infinite bus systems) cause to change in its internal emk (E_f) and just only reactive power of the machine is effected. Active power of the machine stay at constant value [4]. Machine active power can be calculated in two ways. From the tap conditions of the machine, active power can be written for the first one as below:

$$P = Vt.Ia.\cos\theta \tag{1}$$

Where,
Vt = terminal voltage
Ia = armature current
$\cos\theta$ = power factor

For the second one, by ignoring the stator resistance equations (1) can be rewritten relative to load angle, synchronous reactance and internal emk by the excitation current.

$$P = \frac{Ef \cdot Vt}{Xs} \cdot \sin\delta \tag{2}$$

Both terminal voltages (Vt) and active powers (P) stay at constant value. From the equations (1) and (2), the constant values k1 and k2 can be written as below:

$$Ia \cdot \cos\theta = k_1$$
$$Ef \cdot \sin\delta = k_2 \tag{3}$$

By the increasing excitation current, the vector E_f and load current will change within the range of k2 and k1, respectively

Let both the new values of Ef and Ia is E_f' and Ia', respectively (figure 1)

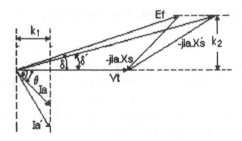

Figure 1. Reactive power control of a synchronous generator

2.2 Active Power Adjustment of a Synchronous Generator

As explained above, active power is not based on the excitation current. When the field current is constant in the machine, E_f voltage is also constant. Thus for the adjustment of the active power, the torque angle should be changed. To make bigger the angle δ as big as $\Delta\delta$, the given torque to generator's shaft should be increased and accelerated in a short time by opening the valve of the water turbine or steam turbine (figure-2).

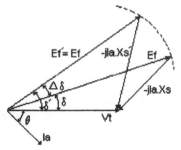

Figure 2. Active power control of a synchronous generator

For the above conditions, the machine gives the reactive power to networks as bellow:

From the tap conditions of the generator

$$Q = Vt \cdot Ia \cdot \sin\theta \tag{4}$$

And depending on the torque angle, terminal voltage, synchronous reactance and internal emk, equation (4) can be rewritten as equation (5) [5]

$$Q = -\frac{Vt}{Xs} \cdot \left(Ef \cdot \cos\delta - Vt\right) \tag{5}$$

3 Design of Fuzzy Logic Controller Supported by Artifical Neural Networks For Active And Reactive Power Control at Synchronous Generators Connected to Infinite Bus System

As known, active and reactive power demand changes in a various time of a day. Thus active and reactive power demand should be kept up with the power generations by adjusting field current (I_f) and torque angle (δ). For our system, fuzzy controller is designed as below :

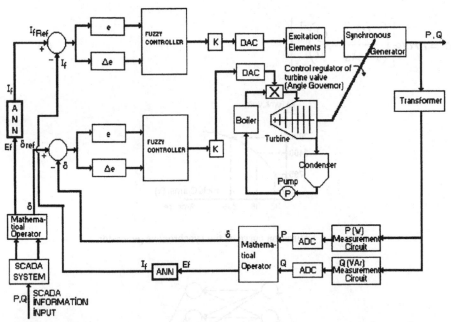

Figure 3. Design of fuzzy logic controller for active and reactive power control at synchronous generators connected to infinite bus system

As seen in the block diagram of the controller, information obtained from SCADA system, measure some values at energy systems, determine which generators should be loaded at which level of active and reactive power. [6]

From the equations (2) and (5) torque angle δ and internal emk E_f are determined as shown equations (6) and (7). As well as this, the data from the saturation curve of the synchronous generator (shown figure 4) is input values of ANN for training. In the software of ANN, mathematical equation of Fast Backpropagation Algorithm was used [7]. The program language is Turbo Pascal and IBM compatible 486-DX 2*66 computer was used to train ANN [8]. Network architecture for Ef → If was given in figure 5 as 1:10:10:1. For the network architecture, the learning error rate is e = 0.0013 % and the test phase of ANN (ANN's output) is tabulated in the table 1.

Consequently the second control variable of the synchronous generator's active and reactive power , is I_f, can found like above.

$$Ef = \frac{P.Xs}{V.\sin\left(arctg\left[\left(\frac{V^2}{P \cdot Xs} - \frac{Q}{P}\right)^{-1}\right]\right)} \tag{6}$$

$$\delta = arctg\left[\left(\frac{V^2}{P \cdot Xs} - \frac{Q}{P}\right)^{-1}\right] \tag{7}$$

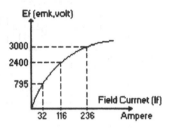

Figure 4. Saturation curve of the synchronous generator

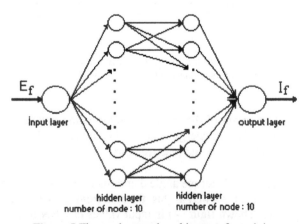

Figure 5 The used network architecture for training

The data active and reactive power demands obtained from SCADA system is used as an input value of mathematical operator for finding E_f and δ. From the E_f, I_f will be determined by using ANN. The equations (6 and 7) is programmed into mathematical operator for finding $E_f \xrightarrow{ANN} I_f$ and δ. These values (I_f and δ) are become the reference input of the block diagram of the systems by relative to load changing.

Table 1: Some values of I_f from the ANN's output versus E_f

Field current (If) (Ampere)	EMK (Ef) (Volt)
32	795
41.5	1010
54	1255
61	1460
116	2400
131.5	2530
149.5	2675
180	2800
236	3000

3.1 Distinguished Membership Functions and Fuzzy Rules

Figure (6) shows the governor of the exciter and torque angle of a generator. As seen in the figure (6) the characteristic cycle versus load is linear (figure - 5)

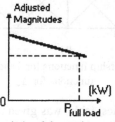

Figure 6. The governor characteristic of the exciter and torque angle of a generator

By searching the figure (1),(2) and (6), membership functions relative to error and changes in error were distinguished as triangular are shown in figure $7_{(a-b)}$ and $8_{(a-b)}$.

Error and change in error are determined as below [9]:

e = Reference value - system output
ce = Previous error - the present time error

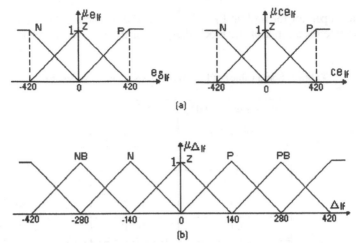

(a)

(b)

Figure 7. (a) Input membership functions for If and C_{If} (b) Output membership functions for Δ_{If}

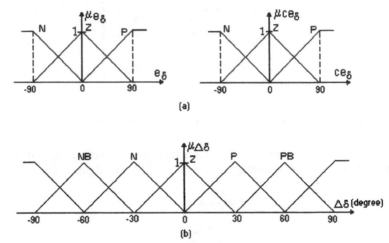

(a)

(b)

Figure 8. (a) Input membership functions for δ and C_δ (b) Output membership functions for Δ_δ

FAM (fuzzy associative memories) table was given in table 2. In the table, P,Z,N,PB and NB are represent the positive, zero, negative, positive big and negative big respectively.

Table 2. Fuzzy rules

e\ce	P	Z	N
P	PB	P	Z
Z	N	Z	P
N	Z	N	NB

3.2 Simulation Results

Lets search the adjustments of active and reactive power of synchronous generator were considered in two time slice. For the first period of the time, lets adjust the both excitation current from zero to 250 A and torque angle δ from zero to 15 degree. As well as this, for the second period of the time, lets adjust the both excitation current from 250A to 200A and torque angle δ from 15 degree to 12 degree. The software of the programme was written in Q - basic programming language and simulation results were given in figure (9a-b) and (10a-b)

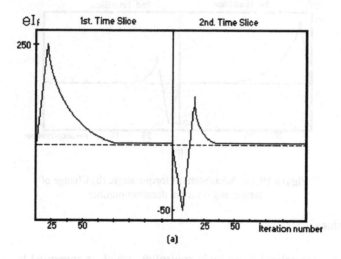

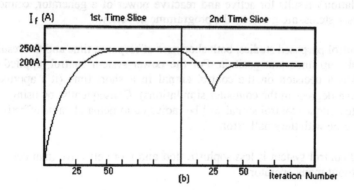

Figure 9. (a) Change of excitation current versus iteration number
(b) Adjustment of field current

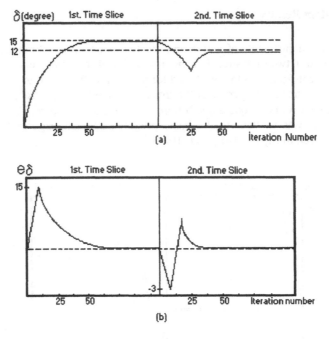

Figure 10. (a) Adjustment of torque angle (b) Change of
torque angle versus iteration number

4 Conclusions

In this study, generalized fuzzy logic controller, which is supported by ANN, was resembled by writing its software in Q - basic programming language. As well as this, simulation's results for active and reactive power of a generator, connected to infinite bus systems, are given by the programme as bellow :

- In the control purposed studies, both shortness of response time and the result value of error take up important position. For the power system control, applied control system takes a decision on the control signal in a short time (it's approximately about at μsn degrees in the computer simulations). Consequently, by using this new approach technique, control signal will be adjusted to demand value effectively in a very short time with the small error.

- Proposed control system is less sophisticated and more effective than conventional system without many operator.

- Due to more quick and more sensitive control actions can be fulfilled by the method, it cause destroying the system stability effects are to be disappeared. So we can say that the method has a effect of improvement for the getting better power system stability.

- At the conclusion of control operation by fuzzy logic controller, negativeness sourced by the operator is absolutely annihilated.

If we search the system at the point of cost, we can come to a conclusion as follow :

- The additional cost sourced from the operation of the system with the error is disappeared by the method.

- Because of shortness of response time in the developed method, occurring additional cost during the time of control actions by the operator is annihilated. Thus as soon as the undemand conditions occur, system will continue to operate in the demand limits. Consequently, minimum cost and maximum efficiency compared with the conventional system are provided by the method.

- Because of system's above feature, we can say that developed control system for active and reactive power of a generator, connected to infinite bus systems is the optimum control type.

References

1. Procky and E.H. Mamdani, A linguistic self-organising process controller, Automatica, Vol. 15, No.1 1979, pp. 15-30

2. Nagrathy, I., J., Kathari D., P., Modern Power System Analysis, Mc Grawn Hill, New Delhi, 1982.

3. Sarıoğlu, K. , Foundation of Electrical Machines (Synchronous Machines) İ.T.Ü. Electric-Electronic Faculty, İstanbul, 1993

4. Neal J., B., Mark G. Lauby, Power System Stability and Control, Newyork 1993.

5. Carson, W., T., Balu, Neal J.,B., Power System Voltage Stability Mc Grawn Hill, Newyork, 1993.

6. Uçan, B., A SCADA/EMS Applications, 5th Electrical Engineering National Symposium, TURKEY, 1995

7. Karayannis, N.B. And Venetsanopoulas A.N., "Fast Learning Algorithms for Neural Networks", IEEE Trans. Cir. and Sys.-II Analog and Digital Systems Proc. Vol.39, no.7, Page 453-474, July 1992.

8. Karlık, B., "Mio Electric Control by Using Artificial Neural Networks for Multi-functional Prothesis", Doctora Thesis, Page 36 , 38 ve EK , Y.T.Ü. TURKEY, 1994.

9. Timothy J.R., Fuzzy Logic With Engineering Applications, U.S.A, 1995

Design of Adaptive Fuzzy Controller for Bolu Highway Tunnel Ventilation System

Ercüment Karakaş[1], Hasan Külünk[1], Engin Özdemir[1], Ercan Ölçer[2], Bülent Karagöz[3]

[1] Department of Electrical Education, University of Kocaeli, 41100, Izmit, Turkey
[2] Department of Computer Engineering, University of Kocaeli, 41100, Izmit, Turkey
[3] Department of Electronics & Com. Engineering, University of Kocaeli, 41100, Izmit, Turkey
E-mail: curgu-ed@yunus.mam.tubitak.gov.tr

Abstract. The purpose of tunnel ventilation control is to provide a safe and comfortable environment for the users. The tunnel ventilation is optimized by controlling jet fans and dust collectors installed inside the tunnel. The jet fans blow polluted air from inside the tunnel toward air exit ports. The dust collectors remove soot and smoke so that pollutant concentration inside the tunnel is measured by CO (carbon monoxide) meters. Since this is a process involving many elements which are difficult to quantify exactly, in this paper predictive fuzzy control was introduced to solve this problem. By introducing predictive fuzzy control it was made possible to greatly reduce electric power consumption while keeping the degree of pollution within the allowable limit.

1 Introduction

The necessary and sufficient amount of ventilation air flow for the traffic conditions in the tunnel at a given moment must be provided with minimum electric power consumption. In particular, increasing ventilation efficiency in tunnels is important to control operating costs.

The dust collectors remove soot and smoke so that pollutant concentration inside the tunnel is measured by CO (carbon monoxide) meters, while the numbers of large and small vehicles and their speeds are measured by traffic flow meters.

The purpose of tunnel ventilation control is to provide a safe and comfortable environment for the users. The tunnel ventilation is optimized by controlling jet fans and dust collectors installed inside the tunnel. The jet fans blow polluted air from inside the tunnel toward air exit ports. The dust collectors remove soot and smoke so that pollutant concentration inside the tunnel is measured by CO (carbon monoxide) meters. Since this is a process involving many elements which are difficult to quantify exactly, in this paper predictive fuzzy control was introduced to solve this problem. By introducing predictive fuzzy control it was made possible to greatly reduce electric power consumption while keeping the degree of pollution within the allowable limit.

The main feature of a fuzzy controller is that it can convert the linguistic control rules based on expert knowledge into automatic control strategy. So it can be applied to control the systems with unknown or unmodelled dynamics.

2 Tunnel Ventilation System

Bolu Tunnel is located on the Istanbul-Ankara highway. It is 250 km far away from Istanbul. Tunnel specifications are shown in Table 1.

Table 1. Tunnel Specifications.

Right carriageway	3311 m
Left carriageway	3363 m
Max. Gradient	2 %
Number of lanes per tube	3
Max. permissible traffic speed	90 km/h

Emissions from cars are determined not only by the way they are build but also by the way they are driven in various traffic situations. Various gases are emitted by combustion engine. They consist largely of nitrogen (N_2), carbon dioxide (CO_2), steam (H_2O) and particals. In addition, a number of harmful substances are hydrocarbons (HC) carbon monoxide (CO), lead, sulfur dioxide (SO_2). Because of these dangerous gases, it is necessary to provide fresh air in longer tunnels. The fresh air which is used to lower the concentration of CO also serves to improve visibility.

The purpose of ventilation is to reduce the noxious fumes in a tunnel to a bearable amount by introducing fresh air. Every tunnel has some degree of natural ventilation. But a mechanical ventilation system should have to be installed. In order to create air stream, fans are installed on the ceiling or side walls of the tunnel. The fans take in tunnel air and blow it out at higher speed along the axis of the tunnel. 16 jet fans in the right carriageway and the six jet fans in the left carriageway is mounted in the tunnel.

In each tunnel tube two locations for measuring the carbon monoxide, dust particle concentration and traffic volume. From the measuring units all essential data is transmitted to the ventilation control system.

3 Design of Adaptive Fuzzy Logic Controller

The design and industrial implementation of automatic control systems requires powerful and economic techniques together with efficient tools. In order to solve a control problem it is necessary to first describe somehow the dynamic behaviour of the system to de controlled. Traditionally this is done in terms of a mathematical model. However, it is well known that mathematical modelling of a plan is always forced with the problem of uncertainty, i.e. there is always a discrepancy between the mathematical model of a plant and the plant's actual behaviour. Control engineer has to find the simplest and the cheapest solution which fulfils the performance requirements within given tolerances in the face of the existing modelling uncertainties of the plant.

Fuzzy controllers are nonlinear and so they can be designed to cope with a certain amount of process nonlinearity. However, such design is difficult, especially if the controller must cope with nonlinearity over a significant portion of the operating range of process. Also, the rules of the fuzzy controller the not contain a temporal component, so they can not cope with process changes over time. So there is a need for an adaptation mechanism for fuzzy controller as well.

Adaptive controllers generally contain to extra components on top of the standard controller it self. The first is a *process monitor* that detects changes in the process characteristics. The second component is the *adaptation mechanism*. It uses data passed to it by the process monitor to update the controller parameters and so adapts the controller.

Process monitor looks for changes in process characteristics and the adaptation mechanism alters the controller parameters on the basis of any detected changes. The adaptation mechanism is specifically designed for changing the parameters of adaptive Fuzzy Logic Controller (AFLC).

Most of real-world processes that require automatic control are nonlinear in nature. That is their parameter values alter as the operating point changes, over time, or both. As conventional control schemes are linear, controller can only be tuned to give good performance at a particular operating point or for a limited period of time. The controller needs to be retinued if the operating point changes, or retained periodically if the process changes with time. This necessity to retinue has driven the need for adaptive controllers that can automatically retinue to match the process characteristic.

A non-adaptive fuzzy logic controller (FLC) is one in which these parameters do not change once the controller is being used on-line. If any of these parameters are altered on-line, it is called the controller an AFLC. AFLC that modify the rules will also be called self organizing controllers. They can either modify existing set of rules, or they can start with no rules at all and learn their control strategy as they go.

3.1 Membership Function Tuning

The structure of the controller is shown in Figure 1. The aim is to maintain a single process-state variable at set point. The controller is a PD like AFLC with the inputs being the error and change of error, and its output being the required change in the controller variable. The performance monitor uses a set of performance criteria to assess the AFLC while the controller is operating on-line.

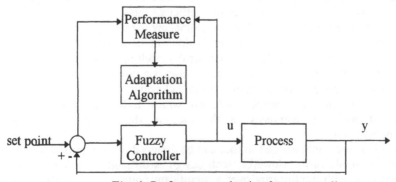

Fig. 1. Performance adaptive fuzzy controller.

The control logic structure is shown in Figure 2.

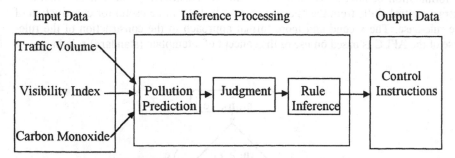

Fig. 2. Relationship among control functions in a tunnel ventilation control system.

Adaptation action is concerned with improving the controller performance when the process is regularly below set point. This is done by increasing the degree to which values of the error are recognized as negative (below the set point). Modification of the membership functions to achieve this is illustrated in Figure 3. Adaptation algorithm for minimizing control variations is shown in Figure 4.

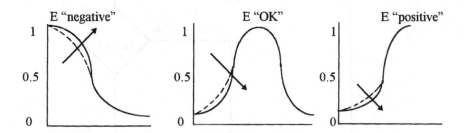

Fig. 3. Adaptation action.

The aim of the adaptation process is to provide quick controller adaptation without causing instability or oscillations. In addition, the adaptation procedure need to be carried out at every sampling time.

Inputs of the fuzzy control system are visibility index (VI) that includes dust and smoke penetration rate, and carbon monoxide ratio (CO) in air, and traffic volume that is measured by traffic counter that is placed on highway entrance. Output is control instruction for ventilation equipment that includes jet fan and dust removal. First VI and CO values are measured, and degree of pollution inside tunnel predicted. Then traffic volume is measured and the effect of air flow driven by passing vehicles on degree of pollution is predicted and control instruction is send. Control range that includes inputs and outputs is well covered with seven membership functions.

The construction of the rule-base is the crucial and the most difficult aspect of AFLC design. It is also one reason for criticism toward fuzzy logic control because, in general, there are no systematic tools for forming the rule-base of the AFLC. There are two notable methods. The first is based on intuitive knowledge and experience-the

AFLC is designed as a simple expert system. Different sources of knowledge, resulting in formulation of alternative rule-bases, can be considered. Usually it is difficult to extract control skills from the operator in a form that can be useful for construction of the rule-base. The second and more formal approach to the construction of the rule-base of the AFLC is based on use of the concept of a template (standard) rule-base.

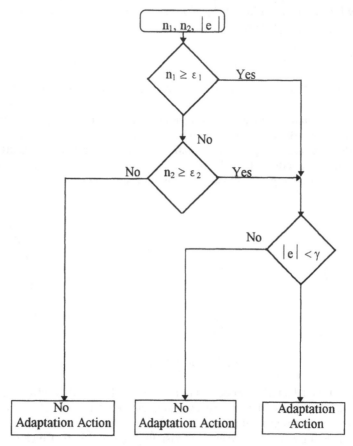

Fig. 4. Adaptation algorithm to reduce the number of control variations.

4 Simulation Results

The block diagram of the proposed adaptive fuzzy controller model for an induction jet fan motor drive is shown in Figure 5 which consist of the induction jet fan motor driven by an inverter, a dynamic load, and the adaptive fuzzy controller.

In order to validate the control strategies as discuss about, digital simulation studies were made. Table 2 shows the parameters of the drive system used for simulation study.

Table 2. Parameters of motor model.

R_1 (Ω)	0.0217
R_2 (Ω)	0.0329
L_1 (H)	0.0003235
L_2 (H)	0.003686
L_m (H)	0.01351
P	4
J (kg m^2)	11.4
L_t (H)	0.00017012

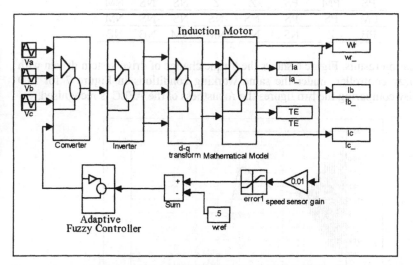

Fig. 5. Adaptive fuzzy controller model for an induction jet fan motor drive.

The derivation of the fuzzy control rules is based on the following criteria:
*   If the output of the converter is far from the set point, the change of duty cycle must be large so as to bring the output to the set point quickly.
* If the output of the converter is approaching the set point, a small change change of duty cycle is necessary.
* If the output of the converter is near the set point and is approaching it rapidly, the duty cycle must be kept constant so as to prevent overshoot.
*   If the set point is reached and the output is still changing, the duty cycle must be changed a little bit to prevent the output from moving away.
* If the output is above the set point, the sign of the change of duty cycle must be negative, and vice versa.
*   If the set point is reached and the output is steady, the duty cycle remains unchanged.
According to these criteria and after controlling several times by conventional controllers, fuzzy algorithm is derived and expressed by the Table 3 which gives the inferred linguistic value of 't_{on}' for any pair of linguistic values of "ce" and "e". Each universe of discourse is divided into seven fuzzy subsets: PB (Positive Big), PM

(Positive Medium), PS (Positive Small), Z (Zero), NS (Negative Small), NM (Negative Medium), and NB (Negative Big).

Table 3. Fuzzy controller rules.

e\ce	NB	NM	NS	Z	Ps	PM	PB
NB	PB	PB	PM	PS	PS	PS	Z
NM	PB	PM	PM	PS	PS	Z	NS
NS	PB	PM	PS	Z	Z	NS	NS
Z	PM	PS	Z	Z	Z	NS	NS
PS	PM	PS	Z	Z	NS	NS	NM
PM	PS	PS	NS	NS	NS	NS	NM
PB	Z	NS	NS	NM	NM	NB	NB

From simulation results, Figure 6 shows the torque change of drive system driven with adaptive fuzzy controller, under the same working conditions in Figure 7 for PID controller. By comparing the two figures the robustness of the AFLC is recognised.

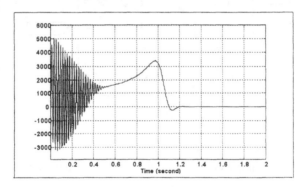

Fig. 6. Adaptive Fuzzy Controller step response of drive system torque change.

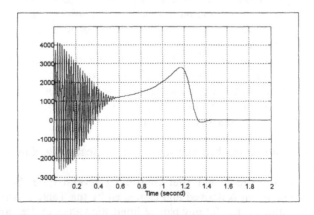

Fig. 7. PID Controller step response of drive system torque change.

5 Conclusion

Ventilation control is based on this sensor information. The amounts of pollutants in exhaust gas, air flow driven by the vehicles and degree of pollution inside the tunnel are predicted, and optimized operation commands are given to the jet fans and dust collectors. "Optimum" means that pollutant concentration is kept within the allowable limit, and at the same time electric power consumption is minimized. In the past, this control was performed using a quantitative numerical model; but the model failed to accurately account for a number of phenomena including turbulence inside the tunnel and emission of pollutants from vehicles, making it difficult to obtain optimum operation in which the electrical power consumption is minimized.

References

1. K. Jamshidi, V. Subramanyam: Self organising fuzzy controller for CSI fed induction motor. Proceed. of the 1995 PEDS conference, Part 2, 744-748 (1995)
2. D. T. Pham, D. Karaboga: Design of an adaptive fuzzy logic controller. Proceeding of the 1994 IEEE Int. Conf. on Systems, Man and Cybernetics, 437-442 (1994)
3. A. Kandel, G. Langholz: Fuzzy Control Systems. CRC Press 1994, pp. 295-314
4. G. Gateau, P. Maussion, J. Faucher: Investigations on adaptive fuzzy controllers. IEEE Conference on Control Applications, 354-359 (1995)
5. T. J. Procyk, E. H. Mamdani: A linguistic self-organizing process controller. Automatica 1979. pp 15-30
6. Y. Kung and C. M. Liaw: A fuzzy controller improving a linear model following controller for motor drives. IEEE Trans. on Fuzzy Systems, vol.2, no.3, 194-197 (1994)

Layered Neural Networks as Universal Approximators

I. Ciuca* J. A. Ware**

*Research Institute for Informatics Bucharest. Email: ciuca@u3.ici.ro
**Glamorgan University. Pontypridd. United Kingdom. Email: jaware@glamorgan.ac.uk

Abstract. The paper considers Ito's results on the approximation capability of layered neural networks with sigmoid units in two layers. First of all the paper recalls one of Ito's main results. Then the results of Ito regarding Heaviside function as sigmoid functions are extended using a signum function. For Heaviside functions a layered neural network implementation is presented that is also valid for signum functions. The focus of paper is on the implementation of Ito's appoximators as four layer feed-forward neural networks.

1. Introduction

The use of three-layered neural networks as universal approximators of continuous functions - with sigmoid units in the hidden layer - has been considered in detail by Cybenko[1]. Funahashi[2]. Hecht-Nielsen[4] Hornik[6]. Hornik et al.[7]. Ito [9].[10].[11].[12].[13] and Stinchcombe and White[17]. While the use of four-layered neural networks as universal approximators of continuous functions that have two sigmoid unit layers have been investigated by Hecht-Nielsen[4]. Funahashi[2]. Girosi and Poggio[3] and Kurkova[15].[16] The theory on networks of this type has been based on Kolmogorov's representation theorem of continuous functions f (x) given by the expression:

$$f(x) = \sum_{i=1}^{2n+1} \phi_i (\sum_{j=1}^{n} \psi_{ij}(x_j)) \qquad (1)$$

where x_j is the jth component of the input $x \in R^n$ and where the functions ϕ_i and ψ_{ij} are continuous functions. each of which have only one variable. A formula of this form can be realised by a four-layered neural networks with sigmoid units on the first and second hidden layers. if several first layer units are provided for each component x_j - as can be seen from Fig 1 for the case when i =1,..., 2n+1 and j=1,..., n

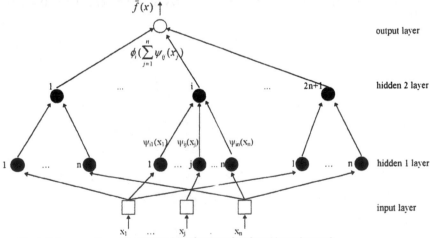

Fig. 1. Kolmogorov theorem implemented as a four-layer feed-forward neural network

This implementation requires (2n+1).n nodes in first hidden layer and 2n+1 nodes in the second hidden layer. to ensure that every node has only a single activation function. However, to realise this implementation it would be necessarily for functions ϕ and ψ to have particular forms.

Funahashi[2] and Kurkova [16] - using Kolmogorov theorem - have obtained an approximation function $\bar{f}(x)$ of f(x) realised by a four-layered neural network with sigmoid units in the first and second hidden layers. if several first layer units are provided for each component x_j

Kurkova[16] has recently estimated the numbers of units in the respective sigmoid layers of the four-layered neural networks using Kolmogorov's representation theorem [14]. Ito[13] obtained an approximation formula of type (1) and a bound on the number of sigmoid units using not Kolmogorov's representation theorem but an elementary method.

Implementing the expression (1) as a neural network means that for every node there is only one output. while in the expression (1) Ψ_{ij} is in fact of the form $\Psi_{ij}(x_j) = \varphi_j(x_j) a_{ij}$ with φ_j being the output function (activation function) for the node j.

2. Ito's main theorem

Ito[13] calls a function h sigmoid if it is monotone increasing. $h(t) \to 1$ as $t \to \infty$ and $h(t) \to 0$ as $t \to -\infty$. Using the notation H for Heaviside function: $H(x) = \begin{cases} 1 & x \geq 0 \\ 0 & x < 0 \end{cases}$ it is obvious that H is a sigmoid function.

Let the unit Hypercube be noted as $E^d = [0,1]^d$ in the d-dimensional Euclidean spaceR^d.

For an integer m Ito used the notation $Z_m^d = \{0,...,m-1\}^d$ and for a point $k \in Z_m^d$, k = (k_1 ..... k_d).

he makes the notation $x^{(k)} = (1/m)(k_1,...,k_d)$.

Let f be a function defined on E^d and let the notation be

$$\varepsilon_f(\delta) = \sup\{| f(y) - f(x)| \mid x, y \in E^d . |y_i - x_i| \leq \delta, i=1.....d\} \qquad (2)$$

For k = (k_1 ..... k_d). Ito defined the function I_k^H as having the expression:

$$I_k^H(x) = H\left[\sum_{i=1}^d H\left(x_i - \frac{1}{m}k_i\right) - d + \frac{1}{2}\right] \qquad (3)$$

This function is the indicator function of a set $\prod_{i=1}^d [(1/m)k, \infty)$ There are m^d such functions.

With these notations. Ito (13) presented the following main result:

Theorem 1.

For any continuous function f defined on E^d and for any positive integer m. there are constants a_k, $k \in Z_m^d$ for which

$$\bar{f}(x) = \sum_{k \in Z_m^d} a_k H\left[\sum_{i=1}^d H\left(x_i - \frac{1}{m}k_i\right) - d + \frac{1}{2}\right] \qquad (4)$$

satisfies

$$|f(x) - \bar{f}(x)| \leq \varepsilon_f\left(\frac{1}{m}\right) \text{ on } E^d. \qquad (5)$$

It is obvious that (4) is of the form (1) with $\phi_i(x) = a_i H(x)$. i=1..... m^d

and $\psi_{ij}(x_j) = \psi_{k_j} = H(x_j - \frac{1}{m}k_j) - 1 + \frac{1}{2d}$ with $k_j = 0.....$ m-1 and j=1..... d

Using Theorem 1. Ito [13] estimated the numbers of units in the first and second sigmoid layers as being dm and m^d - compared with Kurkova[16] results of dm(m+1) and $m^2(m+1)^d$.

3. Some considerations regarding Ito's results

To get a neural network representation of the approximation function $\bar{f}(x)$. let k = (k_1 ..... k_d). be interpreted as m-radix representation of k. that is $k = \sum_{i=1}^d k_i m^{i-1}$ where now k = (k_1 ..... k_d) takes the values. in m-radix representation. from k=0 to k=m^d-1.

This gives the following neural network implementation for the approximation $\bar{f}(x)$ given by expression (4) as presented in Fig 2. The schema is a four-layered feed-forward neural network. Every input x_j goes into a proper group of m nodes in the first hidden layer. To a node $k = (k_1 ..... k_d)$ in second hidden layer are connected. with weights equal one. a number of d nodes from first hidden layer. one node from each group. namely $H(x_1 - k_1 / m)....H(x_d - k_d / m)$.

The two hidden layers are sigmoid layers while the output layer is linear. having transfer function given by the identity function $f(x) = x$.

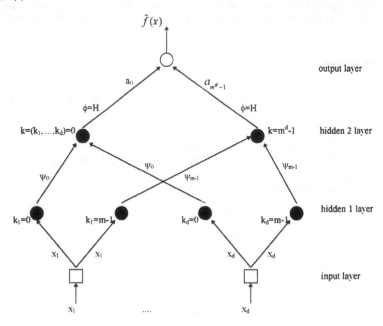

Fig. 2 Four-layer neural network implementation with Heaviside functions in the two hidden layers

Obviously the m functions ψ_{k_j} each of only one variable in the first hidden layer with $k_j=0..... m-1$ have expressions given by $\psi_{k_j}(x_j) = H(x_j - \frac{1}{m}k_j) - 1 + \frac{1}{2d}$. The implementation therefore respects the restriction of neural networks. every node must have only one output.

Fig 3 shows an implementation with d and m^d nodes in first and second hidden layer that agrees with Kolmogorov's representation (1) but does not agrees with the constraint of a neural network where every node must have only one output. Here. a node j in first hidden layer has m activation functions $H(x_j - \frac{1}{m}k_j)$ corresponding to the values $k_j=0..... k_j=m-1$ taken by j-th component of every node $k=(k_1..... k_j..... k_d)$ in the second hidden layer. Obvious this is a false implementation.

Considering the term d-1/2 as a bias in the second hidden layer. another implementation of a feed-forward neural network with two hidden layer of Ito's approximator is shown in Fig 4. In this case the functions ψ would have a typical form of Heaviside function.

414

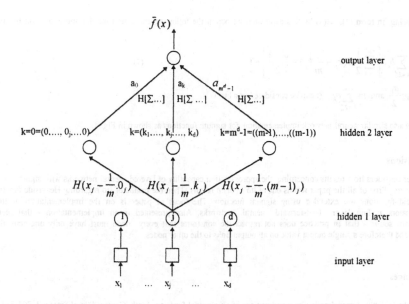

Fig. 3 A false implementation : 4-layer neural network of Ito's approximators with d and m^d-1 nodes in first respective second hidden layer.

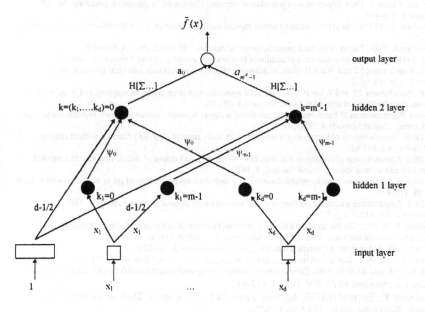

Fig 4 Four-layer neural network implementation of Ito's approximators using bias on second hidden layer

Using the relation between Heaviside function H(x) and signum function, noted as S(x), given by expression

$$H(x) = \frac{1 + S(x)}{2} \qquad (6)$$

415

and replacing. in form (4). H(x) by this expression we obtain the following expression for the approximation function $\bar{f}(x)$:

$$\bar{f}(x)= \sum_{k \in Z_m^d} b_k S \left[\sum_{j=1}^{d} \frac{1}{2} S(x_j - \frac{1}{m}k_j) - \frac{d-1}{2} \right] + t \qquad (7)$$

where $b_k = \frac{a_k}{2}$ and $t = \sum_{k \in Z_m^d} b_k$ (t can be considered as a bias)

There are also similar neural network implementation for signum functions as shown in Fig. 4.

Conclusions

The paper considers Ito's results concerning the approximation capability of layered neural networks with sigmoid units in two layers. First of all the paper recalls one of Ito's main results. Then the results of Ito regarding Heaviside function as sigmoid functions are extended using signum function. The focus of paper is on the implementation of Ito's approximators as four layer feed-forward neural networks. Also presented is an implementation - that seems theoretically sound - that in practice does not respect the constraint that every node must have only one activation function and therefore a single output value on all output links to the other nodes.

References

1. Cybenko G. 1989. Approximation by superpositions of a sigmoidal function. *Math. Control Signal System* 2. 303-314
2. Funahashi K. 1989, On the approximate realisation of continuous mapping by neural networks. *Neural Networks* 2. 183-192
3. Girosi F. and Poggio T. 1989. Representation properties of network: Kolmogorov's theorem is irrelevant. *Neural Computation* 1. 456-469
4. Hecht-Nielsen R. 1987. Kolmogorov's mapping neural network existence theorem. *IEEE First Conf. Neural Networks* III. 11-13
5. Hecht-Nielsen R. 1989. Theory of the back propagation neural network. '89 *IJCNN Proc.* I. 593-605
6. Hornik K 1991 Approximation capabilities of multilayer feedforward networks. *Neural Networks* 4. 251-257
7. Hornik k., Stinchcombe M. and White H. 1989. Multilayer feedforward networks are universal approximators. *Neural Networks* 2. 359-366
8. Hornik K., Stinchcombe M. and White H 1990. Universal approximation of an unknown mapping and its derivatives using multilayer feedforward networks. *Neural Networks* 3. 551-560
9. Ito Y 1991a. Representation of Functions by superposition of a step or sigmoid function and their applications to neural network theory. *Neural Networks* 4. 385-394.
10. Ito Y. 1991b. Approximation of functions on a compact set by finite sums of a sigmoid function without scaling. *Neural Networks* 4. 817-826
11. Ito Y. 1992. Approximation of continuous functions of R" by linear combinations of shifted rotations of a sigmoid function with and without scaling. *Neural Networks* 5. 105-115.
12. Ito Y. 1993. Approximations of differentiable functions and their derivatives on compact set by neural networks. *Math. Scient.* 18. 11-19
13. Ito Y. 1994. Approximation capability of layered neural networks with sigmoid units on two layers. *Neural Computation* 6. 1233-1243
14. Kolmogorov A. N. 1957. On the representations of continuous functions of many variables by superpositions of continuous functions of one variable and addition. *Dokl. Akad. Nauk* USSR **114** (5). 953-956.
15. Kurkova V 1991. Kolmogorov's theorem is relevant. *Neural Computation* 3. 617-622
16. Kurkova V 1992. Kolmogorov's theorem and multilayer neural networks. *Neural Networks* 5. 501-506
17. Stinchcombe M. and White H. 1989. Universal approximation using feedforward networks with non-sigmoid hidden layer activation functions. '89 *IJCNN, Proc.* I, 613-617.
18. Cardaliaguet P., Euvrard G,1992, Approximation of a Function and its Derivatives with a Neural Network, *Neural Networks*, Vol 5 pp 207-220

Fuzzy Backpropagation Training of Neural Networks

Michael Gerke and Helmut Hoyer

Chair of Process Control
FernUniversität Hagen
D-58084 Hagen, Germany
Tel.xx49-2331-987-1105
FAX xx49-2331-987-354
Email michael.gerke@fernuni-hagen.de

Abstract. Adaption of parameters during neural network training according to the actual shape of the error surface is supposed to be a powerful instrument to enforce convergence and to decrease time consumption of neural network training.

This paper presents an analysis how fuzzy adaption of training parameters (learning rate and momentum) could accelerate backpropagation learning in feedforward networks.

It summarizes first experiences of this approach if applied to a neural network simulator for pattern recognition, especially how to set up an appropriate fuzzy rulebase and how to choose efficient fuzzy sets.

1 Introduction

Learning through error-backpropagation in multilayer feedforward network architectures has been applied to a variety of theoretical and practical problems including pattern recognition.

Backpropagation represents supervised gradient-descent learning algorithms. It performs a gradient descent search in weight space of the neurons. An error function E is intended to be minimized, which is calculated as the sum of squared differences between the actual outputs o of the neural network and the desired output t during each iteration of the algorithm.

Modifications in weight space are performed according to:

$$\Delta w_{ij}(n+1) = \varepsilon\, E(n) + \alpha \Delta w_{ij}(n)$$

with learning error:

$$E(n) = \delta_j\, o_i$$

The changes in weights after iteration n are denoted by $\Delta w_{ij}(n)$, implicitly nominating the vector of weights $w_{ij}(n)$; the output of an individual neuron i is given by o_i, whereas the training error of a neuron is δ_j.

Due to the fact, that there is no desired output t_j available for hidden layers of the network, calculation of the training error is different for hidden layers and output layer.

Output layer:
$$\delta_j = (o_j + \beta) (1 - o_j) (t_j - o_j)$$

Hidden layer:
$$\delta_j = (o_j + \beta) (1 - o_j) \sum_j \delta_j w_{ij}$$

Usually training parameters *learning rate* ε, *momentum* α and *offset* β are constant values chosen by experience. This is of course in no way optimal, because stepping through different topological regions of the error surface requires different optimal training parameters for efficient training with respect to convergence speed.

Although there is no unified approach to adapt learning parameters for gradient descent search of global minima in cluttered error surfaces, it is quiet easy to set up a fuzzy decision table how parameters could be adapted in tendency, if the actual value of the error function E and the change in error ΔE during the last two iterations are known.

An approach to accelerate the convergence of the learning algorithm based on fuzzy control of training behaviour is presented here. It is based on ideas presented in [JACOBS 1988] and [ARAB 1992]. It demonstrates remarkable effects on convergence speed for different fuzzy sets and decision rulebases in comparison to conventional learning. As a result appropriate fuzzy sets and a decision rulebase are given for a single pattern recognition problem.

An analysis of its general applicability for a class of divergent pattern recognition problems is carried out in parallel. Non-pattern specific efficiency is necessary for fuzzy controlled training of neural networks, to justify the efforts spent in deriving the fuzzy parameter adaption.

An alternative approach for parameter adaption is introduced by [RIED 1993], where neural weights are updated based on partitial derivation of error function $\frac{dE}{dw_{ij}}$. In consequence, this formal approach is quiet similar to fuzzy adaption. However, fuzzy adaption is more intuitive and more accesssible for fine-tuning.

2   Standard Backpropagation

To have a valid reference for fuzzy control of training parameters, a set of 10 patterns (alphabetic letters from 'a' to 'j') has been presented to a standard backpropagation learning algorithm. Simulations have been carried out for constant learning rates ε in a range from 0.10 to 2.00, in combination with constant momentum α in a range from 0.00 to 1.00, which as a result leads to the total number of iterations, until the error function has reach a predefined minimum. To avoid improper conditions of comparison, two different randomly chosen, but fixed initializations for neural weights were used.

Each series of simulation leads to a 3D-surface of learning cycles (numbers of iteration) which are necessary to train the network. Learning rate and momentum at the base axes of the 3D surface give initial values for a specific simulation. Figure 1 gives the number of iterations necessary with standard backpropagation for one of the network weight initializations under consideration. In this particular case learning rate and momentum remain unchanged throughout each simulation (no fuzzy adaption).

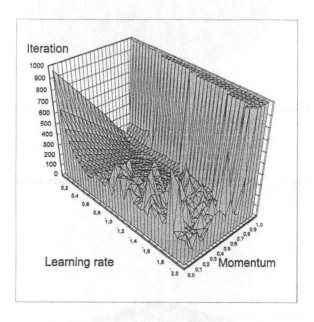

Fig. 1. Convergence without fuzzy control, alphabetic letters

The typical quality of this surface is a channel-like area of fast learning for momentum values of $0.7 < \alpha < 0.8$ over a wide range of learning rate values ε for this example. The 3D-surface for a second random initialisation is quiet similar, whereas 3D-surfaces of different pattern sets could be different. We used other sets of geometric patterns (10 digits, 10 japanese letters), and we found different 3D-surfaces of learning cycles in figure 2 and 3, but with an overall tendency to develop a valley of fast learning behaviour. However, the only conclusion to be drawn from this fact is, that for typical problems there is a contiguous region in ε, α parameter space with good convergence speed.

Assuming that this region in (ε, α)- parameter space is unknown in advance, and noone spends too much time to find it, it can be expected that an implemented parameter adaption speeds up convergence in general, whereas choosing constant parameters by hand depends on good luck.

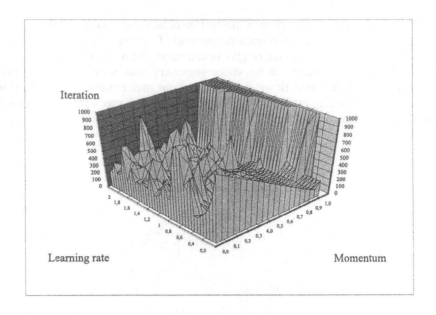

Fig. 2. Convergence without fuzzy control, numerical digits

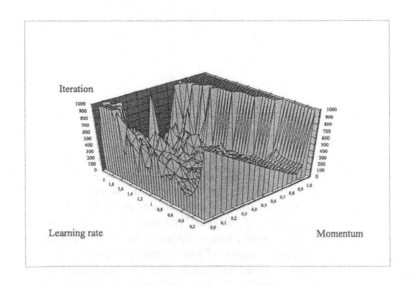

Fig. 3. Convergence without fuzzy control, japanese letters

3 Fuzzy Adaption of Learning Parameters

3.1 On the choice of fuzzy sets

We fuzzified the knowledge about the actual state of learning during each step
of iteration in terms of learning error $E(n)$ and its tendency $\Delta E(n)$ (change in
error during last two iterations), according to figure 4 and 5.

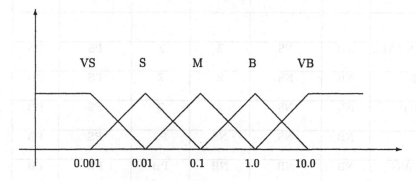

Fig. 4. Fuzzification of learning error $E(n)$

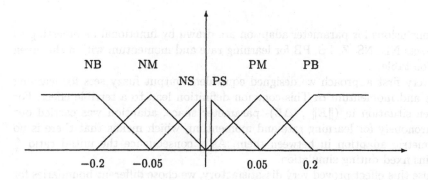

Fig. 5. Fuzzification of error changes $\Delta E(n)$

The given fuzzy sets incorporate some basic knowledge which has been derived from early simulations:

- Fuzzification of learning error (squared sum of errors between desired and
  actual output) should be logarithmic to cover a wide range of error values.
- The center fuzzy set for error changes would be usually called ZERO, but
  due to the fact that even small tendencies have to be treated, this set has
  to be split into two fuzzy sets NEGATIVE SMALL (NS) and POSITIVE
  SMALL (PS). These tendencies require different actions.

Based on fuzzified learning error and change of error, an inference rulebase is set up to conclude appropriate adaption of training parameters.

A first heuristic approach leads to a decision table for learning rate ε and momentum α, which probably looks like table 1:

ΔE	Neg. Big	Neg. Medium	Neg. Small	Pos. Small	Pos. Medium	Pos. Big
$\|E\|$						
VERY SMALL	NB	NS	Z	Z	PS	PB
SMALL	NB	NS	Z	Z	PS	PB
MEDIUM	NB	NS	Z	Z	PS	PB
BIG	NB	NS	NS	PS	PS	PB
VERY BIG	NB	NB	NB	PB	PB	PB

Table 1. Decision table to define fuzzy parameter adaption

Conclusions for parameter adaption are drawn by functional membership to fuzzy sets NB, NS, Z, PS, PB for learning rate and momentum within the given decision table.

In a very first approach we designed equivalent output fuzzy sets for learning rate ε and momentum α. This common definition lead to a terrible effect. For a given situation in ($\|E\|$, ΔE)- parameter space, adaption was carried out synchronously for learning rate and momentum, which means that there is no asymmetric adaption in between them. As a consequence the initial ratio $\frac{\varepsilon}{\alpha}$ remains fixed during simulation.

Because this effect proved very dissatisfactory, we chose different boundaries for the output fuzzy set *learning rate* given in figure 6 and for the output fuzzy set *momentum* presented in figure 7. [1]

Fuzzy set *learning rate* NB is restricted to -0.04 at its left boundary and fuzzy set *learning rate* PB is restricted to 0.02 at its right boundary to avoid excessive adaption during simulation.

Fuzzy sets *momentum* NB and PB are restricted to -0.05 and 0.02 as outmost boundaries.

---

[1] The same motivation made us change the boundaries of input fuzzy set *change of error* slightly for derivation of an appropriate adaption of training parameter momentum. The given boundary values in figure 5 are valid for the adaption of learning rate only, whereas boundary value 0.05 is decreased to 0.02 for the momentum.

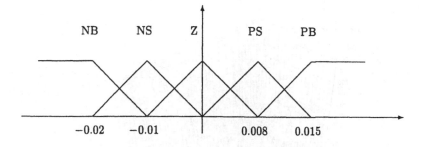

Fig. 6. Fuzzified conclusions (adaption of learning rate)

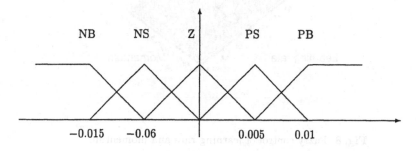

Fig. 7. Fuzzified conclusions (adaption of momentum)

The actual degree of membership to these fuzzy sets form a composite membership function, which requires *defuzzification*. Applying the center of gravity (C.O.G.) defuzzification method to it proposes the crisp modifications of training parameters.

As an intermediate result we present fuzzy adapted backpropagation for the alphabetic pattern set in figure 8. The design of fuzzy sets depends on experiences mentioned above for this serie of simulations, whereas the decision rulebase is still heuristic.

Consequent adaption and verification through simulation leads to the following 3D-surface of iterations necessary to reach the predefined minimum error value, with absolutely comparable preconditions as in the case of standard backpropagation.

Although the assumptions made above during the definition of fuzzy sets sound consistent the resulting 3D-surface is very cluttered. Parameter adaption has not proven to be efficient here, if compared with our initial results for standard backpropagation in figure 1. However, the results of figure 8 are still promising. Many combinations of parameter initializations for learning rate and momentum lead to fast learning, but in some special cases convergence is very poor. These cases lead to singular peaks in 3D. A closer analysis of simulations connected with these peaks shows, that in these cases the absolute error $\|E\|$ becomes oscillatory for a long period of iterations.

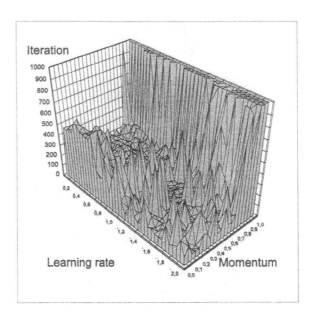

Fig. 8. Fuzzy control of learning rate and momentum

3.2 An advanced inference rulebase

A redesign of our heuristic rulebase is necessary to get rid of the oscillatory behaviour. Based on previous experiences here are some verbal rules which could help to optimize the rulebase:

- If the actual change in error is ZERO and the error itself is less than 0.01, the learning rate should be reduced. Otherwise adaption of ε could cause training oscillations.
- If changes in error are ZERO, the momentum value α should not be increased; in contrary, if the error itself is large, the momentum has to be slightly reduced, because obviously the training behaviour is unsatisfying.
- If error changes occur, which are PB, the consequent increase of parameters ε and α is not of the same amount as in the opposite case of error changes NB, which lead to a reduction of parameter values. Otherwise the training parameters for a particular error value become dependent on initial parameters.

To follow these ideas, here are the redesigned rulebases, which also include a certain fine-tuning; rulebases for adaption of ε and α have become slightly different:

Simulation series with alphabetic patterns in figure 9 clearly present advantages of fuzzy parameter adaption during neural network training. The 3D-surface has been flattened through wide ranges of initial start values ε and α.

ΔE	Neg. Big	Neg. Medium	Neg. Small	Pos. Small	Pos. Medium	Pos. Big
$\|E\|$						
VERY SMALL	NB	NS	NS	Z	Z	PS
SMALL	NB	NS	Z	Z	Z	PS
MEDIUM	NB	NS	Z	Z	PS	PS
BIG	NB	NS	NS	PS	PS	PB
VERY BIG	NB	NB	NS	PS	PB	PB

Table 2. Decision rulebase for learning rate ε

ΔE	Neg. Big	Neg. Medium	Neg. Small	Pos. Small	Pos. Medium	Pos. Big
$\|E\|$						
VERY SMALL	NB	NS	NS	Z	Z	PS
SMALL	NB	NS	NS	Z	Z	PS
MEDIUM	NB	NS	Z	Z	PS	PS
BIG	NB	NS	Z	Z	PS	PB
VERY BIG	NB	NB	NS	Z	PS	PB

Table 3. Decision rulebase for momentum α

In comparison with 3D simulation without parameter adaption, as given in figure 1, we derived a large central plateau of high speed of learning.

Additionally, in all cases where adaption starts with relatively high values for momentum α the time-consumption during training has been reduced largely.

To avoid a lack of generality, the given fuzzy parameter adaption has been applied to the numerical pattern set and to the japanese letters, too. The results are given in figures 10 and 11.

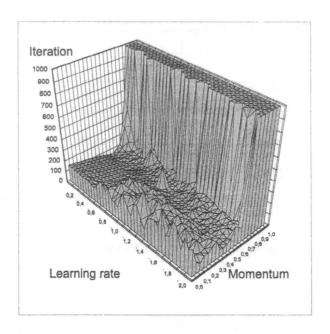

Fig. 9. Fuzzy control of learning rate and momentum

4   Conclusions

It has been shown by neural simulation, that a fuzzy approach for parameter adaption increases the speed of convergence for multilayer feedforward networks with backpropagation training algorithms. Analysis of efficient parameter adaption has been made for wide ranges of initial values of learning rate ε and momentum α.
In all cases fuzzy adaption of parameters seemed to be an appropriate method to speed up network training. This is an advantage during offline training and an important precondition for feasability of online training.

Even though fuzzy adaption has proven to speed network convergence during training for three examples under consideration, this is no prove in general.
However there is no method available for a general prove. We expect a large class of audio-visual pattern sets and applications where the method works and probably an even larger class of problems where its performance is poor and does not justify parameter adaption.

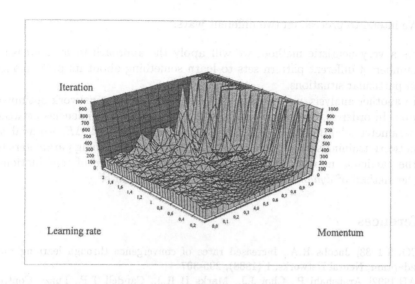

Fig. 10. Convergence with adaptive fuzzy control, numerical digits

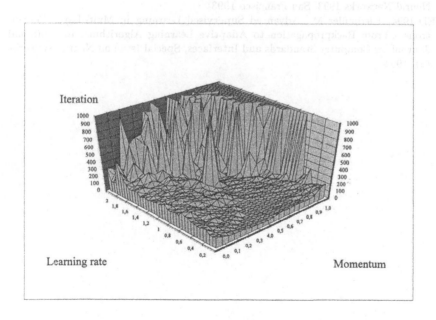

Fig. 11. Convergence with adaptive fuzzy control, japanese letters

We intend to proceed on two different ways:

- As a very heuristic method, we will apply the approach to an increasing number of different pattern sets to learn something about its performance in particular situations.
- In another analysis, we turn our attention to finite state network optimization. In order to learn something about systematical dependencies between parameter adaption in ε, α and consequences in $|E|$ and ΔE, we analyse network training step by step, applying a finite set of learning parameters to the particular network state. To handle the large amount of combinations, the method of *dynamic programming* seems appropriate.

References

[JACOBS 1988] Jacobs R.A., Increased rates of convergence through learning rate adaption, Neural Networks, 1 (1988), 295-307

[ARAB 1992] Arabshahi P., Choi J.J., Marks II R.J., Caudell T.P., Fuzzy Control of Backpropagation, IEEE International Conference on Fuzzy Systems 1992, San Diego 1992

[RIED 1993] Riedmiller M., Braun H., A Direct Adaptive Method for Faster Backpropagation Learning: The RPROP Algorithm, IEEE International Conference on Neural Networks 1993, San Francisco 1993

[RIED 1994] Riedmiller M., Advanced Supervised Learning in Multi-Layer Perceptrons - From Backpropagation to Adaptive Learning Algorithms, International Journal on Computer Standards and Interfaces, Special Issue on Neural Networks (5), 1994

Fuzzy and Neural Network Controller - A New Tool for Stability Analysis

Andreas Simon

Institut für Regelungstechnik,
Technische Universität Braunschweig

Abstract. Today, in important fields of application fuzzy and neural network controller are often not used because of the lack of a stability proof for such non-linear control systems. Therefore, we introduce a numerical algorithm which examines the stability of a closed-loop system in the Lyapunov sense without any analytical description of the plant or of the controller. We developed a software tool using this algorithm and we have accumulated many experiences in implementation and application of this method.

1 Introduction

We want to examine the behaviour of a closed-loop control system. This system consists of a plant and a controller as shown in fig.1. The control loop is a dynamical system with a state vector $\mathbf{x}$. The components of $\mathbf{x}$ are the state variables of both the plant and the controller. If the controller is static like a fuzzy controller it contains no state variables. In this case the state of the control system is identical to the state of the plant.

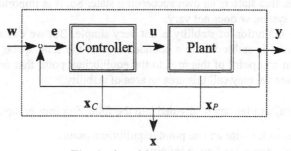

Fig. 1. closed-loop system

The behaviour of the closed control loop can be expressed by the following equations:

$$\mathbf{x}(k+1) = \mathbf{f}(\mathbf{x}(k), \mathbf{w}) \qquad (1)$$
$$\mathbf{w} = const.$$
$$\mathbf{y}(k) = \mathbf{g}(\mathbf{x}(k), \mathbf{w}) \qquad (2)$$

or respectively in the continuous case:

$$\dot{\mathbf{x}} = \mathbf{f}(\mathbf{x}, \mathbf{w}) \tag{1a}$$
$$\mathbf{y} = \mathbf{g}(\mathbf{x}, \mathbf{w}) \qquad \mathbf{w} = const. \tag{2a}$$

(1), (1a) describe all trajectories in the state space. In the discrete case a trajectory appears dotted as a chain of time-equidistant states (fig. 2).

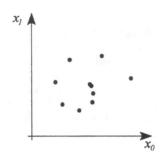

Fig. 2. time-discrete trajectory

2 When is a system stable?

In the Lyapunov sense, stability is the property of an equilibrium point. This point is a state in the state space which solves

$$\mathbf{f}(\mathbf{x}(t), \mathbf{w}) = \mathbf{0},$$

in other words, this state is its own succeeding state. So, it is important that the input of the control system **w** does not vary.

The precise definition of stability is not very simple. But we can say here that the equilibrium point is stable within a bounded area of the state space, if there exists a trajectory from any point of this area to the equilibrium point that does not leave the given area. Then we may call this area an area of stability.

The types of trajectories can be divided into the following four groups:

- stable trajectories with an end point (equilibrium point)
- instable trajectories leading to infinity
- trajectories of a limit cycle
- and trajectories leading to a limit cycle

With this, the following three questions arise in order to estimate the stability of the system:

- Where are equilibrium points in the state space?
- Which extension do the corresponding areas of stability have?
- Where are limit cycles?

3   The Algorithm of the State Space Investigation

Since the memory and speed of a computer is limited, the following three steps are necessary to handle the stability analysis with a computer:

1.) discretize the time (sampling)
2.) discretize the state space
3.) limit the area of investigation to a finite size

The first step is a well-known procedure from discrete-time control theory. Therefore, we do not go into details here. In the second step, we replace the continous state space by an equidistant grid of discrete states. These points we call D-points in the following. The question of an optimal size and resolution of the discrete state space will be treated in chapter 4.

As in [4] the stability analysis can be divided into the following steps:

1.) Computing the succeeding D-point for every D-point,
2.) searching all equilibrium points in the area of investigation,
3.) computing the areas of stability of every equilibrium point,
4.) finding instable trajectories
5.) and searching limit cycles.

In the first step, we compute the succeding D-point for every D-point in the area of investigation. Since a D-point is also a system state we can compute the succeeding system state for a D-point but this succeeding state will usually be different from any other D-point. A solution for this problem is to define for any D-point a part of the state space as its cell. With this, for any D-point we compute its succeeding state, then the succeeding state of the succeeding state, and so on until the first state is in the cell of another D-point (fig. 3). This D-point is defined as the succeeding D-point for the first one. This method is well-known as 'cell-to-cell mapping' [1, 2].

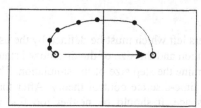

Fig. 3. discrete state transition

This is done for every D-point of the discretized state space. As a result, we get a vector field that describes the dynamical behaviour of the system in the discretized state space. Based on this information, we can easily detect equilibrium points: An equilibrium point has itself as succeeding state. This means, that all succeeding states of the D-point stay inside its cell.

With this, we can compute all D-points which belong to trajectories reaching a given equilibrium point. With the information of the vector field, we start at an equilibrium point and trace backwards all D-points that lead to the equilibrium point. Doing this, we obtain the stable workspace of the control system around an equilibrium point. This area usually can not be expressed in mathematical terms because it has an irregular shape. Therefore, we show to the user the biggest area which can be expressed as

$$\left\{ \mathbf{x} \middle| x_{i,LB} < x_i < x_{i,UB} \text{ for all } i \right\}$$

with $x_{i,LB}$ and $x_{i,UB}$ the lower and the upper bounds for each coordinate, as the stability area of an equilibrium point.

For detecting limit cycles and trajectories to limit cycles, it is useful to first find unstable D-points. One group of unstable points are D-points whose succeeding D-point is outside of the area to be treated, although there might exist a trajectory from this point to an equilibrium point. Since we do not know the type of trajectory, we define this D-point as unstable. The second group are unstable points which belong to trajectories that reach the D-points just described. These trajectories are computed in opposite direction in the same way as the common trajectories to the equlibrium points are computed.

Now, there are two types of trajectories left: limit cycles and trajectories to limit cycles. To detect limit cycles, we look at the succeeding point of every D-point and at the successor of the succeeding point and so on until one D-point of the trajectory will be reached for the second time or a succeeding point is already classified.

In the last case, the trajectory leads to a limit cycle. In the first case, all D-points of the closed loop are classified as a limit cycle and the remaining points as a part of a trajectory to a limit cycle.

4 Tools to support the user in finding the right parameters for the analysis

There are three parameters left which must be defined by the user: The step size of the simulation and the resolution and the size of the discretized state space.

First, we have to determine the step size of the simulation. This is done according to the well-known rules of time-discrete control theory. After that, we choose the right resolution of the state space. It should be neither too fine in order to economize computer memory and calculation time nor too rough in order to minimize the error of discretization. This parameter also depends on the chosen step size for the simulation as shown in fig. 4. One way for supporting the user in finding the right value is to examine a large number of arbitrarily chosen state transitions. The smallest and the largest length of transition provide a lower and upper bound of possible resolution. But the most interesting value is surely the average length of a discrete state transition. The resolution of the state space should be of about this size.

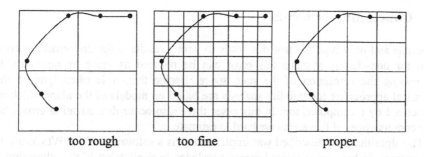

| | | |
| too rough | too fine | proper |

Fig. 4. resolution of the state space

Now, we have to choose the area of investigation. Since it is difficult to estimate the range of the state variables in practice and it is much easier to appreciate 'visible' variables like the output of the plant, we introduce a method to derive the range of the state variables from the control system input and output.

We assume that the control system input $\mathbf{w}$ and the range of the ouput $[y_{a,i}, y_{b,i}]$ is given. The problem to solve is, finding all D-points that lead together with the control system input to output values in the given range. Since the function $g(\mathbf{x},\mathbf{w})$ cannot usually solved for $\mathbf{x}$, we take a large number of arbitrarily chosen D-points and check if they belong to

$$\left\{\mathbf{x} \middle| g_i(\mathbf{x},\mathbf{w}) \in [y_{a,i}, y_{b,i}] \text{ for all } i\right\} \text{ with } \mathbf{x} = \begin{pmatrix} \mathbf{x}_C \\ \mathbf{x}_P \end{pmatrix} \tag{3}$$

The area of the discrete state space which only contains D-points that solve (3) is the desired area of investigation.

If the controller is a dynamical one with his own state vektor $\mathbf{x}_C$ and if we know the range of the controller output $[u_{a,i}, u_{b,i}]$ we can split our computation into a controller part and a plant part:

$$\left\{\mathbf{x}_C \middle| g_{C,i}(\mathbf{x}_C, \mathbf{e}) \in [u_{a,i}, u_{b,i}] \text{ for all } i\right\} \text{ with } e_i \in [w_i - y_{b,i}, w_i - y_{a,i}] \tag{4}$$

$$\left\{\mathbf{x}_P \middle| g_{P,i}(\mathbf{x}_P, \mathbf{u}) \in [y_{a,i}, y_{b,i}] \text{ for all } i\right\} \text{ with } u_i \in [u_{a,i}, u_{b,i}] \tag{5}$$

Hence, we have splitted the n-dimensional state space into an s-dimensional and a t-dimensional state space where $s+t=n$. Thus, we have reduced the number of D-points for finding the state ranges in an effective way.

5 Concluding remarks

The one and only type of error that leads to wrong results is the discretization error. But for non-chaotic systems this error can be reduced as much as necessary by increasing the resolution of the state space. Beside this, it is often ignored that classical approaches for stability analysis are based on models of the plant which are generated by a computer, too. In this cases there also occur discretization errors, but the consequences of this errors are hard to estimate.

The algorithm just described was implemented as a software tool for WINDOWS 95. Therefore, we have accumulated more knowledge in application of this algorithm as we can describe here briefly.

We also want to thank the software specialists of the TRANSFERTECH company in Braunschweig for their valuable contributions in long discussions while developing a commercial version of this algorithm.

References

1. Y.Y. Chen, T.C. Tsao: A Description of the Dynamical Behaviour of Fuzzy Systems. IEEE Transactions on Systems, Man, and Cybernetics 19, 745-755 (1989)

2. C.S. Hsu: A Theory of Cell-to-Cell Mapping Dynamical Systems. Journal of Applied Mechanics 47, 931-939 (1980)

3. C.S. Hsu, R.S. Guttalu: An Unravelling Algorithm for Global Analysis of Dynamical Systems: An Application of Cell-to-Cell Mappings. Journal of Applied Mechanics 47, 940-948 (1980)

4. K. Michels, H.-G. Nitzke: Numerical Stability Analysis for Fuzzy Control (in German). Proceedings of *Fuzzy-Neuro-Systeme '95*, Darmstadt (1995)

Evolution Strategies Applied to Controls on a Two Axis Robot

R. Richter, W. Hofmann

TU Chemnitz
Department Electrical Machines and Drives
09107 Chemnitz
Germany

Phone: Germany / 371 / 531 3346
Fax: Germany / 371 / 531 3324
Email: ralf.richter@e-technik.tu-chemnitz.de

Abstract. The paper will focus on modern control design techniques applied on a two axis robot with synchronous servo drives. PID and fuzzy controllers are designed by means of Evolution Strategies. The implementation will also run on a 80C166 microcontroller board in connection with a DSP extension (ADSP 2111).

Introduction

Nowadays in industrial servo drive applications the brushless synchronous machine [1] is the dominate solution. The main reason for using brushless machines is the robustness and serviceability cause of dropped commutation at DC machines.

The goal of the work is to provide a better way to get an improved motion dynamics of such servo drives by discrete control. To avoid disturbance and nonlinear effect's like friction a structure-optimal controller (General Linear Controller - GLC [5]) and nonlinear concepts such as fuzzy control [4] are implemented on speed control. Special concern was on the minimizing overshoot on edges and the velocity error when both axis were in motion. A common way to meet these expectations is to use optimization algorithms like gradient decent methods, stochastically search methods or evolutionary search methods during the design process or during the online or off-line tuning phase.

The research is also part of a project which focused on the use of Evolution Strategies in the field of Fuzzy Control.

The paper presents first a view on the plant and the hardware that has to be controlled, then gives a short look on the model estimation process, introduce Evolution Strategies as a way of parameter optimization and finally presents the controller designs, simulations and results on the plant.

The Plant under Investigation and Hardware used

All research was done on a two axis robot, similar to a plotting system. Every axis is driven by a synchronous servo drive (see figure 1) and contains an incremental sensor to measure the speed signal. No further sensors are available for control purposes.

From the hardware point of view, there are two ways of controlling the drives, the first one is a PC-internal IO-Board mounted on a Pentium-PC and the second one is a stand alone 80C166 based microcontroller device with multiple input/output channels. The basis controller C166 board is speeded up by a DSP-extension with a ADSP 2111 to meet the millisecond sample times of electromechanical drives. The stand alone device is connected by a serial link (RS 232) to a PC. The link enables the exchange of measured data as well as the exchange of the controller software.

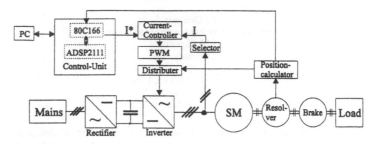

Figure 1: Scheme of one axis

In the present stage most of the implementations were done on the PC version which provides a better opportunity to make changes in the control structure and parameters in an easy manner. The final goal is the implementation on the C166 hardware which runs as an industry relevant stand alone solution.

Approach

The goal is to implement speed control for both axes on the robot. Based on an experimental model estimation in respect of stationary and dynamically behavior a general linear controller [5,6] for comparison reasons and a PID algorithm as well as a fuzzy controller [4] had been designed, simulated and implemented.

To improve the controller performance on common benchmark tests like sharp edges and circle trajectories, and to improve the robustness of the model-based designed controllers, the PID and Fuzzy controller parameters had been tuned with Evolution Strategies.

The introduction of Evolution Strategies in the field of controls that includes a lot of optimization steps like model estimation, fitting model parameter to real curves and regulator parameter design was the main focus of the work. More general off-line and online parameter tuning can be handled well by this type of optimization strategies.

Experimental Model Estimation

In order to apply two linear control concepts like PID and GLC, the first choice was to estimate a linear discrete plant model. This means a significant simplification in respect to friction forces but also demonstrates the capability of the control algorithms.

The discrete time model common used as a discrete transfer function which presents the input-output relation of a system typically has the form

$$G(z) = \frac{Y(z)}{U(z)} = \frac{b_1 z^{-1} + b_2 z^{-2} + \dots + b_m z^{-m}}{1 + a_1 z^{-1} + a_2 z^{-2} + \dots + a_n z^{-n}} z^{-d} \qquad (1)$$

and can be estimated in two ways,
    - open-loop by using least square method on relevant activated system signals
    - closed-loop by using a P-regulator with time delay.

After examine different step responses and using the approximation that the sample time results from 10% of the settling time the set up of the sample times T_r was done at 2 milliseconds [10]. Afterwards both estimation algorithms had been performed and led to almost equal models. The model turned out to fit well with two real poles in the unit circle.

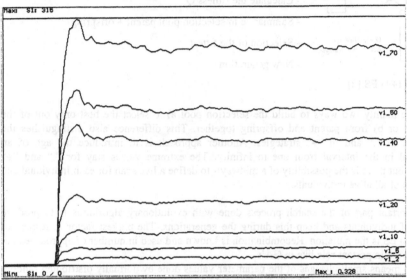

Figure 2: Step responses of the x-axis from 2 to 70 % magnitude

Evolution Strategies (ES)

ES belongs to the class of probabilistic optimization algorithms as well as Genetic Algorithm (GA), Evolutionary Programming (EP) and Simulated Annealing (SA).

These algorithms are based on an arbitrarily initialized population of search points, which by means of randomized processes of selection, mutation and recombination evolves toward a region of better performance. Every search point is represented by a set of parameters - controller parameters in this case - and it's fitness value - the control performance index. A search point in terms of evolution is called an individual which is summarized in a population. To every individual belong also a set of strategy parameters like step size, step size gains, age and so on.

The figure 3 shows the flow of the individuals during a generation on a (1+1) strategy.

Generation series:

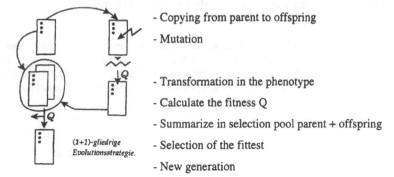

- Copying from parent to offspring

- Mutation

- Transformation in the phenotype

- Calculate the fitness Q

- Summarize in selection pool parent + offspring

(1+1)-gliedrige
Evolutionsstrategie.

- Selection of the fittest

- New generation

Figure 3: (1+1) ES [2]

There are mainly two ways to build the selection pool a) to select the best only out of the offspring or b) from parent and offspring together. This difference also distinguishes the strategies in "," and in "+" strategies. Another approach is to introduce the age of an individual in the interval from one to infinity. The extreme values stay for "," and "+" strategy, but there is the possibility of a midway - to define a live span for each individual and select out of all alive individuals.

The important part of the search process done with evolutionary algorithms is to produce variety of individuals and keep this during the generations. The biggest factor in respect of variety on ES is the mutation. Recombination is known and used in number of approaches but plays not the important role like in GA's.
Mutation means small changes of the parameter values done by normally distributed random generator with step size σ also called standard deviation and a mean or expectation of zero - $N(0, \sigma)$. To keep the variety in the search process the concept of self-adaptation of strategy parameters was introduced by Schwefel and Rechenberg. The simplest adaptation algorithm is the so called mutative step size control (MSC), which proceeds by multiplying a constant factor or the inverse of the factor with equal probability to the step size. Much more complex mechanism such the correlated mutation by Schwefel exist, but only work on a basis of huge population sizes which cost high computational effort to compute the objective function.

For the parameter optimization of controllers, the $(\mu/p,\lambda)$-ES with mutative step-size control turned out as a suited algorithm and a compromise which was used during this work. In further research a higher level strategy which is known as multilevel strategy will be used. This multilevel strategy provides additional an evolution on the population level. Every population development runs γ - generations by there self's and after this information exchange take place. The advantage of this way is an undisturbed evolution which might come up with better convergence characteristics. The general notation for the multilevel evolution is

$$[\mu'/p' \overset{+}{\_} \lambda` (\mu/p \overset{+}{\_} \lambda)^\gamma] \text{-} ES \tag{2}$$

and was introduced by Rechenberg [see also 2].

Objective Function

A general optimization problem with the assumption of n-dimensional, continuous parameter space is written in the form

$$f^* = f(\underline{x}^*) = \min\{f(\underline{x}) \mid \underline{x} \in R^n\}, \tag{3}$$

where f^* is the optimal - the minimum value of the objective function $f(\underline{x})$. R^n is the n-dimensional real space that can be constraint.
The fitness value or objective function in respect of controls represents the quality of dynamic behavior and is measured in most of the cases as an integral criterion

$$\text{i.e. ISE} \int_0^\infty e^2(t)dt \tag{4}$$

where e is the error signal. In a nonlinear environment it's necessary to build the criterion over a certain number of input responses for instance input changes with small and with large amplitude values to cover the nature of nonlinearity. In addition to the error signal the influence of the controller output signal - the control effort - is taken into account when limited gains are given. That's the case in almost all real cases. So combinations of absolute and quadratic terms as well as error signal and control effort with time weights are common

$$\text{i.e.} \int_0^\infty ae^2(t) + (1-a)\left|\frac{du}{dt}\right|dt, \text{with } a \text{ in } [0,1]. \tag{5}$$

For linear systems the closed-loop frequency-response curve in respect to disturbances provides also a well-suited criterion to estimate the quality of control. In our case the GLC is designed by means of this criterion. A side effect of the frequency domain design is the guaranty of stability in the desired bandwidth.

Design of PID Controller

The PID controller with the following transfer function and structure (Fig. 6)

$$G(z) = \frac{U(z)}{E(z)} = \frac{q_0 + q_1 z^{-1} + q_2 z^{-2}}{1 - z^{-1}} \tag{6}$$

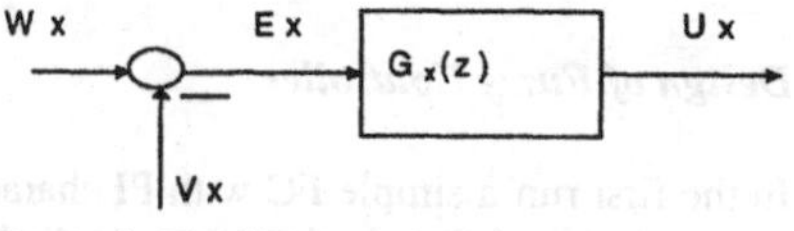

Figure 4: Feed Forward Controller

was designed with a (3,10) ES by minimizing the ISE criterion. After fifty generations the cancellation criterion was reached. The simulation is shown in figure 4.

The input of the drive axis X was a step series that needs under steady state conditions about 15 % of the possible control effort. This test signal described as Wx in the figures and is used in the simulation as well as in the real test for all controller types. The time scale in the simulation was 0.24 seconds. The profile of the step series imitates a forward - backward motion of the robot in every axis.

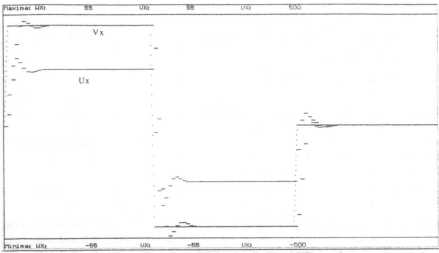

Figure 5: Simulated results of the plant model controlled by a PID regulator

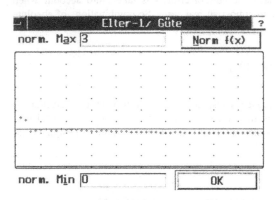

Figure 6: Best fitness values over generations

The tuned PID regulator acts smoothly and shows a response with a maximum overshoot of 3.5 % and a settling time of 14 milliseconds. The fitness value at the initial point (hand tuned) with $f=2.6$ decreased during the optimization to $f= 0.88$, as shown in the figure 6 bellow. As mentioned before the ES (3,10) produced out of 3 parents 10 offspring and select the 3 fittest as new parents. After the first four generations after 40 simulation runs the main improvement was done.

Design of Fuzzy Controller

In the first run a simple FC with PI characteristic had been set up (figure 7). PI characteristic means the regulator use the error and the integrated error signal as input signals and calculates the control signal. A 5 by 5 symmetric rule base was used also because of the time restrictions of the real-time procedure.

The input signals had been adjusted with the input coefficients K_{in} to fit to the normalized input membership function. These membership functions consist of triangles and z- / s-forms on the left respectively right edge. The output membership function was built by singletons. Only the position of the singletons where adjustable by the optimizing process, which is in that case a constraint one. The constraints ensure the correctness of the logic.

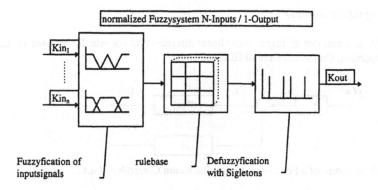

Figure 7: General scheme of a FC regulator

The settling time of the designed controller is in the simulation larger than it was at the PID regulator. The control effort is close to an aperiodic function , which may turn out in real process as an advantage. The overshoot is very small approximately 7%.

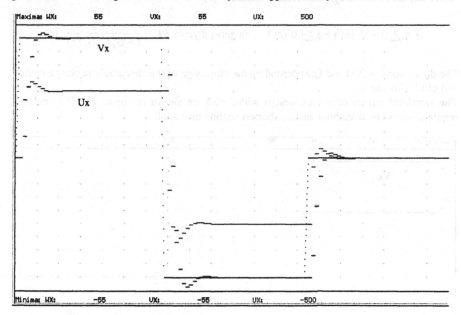

Figure 8: Simulated results of the plant model controlled by a FC regulator

As mentioned before the tunable parameters of the controllers had been the position of the output singletons. That means only four parameters had been tuned because of symmetry reasons. Because of the nonlinearity of the FC the coefficients K_{in} and K_{out} have to calibrate to entire control range, otherwise special changes in the desired signal (much smaller or larger than the optimized) will create problems.

Design of GLC Controller

The GLC is a member of higher order linear discrete regulator which was used in this paper for comparison. On the structure of GLC controller

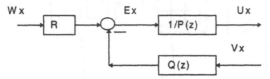

Figure 9: Structure of a Feed Forward and Backward Controller - GLC

it's to be seen, that this regulator is able to handle input and disturbance signals separately. The controller has a discrete term in the backward branch. The design procedure was performed with Matlab which uses gradient methods for parameter optimization. The objective function was constructed in frequency domain [6] and combines the deviation from the magnitude of the ideal input frequency response function abs($G_w(j\omega)$)=1 and a weighted term from the so called dynamic control-factor RF [5].

$$ f = \sum_i (1 - |G_w(\omega_i)| + \lambda \sum_i |RF(\omega_i)| \quad , \text{ in general yield } RF = \frac{1}{1 + G_R(z)G_P(z)} \cdot (6) $$

The dimensions of P(z) and Q(z) depend on the dimension of the characteristic plant equation and of it's time delay.
The simulated curves of a GLC-design with λ=0.5 are shown in figure 10. The designed regulator shows no overshoot and the shortest settling time at all.

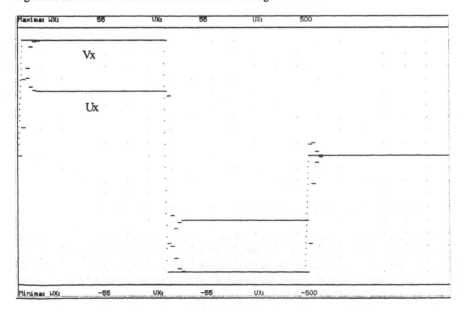

Figure 10: Simulated results of the plant model controlled by a GLC regulator

Reached Results on the Plant

The figure 11 shows controlled speed signals with the implemented controller algorithms in comparison. The time scale is zoomed to make the differences clear the magnitude is the same as used in the simulation - 50.

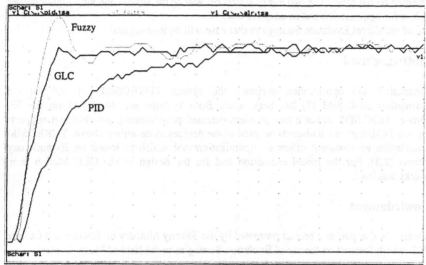

Figure 11: Speed signals on the X-axis

Still the best design at that stage is the GLC with 6 parameters to be tuned. This design comes along with most time consuming design in the frequency domain.
The Fuzzy design is able to follow the GLC in respect of the rice time but is not able to avoid significant overshoot, which stands in contrast to the simulation. The problem might be solvable by tuning the input membership functions as well. Even a bigger number of memberships itself lead to an improvement. The PID regulator as a conventional regulator gives reasonable results without overshoot. The design was done with 3% overshoot.

Conclusion

The ES as a member of the probabilistic optimization algorithms is well suited for parameter estimation (not shown on the paper) and for controller tuning based on models or even direct on real plant under the assumption of a stabile controller structure. The optimization gives the desired improvements.

Furthermore it could be shown, that controllers with a higher degree of freedom reach an improvement in the performance index in comparison to common PID techniques.

Further Work

First of all, the used Fuzzy controller will be further investigated to come up with a design which do not have overshoot.
Two different fuzzy structures and an artificial neuronal network controller will be practically implemented on the robot. The two Fuzzy structures will be the simple one with higher degree of freedom for optimization and an adaptive one.
The use of multilevel Evolution Strategy in that case will be investigated.

The Software used

For simulation and optimization purposes the system TURBOSIM [9] and for the implementation DDC-SIM [7] has been used. Both systems are developed on the EE Department. DDC-SIM makes a free problem-oriented programming possible and supports controls via PC-integrated IO-boards or stand alone devices as described above. TURBOSIM as a simulation environment offers an optimization-tool which is based on Evolutionary Algorithms [2,3]. For the model estimation and for the design of the GLC Matlab from Mathworks was used.

Acknowledgment

The research is also part of a project promoted by the Saxony Ministry of Science and Culture (SMWK) which focused on the use of Evolution Strategies in the field of Fuzzy Control.

References

1. R.Schönfeld: Grundlagen der automatischen Steuerungen.Springer, Berlin, 1995
2. I. Rechenberg: Evolutionsstrategie: Optimierung technischer Systeme nach Prinzipien der biologischen Evolution. Frommann-Holzboog. Stuttgart, 1994
3. H.P. Schwefel: Evolution and Optimum Seeking. Wiley. New York, 1994
4. M. Koch, Th. Kuhn, J.Wernstedt: Fuzzy Control. Oldenbourg. Müchen 1996
5. R.Isermann: Digitale Regelsysteme I. Springer-Verlag. Berlin, 1988
6. W.Ehrlich: Entwurf von Allgemeinen Linearen Reglern (ALR), Vorlesungsskirpt
7. R. Neumann: DDC-SIM ein Soft- und Hardwarekonzept für Echtzeitanwendungen. Chemnitz,1996
8. Gerätehandbuch Servomation. Elektrische Servoantriebe GmbH, Stadtbergen, 1990
9. R.Richter: Handbuch TURBOSIM V4.0, Chemnitz, 1996
10. I.Golle: Moderen Konzepte der Lageregelung, Chemnitz, 1995
11. R.Isermann: Identifikation dynamische Systeme I/II. Springer-Verlag. Berlin, 1988

Purposeful and Reactive Navigation Based on Qualitative Path Planning and Fuzzy Logic

Eleni Stroulia, Daniel Schneider, and Erwin Prassler

Center for Applied Knowledge Processing
Helmholtzstr. 16
89081 Ulm, Germany

Abstract. An important problem in robotics research is to develop methods for safe, robust, and efficient navigation. To be able to navigate purposefully in its environment, a robot has, first, to have some a priori knowledge about its environment, so that it can plan the paths necessary for accomplishing its tasks, and, second, to be able to react to any unforeseen obstacles that may arise it its way during the execution of its paths. Since real environments are usually complex and dynamic, the robot cannot have complete and correct a priori knowledge about them. To aggravate the situation, the robot's sensors usually deliver noisy readings, so even its runtime perceptions are not absolutely correct. Towards addressing the problem of purposeful yet reactive navigation without assuming either precise a priori knowledge or noise-free runtime sensory input, we have developed a navigation method integrating a qualitative path planner and a fuzzy controller. This method has been implemented and evaluated in the context of DAVID, a Nomad 200 robot.

1 Introduction

The deployment of an autonomous robot in a real environment gives rise to a set of research problems, one of which, is its safe, robust, and efficient navigation. On one hand, purposeful navigation must service the high-level goals of the robot, and therefore the robot must have some knowledge about the layout of its environment in order to plan the paths that it needs to traverse to accomplish its tasks. On the other, navigation has to be reactive since real environments are dynamic and unknown obstacles may appear in its proximity while traversing these paths. A major problem with both these two knowledge requirements is that they are quite difficult to meet. Since any realistic environment is dynamic and may be quite complex, endowing the robot with precise a priori knowledge about it implies a prohibitively costly environment-modeling effort. In addition, the robot's sensors, responsible for delivering its runtime perceptual input, are usually noisy, and therefore, the information they provide is also not precise. Due to these problems, the integration of information at the task and at the perceptual level for the design of a robust, purposeful, and reactive navigation method becomes a major research issue.

To date, a lot of research has been devoted to developing qualitative methods for path planning [1, 3]. Qualitative path-planning methods offer several advantages when compared to quantitative ones. They have lower environment-modeling requirements, they, sometimes, mimic human path planning and are therefore more explainable, and they efficiently produce good, if not necessarily optimal, paths.

On the other hand, several fuzzy-logic methods have been developed for reactive navigation [4, 5]. In general, a fuzzy behavior, B, is specified in terms of (i) the context in which it is relevant, and (ii) a set of rules "$If A_i then C_i$" specifying the preferred values for the robot's controls, C_i, in the state A_i, from the point of view of B. A fuzzy behavior may be designed to suggest controls to fulfill low-level navigation requirements (e.g., obstacle avoidance) or to accomplish motion of a more purposeful type (e.g., wall following). Thus, in principle, a blending of a set of different behaviors may result in navigation sensitive both to the requirements of the robot's current task and to the type of environments which it needs to navigate. As compared with another major paradigm of reactive-navigation methods, i.e., potential fields, fuzzy reactive navigation offers the advantages of explainability of the produced trajectories and higher resilience to the problem of getting stuck in local minima.

Qualitative path-planning methods, alone, are not capable for guiding a robot in navigating its environment, since they cannot control the robot's motion at the level of deciding its speed and orientation. On the other hand, fuzzy-logic methods, although effective in controlling the robot's motion, are, in general, insufficient for planning paths to accomplish the robot's tasks. In the DAVID project, we have been investigating the synergy between these two approaches. In this paper, we describe a method, we have developed, which demonstrates (i) that qualitative knowledge alone, about the environment is sufficient with respect to navigation, (ii) that a small set of fuzzy behaviors, each designed for navigating a different type of space, may suffice for reactive navigation in indoor environments, and (iii) that these behaviors can be organized in a way that a qualitative planner can schedule the robot's behaviors appropriate for its navigation task.

We have developed a qualitative path planner and we have integrated it with a fuzzy method for reactive navigation. The qualitative path planner has a model of the robot's world in terms of spaces (i.e., rooms, corridors, open areas), qualitative locations, (i.e., staircases and elevators), their orientation (i.e., east west etc), and the connections between them. Given an initial and a destination location, the planner produces a qualitative path consisting of a sequence of spaces that the robot must traverse in order to reach from the initial location its destination. Each different type of space indexes a set of fuzzy behaviors which are appropriate for navigating spaces of that type and for perceiving the landmarks signaling their beginning and end. Thus, each path segment can be "translated" to a set of behaviors appropriate for re-orienting the robot into the space the segment refers to, and for traversing it. During the navigation of a particular segment, only the behaviors relevant to navigating through it and perceiving the landmarks expected in it are active. DAVID, a Nomad 200 robot uses this method to navigate in an office environment. The whole process is configured and monitored by AUTOGNOSTIC [7], a high-level executive system.

2 The Overall System Architecture

Our experimental platform, DAVID, is a Nomad 200 robot system developed by Nomadic Technologies. The system has a three-wheeled synchrodrive with all wheels steered in the same direction. In its standard configuration the robot vehicle is equipped with a ring of 16 sonar sensors, a ring of 16 infrared sensors, and two rings of tactile sensors. As an additional navigation sensor a SICK laser radar is mounted on top of the vehicle providing a 180° 2D depth profile of the robot's surrounding with an angular resolution of .5 degrees and a range resolution of 3cm. To identify specific objects in its surroundings, the robot is equipped with a color video camera mounted on a pan and tilt unit. The onboard computer is a 486 processor with 16 MB of RAM running under the Linux 1.2.12 operating system. For elementary manipulation tasks the robot vehicle can be equipped with a 2 DOF arm consisting of a vertical prismatic joint and a simple two-finger gripper.

The overall robot architecture we have developed in the context of the DAVID project is shown in Figure 1 below. It consists of two main elements, shown in boldface boxes in the Figure. The first element is AUTOGNOSTIC, [7], a deliberative reasoner responsible for planning a task structure appropriate for accomplishing the robot's current mission, monitoring its execution and adapting it when it fails. The second is DAVID itself, the physical robot, responsible for executing the elementary tasks that constitute the leaves of the task structure.

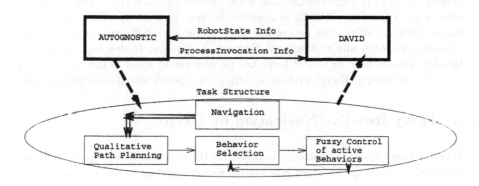

Fig. 1. The System Architecture.

AUTOGNOSTIC has a library of the elementary tasks that DAVID is able to accomplish, that is the elementary functionalities it is able to deliver. Given a new mission for the robot, for example, to go from one office to another, AUTOGNOSTIC, based on its knowledge of DAVID's elementary tasks, configures a task structure appropriate for accomplishing this current mission. A task structure is a hierarchical task network, as the one shown inside the oval at the bottom of Figure 1. Its root is the robot's mission, in this case the *navigation* from one office to the other, and its leaves are the specific elementary tasks in which it gets recursively decomposed, in this case *path planning, behavior selection* and *fuzzy control*. The task structure also specifies the control of processing among the tasks, shown with the different types of arrows inside the oval. In this example, the *navigation* task *gets decomposed* (denoted by a double arrow) into the three abovementioned subtasks, that is, this task is accomplished when these three simpler subtasks have been accomplished. *Path planning* is the first of the three to be performed, where *behavior selection* and *fuzzy control* follow in a loop (denoted by single arrows). Once a task structure is configured, AUTOGNOSTIC starts to execute and monitor it. When a complex task, such as *navigation*, is encountered, AUTOGNOSTIC simply pushes down in its task stack and invokes the first of its subtasks. As soon as a leaf task needs to be performed, AUTOGNOSTIC sends the information relevant to the invocation of the procedure accomplishing the task to DAVID. DAVID, in turn, executes the procedure and communicates back to AUTOGNOSTIC its output. AUTOGNOSTIC evaluates whether or not this output is correct, given its knowledge of the functionality that the invoked procedure should deliver. If indeed the results are correct, it continues the task-structure execution with the next subtask, otherwise, it suspends it and proceeds to adapt the failing task structure.

AUTOGNOSTIC is implemented in Lucid Common Lisp, and is connected through a TCP/IP socket connection with DAVID. In this setup, AUTOGNOSTIC acts as the Client and DAVID acts as the Server. This set up offers two advantages. First, it allows the processing load to be shared between the machine onboard the robot and another one. Second, it allows us to develop functionalities for DAVID both in C and Lisp. Let us now see in some detail how exactly the AUTOGNOSTIC-DAVID system accomplishes specifically its navigation task.

3 Fuzzy Reactive Navigation in DAVID

DAVID's fuzzy reactive controller is based on the concept of *behaviors*, each one designed to accomplish either a low-level motion goal or a perceptual goal. A fuzzy behavior, B, is specified in terms of a set of rules *If A_i then C_i* specifying the preferred values for the robot's controls, C_i, in the state A_i, from the point of view of B. The controlled variables in the case of DAVID are its speed and its orientation. In addition, each fuzzy behavior is associated with a *context*, that is, a situation, characteristic of several states, in which it is relevant. The context of the behavior can be thought of as yet another set of conditionals, $Cntxt_i$, that should be added in the evaluation of the behavior's rules.

The fuzzy control process consists of the following five steps: (i) *perception* of the local environment through the robot's sensors, (ii) *fuzzification* of the perceived values in order to update the fuzzy variables defining the robot's state, (iii) *inferencing*, i.e., evaluation of the rules composing the active fuzzy behaviors, (iv) *synthesis* of all the fuzzy values suggested for the controls by the rules of each active behavior into a single fuzzy value proposed by each behavior, and (v) *defuzzification* of the fuzzy values inferred by all the behaviors into the actual values used for the robot's control variables. DAVID uses the Max-Prod rule for the fuzzification step and the max T-norm rule for the synthesis step.

Since each fuzzy behavior is designed to suggest controls to fulfill a quit specific goal of the robot, whether it is a navigation or a perceptual requirement (e.g., obstacle avoidance or recognition of an open door), complex tasks in complex environments can only be accomplished by the blending of a large number of such behaviors. An important problem that arises from blending a variety of behaviors is that, since they are designed for different purposes, they may suggest very different controls, which during the averaging process of synthesis may result to a control value inappropriate for any of the currently active purposes of the robot. To avoid such conflicts, it is preferable that only few behaviors are active at nay point in time. Therefore, the notion of context was conceived in order to manage the number of active behaviors.

In DAVID, we have added an additional level of filtering the robot's active behaviors, controlled part by the type of space the robot is traversing in the current segment. The basic intuition behind this decision is that, different types of spaces impose different challenges to the robot which can be better addressed by behaviors specifically tailored to them. Thus, for each different type of area in DAVID's environment, we have developed a set of *motion behaviors* appropriate for navigating areas of this type. For example, to navigate through a corridor, a long and narrow space, DAVID has to be careful not to hit its walls. For that purpose, DAVID's *follow_corridor* behavior consists of a set of rules whose purpose is to enable DAVID to maximize its distance from the walls on its left and right side. In addition, while moving within a space, DAVID has to be able to localize itself inside it. To that end, we have developed a set of *perceptual behaviors*, each one appropriate for recognizing a particular type of landmark that can be naturally encountered within an office environment. For example, while moving through corridors, DAVID's *find_door_direction?* behaviors are responsible for recognizing the doors of the offices to which the corridor leads, and keeping track of their number.

To date, DAVID has six (6) motion behaviors and (5) perceptual ones. For its motion behaviors, it is using its short-range, infrared sensors. For its perceptual ones, it uses its laser scanner, which is more precise and longer range sensor. Each laser scan consists of 360 values in a range from 0 to 6000, thus covering a 180-degree area, at which the laser is pointing, with a precision of 0.5 degrees. For each different landmark to be recognized, we have developed a specific pattern of a typical scan that the laser would deliver, were it pointing to such landmark. Therefore, to recognize a landmark in a given scan, the scan is searched for the landmark pattern. To make the recognition process more robust, and to enable the robot to recognize a landmark from a variety of positions (and not only when the laser is pointing directly at it) the pattern values are also represented as belong in fuzzy sets. Clearly, DAVID's perceptual behaviors are more computationally intensive than its motion ones since they involve a pattern-matching process per cycle. In general, a fuzzy-control cycle in DAVID lasts 40000 microsecs, but it can rise if several perceptual behaviors are active at the same time.

Having thus designed DAVID's fuzzy behaviors, during the execution of each path segment, only the motion and perceptual behaviors associated with the particular type of space that the segment belongs in, are active.

4   Navigating in an Office Environment

As AUTOGNOSTIC-DAVID receives its mission, it first plans a complete path from the initial to the goal area. This path consists of a set of areas that need to be traversed. Then, each segment of this path is successively "translated" into a set of behaviors sufficient for traversing it, and the fuzzy controller is activated to execute them. The successful execution of each segment is signaled by the controller when the activation levels of all the behaviors associated with the current segment fall below a certain threshold level. The whole process stops when all the segments of the path have been executed.

Path Planning DAVID's high-level, qualitative planner is an adaptation of ROUTER [2], a path planner originally developed for navigating within a city neighborhood. We have developed IN-ROUTER to navigate within an office building in a manner similar to which ROUTER navigates city streets. IN-ROUTER has a model-based method and a case-based method for planning paths. When presented with a problem, it chooses one of them based on a set of heuristics which evaluate the applicability and utility of each method on the problem at hand.

IN-ROUTER's model of its environment consists of a set of areas in an building, i.e., offices or corridors, and the connectors, i.e., doors or passages, between them. The areas are grouped into "neighborhoods" which are organized in a space-subspace hierarchy. At the top level of this hierarchy are the buildings known to DAVID, which are then decomposed into their floors, which get decomposed into wings, which, in turn, consist of rooms, open spaces, and corridors. When IN-ROUTER uses the model-based method to solve a problem, it first locates the initial and goal areas in its space hierarchy. If the two areas belong in the same wing, that is in the same lowest-level neighborhood, IN-ROUTER performs a local search to find a path between them. Otherwise, it decomposes the overall problem into a set of subproblems, each of which can be solved locally within some space-type level. In then synthesizes the partial solutions to find the desired path and stores it in memory for future use.

The path produced by IN-ROUTER specifies two types of information. First, it specifies the spaces the robot has to traverse to go from its initial location to its destination location, and second, the connectors it has to recognize between them in order to switch between one space and the next. The types of the different spaces contained in the path will be used to decide the fuzzy behaviors that will be activated to control the motion of the robot, and the connectors between them will be used to decide the landmark-recognition behaviors that will be activated to enable the robot to locate itself in its environment during the execution of the path.

Let us now describe IN-ROUTER's planning in the context of a particular experiment. DAVID was presented with the task of going from "office 2.1" to "office 2.21". As shown on the right of Figure 2, these two offices are located in the opposite corners of the second floor of our building, and therefore in two different wings in IN-ROUTER's model of the building. IN-ROUTER identifies that a common area between the two wings is the "Free Space" in front of the library. Therefore it plans a path between "office 2.1" and this area, and between this area and "office 2.21". The complete path suggests that to go from one office to the other, DAVID has to get out of "office 2.1", go to end of the "Alann-Vips corridor", traverse the open space between the elevator and the library, and then between the library and the washroom, traverse the "David-lab corridor", and turn into "office 2.1". The first column of the table on the left of the Figure 2 shows the actual spaces that DAVID has to traverse and (in parentheses) the different connectors it has to recognize between them.

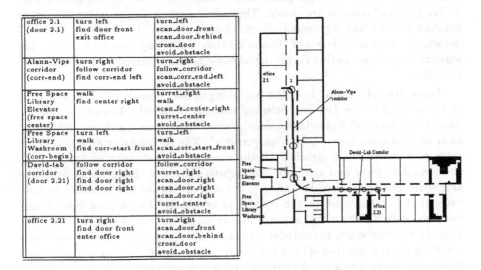

office 2.1 (door 2.1)	turn left find door front exit office	turn_left scan_door_front scan_door_behind cross_door avoid_obstacle
Alann-Vips corridor (corr-end)	turn right follow corridor find corr-end left	turn_right follow_corridor scan_corr_end_left avoid_obstacle
Free Space Library Elevator (free space center)	walk find center right	turret_right walk scan_fs_center_right turret_center avoid_obstacle
Free Space Library Washroom (corr-begin)	turn left walk find corr-start front	turn_left walk scan_corr_start_front avoid_obstacle
David-lab corridor (door 2.21)	follow corridor find door right find door right find door right	follow_corridor turret_right scan_door_right scan_door_right scan_door_right turret_center avoid_obstacle
office 2.21	turn right find door front enter office	turn_right scan_door_front scan_door_behind cross_door avoid_obstacle

Fig. 2. A complete experiment: Going from "office 2.1" to "office 2.21". Left, the qualitative path plan and the behaviors activated for the traversal of each one of its segments, and right the trajectory.

Behavior Selection The path produced by IN-ROUTER describes which spaces the robot needs to traverse in order to accomplish its task, and which landmarks it needs to recognize in order to switch between them. This path is extended to include directions for the turns the robot has to make to traverse this space sequence given its current orientation. Each step of this elaborated path is then transformed into a set of behaviors appropriate for executing it. Finally, the segments of the elaborated path are translated to sets of behaviors appropriate for accomplishing them. To ensure that DAVID is able to react to obstacles (i.e., static furniture, or moving people) in its proximity the *avoid_obstacle* behavior is active in all the different segments.

The second column of the table in Figure 2 shows the extension of the qualitative path produced by IN-ROUTER in the above example, and the third one the behaviors activated for the execution of each segment. The behavior *avoid_obstacle* is always active since, irrespective of the type of space in which the robot moves, it has to avoid collision with any object in its trajectory.

Fuzzy-Control Navigation Once the behaviors appropriate for a segment have been identified, the fuzzy controller is invoked to execute them. Let us now discuss the evolution of the behaviors' activation and the controls' values during the traversal of our example path. In the map, shown on the right of Figure 2, five points of interest are shown. These five points roughly depict the points between the different segments in the path generated by IN-ROUTER. Figure 3 depicts the activation levels of DAVID's behaviors during this experiment. The vertical dashed lines in this Figure depict these points occurred in the experiment time line.

From start to point 1: In the beginning of the experiment, DAVID is close to the door of "office 2.1" and oriented towards the south. To exit the office it has first to turn left towards the door, then to recognize the door, and finally, to cross it to enter the "Alann-Vips corridor". The successful completion of the cross_door maneuver is signified by the surge of activation exhibited by both the *scan_door_front* and the *scan_door_behind* behaviors. Both these behaviors are active during the whole segment, but the context of the second one indicates that it contributes to the control process only after a door has already been recognized. Therefore, as soon as the surge of *scan_door_front* occurred, *scan_door_behind* starts to contribute to the control process. This last surge also results in the invalidation of both behaviors' contexts, which in turn, results that the control be returned to AUTOGNOSTIC, which proceeds to the next segment of the path to be executed.

From point 1 to point 2: To traverse the corridor a set of three behaviors are needed: DAVID had to turn right, so as to orient itself along the corridor, and then to follow it until it recognizes its end. Following the corridor means for DAVID moving along the long axis of the corridor while staying in the middle of the area defined by the two walls on its left and right. The end of the corridor is recognized by the increase (or, decrease) of the width of the area in which DAVID finds itself. The end of this segment was signaled by a surge of activation of the *scan_corr_end_left* behavior.

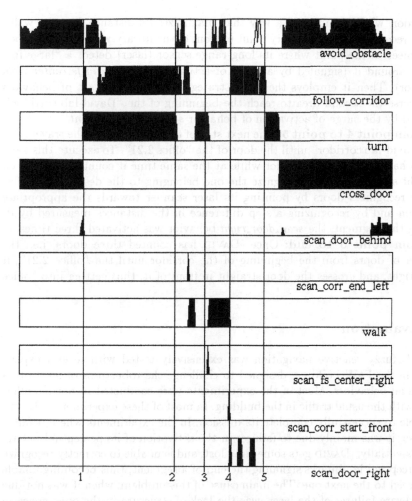

Fig. 3. The activation levels of DAVID's behaviors during the experiment.

From point 2 to point 3 to point 4: As it can be seen in the map of Figure 2, the two main corridors meet at a point where the staircase leading to the first floor of the building is located. This is an especially dangerous area for the robot, because it is very unstructured and also because it has to steer safely away from the stairs. In order to enable DAVID to traverse this unstructured environment between the staircase area and the library safely, we have conceptualized it as consisting of two separate areas. One area is close to the north-south "Alann-Vips corridor", and the other is close to the east-west "David-lab corridor". The two areas meet at an abstract point conceptualized in front of the library, and far away from the staircase. Thus, as DAVID finds itself in the first area, it orients itself towards the right wall, in contrast with its behavior while following a corridor, when it moves in a way that maximizes its distance from its walls. DAVID recognizes the imaginary point in front of the library where the two open areas meet as the point where its long-range sensor (laser) detect a "large free space" around it (signaled by a surge of activation of the *scan_fs_center_right* behavior). Then it employs the same strategy, i.e., the same set of behaviors to traverse the second area to reach the beginning of the "David-lab corridor", signaled by the surge of activation of behavior *scan_corr_start_front*.

From point 4 to point 5: The next step of the path involves the traversal of the "David-lab corridor" until the door of the "office 2.21". To execute this step, DAVID has to follow the corridor while at the same time it counts the doors on its right side, in order to recognize the one belonging to the destination office. DAVID recognizes doors by pointing its laser scanner towards the appropriate direction and by recognizing a step difference in the distances measured by it. During this segment, the *scan_door_right* behavior was activated three times.

From point 5 to end: Once DAVID has counted three doors, i.e., the number of doors from the beginning of the corridor until the "office 2.21", it turns right, and crosses the door straight in front of it, thus getting into "office 2.1".

5 Evaluation

DAVID's fuzzy reactive navigation was extensively tested with several experiments in the FAW building, designed to challenge the robustness of most of its individual behaviors. Most of the experiments were conducted in normal office hours with the usual traffic in the building. In most of these experiments, DAVID was able to successfully complete its mission. In the experiments where it failed to do so, it was mainly due to failures in the activation of its perceptual behaviors. Essentially, DAVID gets sometimes lost, and is unable to correctly recognize the perceptual landmarks signaling the end of a segment, thus becoming unable to proceed to the next one. The main cause of this problem, when it was not due to hardware failures of the laser, was the lack of structure in the open space in the area of the library. In this area, DAVID found most difficulties in recognizing and orienting itself towards the corner of the space opposite from the stairs. We have modified the *walk* behavior into a somehow wall-following behavior and the results have since improved.

In addition to the real-world experiments, DAVID was tested with two sequences of experiments in simulation mode [6] where it was asked to wander around in an area, with dense static and moving obstacles. In these experiments, we varied the density and the speed of the obstacles in order to test the robustness of the *avoid_obstacle* behavior.

5.1 Related Research

Another qualitative path planner, descendant of ROUTER, has also been integrated in Stimpy, a robot implemented in the context of the AuRA architecture [1]. The main difference between Stimpy and DAVID is that Stimpy's navigation is based on potential fields, and as a result at that level it is using numerical information. The main difference between the blending of fuzzy behaviors and the synthesis of vectors in a potential field is that contribution of each active behavior to the robot's controls' values depends, at any point in time, on the degree to which the context of this behavior is fulfilled. Instead, in potential-fields approaches the overall force applied to the robot is a linear superposition of all the potential fields acting on the robot's current position. Thus, fuzzy-behavior blending tends to deliver controls' values more "sensitive" to the robot's state and tends to be more immune to the local-minima problem from which potential-fields approaches often suffer. This advantage was part of our motivation in investigating fuzzy logic as the basis for reactive navigation.

In comparison with other work on fuzzy reactive navigation, we should mention that unlike [4, 5] our approach does not require any metric information at all. In these other approaches, some of the behaviors (i.e., corridor following) depend on metric information about the environment (i.e., length or width of corridor). In addition, they have to "anchor" the robot's perception during execution with this information. Our qualitative approach offers two advantages, since, first, its knowledge requirements are lower and therefore simpler to be met, and second it limits the needs for coordination between internal knowledge and perceptual input during execution.

Another difference between [4] and our work is that, all the behaviors which will be needed at some point in the execution of the planned path are presented to the fuzzy controller together in the beginning. In contrast, in our work, the fuzzy controller is given only the behaviors necessary for the next path segment to be executed. Thus, because only a limited set of behaviors is active at each point in time, AUTOGNOSTIC's monitoring element has a more fine-grained feedback in case of failures during navigation, and therefore, more information to use for recovery.

6 Conclusions and Future Work

From our experiments with DAVID's fuzzy-navigation module we would like to draw out the following conclusions:

(1) Qualitative information suffices for navigation in indoor environments. With the exception of navigating from one precise point to another, DAVID is able to navigate throughout the whole second floor of our building.

(2) A compact (DAVID has 11 motion and perceptual behaviors) set of fuzzy behaviors suffices for robustly and efficiently navigating a variety of types of areas in an office environment.

This set of behaviors, our sets of experiments show, suffices for maneuvers such as crossing a door carrying a waste basket, a fact that significantly limits the free space between the robot and the door, and safely traversing a corridor with big obstacles, such as tables, in it.

(3) Qualitative planning and fuzzy navigation provide a good match for robust navigation without coordinates.

IN-ROUTER's hierarchical, qualitative planning and DAVID's fuzzy navigation constitute an example of this match. Both modules have similar knowledge requirements in terms of environment modeling, and in addition, to providing a plan executable by the fuzzy navigator, IN-ROUTER also provides to some degree a "context" for each execution segment.

In closing, we would like to mention that we intend to investigate the issue of extending the self-monitoring capabilities of the AUTOGNOSTIC-DAVID system by enriching the feedback that the fuzzy controller can return to AUTOGNOSTIC. By examining the evolution of the activation levels of the currently active behaviors, the fuzzy controller is able to recognize pathological situations. For example, in a case where DAVID is trying to cross a door blocked by an obstacle, the activation level of the *cross_door* behavior is periodically rising and falling. The activation trace can be monitored for such pathological cases, which can then be signaled to AUTOGNOSTIC's monitoring.

References

1. Ali, K.; Goel, A.; Stroulia, E.: Some Experimental Results in Multistrategy Navigation Planning, Tech. Report GIT-CC-95-51 Georgia Inst. of Technology, 1995.
2. Goel, A.; Callantine, T.; Shankar, M.; Chandrasekaran, B.: Representation, Organization, and Use of Topographic Models of Physical Spaces for Route Planning. In *Proceedings of the Seventh IEEE Conference on Artificial Intelligence Applications*, Miami Beach, Florida, pp. 308-314, IEEE Computer Society Press, 1991.
3. Levitt, T.S.; Lawton, D.T.; Chelberg, D.M.; Nelson, P.C.: Qualitative Landmark-Based Path Planning and Following. In *Proceedings AAAI-87*, Seattle, July 1987, pp. 689–694.
4. Saffiotti, A.; Ruspini, E.H.; Konolige, K.: Using Fuzzy Logic for Mobile Robot Control. In: International Handbook of Fuzzy Sets and Possibility Theory. D.Dubois, H. Prade and H.J. Zimmermann Eds. Kluwer Academic, forthcoming 97.
5. Surmann, H.; Huser, J.; Peters, L.: Guiding and controlling mobile robots with a fuzzy controller. In: Proc. Fourth IEEE International Conference on Fuzzy Systems, Yokohama, Japan, 1995.
6. Schneider, D: Robuste Bewegungsstrategien Autonomer Mobiler Roboter, FAW Tech. Report, 1996.
7. Stroulia, E: *Failure-Driven Learning as Model-Based Self-Redesign*. PhD Thesis, Tech. Report GIT-CC-95-38 Georgia Inst. of Technology, 1994.

Application of the Approximate Fuzzy Reasoning Based on Interpolation in the Vague Environment of the Fuzzy Rulebase in the Fuzzy Logic Controlled Path Tracking Strategy of Differential Steered AGVs

Szilveszter Kovács

Computer Centre, University of Miskolc
Miskolc-Egyetemváros, Miskolc, H-3515, Hungary
e-mail: szkszilv@gold.uni-miskolc.hu

László T. Kóczy

Department of Telecommunication an Telematics,
Technical University of Budapest
Sztoczek u.2, Budapest, H-1111, Hungary

Abstract. In most of the practical applications the concept of vague environment [1] gives a simple way for fuzzy approximate reasoning. If the fuzzy partitions (used as primary sets of the fuzzy rulebase) can be described by vague environments [1], the primary fuzzy sets of the antecedent and the consequent parts of the fuzzy rules can be characterised by points in their vague environments. So the fuzzy rules themselves can be characterised by points in their vague environment too. It means, that the question of approximate fuzzy reasoning can be reduced to the problem of interpolation of the rule points in the vague environment of the fuzzy rulebase relation [2,3]. In this paper an approximate fuzzy reasoning method based on *rational interpolation* in the vague environment of the fuzzy rulebase will be introduced, and as an example of a practical application of the method, a path tracking control strategy for differential steered AGVs (Automated Guided Vehicle) [4] implemented on such a fuzzy logic controller will be introduced.

1 The Vague Environment and the Fuzzy Partition

The concept of vague environment is based on the similarity or indistinguishability of the elements. Two values in the vague environment are ε-distinguishable if their distance is grater then ε. The distances in vague environment are weighted distances. The weighting factor or function is called *scaling function* (or factor) [1].

For finding connections between fuzzy sets and a vague environment we can introduce the membership function $\mu_A(x)$ as a level of similarity **a** to x, as the degree to which x is indistinguishable to **a** [1]. So the α-cuts of the fuzzy set $\mu_A(x)$ is the set which contains the elements that are $(1-\alpha)$-*indistinguishable* from **a** (see fig.1.):

$$\delta_s(a,b) \le 1 - \alpha \ ,$$

$$\mu_A(x) = 1 - \min\{\delta_s(a,b),1\} = 1 - \min\left\{ \left| \int_a^b s(x)dx \right|, 1 \right\}$$

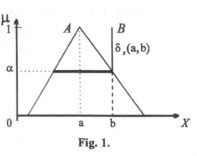

where $\delta_s(a,b)$ is the vague distance of the values a, b, and $s(x)$ are the scaling function of the universe X.

Fig. 1.

It is very easy to realize (see fig.1.), that this case the vague distance of points a and b ($\delta_s(a,b)$) is basically the *Disconsistency Measure* (S_D) of the fuzzy sets A and B (where B is a singleton):

$$S_D = 1 - \sup_{x \in X} \mu_{A \cap B}(x) = \delta_s(a,b) \qquad \text{if } \delta_s(a,b) \in [0,1]$$

where $A \cap B$ is the min t-norm, $\mu_{A \cap B}(x) = \min[\mu_A(x), \mu_B(x)] \ \forall x \in X$.

The main difference between the disconsistency measure and the vague distance is, that the vague distance is a crisp value in range of $[0,\infty]$, while the disconsistency measure is limited to $[0,1]$. That is why it is useful in interpolate reasoning with insufficient evidence too.

So if it is possible to describe all the fuzzy partitions of the primary fuzzy sets (the antecedent and consequent universes) of our fuzzy rulebase, and the observation is a singleton, we can calculate the "extended" disconsistency measures of the antecedent primary fuzzy sets of the rulebase and the observation, and the "extended" disconsistency measures of the consequent primary fuzzy sets and the consequence (we are looking for) as vague distances of points in the antecedent and consequent vague universes.

For generating a vague environment of a fuzzy partition we have to find an appropriate scaling function, which describes the shapes of all the terms in the fuzzy partition. A fuzzy partition can be characterised by a vague environment if and only if the membership functions of the terms fulfills the following requirement [1]:

$$s(x) = |\mu'(x)| = \left| \frac{d\mu}{dx} \right| \quad \text{exists iff} \quad \min\{\mu_i(x), \mu_j(x)\} > 0 \ \Rightarrow \ |\mu'_i(x)| = |\mu'_j(x)| \ \forall i,j \in I,$$

where $s(x)$ is the vague environment we are looking for.

Generally the above condition is not fulfilling, so the question is how to describe all fuzzy sets of the fuzzy partition with one "universal" scaling function. For this reason we propose to use the *approximate scaling function* [2,3].

The *approximate scaling function* is an approximation of the scaling functions describes the terms of the fuzzy partition separately.

Supposing that the fuzzy terms are *triangles*, each fuzzy term can be characterised by two constant scaling functions, the scaling factor of the left and the right slope of the triangle. So a triangle shaped fuzzy term can be characterised by three values (by a triple), by the values of the left and the right scaling factors and the value of its core point (e.g., fig.2.). For generating the approximate scaling function we suggest to interpolate the neighbouring scaling factors (e.g., fig.2.).

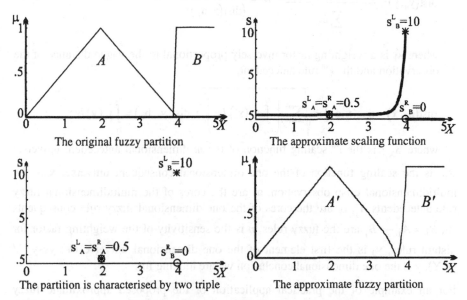

Fig. 2. Approximate scaling function generated by non-linear interpolation, and the original fuzzy partition (*A,B*) as the approximate scaling function describes it (*A',B'*)

2 Approximate Reasoning Based on Vague Environment

If the vague environment of a fuzzy partition (the scaling function or the approximate scaling function) exists, the member sets of the fuzzy partition can be characterised by points in the vague environment. (In our case the points are characterising the cores of the terms, while the shapes of the membership functions are described by the scaling function.) If all the vague environments of the antecedent and consequent universes of the fuzzy rulebase are exist, all the primary fuzzy sets (linguistic terms) used in the fuzzy rulebase can be characterised by points in their vague environment. So the fuzzy rules (build on the primary fuzzy sets) can be characterised by points in the vague environment of the fuzzy rulebase too. This case the approximate fuzzy reasoning can be handled as a classical interpolation task. Applying the concept of vague environment (the distances of points are weighted distances), any interpolation, extrapolation or regression methods can be adapted very simply for approximate fuzzy reasoning [2,3].

For example we can adapt the *rational interpolation* [2,3]. This method generates the conclusion as a weighted sum of the vague consequent values, where the weighting factors are inversely proportional to the vague distances of the observation and the corresponding rule antecedents:

$$\text{dist}(y_0,y)=\frac{\sum_{k=1}^{r} w_k \cdot \text{dist}(y_0,b_k)}{\sum_{k=1}^{r} w_k}, \quad w_k=\frac{1}{\left(\text{dist}(\mathbf{x},\mathbf{a}_k)\right)^p},$$

where w_k is a weighting factor inversely proportional to the vague distance of the observation and the kth rule antecedent,

$$\text{dist}(\mathbf{a}_k,\mathbf{x})=\text{dist}(\mathbf{x},\mathbf{a}_k)=\sqrt{\sum_{i=1}^{m}\left(\int_{a_{k,i}}^{x_i} s_{X_i}(x_i)dx_i\right)^2}, \quad \text{dist}(y_0,b_k)=\int_{y_0}^{b_k} s_Y(y)dy,$$

where s_{X_i} is the ith scaling function of the m dimensional antecedent universe, s_Y is the scaling function of the one dimensional consequent universe, $\mathbf{x}$ is the multidimensional crisp observation, $\mathbf{a}_k$ are the cores of the multidimensional fuzzy rule antecedents A_k, b_k are the cores of the one dimensional fuzzy rule consequents B_k, $R_i = A_i \rightarrow B_i$ are the fuzzy rules, p is the sensitivity of the weighting factor for distant rules, y_0 is first element of the one dimensional universe (Y: $y_0 \leq y$ $\forall$ $y \in Y$), y is the one dimensional conclusion we are looking for.

For an example of the practical application of the proposed approximate fuzzy reasoning method a real path tracking control strategy for differential steered AGVs (Automated Guided Vehicle) [4] will be introduced.

3 The Guide Path Controlled AGV

The Automatically Guided Vehicle (AGV) is a typical element of the group of materials handling equipment. A popular way of AGV guidance is based on the guide path method. This is because of the simple structure of the guidance system [5]. The guide path is usually a painted marking or a passive or active wire (guidewire) glued onto or build into the floor. The goal of the steering part of the guidance system of the AGV is to follow the marking of the guide path. The guiding system senses the position of the guide path by special sensors tuned for the guide path. Usually an AGV has one or two guide path sensors. The number of the sensors depends on the wheel configuration of the AGV. The AGVs without fixed directional wheels in their wheel configuration (can be moved to arbitrary direction) usually have two guide path sensors, the AGVs with at least one fixed directional wheel have only one. This is because of the restricted possible moving directions of the fixed directional wheel. An AGVs with at least one fixed directional wheel can run only on a path curve has its momentary centre on the line fits the axe of the fixed directional wheel. In the further part of this article we would like to concentrate on the path tracking strategy of a differential steered AGV which has fixed directional wheel (fig.4.).

The guide point is a point of the AGV determined by the guide path sensor. The goal of the steering control is to follow the guide path with the guide point. The main problem of the guide point based path tracking strategy, that on straight path the path tracking error (the distance of the guide path and the driving centre of the AGV) decreasing relatively slow (see fig.3.):

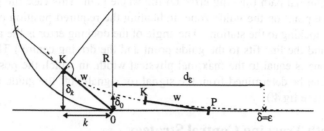

Fig.3. Trajectory of the driving centre of a differential steered AGV using the guide point based path tracking strategy (see fig.4. for the notation)

The quick convergence of the trajectory of the driving centre to the guide path (the quick convergence of the path tracking error to zero) is very important in determining the possible positions of the docking points of the AGV. Quicker the convergence of the path tracking error to zero, quicker the docking accuracy specified for the docking point can be reached, so the minimal distance needed between the last curve of the guide path and a docking point (minimal docking distance) is shorter too. Using the concept of guide point path tracking strategy gives no freedom for the guidance system in choosing trajectory with quicker convergence to the guide path.

For finding a better trajectory we suggest to use the concept of *guide zone* [4] instead of the guide point. The guide zone is an extension of the guide point. Basically the guide zone is a new interpretation of the guide point. Using the concept of guide zone, the signal of the guide path sensor is interpreted as a distance between the guide path and the guide point, instead of the meaning of an error value.

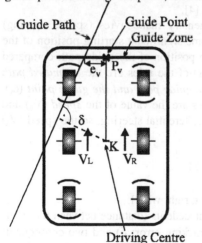

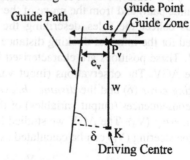

δ path tracking error
e_v distance of the guide path, guide point
d_s width of the guide zone
w distance of the guide point, driving centre
P_v guide point, K driving centre

Fig.4. Differential steered AGV with guide zone

The *guide zone* is a section of the AGV determined by the guide path sensor (or raw of sensors). The goal of the steering control is to follow the guide path by the guide zone with minimal path tracking error on the whole path. This case the guide point is a reference point on the guide zone, indicating the required position of the guide path during docking to the station. (The angle of the docking error is the angle of the guide path and the line fits to the guide point and the driving centre.) The width of the guide zone is equal to the maximal physical width, in which the position of the guide path can be determined from the signal (or signals) of the guide path sensor (or sensors) (see fig.4.).

4 The Path Tracking Control Strategy

The simplest way of defining a path tracking strategy is based on collecting the operator's knowledge. This case we would like to find a better guidance strategy, compared to the single guide point based method, for reducing the minimal distance needed from the last curve of the path to a docking station (minimal docking distance). The control of a differential steered AGV is very similar to the control of a car. The base of our guidance strategy is very simple: keep the driving centre of the AGV as close as it possible to the guide path, than if the driving centre is close enough to the guide path, simply turn the AGV into the docking direction.

For defining the strategy, we have to examine the observations we need for the guidance system. The above simple strategy we would like to describe needs only two observations: The distance between the guide path and the driving centre (path tracking error), and the distance between the guide path and the guide point. Using the guide zone, we can determine the distance of the guide path and the guide point, but we have no information on the path tracking error. We suggest to calculate the estimated momentary path tracking error from the previous (e_{vo}) and the current value (e_v) of the distance between the guide path and the guide point (measured by the guide path) and from the move of the AGV [4].

Let us collect the rules describing the momentary manoeuvres (speed, steering) needed for the minimal docking distance in some significant starting position of the AGV. These positions are characterised by the position of the guide path compared to the AGV. The observations (input variables) of the rules are the *estimated path tracking error* (δ) and the *distance between the guide path and the guide point* (e_v). The consequences (output variables) of the rules are the value of the *speed* (V_a) and the *steering* (V_d). The AGV we studied has a differential steering, so the speed (V_a) and the steering (V_d) can be calculated as:

$$V_d = V_L - V_R \text{ steering}, \qquad V_a = \frac{V_L + V_R}{2} \text{ speed}.$$

where V_L, V_R is the contour speed of the left and right wheel.

So the rules we are collecting will have two antecedents (distance between the guide path and the guide point: e_v, estimated path tracking error: δ) and two consequents (steering: V_d, speed: V_a). Practically it means two rulebases with rules has only one consequent, one rulebase for the steering consequence and one for the speed. The structure of the FLC controlled guidance system of a differential steered AGV is shown on fig.5.

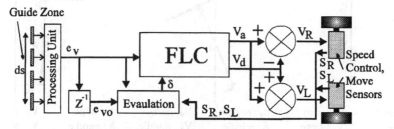

Fig.5. Structure of the guidance system of a differential steered AGV

For the simplicity of converting human knowledge to rules, the values we are using in the rulebases are linguistic terms. Let the ith rules have the following form:

$R_{Vd,i}$ (rules of the steering): $R_{Va,i}$ (rules of the speed):

  If $e_v = A_{1,i}$ **And** $\delta = A_{2,i}$, **If** $e_v = A_{1,i}$ **And** $\delta = A_{2,i}$,

   Then $V_d = B_i$.   **Then** $V_a = B_i$.

e.g.,

  R_{Vd}: If $e_v = NM$ **And** $\delta = PS$ **Then** $V_d = Z$.

(**If** the *distance between the guide path and the guide point* (e_v) is *Negative Middle* **and** *estimated path tracking error* (δ) is *Positive Small* **then** the *steering* (V_d) is *Zero*)

The whole rulebase describing the rules of the momentary steering actions (V_d) and the momentary speed (V_a) is the following:

R_{Vd}:

$e_v =$							
	NL :	NM :	NS :	Z :	PS :	PM :	PL :
$\delta =$	NL :	PM	PS	Z	Z	NL	NL	NL
	NM :	PL	PS	PS	PS	PS	Z	NL
	NS :	PL	PM	PS	PS	Z	Z	NL
	Z :	PL	PM	PS	Z	NS	NM	NL
	PS :	PL	Z	Z	NS	NS	NM	NL
	PM :	PL	Z	NS	NS	NS	NS	NL
	PL :	PL	PL	PL	Z	Z	NS	NM

R_{Va}:

$e_v =$							
	NL :	NM :	NS :	Z :	PS :	PM :	PL :
$\delta =$	NL :	M	S	S	S	S	Z	Z
	NM :	S	M	M	M	M	M	S
	NS :	Z	S	L	L	L	M	S
	Z :	S	M	L	L	L	M	S
	PS :	S	M	L	L	L	S	Z
	PM :	S	M	M	M	M	M	S
	PL :	Z	Z	S	S	S	S	M

For example:

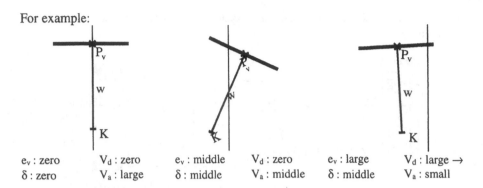

e_v : zero	V_d : zero	e_v : middle	V_d : zero	e_v : large	V_d : large →
δ : zero	V_a : large	δ : middle	V_a : middle	δ : middle	V_a : small

5 FLC Based on Compositional Rule of Inference

Our rulebases is complete, all the possible combinations of observations (in linguistic terms) are covered by rule antecedents. In case of building complete fuzzy partitions for the antecedent linguistic terms we can use the classical *min-max compositional rule of inference* in the fuzzy logic controller.

For building a fuzzy logic controller, we have to complete these rulebases with the meanings of the linguistic terms we have used. In a fuzzy logic controller the linguistic terms are characterised by fuzzy sets (primary fuzzy sets). We have generated these fuzzy sets by a tuning process based on expert's knowledge. The tuning process was optimized the core positions of the primary fuzzy sets for getting the shortest docking distance on a trial guide path (fig.7.) using a simulated model of an existing AGV. The fuzzy partitions of the primary fuzzy sets we have got after the tuning process are shown below:

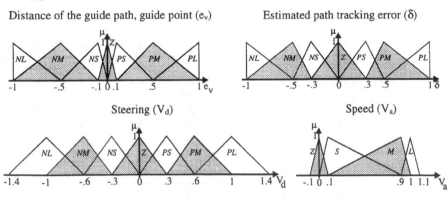

The shortenings we used during the definitions of the primary fuzzy sets are:
N: Negative, *P*: Positive, *L*: Large, *M*: Middle, *S*: Small, *Z*: Zero

Using the classical method of the max-min composition (Zadeh) for the fuzzy rule inference and the centre of gravity method for defuzzification we have got the following control surfaces:

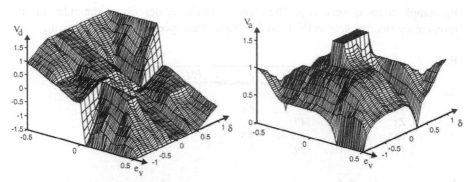

Fig.6. Control surface of the steering (V_d) and the speed (V_a)

For checking the performance of the fuzzy logic controller (FLC) based on CRI, we have worked out a simulated model for a differential steered AGV and an implementation of the FLC. Using this model we have approximated the minimal docking distance of the simulated AGV. The trial guide path we used, and the minimal docking distances in function of the guide path radius are shown on fig.7. For comparing the results of the minimal docking distances to the single guide point based steering system, its calculated results are shown on Fig.7. too.

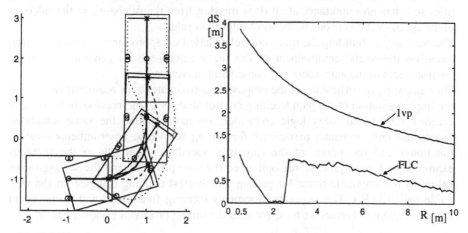

Fig.7. Minimal docking distances (dS) calculated for the optimal single guide point based steering system (1vp) and the simulated results of the AGV using the CRI based FLC path tracking strategy (FLC) in function of the trial guide path radius (R)

6 FLC Based on Interpolation in the Vague Environment

For showing the efficiency of the proposed approximate fuzzy reasoning method, the size of the rulebase, describing the path tracking strategy, is reduced dramatically. All the unimportant rules, rules concluded from the other rules, are removed from the rulebase. It means, that this rulebase contains the most important rules only, so its completeness is necessary. The reduced rulebase, describing the rules of the momentary steering actions (V_d) and the momentary speed (V_a) is the following:

R_{Vd}:

$\delta =$	$e_v =$ NL :	NM :	Z :	PM :	PL :
NL :				NL	
NM :	PL		PS	PS	NL
Z :		PL		NL	
PM :	PL	NS	NS		NL
PL :		PL			

R_{Va}:

$\delta =$	$e_v =$ NL :	NM :	Z :	PM :	PL :
NL :					Z
NM :					
Z :	S		L		S
PM :					
PL :	Z				

It is interesting to notice, that while in the rulebase of the steering (V_d) the conclusion of the rule e_v:zero and δ:zero can be concluded from the surrounding rules so it has no importance at all (it is missing from the rulebase), in the rulebase of the speed (V_a) this is one of the most important rules.

The next step of building the fuzzy controller based on approximate fuzzy reasoning based on the vague environment of the fuzzy rulebase is to generate the vague environments of the antecedent and consequent universes.

For comparing the efficiency of the proposed approximate fuzzy reasoning method in the implementation of the path tracking control strategy to the previously introduced classical CRI based fuzzy logic controller, we have applied the same simulated model and environmental parameters for tuning the vague environments (scaling functions) and the points of the linguistic variables. Similarly to the previous example, the tuning process was optimized the core positions and the scaling factor values of the linguistic terms for getting the shortest docking distance on the trial guide path (fig.9.). The vague environments (scaling functions) of the antecedent and consequent universes we have got after the tuning process are shown below:

Distance of the guide path, guide point (e_v)   Path tracking error (δ)

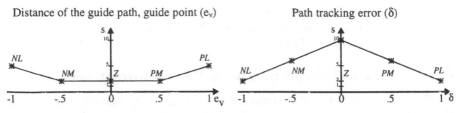

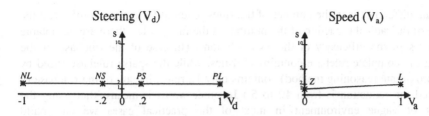

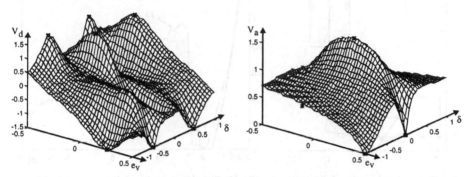

Applying the proposed approximate fuzzy reasoning method based on rational interpolation (introduced in point two) we have got the following control surfaces (for some kind of analogy to the center of gravity defuzzification, the sensitivity factor is chosen to two ($p=2$)):

Fig.8. Control surface of the steering (V_d) and the speed (V_a) (the rule points are signed by *)

7 Comparing the Simulated Results

For comparing the efficiency of the simulated fuzzy logic controllers based on the classical CRI and the proposed approximate fuzzy reasoning method, during the tests we have applied the same simulated model and environmental parameters. The results, the minimal docking distances on the trial guide path in function in function of the path radius are summarized on fig.10.

Summarizing our results, on the trial guide path we have used, there are no significant differences in minimal docking distances of the two simulated implementations of the FLC (classical CRI and the proposed approximate fuzzy reasoning) path tracking strategies. In both cases these results are always better than the minimal docking distance calculated for the optimal, single guide point based steering system (fig.7,10). So using any of our path tracking guidance systems, the minimal distance needed for a possible station from the last curve of the guide path can be reduced considerably.

This is the conclusion we were expected. Both rulebases, the complete (for CRI) and the sparse (for approximate reasoning) one, were fetched from the same "expert knowledge" describing the same path tracking strategy. So the simulated results of the two solutions should not be differ dramatically from each other.

467

The main difference is in the number of the rules required for getting similar results. In spite of the radical reduction of the number of the fuzzy rules, there are no notable differences in the efficiency of the two solutions. (In case of the rulebase of the steering the complete rulebase contains 49 rules, while the sparse rulebase (used by the approximate reasoning method) contains only 12 rules, in case of the rulebase of the speed the reduction is from 49 to 5.) In other words it means, that using the concept of vague environment in most of the practical cases we can build approximate fuzzy reasoning methods simple enough to be a good alternative of the classical Compositional Rule of Inference methods in practical applications.

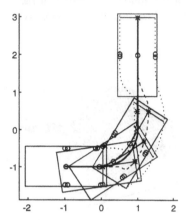

Fig.9. Simulated results of the AGV using the approximate fuzzy reasoning based FLC path tracking strategy

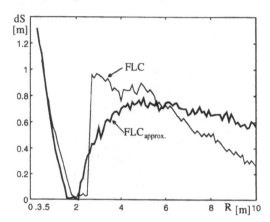

Fig.10. Simulated results of the minimal docking distances (dS) using the CRI based FLC (FLC) and the approximate fuzzy reasoning based (FLC_approx.) path tracking strategy in function of the trial guide path radius (R)

References

1. Klawonn, F.: Fuzzy Sets and Vague Environments, Fuzzy Sets and Systems, 66, pp207-221, (1994).
2. Kovács, Sz., Kóczy, L.T.: Fuzzy Rule Interpolation in Vague Environment, Proceedings of the 3rd. European Congress on Intelligent Techniques and Soft Computing, pp.95-98, Aachen, Germany, (1995).
3. Kovács, Sz.: New Aspects of Interpolative Reasoning, Proceedings of the 6th. International Conference on Information Processing and Management of Uncertainty in Knowledge-Based Systems, pp.477-482, Granada, Spain, (1996).
4. Cselényi, J., Kovács, Sz., Pap, L., Kóczy, L.T.: New concepts in the fuzzy logic controlled path tracking strategy of the differential steered AGVs, Proceedings of the 5th International Workshop on Robotics in Alpe-Adria-Danube Region, p.6, Budapest, Hungary, (1996).
5. Hammond, G.: AGVS at Work - Automated Guided Vehicle Systems. Springer-Verlag, p.232., (1986).

An Approach for Data Analysis and Forecasting with Neuro Fuzzy Systems – Demonstrated on Flood Events at River Mosel

M. Stüber and P. Gemmar

Department of Applied Computer Science, Fachhochschule Trier, 54208 Trier

Abstract. The development and usage of soft computing systems for forecasting of water level progress in case of flood events at river Mosel are presented. The practical situation and its requirements are explained and two different system approaches are discussed: a) a neural network for supervised learning of the functional behavior of time series data and its approximation, and b) a fuzzy system for modeling of the system behavior with possibilities to exploit expert information and for systematic optimization. Advantages and disadvantages of both concepts are described and emphasis is laid on the structural development of the fuzzy system. Both systems have been tested and satisfying results are shown with practical data.

1 Introduction

Events of flooded rivers represent a threat for adjoining regions. They can cause tremendous harms depending on their temporal progress. Within 13 months, in December 1993 and January 1995, there have been two extreme events of floods at river Mosel for example, with considerable harms for the surrounding region at middle and lower river section between Trier and Koblenz. In order to plan and timely decide about according measures for protection of people and buildings, it is necessary to have rather accurate and long term forecasting of water level progress. Therefore, e.g. a dense grid of measurement points for water levels, and precipitate has been established in the adjoining regions of river Mosel. Based on this data the progress of water levels has to be estimated during flood events. In practice, precise forecasting of water levels within next 6 hours is required for taking according precautions.

2 Problem and Objectives

Up to now, forecasting is mainly done by experts, whose estimates from actual data are typically carried out manually. They can consider the actual local geological, topological, and meteorological situation to some extend, however estimation is mainly based on their long term experience - and intuition sometimes. A formal description or direct transformation of their approach is hardly possible. On one side, a comprehensive mathematical model of the influencing variables and functional relations is not available till now and can hardly be developed due to the yet unfixed complexity of the hydrological system. On the other side, the lacking of knowledge about quantitative and temporal dependencies of the type and behavior of flood process in the catchment area results in the practical situation that only a small part of available data are being used for level forecasting.

An approach for short term level forecasts is described by the so called pre-level procedure. Here, the available data about levels in up-river sections are used to estimate the passage time and extend of the flood wave in down-river sections. In the following two soft computing approaches are described for a 6 hours forecast of two levels at river Mosel (level at Trier and Cochem, res.) using a neural network and a fuzzy system. For both cases, there are data of 14 previously occurred floods available to be used for building a system model. Up-river levels (at Trier and Perl) and levels of tributary rivers (at Kordel, Platten, Fremersdorf, Bollendorf, Prümzurlay, see Fig. 1) are provided on a one-hour time basis. Fig. 2 shows the progress of flood event in January 1995 with discharge of level Trier to be forecasted.

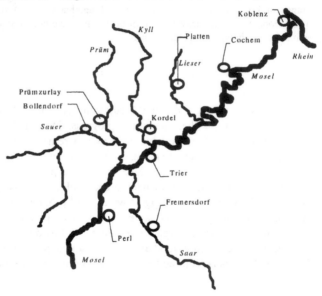

Fig. 1: River Mosel with tributaries

Main goal of the work carried out has been the investigation and development of systems based on Neural Networks and/or Fuzzy Logic for automatic forecasting of the water levels during flood events. Forecasts should estimate the progress of selected levels within next six hours. First, the forecasting problem was investigated using neural networks [sal]. This approach bears the advantage that neither special knowledge from experts nor knowledge about a system model has to be provided or explicitly to be included.

The system to be modeled can be looked upon as a time series consisting of sequences of patterns whose elements are standing in a temporal relationship. This

470

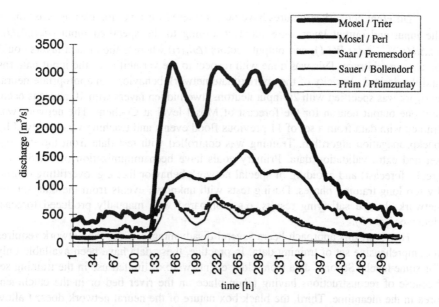

Fig. 2: Flood event January 1995
Discharge level Trier, up-river level Perl, and levels of tributaries Saar, Sauer, Prüm,
critical out-shore level at Trier is arrived at 1230 m³/s equaling a water level of 6,5m

relationship determines both the pattern and its position in the time series. Using a neural network for processing of such data the output of the network must depend on the actual input and previous inputs too. If one wants to use state-free feedforward neural networks, one has to provide specific input patterns showing the temporal dependencies to the network. In the given task a time window of size W is laid over the data series and shifted along the time axis during the training and recalling phase (see Fig. 3). Different schemes for input vectors were defined using actual discharge values $d(t)$ of river levels and derivatives (discrete differences) $\Delta d(t)$ of discharges in order to eliminate dependencies from absolute levels.

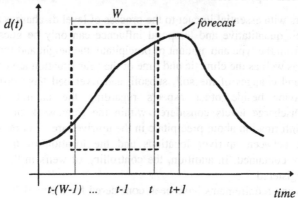

Fig. 3: Times series with input window of size W

Different network structures have be considered for level forecasting. Obviously the input and output layer were fixed according to the specified input $(d_i(t), \Delta d_i(t-1), \Delta d_i(t-2), ..., \Delta d_i(t-W+1))$ and output vectors $(\Delta d_o(t))$ whereas the hidden layers could be chosen arbitrary. This was done with respect to the variability of the input patterns and the desired plasticity of the approximate network behavior. An appropriate neural network was specified with 20 input neurons, two hidden layers with 10 neurons each, and one output neuron for the forecast of Mosel level at Cochem. The network was trained with data from a set of 11 previous flood events and teaching was achieved by backpropagation algorithm. Training was controlled with test data from the training set and extra validation data. Primary goals have been minimization of RMS error (real - forecast) and avoiding of special network behavior like e.g. overfitting caused by too long training phases. During tests with unknown events from the data set the network showed satisfying results, e.g. compared with manually produced forecast data (see Tab. 3).

However, this approach bears several disadvantages. First, the network requires a comprehensive set of training data. In practice, these data have been available only for some levels. Second, data from elder events were of limited use in the training set because of reconstructions having taken place on the river bed or in the catchment area in the meantime. Third, the black box nature of the neural network doesn't allow to extract acquired model information and therefore cannot be systematically improved with regard to the specific application. This means for example, that the system cannot work correctly in cases of extreme situations that have not be shown in the training data but do not injure the nature of the function approximation.

Alternatively, a Fuzzy Model for the forecasting of water levels (for Trier and Cochem) was investigated and developed [stu]. This approach allows the consideration of knowledge in the system description. In addition, this technique gives the opportunity to analyze directly the used system model and according input data in the development phase and also during system tuning. By this means, practical evaluations and conclusions about the actual problem situation can be achieved.

3 Fuzzy System Approach

Many parameters with essential impact to the process of level discharges are known in general, but their quantitative and temporal influence can only be guessed even by experts. Aside from the type and amount of precipitate, the height and topology of the catchment area as well as the climatic and time of the year situation are of importance. The condition and changes of the soil, subsoil, and river bed have influence to the discharge behavior beside other aspects regarding the natural and artificial environment. Discharge levels considered within the framework of this approach implicitly bear information about precipitate in the up-river area, but rain fallen in the catchment area between up-river locations and the location of the level to be forecasted is not contained. In addition, the controlling of weirs in this intermediate section is not registered.

The following requirements have been considered for the modeling of the fuzzy system. From the investigation of the problem and from interviews with experts it was not possible to derive specific hints for the partitioning of the input and output variables and for the defining of according membership functions. The time delay of waves between Trier and Cochem lies in the range of 11 and 17 hours according to informations from experts. A dependency between varying time delays Δt of flood wave and progress of flood at down-river levels is assumed. However, its effect is not clearly known for river Mosel. The time delay or passage time is dependent from the distance of the according level locations, the height of the discharge level, the shape of the flood waves, and meetings with waves from tributaries a.s.o.

A general description of the dependencies of the forecasted discharge (d_o) and its changes (Δd_o) during the next six hours are given by equation (1).

$$d_o(t) = d_o(t-6) + \Delta d_o(t)_{w=6} \tag{1}$$

The fuzzy description of discharge changes is given by (2):

$$\Delta d_o(t)_{w=6} = f(\Delta d_1(t - \Delta t_1)_{w=3,6,9}, \Delta d_2(t - \Delta t_2)_{w=3,6,9}, \cdots$$
$$\cdots, \Delta d_n(t - \Delta t_n)_{w=3,6,9}) \tag{2}$$

($w \in W$, input window; d_i, $i=1...,n$, input levels)

At this stage of problem analysis it could be stated that expert knowledge gained from many years of experience could only be used to a small extend. For this reason, the rule base will bear little pre-knowledge about the problem. A system model with approximate functional behavior based on previous events leads to limits in the extreme results the system can provide. Future events would be forecasted with increased error values in such cases.

3.1 Fuzzy System Design

A fuzzy model based on *Sugeno* approach [sug, tag] was chosen as an appropriate solution for the given problem. The selected approach allows a piece-wise approximation of non-linear functions. In a first step the input space can be partitioned into coarse intervals. Subspaces which yield insufficient results with respect to predefined quality measures can be subdivided further. The optimization of single subspaces by different weighting of input variables leads to distinct rule sets. This can be equated with an acquisition and adaptation of expert information.

The function approximation can be improved by choosing the available functional degree of freedom to a large extend. The forecast can be evaluated from a linear combination of input variables. Partitioning of the output space with membership functions can be omitted. The influence of input variables on the result can be extracted directly from the rule base and therefore alleviates considerably the optimization of parameters. Optimization can so be performed manually or automatically during the modeling phase. Input variables can be used in output terms without being contained in the premise. This leeds to a smaller rule base for the given problem space. Weighting of input variables in the conclusions allows direct formulation of their influences with respect to the output. This is advantageous compared e.g. with a system approach of *Mamdani* type.

The i-th control rule is of the form

$$R^i: \qquad x_1 \text{ is } A_1^i, \quad x_2 \text{ is } A_2^i, ..., x_n \text{ is } A_n^i$$

$$\rightarrow y^i = p_0^i + p_1^i x_1 + ... + p_n^i x_n \tag{3}$$

where the A_j^i are fuzzy variables and y^i is the output of the i-th control rule determined by a linear equation with coefficients p_j^i. The membership function of a fuzzy set A is simply written $A(x)$ and is composed of triangle functions. If the inputs $x_1, ..., x_n$ are given, the truth value w^i of the premise of the i-th rule is calculated as

$$w^i = \prod_{j=1}^{n} A_j^i(x_j) \tag{4}$$

and the output y is inferred from m rules by taking the weighted average of the y^i:

$$y = \sum_{i=1}^{m} w^i y^i \bigg/ \sum_{i=1}^{m} w^i \tag{5}$$

3.2 Structural Identification

The structural identification and optimization of the fuzzy system can become rather demanding following the approach after *Tagaki*. In practical applications at least a semi-automatic optimization process is required. For this reason a slightly changed strategy was chosen for the given task. This strategy is characterized mainly by the following steps.

1. Discharge values ($d(t)$) serve as input and output variables for the fuzzy system. They can be calculated from flood (or water) levels via transformation tables.
2. The set of available flood data (events) is devided into two sets: one for system modeling - selected regarding characteristic features - and one for system testing.
3. Input variables which probably can influence passing times Δt of waves are subdivided in membership functions. Since there was no support by expert knowledge a clustering algorithm was used. For this, the summit of waves of input levels and the output level can be used in the feature vector.
4. Independent models are then generated for all input variables determined according to 1. For each model a rule base is defined containing as much rules as the input variable possesses membership functions. The premises contain one membership function. The conclusions describe the changes of the discharge value $\Delta d(t)$ within 6 hours in advance to the estimated passage time of the wave.
5. The passage times Δt and the parameters p_r^s contained within the conclusions are optimized according to criteria (6) so that the total error between forecasted (Δd_{fore}) and real discharge change (Δd_{real}) yields a minimum.

$$E = \min\left\{\left(E_{abs}\right)_q\right\}; \qquad q = 1, .., N; \qquad N \in \mathbb{N} \tag{6}$$

N: number of parameter sets (Δt, p_r^s)

$$E_{abs} = \sum_{j=1}^{k} \sum_{i=1}^{n} \left| \Delta d_{real} - \Delta d_{fore} \right| \tag{7}$$

with k elements of modeling set and each with n hourly data

6. Input variables showing dependencies of the discharge value from passing time are collected in premises and then the maximum number of rules is built. The conclusion then contains all discharge changes of all input levels. The parameters than are optimized.
7. If there are discharge changes of several levels (river locations) in a conclusion the passing times are changed until a minimum error value is achieved. After every changing of a passing time an optimization of parameters is performed. These steps are repeated until the required forecasting accuracy is reached or changes don't result in essential improvements.
8. Further improvements of forecasting results can solely be achieved if new rules are inserted for regions with maximum error values. Again this step is repeated together with optimization steps as mentioned above until there is no further improvement.
9. The influence of the input variables to the passing times can be recognized within the rule base. Assumptions can be made, that these dependencies exist also for other (higher) discharge values. Based on this additional rules are built for discharge levels that are not available or yet displayed in the training set.
10. The discharge values are transformed to (flood or water) level data.

By this means, an interpretation of acquired knowledge is made possible relative simply. From the interpretation of the rule base new rules can be derived, which can describe extreme situations beyond events hitherto and can forecast such situations with higher accuracy.

4 Practical Results

Two systems have been investigated for flood level forecasting at river Mosel. A neural network was teached with training data from 11 flood events. A fuzzy system was modeled based on hourly recorded data of two different flood events. A forecast of the level at time $t + 6$ hours (so called 6-hours forecast) was defined as central task. The quality measure was defined by the sum of the absolute forecast error (compared to known level progress) beyond a critical level (so called out-shore-threshold). As minor requirement a measure was considered, which provides a better forecast than that produced manually or with the neural network.

Tab. 1 and Tab. 2 show result data of the fuzzy system and Tab. 3 lists according data for the neural network. One recognizes, that quality of forecasting is quite as good for the validation data as for the test data (shaded lines). The fuzzy system forecasted summit levels for flood in 1993 nearly accurately in spite of the fact, that extreme values (see max. flood levels) were not available in the modeling set (Fig. 4). A system comparison (level Cochem) shows, that the fuzzy system approach already was able to produce satisfying results only with two flood events in the modeling set. This system also has better behavior at critical phases of increasing flood levels. Fig. 5 shows manually produced forecasts and Fig. 6 show result of the fuzzy system for the same event.

Legend for Tables: *AAEO*: averaged absolute error between forecasted and real flood level above out-shore threshold, *MES*: maximum error between forecasted and real flood level at summit point, *MAEO*: maximum absolute error between forecasted and real flood level above out-shore threshold; flood events used in modeling data set are in shaded lines

Flood event	*AAEO*	*MES*	*MAEO*	max. flood level
January 1995	8	11	52	1033
December 1993	6	3	61	1128
January 1991	6	8	31	840

Tab. 1: Result flood level forecast using fuzzy model (level Trier, all values in cm)

Flood event	*AAEO*	*MES*	*MAEO*	max. flood level
January 1995	7	-6	45	947
December 1993	6	-24	34	1034
January 1993	6	-14	36	840
January 1991	7	-4	39	750
February 1990	6	-8	31	802

Tab. 2: Result flood level forecast using fuzzy model (level Cochem, all values in cm)

Flood event	*AAEO*	*MES*	*MAEO*	max. flood level
January 1993	10	-18	35	840
February 1990	10	-3	67	802

Tab. 3: Result flood level forecast using neural network (level Cochem, all values in cm)

5 Conclusion

The two systems designed for flood level forecasting at river Mosel proved their capabilities for modeling complex system environments. The neural network delivered acceptable results for the approximation of times series functions without the need of specific knowledge about the problem. It is dependent on and also limited to sufficient training data. In comparison the fuzzy system displayed very promising features. The system achieved better results even with only few data sets for system modeling.

Without pre-knowledge about e.g. hydrological dependencies a rule base was established by structural system optimization. The rule base was subsequently confirmed by interpretation through experts. A special forecast problem could be modeled and transformed into a simple mathematical description. From the rule base dependencies for extreme situations could be derived and described by which yet unknown situations could be forecasted. The analysis of the system structure delivered detailed information about influences of system variables which have been assumed but could not be verified by the experts before.

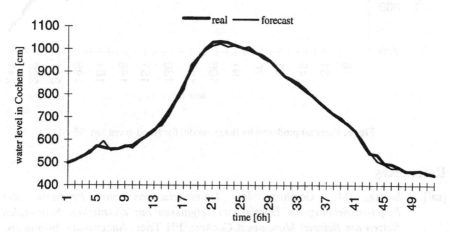

Fig. 4: Flood event Dec. 93 with summit level exceeding modeling data set

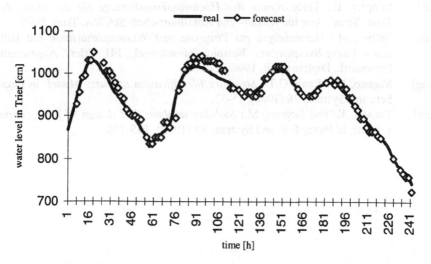

Fig. 5: Forecast produced by experts for flood event Jan. 95[sti]

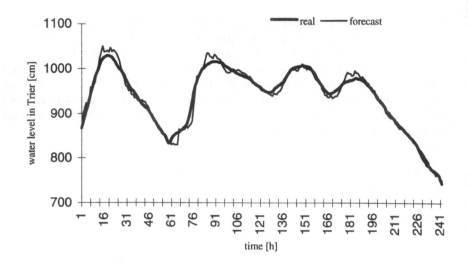

Fig. 6: Forecast produced by fuzzy model for flood event Jan. 95

References

[sal]   Salzig, M., Gemmar, P.: *Untersuchungen zur Prognose der Pegelentwicklung bei Hochwasserereignissen mit künstlichen Neuronalen Netzen am Beispiel Moselpegel Cochem,* FH Trier, Angewandte Informatik, Bericht 9602, 1996.

[sch]   Schmitt, K.-H.: *Der Fluß und sein Einzugsgebiet,* Franz Steiner Verlag, 1984

[sti]   Stippler, E.: *Verbesserung der Hochwasservorhersage für die Mosel bei Trier,* Staatl. Amt für Wasser- und Abfallwirtschaft StAWA, Trier, 1995

[stu]   Stüber, M.: Datenanalyse zur Prognose und Wissensakquisition mit Hilfe eines Fuzzy-Systems am Beispiel Moselpegel, FH Trier, Angewandte Informatik, Diplomarbeit, 1996.

[sug]   Sugeno, M.; Kang, G.T.: *Structure Identification of Fuzzy Model*; in Fuzzy Sets and Systems 28 (1988), 15-33.

[tag]   Tanaka, K. and Sugeno, M.: *Stability analysis and design of fuzzy control systems*; in Fuzzy Sets and Systems 45 (1992), 135-156.

Identification of Waste Water Treatment Plant Using Neural Networks

I. I. Voutchkov, K. D. Velev

Department of Automation, Higher Institute of Chemical Technology,
Sofia 1756, Bulgaria

Abstract: Identification of non-linear processes has always been a problem, as long as the mathematical model structure hardly can be known in advance. In the present paper, conventional Recursive Least Squares estimation is compared with Neural Network identification. Being more flexible and undependable on the model structure, this approach can approximate large variety of relationships. Both strategies are compared identifying a Waste Water Treatment Plant, which posses very strong non-linear properties.

1. Introduction

Mathematical modelling and identification are an essential part of the investigation of real world processes and have been object of considerable attention last few decades. Using mathematical description of any considered process, it can get studied, improved and controlled.

Computer hardware and facilities, mathematical methods and strategies, which are the most common used identification tools, have become quite powerful and sophisticated. Nevertheless there is much yet to be desired and developed.

An identification approach have to posses the following features to be used as a reliable strategy:

- Adequate relationship approximation. The prediction error should not exceed the standard deviation value of the output noise.
- Fast prediction error reduction, i.e. fast convergence.
- Reliable numerical method.
- Suitable for as much types (linear/non-linear, with/without deadtime, noise corrupted, noiseless, etc.) of processes as possible.

Recursive Least Squares (RLS), Maximum Likelihood, Instrumental Variable are well known and studied identification methods, and are often used in theory and in practice. However they do not match all the requirements mentioned above. As long as all of them are model based, they could not cope processes with completely unknown model structure, such as most of the real-life relationships are. The number of non-linearities is infinite and the model types suggested in those cases, such as Hammerstein, Volterra, Wiener, [1, 2], etc. can not always give reasonable approximation.

Neural networks appeared to be very powerful tool when applied to any kind of relationships. Moving away from the exact structures and definitions, they can be suitable for a large range of relationships. In this paper, the neural networks' application to a Waste Water Treatment Plant, as identification approach will be shown.

2. Waste Water Treatment Plant Description

A waste water treatment process generally consists of two stages: primary and secondary treatment. Since the primary treatment includes only rough and mechanical purification, the secondary water treatment is usually the final and most important stage of a waste water treatment. Its aim is to reduce the oxygen demand of an influent waste water (S_0) to a given level of purification (S).

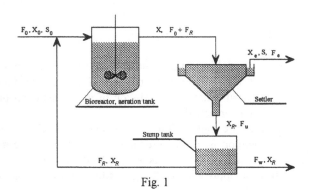

Fig. 1

A schematic diagram of the plant to be controlled is shown in Fig. 1. Polluted water is continuously fed with flow rate F_0, biomass X_0 (which can be neglected $X_0 = 0$), and concentration of Biochemical Oxygen Demand (substrate) S_0. The stirrer in the aeration tank promotes the aeration process and in that it is often combined with a surface aerator which contacts the mixed liquor with the ariel oxygen. Biomass X in the aeration tank is mixed with the incoming sewage, which then grows using it as a food and then being separated in the settler. The sludge is heavier than the water and settles down, afterwards being removed from the settler with flow rate F_u and biomass X_R. Part of it (F_R) is recycled and the rest (F_w) is provided for final sludge treatment. In practice wasted sludge is dried and if it does not contain heavy metal components it can be used as fertiliser.

The liquid which is on the top in the settler is substantially clean water, it overflows into special channels and is removed from the system with substrate concentration S, flow rate F_e and biomass X_e which must be close to zero.

In this analysis all the biological reaction is considered to be taking place in the aeration tank whilst the settler - thickener (secondary settler) is considered to be a purely physical separation device. This is a reasonable simplification since even though the microbial floc is in contact with some substrate the low oxygen concentration prevents much growth taking place.

3. Using Neural Networks as Identification Tool

Analysis of the literature on wastewater treatment [3, 4] shows that the relationship between the concentration of substrate S (denoted by y), the recycle flow rate F_R (denoted by u_1) and the sludge flow rate F_u (denoted by u_2) is significantly nonlinear.

However some attempt to identify the plant using linear and nonlinear ARX type model was made. As an identification input signal input signal a randomly generated, 3-level sequence was used. The data were down loaded from a real water treatment plant's daily journal. Plots of the observed and predicted outputs are shown in Fig. 2

As it can be seen, the ARX model gives a very bad accuracy so that some more flexible and advanced identification tool for modelling this obviously non-linear process, needed. As such appeared to be Neural Networks strategy.

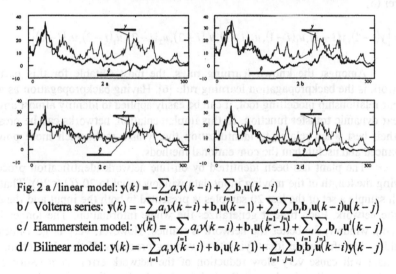

Fig. 2 a / linear model: $y(k) = -\sum_{i=1} a_i y(k-i) + \sum_{i=1} b_i u(k-i)$

b/ Volterra series: $y(k) = -\sum a_i y(k-i) + b_1 u(k-1) + \sum\sum b_i b_j u(k-i)u(k-j)$

c/ Hammerstein model: $y(k) = -\sum a_i y(k-i) + b_1 u(k-1) + \sum\sum b_{i,j} u^i(k-j)$

d/ Bilinear model: $y(k) = -\sum_{i=1} a_i y(k-i) + b_1 u(k-1) + \sum_{i=1}\sum_{j=1} b_i b_j u(k-i)y(k-j)$

Having in mind the non-linear nature of the process, a non-linear structure of the network has been chosen (Fig. 3). It is a three layer network of type $N^3_{8,13,6,1}$, ($R = 8$; $S1 = 13$; $S2 = 6$; $S3 = 1$) having two sigmoid and one linear layers.

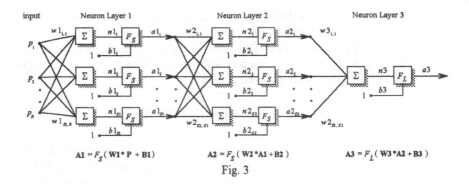

$$A1 = F_S(W1*P + B1) \qquad A2 = F_S(W2*A1 + B2) \qquad A3 = F_L(W3*A2 + B3)$$

Fig. 3

This structure has been chosen after trying with a $N^2_{8,S1,1}$ network. However to reduce the network error to a satisfactory level 72 neurons ($S1$ = 72) were needed, which was incredibly time consuming. Smaller value for $S1$ was not enough, and the network error reached a level, after which it couldn't be reduced any more (the phenomenon is described as "underfitting" in [5]). After the extension to a three layer network, the time for training dropped rapidly, and the level reached a satisfactory error value of 10^{-3}. As network inputs past values for y, u_1 and u_2 were used, as written in (1). The network output was the present value of the treated water (S)

$$\mathbf{P} = \left[y(i-1), y(i-2), u_1(i-1), u_2(i-1), u_1(i-2), u_2(i-2), u_1(i-3), u_2(i-3) \right] \qquad (1)$$

Amongst the known learning rules, the most suitable for this particular network is the backpropagation learning rule [6]. Having backpropagation as a non-linear relationship modelling tool, it can be easily applied to identify almost any non-linear dynamic transfer function. In this implementation, networks can be presented in their best when related to identification theory as a tool, much more powerful, advanced and useful than the conventional methods.

The plant has been identified by on-line network identification procedure, having the length of the moving input data window equal to 20 (Q = 20). That is on each sample a set of the last 20 samples is presented to both the input and the output of the network. Each time it generalizes the whole information. The longer is that moving window, the better the network is trained. However, very large values of Q could be harmful, when identifying time varying processes. If the plant changes the old data will cause very slow reduction of the network error. A network can be considered as well trained when the training epochs (epoch corresponds to iteration in the training procedure needed to reduce the level under a pre-defined level [5]) drop to zero, i.e. the network has learned the process and no more training is needed.

Training stage is shown in Fig. 4. As plant input a randomly generated, 3-level, normaly distributed sequence (RGS) was used, so that

$$F_R = 12.5 + RGS[l/h]$$
$$F_U = 13.5 + RGS[l/h]$$

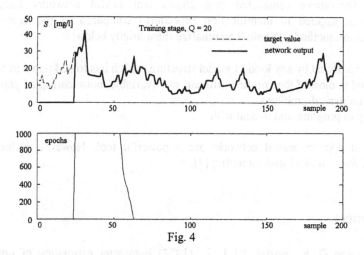

Fig. 4

The network was trained for 48 samples, after which, the epochs needed for reducing the error dropped to zero (at sample 62). At this moment the network can be considered as trained. Its testing is shown in Fig. 5, using newly generated RGS sequence. The result is much better, compared to the models used in Fig. 2. The error plot in Fig. 5 confirms that the so trained neural network can be used as an adequate description for the considered waste water treatment plant. Comparing the error level to the one observed from the ARX description, it can be seen, that Neural Networks approach is about 1000 times more accurate.

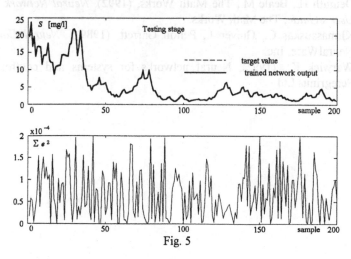

Fig. 5

4. Conclusion

The above considered case shows that neural networks have some advantages, applied to identification procedures compared to the conventional identification methods. We shall summarize them briefly below:

- Not restricted by any kind of model structure, which has to be known in advance
- Fitted to most of the non-linear/linear, multivariable, noise corrupted plants
- No numerical problems
- Easy to program and to deal with

It appears neural networks are a powerful tool. However, it has to be handled with care and understanding [7].

References

1.	Grag D. P., Boziuk III J. S., (1972) Parameter estimation of non-linear dynamical systems, *Int. J. Control*, vol. 15, No. 6, *pp 1121 - 1127*
2.	McCabe S., Davies P., Siedel D., (1991) On the use of nonlinear autoregressive moving average models for simulation and system identification, *ACC*, v.1, Boston
3.	Sundstorm W. D., Klei H. E., (1979) *Wastewater treatment*, Prentice-Hall, Inc.
4.	Horan J. N., (1990) *Biological wastewater treatment systems. Theory and Operation*, John Wiley & Sons
5.	Demuth H., Beale M., The Math Works, (1992), *Neural Network Toolbox User's Guide*, The Math Works.
6.	Klimasauskas C., Guiver J., Pelton Garrett, (1989), *Neural Computing*, NeuralWare, Inc.
7.	Warwick K., (1992), Neural networks for systems and control, Peter Peregrinus Ltd.

Fuzzy Logic Based Control of Switched Reluctance Motor to Reduce Torque Ripple

Adnan Derdiyok[*]   Nihat İnanç[*]   Veysel Özbulur[**]   Halit Pastacı[***]   M. Oruç Bilgiç[**]

* Sakarya University Department of Electronics Engineering Adapazarı , TURKEY.
** Tübitak MRC CAD/CAM Robotics Department Gebze Kocaeli ,TURKEY.
*** Yıldız Technical University Department of Electronics Engineering Istanbul, TURKEY.

Abstract : This paper presents torque ripple reduction of Switched Reluctance Motor (SRM) by using fuzzy speed controller. In this purpose, current control is applied with fuzzy speed controller. Fuzzy controller adjusts value of reference current to keep speed constant. In this study, the nonlinear model of SRM is used and the motor used in simulation is a 8/6 SRM with C-Dump converter.

1. Introduction

In recent years, high interest has been shown on the use of switched reluctance motors (SRM) in many application areas because of simple motor construction and unidirectional power inverter supply. It has no brushes and no rotor windings and its stator has simpler concentrated windings. The motor windings are in series with each main power switching device. The motor is operated with controlled current and controlled conduction angles. In SRM, the motor torque is proportional to the square of the energized winding current, the inverter needs not to feed bidirectional currents. Hence , the scheme of inverters feeding the SRM is unipolar type where one switch is generally used per phase. The most popular drive circuit for SRM is the C-Dump switching circuit. The switching sequence of the inverter is controlled by rotor position signal provided from a position sensor mounted on motor shaft. Because of advantages indicated above SR Motors have been preferred instead of DC motor and inverter driven induction motor in many variable speed application areas. However, excessive torque ripple, especially at low speeds is still one of the important reasons for SRM not to be preferred in some variable speed drive market.

Several efforts to reduce the torque ripple of SRM are presented in literature. In [1,2] phase current shape is modulated to counteract the torque ripple. A balanced commutator which works for accurate current tracking to reduce torque pulsations is reported in [3]. The method described in [4] optimizes the commutation angle for torque ripple reduction. Torque ripple minimization method given in [5] is based on the estimation of the instantaneous SRM torque from the flux linkages versus current and rotor position characteristic curve via bi-cubic spline interpolation. The application of a variable structure system theory to the speed control of SRM given in [6] results that this control is also effective in reducing the torque ripple of the motor, compensating for the nonlinear torque characteristics.

In this paper, the minimization of torque ripple are presented. In this purpose, current control is applied with fuzzy controller which is used to arrange reference current to keep speed of motor constant. The used fuzzy controller generates resetting rate and incremental control input fuzzily from error (e) and change of error rate (de).

2. The Mathematical Model

The Switched Reluctance Motor used in the simulation is a 8/6 SRM (four phases) with C-Dump converter. The crossection of the SRM is shown in figure 1. The state equations of the system are given below. Here the saturation model of SRM is used and mutual inductances are neglected.

System equations

$$d\lambda n / dt = V_n - R.I_n \tag{1}$$
$$dV_c / dt = (-I_g.u_g + I) / C_d \tag{2}$$
$$dI_g / dt = (V_g - R_g.I_g) / L_g \tag{3}$$
$$d\omega / dt = (T_e - T_L - B.\omega) / J \tag{4}$$
$$d\theta / dt = \omega \tag{5}$$

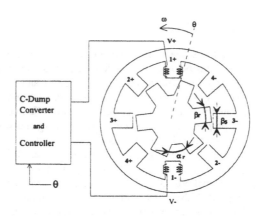

Figure 1. 8/6 crossection of SRM

Motor data :

$\beta_r = \beta_s = 0.4075$ rad, $\alpha_r = 0.6397$ rad,
Nominal power................ : 4 kW / 5.5 Hp,   Speed............................. : 1500 rpm
Resistance per phase........ : 0.45 Ω,   Nominal torque............... : 25 Nm
Viscous friction coef. (B): 0.0008 Nm/(rad/s),   Inertia (j)...........: 0.0053 Nm/(rad/s^2)

In general, the electrical torque for nonlinear system is defined as a function of co-energy (its derivative versus the rotor position) :

$$T_e(\theta,i) = \partial W'(\theta,i) / \partial\theta \tag{6}$$
$$W'(\theta,i) = \int_0^i \lambda(\theta,i).di \tag{7}$$

and the analytical equation which is used to summarize the SRM magnetization characteristics ($\lambda(\theta,i)$) is defined in current-flux linkage plane as follows

$$\lambda(\theta,i) = a(\theta)(1 - e^{p(\theta)i}) \tag{8}$$

$$a(\theta) = \sum_{m=0}^{5} a_m.\cos(6.m.\theta) \tag{9}$$

$$p(\theta) = \sum_{m=0}^{5} p_m.\cos(6.m.\theta) \tag{10}$$

The incremental inductance and phase current of motor are given in equation (11) and (12).

$$L_{inc}(\theta,i) = -a(\theta).p(\theta)e^{p(\theta)i} \tag{11}$$
$$i = Ln(1 - \lambda(\theta,i)/a(\theta))/p(\theta) \tag{12}$$

The equations can be written for each phase by shifting mechanical angle (θ) 15°. The inductances profiles of phases are shown in figure 2.

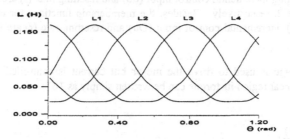

Figure 2. Inductance profiles for the SRM

3. Fuzzy Controller

Fuzzy control systems have been successfully applied to a wide variety of practical problems [7,8,9]. It has been shown that these controllers may perform better than conventional model-based controllers, specially when applied to processes difficult to model, with nonlinearities, and with uncertainties. The fuzzy control is basically nonlinear and adaptive in nature, giving robust performance under parameter variation and load disturbance effect.

A typical fuzzy control describes the relationship between the change of the control Du(k)= u(k)-u(k-1) on one hand, and the error e(k) and its change De(k)= e(k)-e(k-1) on the other hand. Such a control law can be formalized as:

$$Du(k)=F(e(k),De(k)) \tag{13}$$

The actual output of the controller u(k) is obtained from the previous values of control u(k-1) that is updated by Du(k) :

$$u(k) = u(k-1) + Du(k) \tag{14}$$

This type of fuzzy controller is known as fuzzy PI according to the relation between variables e(k) and De(k) on one hand and Du(k) on the other hand. The difference is in the type of relationship. In the case of the PI controller this relationship is linear, while in fuzzy PI it is nonlinear in general [7]. The PI controller (also fuzzy PI) is, however, known to give poor performance in transient response due to the internal integrating operation [10].

In servo motor applications the fast response of the drive is desired [11,12]. That's why the performance of fuzzy PI should be improved to give satisfactory rise time and minimum overshoot in step response. Here, the fuzzy controller system used is similar to that described by Lee [10], and it is a modified type of fuzzy PI controller. It evaluates incremental control input (Du) and, in addition to conventional fuzzy PI, it also evaluates resetting rate (r) of control input applied to the system by error and rate of error change.

The control input is calculated by the following equation :

$$u\ (k+1) = (\ 1 - sqrt(r(k)))^* \ u\ (k) + Du\ (k) \tag{15}$$

The fuzzy rules for resetting rate are constructed to damp overshoot in response by resetting accumulated control input and to make response faster under large incremental control input. Rules for calculating incremental control input (du) and resetting rate (r) are shown in Table 1. and in Table 2. respectively. Besides, the membership functions for error (e), rate of error change (de), incremental control input (Du) and resetting rate (r) are shown in Fig. 3 (a) - (b).

In practice, voltage is used to drive the motor but current is controlled. If the SRM is operated in the linear region its torque can be written simply as [6] :

$$Te = \frac{1}{2} \frac{\partial L_n(\theta)}{\partial \theta} i_n^2 \tag{16}$$

The developed torque is proportional to square of phase currents and the slope of phase inductance. If current of the motor is controlled successfully, torque ripple will be minimized. Here, the fuzzy controller is implemented to arrange the reference current to keep the speed of the motor constant and therefore, to develop ripple free torque . The fuzzy block has two inputs : the error (e) and rate of error change (de) and control output is du.

Table 1. Fuzzy control rules for calculating Du.

$\frac{e}{de}$	NB	NM	NS	ZE	PS	PM	PB
NB	NB	NB	NB	NM	ZE	ZE	ZE
NM	NB	NB	NM	NM	NS	ZE	ZE
NS	NB	NM	NS	NS	ZE	PS	PM
ZE	NB	NM	NS	ZE	PS	PM	PB
PS	NB	NS	ZE	PS	PS	PM	PB
PM	ZE	ZE	PS	PM	PM	PM	PB
PB	ZE	PS	PS	PB	PB	PB	PB

Table 2. Rules for determining resetting rate.

$\frac{e}{de}$	NB	NM	NS	ZE	PS	PM	PB
NB	ZR	SR	BR	AR	BR	SR	ZR
NM	ZR	ZR	MR	VB	MR	ZR	ZR
NS	ZR	ZR	LR	BR	LR	ZR	ZR
ZE	ZR	ZR	ZR	ZR	ZR	ZR	ZR
PS	ZR	ZR	LR	BR	LR	ZR	ZR
PM	ZR	ZR	MR	VB	MR	ZR	ZR
PB	ZR	SR	BR	AR	BR	SR	ZR

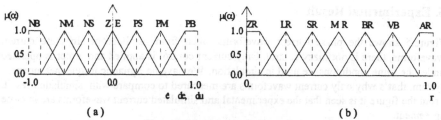

Figure 3. Membership functions for
(a) error e, error rate de, incremental control input du, (b) resetting rate r .

4. Simulation

Usually, the torque control of SRM is made by controlling sum of square of phase currents, which requires generally four current sensors for a 8/6 four phase SRM and increases cost and complexity of measurement circuit. In practice, voltage source is used to drive the motor and current is controlled. This can be done by controlling the total current as

$$i_{ref} = i_1 + i_2 + i_3 + i_4 \qquad\qquad (17)$$

with only one current transducer. The torque variation and corresponding currents which were obtained through simulation are given in Figure.4, there is no fuzzy control.

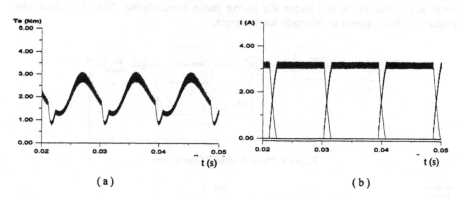

Figure 4. Current control without fuzzy control
(a) Torque variation, (b) Phase currents (i_4, i_1, i_2)

The simulation is achieved with C-Language by using the mathematical model given in equations (1) through (12). Runge-Kutta method is chosen to solve the state equations. Some data for simulation are as follows,

Source voltage...................: V_k = 300 V   C-Dump voltage....................: V_c = 600 V
Phase voltage width...........: θ_{on}= 15°   C-Dump switching frequency.: f_c = 10 kHz
Reference speed.................: ω = 30 rad/s   Reference current....................: I_{ref}= 3.2 A
Load Torque ....: T_L = 2 Nm

5. Experimental Result

Figure.5 shows the phase current waveforms taken from oscilloscope. In these current waveforms, there is not fuzzy control. Only current control is applied and motor is operated under the same conditions we used in simulation. We have no dynamic torque measurement system, that's why only current waveforms are measured to compare with simulation result. From the figure it is seen that the experimental and simulated current waveforms are in close agreement.

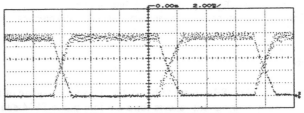

vertical axis 0.7A/div., horizontal axis 2ms/div
Figure 5. Experimental currents waveforms without fuzzy control

6. Proposed Method

In figure 6 the block diagram of system is shown. When fuzzy control is applied in speed loop of the system, the torque ripples and variations are more suppressed as shown in Figure.7. The results show the robustness of fuzzy logic controller by removing low frequency torque ripple and torque dip during phase commutation. The fuzzy controller modulates phase current to minimize torque ripple.

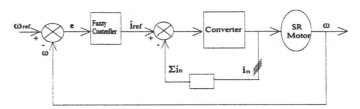

Figure 6. Block diagram of the system

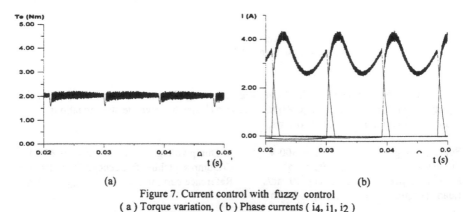

(a)   (b)

Figure 7. Current control with fuzzy control
(a) Torque variation, (b) Phase currents (i_4, i_1, i_2)

7. Conclusion

The Switched Reluctance Motor (SRM) exhibits quite nonlinear characteristics. It is difficult for the classical control algorithm to achieve the desired dynamic performances of the nonlinear SRM. The simulation results show that the fuzzy algorithm has good performance to reduce low frequency torque ripple and torque dip.. Usually, the torque control is made by controlling sum of square of phase currents, which requires generally four current sensors and increases complexity of measurement circuit. The advantage of this method is to use only one sensor to measure sum of the phase currents. However, only knowledge of total current, which needs only one sensor to measure, is enough for fuzzy controller to give satisfied results.

References

1. EGAN, M.G. ,Murphy J.M.D., Kenneally P.F., Lawton J.V. " A High Performance Variable Reluctance Drive Achieving Servomotor Control " Proc. of Motor Con. 1985.
2. LECHENADEC., M. GEOFFORY., B. MOULTON., J. MOUCHOUX.," Torque Ripple Minimisation in Switched Reluctance Motors by Optimisation of Current Waveforms and Tooth Shape with Copper Losses and V.A. Silicon Constrains", Proc. of Int.Conference on Electrical Machines. ICEM294 pp.559, Paris 1994.
3. S. WALLACE., D.G. TAYLOR., "A Balanced Commutator for Switched Reluctance Motor to Reduce Torque Ripple" IEEE Trans. Power Electronics. Vol. 7, No.4, pp.617 October 1992.
4. SCHRAM D.S., WILLIAMS B.W., GREEN T.C. , " Torque Ripple Reduction of Switched Reluctance Motor by Phase current Optimal Profiling ", IEEE Power Electronics Specialists. Conference Record, s 857-860, 1992.
5. MOREIRA, J.C., " Torque Ripple Minimization in Switched Reluctance Motors Via Bi-cubic Spline Interpolation ", IEEE Power Electronics Specialist.
6. BUJA G.S., MENIS, R., VALLA, M.I., " Variable Structure Control of SRM Drive " , IEEE Transactions on Industrial Electronics, Vol. 40, No:1, s. 56-63, February, 1993.
7. KANDEL, A., LANGHOLZ, G., Fuzzy Control System, CRC Press Inc. 1994.
8. KUNG Y.S. , LIAW, C.M. "A Fuzzy Controller Improving a Linear Model Following Controller for Motor Drives " IEEE Trans. on Fuzzy Systems, Vol.2, No.3, Aug. 1994.
9. SOUSA G.C.D. , BOSE, B.K. " A Fuzzy Set Theory Based Control of Phase-Controlled Converter DC Machine Drive " , IEEE Trans. on Industry App. Vol. 30, No.1, January/ February 1994.
10. Lee, J., "On Methods for Improving Performance of PI-Type Fuzzy Logic Controllers.", IEEE Trans. on Fuzzy Sys. vol. 1, No.4, Nov. 1993
11. TSENG.H.C. , HWANG V.H. " Servocontroller Tuning with Fuzzy Logic " IEEE Transactions on Control Systems Tecth. Vol.1. No.4, December 1993.
12. LI, Y.F., LAU, C.C., " Development of Fuzzy Algorithms for Servo Systems ", IEEE Control System Magazine , pp.66 ,April 1989.

Nomenclature

λ_n flux of nth phase, Wb

L_n phase inductance, H

t   time, s

V_n phase voltage of nth phase, V

R_n resistance of nth phase, W

I_n phase current of nth phase, A

V_c C-Dump capacitor voltage, V

I_g inductor current in C-Dump circuit, A

u_g C-Dump converter control signal

i   sum of energy recovery diodes currents, A

C_d C-Dump capacitor, F

V_g voltage across the inductor in C-Dump circuit, V

L_g inductor in C-Dump circuit, H

R_g resistance of L_g, W

w   angular speed, rad/s

q   rotor position, rad

T_e electrical torque, Nm

T_L load torque, Nm

B   coefficition viscous friction, Nm s/rad

J   moment of inertia, kg m^2

Fuzzy Logic Control of High Voltage DC Transmission System

Ercan ÖLÇER[1], Bülent KARAGÖZ[2], Hasan DİNÇER[2]
Engin ÖZDEMİR[3], Ercüment KARAKAŞ[3]

[1] Department of Computer Science Engineering, University of Kocaeli, 41100, Izmit, TURKEY
[2] Department of Electronics & Com. Engineering, University of Kocaeli, 41100, Izmit, TURKEY
[3] Department of Electrical Education, University of Kocaeli, 41100, Izmit, TURKEY
e-mail: curgu-ed@yunus.mam.gov.tr

Abstract. This paper introduces a fuzzy controller of the rectifier side in high voltage direct current (HVDC) system. Especially, rectifier side of the typical point-to-point system has been taken consideration.. In the rectifier, thyristors are used as control devices that can handle high voltage and power. Turn on, turn off, triggering and protections of these devices have lots of problem and are very sensitive. Especially, in changing voltage and frequency, these problems are appeared itself. Therefore, dc current of the rectifier output becomes unstable. Also, device dissipation, transmission losses and effects of environments cause increasing of these problems. In this study, a fuzzy controller is designed to solve these problems. Finally, a comparative study has been performed with and without fuzzy control.

1 Introduction

The economic constraints in many parts of the world could be improved by access to surplus electric power generated elsewhere. One possibility for that is high-voltage dc transmission. As it is realized that this technology is often economical way to interconnect certain power systems situated in different regions or countries or across the sea or that use different frequencies as well as to transmit power over long distances (Figure 1) [1].

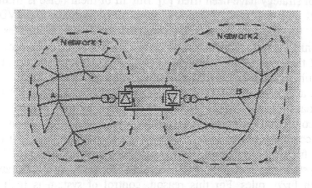

Fig. 1. Interconnection two network by HVDC system

In High Voltage Direct Current (HVDC) energy transmission systems, thyristors are used as control devices that can handle high voltage and power (figure 2). Turn on, turn off, triggering and protections of these devices have lots of problem and are very sensitive. Especially, in changing voltage and frequency, these problems are appeared itself. Device dissipation, transmission losses and effects of environments cause increasing of these problems.

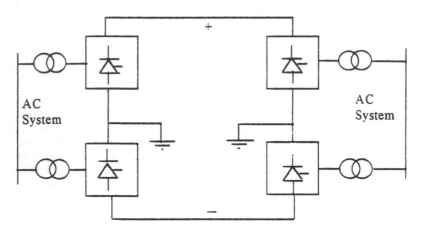

Fig. 2. Bipolar HVDC link.

Under various operating system, the highly nonlinear nature of the HVDC control loops require careful selection of control constants that will accommodate a range of operating conditions. With fixed parameters, the control fails under abnormal circumstances. With large disturbances, performance of adaptive control may degrade the performance [2],[3].

Robust coordinated control scheme has been proposed [4]. Gain scheduling adaptive control strategy have been tried [5]. But all of them noise is limited and uses conventional PI type controller. Fuzzy self tuning PI controller for HVDC links has been provided [6]. Design of a very simple fuzzy logic controller for HVDC transmission links for fast stabilization of transient oscillations. was presented [7] On the other hand these realized as simulation studies and phase sliding and frequency variation of 3-phase input AC signal are rather limited. In this study, to circumvent above problem, a fuzzy control is realized on a prototype circuit of the HVDC system.

Nowadays. practical application of fuzzy logic control increase rapidly. Such types of fuzzy controllers have proven themselves successful in nonlinear plants whose detailed and accurate mathematical descriptions are not available. Particularly. the system, used fuzzy logic control, are complex and nonlinear. HVDC energy transmission systems have second order nonlinear functions, so it is suitable to control the system with fuzzy rules. For this reason. control of system is realized by fuzzy linguistic rules.

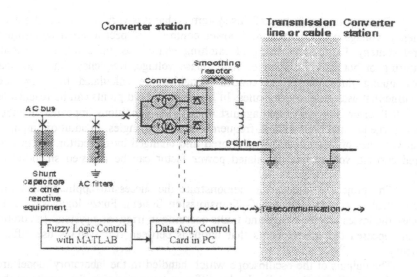

Fig. 3. Fuzzy control block diagram of converter station

2 Realized HVDC System Prototype

In this study, it is realized controlling of a prototype of a HVDC energy transmission system that contains a 25 kVA converter and 25 kVA inverter (Figure 2). A Data Acquisition Card (DAC) is used for information between PC and power electronic devices. A software program was written in C++ language which can send and receive command and information to MATLAB program for control of this card. The current and voltage control of the converter output and frequency control of inverter output are provided by calculating of input and output data in fuzzy logic control subroutine via DAC (figure 3)

Each phase of three phase network, that come to converter, is sampled and sends to PC by DAC. During this sampling period, interrupts that come from circuit are used in fuzzy logic subroutine and suitable triggering signals are send to appropriate thyristors via the optic isolator devices. In this system, instead of zero crossing point, phase crossing points are caught for controlling of power electronic devices. Thus, it is prevented the missing of the turn on and turn off point of thyristors in excessive positions such as decrease of voltage width in network and deviation of frequency.

In some cases, even when the frequency control strategies of two ac systems are the same, an ac interconnection between them will be difficult to establish if their centers of generation are electrically too far apart (high impedance). The phase angle between two remote ac interconnection points tends to fluctuate too much, so that current through an ac interconnection would also fluctuate to dangerously high levels. Fuzzy is used to control amplitude and frequency of line voltage, so fluctuation in line voltage can be limited.

The most important part of this system is that it is very sensitive and quick in frequency, voltage width, current and power control in respect to other conventional control strategy. Particularly, using of catching phase crossing point, it is possible monitoring of not only variation of each phase voltage, frequency; but also phase angles among three phases, so control variables are calculated by fuzzy logic subroutine that uses fuzzy control rules. In this way, desired points can be reached less than $\pm 0.1^0$ error. It is prevented against unexpected conditions appeared in inputs, such as surge low and high voltage, frequency and phase angles. In addition, input and output variables of system can be monitored and changed by the editor. By this way desired current, voltage and calculated power factor can be adjusted smoothly and easily.

The project is intended to demonstrate the successful application of fuzzy logic control to a High Voltage DC Transmission System. Fuzzy logic was used to decrease the losses in line voltage and faults accruing in triggering system. In addition, it is disappeared the control faults that happens between two systems use different frequency.

The outputs of the oscilloscope which handled in the laboratory model are given as follows (Figure 4 and 5). In the different trigger angles, the results of the stabilized and unstabilized network have been taken successfully. Especially, the filtered outputs are in the near of the required values. These results show that even if phase and voltages of 3-phase AC signal are variable, the trigger of the converter will be adaptive and not be false because of monitoring crossing points of all input phases.

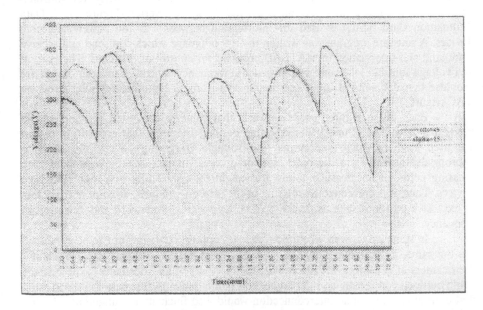

Fig. 4. The outputs of the converter for alpha=15 and alpha=45 in unstabilized AC network with resistance load R=39 ohm

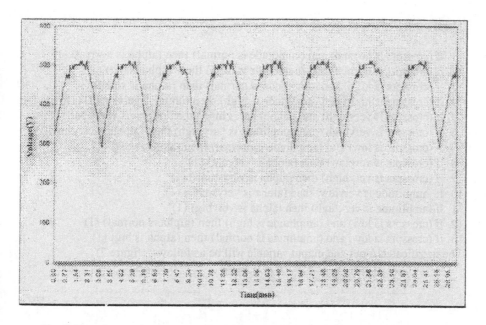

Fig. 5. The outputs of the converter for alpha=0 in stabilized AC network R=39 ohm

3 Fuzzy Control of The System

The fuzzy control system is described as two inputs-one outputs(figure 6)

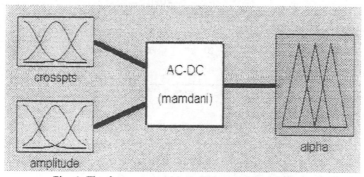

Fig. 6. The fuzzy control block diagram of the converter

On the other hand input variable given as amplitude is feedback that taken from output. The output given as alpha is the value of angle difference that required trigger signal. In the system, Mamdani model is used writing 13 rules of the fuzzy logic.

These rules :

1. If (crosspts is normal) and (amplitude is normal) then (alpha is normal) (1)
2. If (crosspts is normal) and (amplitude is high) then (alpha is big) (1)
3. If (crosspts is high) and (amplitude is normal) then (alpha is small) (1)
4. If (crosspts is high) and (amplitude is high) then (alpha is verysmall) (1)
5. If (crosspts is veryhigh) and (amplitude is high) then (alpha is verysmall) (1)
6. If (crosspts is veryhigh) and (amplitude is veryhigh) then (alpha is verysmall) (1)
7. If (crosspts is low) and (amplitude is normal) then (alpha is big) (1)
8. If (crosspts is verylow) then (alpha is verybig) (1)
9. If (crosspts is veryhigh) then (alpha is verysmall) (1)
10. If (amplitude is verylow) then (alpha is verysmall) (1)
11. If (amplitude is veryhigh) then (alpha is verybig) (1)
12. If (crosspts is low) and (amplitude is high) then (alpha is normal) (1)
13. If (crosspts is low) and (amplitude is normal) then (alpha is big) (1)

The fuzzificated input and output variable will be as follows: (figure 7)

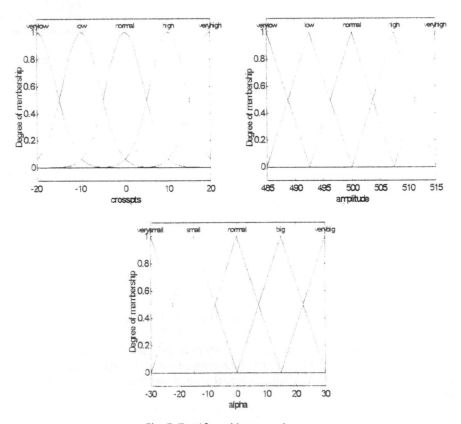

Fig. 7. Fuzzificated inputs and output

In figure 8 and 9, the fuzzy control system has 2-input and 1-output. Crosspts is the amplitude of the crossing point of the two input phases. Amplitude is the DC amplitude of the converter output. Alpha is a new difference value of the alpha. Alpha is the triggering angle of the rectifier thyristors. To find triggering alpha, following calculation is done:

new alpha = old alpha - difference alpha

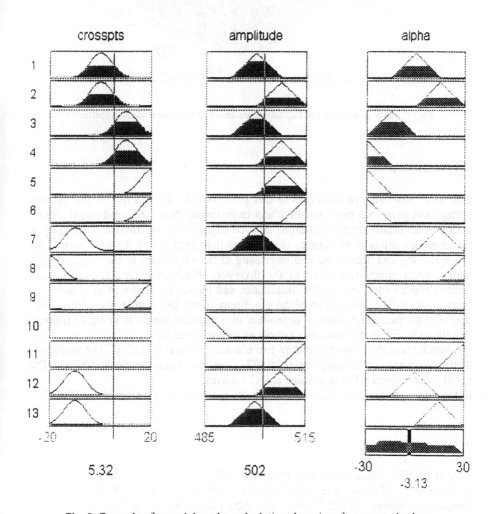

Fig. 8. Example of new alpha value calculations by using fuzzy control rules

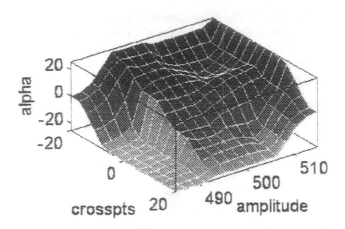

Fig. 9. Fuzzy control output (alpha) dependence on inputs(crosspts,amplitude)

4 Conclusion

The most important specifications of this fuzzy system: the controls of the current, voltage and power are more sensitive and more speedy than other traditional control methods. Especially, to produce alpha delayed triggering of the thyristors catching of the crossings points of the phases is the better than catching of the zero points. So, phase drifting and decreasing and increasing of the amplitudes of the 3-phase input signals don't effects triggering of the thyristors. Furthermore fuzzy control rules remove sudden variations of the amplitudes and phases are removed as possible by adapting to it. For controlling of the amplitude of the DC output, the alpha angle output of the fuzzy logic control dependent on its control inputs is given in figure 8 and figure 9. Figure 8 shows an example as following values:
the amplitude of the crossing point of two phases (crosspts) is 5.32 V. The amplitude of the converter output (amplitude) is 502 V (set DC output value is 500 V). The result : a new alpha difference value is -3.13 degree.
In the future our aim is to design a fuzzy logic control of the inverter of the HVDC.

References

1. ABB " HVDC Research Center" http page, 1996
2. S.Lefebvre, M.Saad, and A.R. Hurteau, "Adaptive control for HVDC power transmission system", IEEE Trans. Power Apparatus Syst., vol. PAS-104, no. 9, pp. 2329-2335, Sept. 1985.
3. W.J.Rugh, "Analytical framework for gain scheduling" IEEE Control Syst. Mag. , vol. II, no. 1, pp. 77-84, Jan. 1991.
4. K.W.V. To, A.K. David, and A.E. Hammad, " A robust coordinated control scheme for HVDC transmission with parallel AC systems." Presented at IEEE'94 WM 061-2 PWRD.

5. J.Reeve and M. Sultan, "Gain scheduling adaptive control strategies for HVDC systems to accommodate large disturbances", IEEE Trans. Power Syst, vol. 9, no. 1, pp. 366-372, Feb. 1994

6. Routray A., Dash P.K. and Panda S. K., "A fuzzy self-tuning PI controller for HVDC links" IEEE Trans. Power Elect., vol. 11, no. 5, pp. 669-679, Sept. 1996

7. Dash, P.K. Liew, A.C. Routray, A., "High performance controllers for HVDC transmission links". IEE Proc. Generation, Trans. And Distr. 141, p. 422- 428, Sept. 1994

Compensation for Unbalances at Magnetic Bearings

Dipl.-Ing. Mathias Paul   Prof. Dr.-Ing. Wilfried Hofmann

TU-Chemnitz
Department Electrical Machines and Drives
09107 Chemnitz
Germany

Abstract: The aim of this paper is to provide a new method to compensate remaining unbalances at magnetic bearings. The compensating signal will be generated by a MLP-network. The main advantage that this new method presents lies in the simplified manner of operation rather than in a qualitative improvement of disturbance compensation.

1. Introduction

Active magnetic bearings have a lot of advantages compared to sliding and ball bearings for high speed applications. Remarkable features of magnetic bearings are the durability and the adjustable stiffness. Many unbalances are measureable by simple means and removable with conventional methods.

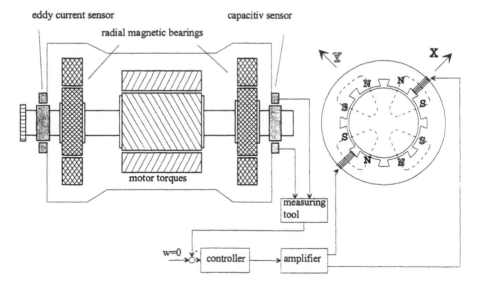

figure 1: system aktive magnetic bearing

Figure 1 shows the complete sytem "activ magnetic bearing". The motor is a common 75 kW three-phase motor. The controller bases on a digital signal processor TMS320-C40. Nevertheless, there are a lot of unwelcome effects for high speed applications. Unbalances cannot be eliminated completely and thus radial forces are still to be considered.

2. Remaining Unbalances at Magnetic Bearings

As mentioned above, remaining unbalances are inevitable for high speed rotors. This causes the shaft-axis to rotate around the central-axis due to centrifugal forces, which are calculated as follows:

$$F = \sum_{i=1}^{n} mi \cdot ri \cdot \omega^2 \qquad \begin{array}{l} m_i \ masses \\ r_i \ radiuses \\ \omega \ speed \end{array} \qquad (1)$$

These centrifugal forces have the form of a sinusoidal disturbance with rotation frequency for the respective axis. The disturbance can neither be influenced nor measured. In conventional compensation for disturbance, the disturbance has to be measureable or at least observeable (figure 2).

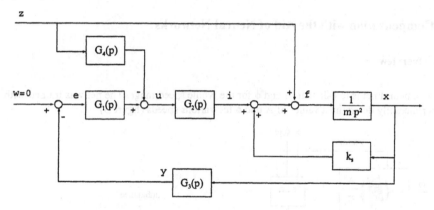

$G_1(p)$ transfer function of controller
$G_2(p)$ transfer function of adjuster and magnetic system
$G_3(p)$ transfer function of measuring tool
$G_4(p)$ transfer function of compensation for disturbance

figure 2: block diagram of conventional compensation for disturbance

The transfer function of disturbance turns out to be:

$$\frac{X(p)}{Z(p)} = \frac{1 + G_2(p) \cdot G_4(p)}{G_1(p) \cdot G_2(p) \cdot G_3(p) - k_s + mp^2} \qquad (2)$$

The aim is to choose $G_4(p)$ such, that for all $p = j\omega_r$ the following equation is true:

$$1 + G_2(p) \cdot G_4(p) = 0 \qquad (3)$$

The disturbance depends on the rotor angle. Therefore the compensating signal can be calculated approximately by:

$$z(t) = A\,sin(\omega_r t + \varphi)$$ (4)

Simple electrical filters were not able to produce satisfactory results /Gü93/. In a usual compensation for unbalances the compensating signal will be generated by a sinus generating set or by a digital signal processor. The position tolerance which is caused by the remaining unbalances can be limited to roughly 10 % of the uncompensated value. The great disadvantage of these two methods of compensation is, that the magnitude A and the angle ϕ have to be provided manually. That is not at all convenient for a drive with alternating working points.

3. Compensation with the Aid of Neural Networks

3.1. Overview

Now, a neural net is to calculate A and ϕ for the whole speed range. The net was trained with experimentally determined values of A und ϕ for various speeds (figure 3)

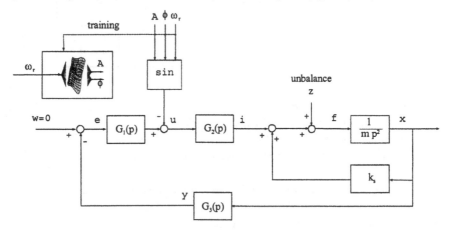

figure 3: block diagram of the compensation for unbalances with neural network

The test set-up, a magnetic beared turbo fan, works in a speed range from 0 to 6000 r.p.m. Considerable unbalances do not occur under approximately 3000 r.p.m. The compensation signal has to be chosen according to the speed such that the tolerance of cyclic running is as small as possible throughout the whole speed range.

The root-mean-square rms of the remaining ripple can be used as a characteristic for the quality of compensation. On the other hand, there is no definite connection between a too high rms and the single compensation parameter. It remains unclear whether the magnitude or the angle or both must be changed to improve the compensation (fig. 4).

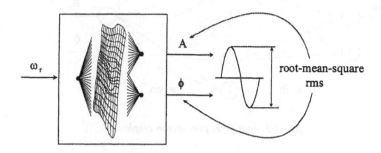

figure 4: connection between the quality of compensation (rms)
and the compensation parameters A and ϕ

The first tests served to find an optimal network topology that would be able to learn all the needed information (magnitude A and angle ϕ) for all speeds. A dense net of magnitudes and angles was determined experimentally and fed to a network.

To learn these two functions an optimal network-configuration is sought after. The number of neurons must be sufficient to store the information. On the other hand with every neuron more the time for training becomes longer. The following section will discuss the particular network characteristics needed and employed.

3.2. Network employed

Multilayer - Perceptron networks combined with the backpropagation algorithm are able to learn every nonlinear function. Therefore the continuous differentiable sigmoid function was used as the activation function:

$$s\left(x\right) = \frac{1}{1 + e^{-cx}} \tag{5}$$

The rotor speed is the input data. Magnitude and angle will be calculated as output data. Figure 5 shows the notational conventions.

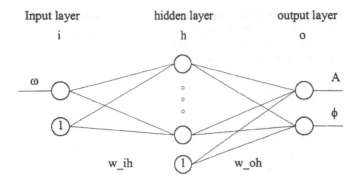

figure 5: multilayer perceptron employed

The neuron output values are calculated:

$$o_j^1 = s\left(\sum_{i=1}^{n+1} w_{ij}^1 \, \hat{o}_i \right) \tag{6}$$

Backpropagation is a simple gradient descent methode. The aim is to search after the minimum of the error function. With further application of the chain rule the weight changes of the neurons are calculated as follows:

$$\vec{\nabla}E = \left(\frac{\partial E}{\partial w_{11}^{1(2)}}, ..., \frac{\partial E}{\partial w_{i+1,h}^{1(2)}} \right) \tag{7}$$

Furthermore, some variations of backpropagation are implemented in order to improve the learning process:

- a **mometum term** $\alpha \, \Delta_m \, w_{ij}(t-1)$ added to the weight change aiming at different learning speeds (on plateaus acceleration and in gorges delayal)

- a **weight decay term** $-d \, w_{ij}(t-1)$ added to the weight change to limit the weights ($0{,}0001 < d < 0{,}0003$)

- a **quickpropagation** algorithm to speed up the training with backpropagation in general

- **flat-spot-elimination** means that a constant value will added to the sigmoid function (some neurons operate near the saturation region of the activation function).

3.3. Simulation Results

The following network configuration was used to learn the training data:
- input layer - 1 input neuron (rotor speed) / 1 Bias
- hidden layer - 15 hidden neuron / 1 Bias
- output layer - 2 output neurons (magnitude, angle)
- 10 training data
- learning rate $\eta = 0,3$
- momentum term $\alpha = 0,09$

If the learning rate or momentum term were higher than indicated above, the weight corrections would soon become so high that the floating point overflowed. After 7768 iterations the square error was below 0.0003, and the ability of the network to approximate became satisfactory well (fig. 6).

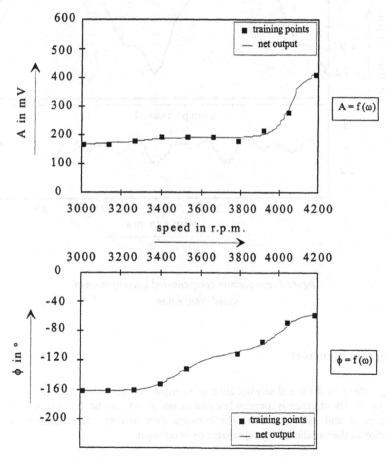

figure 6: training points and calculated net-output after 7768 Iterations

3.4. Measured Results

Figure 7 shows comparison between the uncompensated and compensated behaviour of one bearing-axis. The position tolerance will be limited to roughly 4μm (pic-pic). The measured noise of this capacity measuering tool is approximatly 1,5 μm. That is the lower limit for a sensible compensation for unbalances.

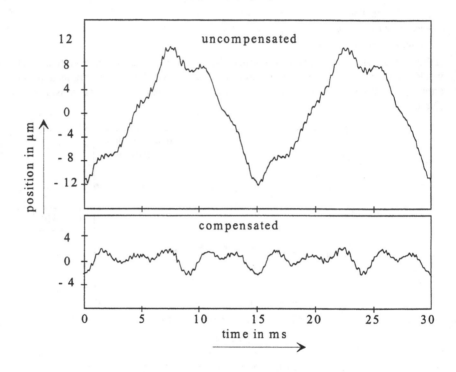

figure 7: comparison compensated-uncompensated
speed 4000 r.p.m.

4. Program Structure

The algorithms for the neural network are implemented in common C. They run with DOS on every system. The structure is completely dynamic not object orientat. It can be stopped at any definite point and continued later. The network data will be saved automatically. The application of the variations of backpropagation is optional.

Furthermore, all intresting information like square error, number of iterations and the input/output values of all neurons are graficaly retrievable.

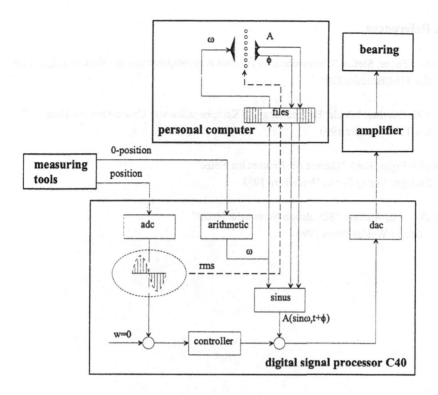

figure 8: hardware structure

The signal processor calculates the speed from of the 0-position data and writes the speed together with the actual magnitude and angle and the rms into files. The neural network, which is implemented on the personal computer, is able to read the files and learns with the aid of these data (figure 8).

5. Conclusions

To provide a real advantage to the conventionell method the network should to need to be fed with only very few values of magnitude and angles. The other values will be learned online during operation. Therefore the rms is used as a quality criterion.The network should learn further online with the help of the quality criterion.

6. Acknowledgment

This project "Magnetisch vollständig gelagerte Monoblockeinheit" is supported by the Deutsche Forschungsgemeinschaft.

7. References

/Fu96 / Fuchs, Stefan "Untersuchungen zur Unwuchtkompensation am aktiven Magnetlager"
DA TU-Chemnitz 1996

/ Gü93/ Günther, Ulrich "Bestimmung und Kompensation von Unwuchten mit Hilfe ....."
DA TU-Chemnitz 1993

/ Ro93/ Rojas, Raul "Theorie der Neuronalen Netze"
Springer Verlag Berlin Heidelberg 1993

/ Ze94/ Zell, Andreas "Simulation Neuronaler Netze"
Addison-Wesley Bonn 1994

Supervised Classification of Remotely Sensed Images Based on Fuzzy Sets

Peter Sinčák*

Laboratory of AI
Department of Cybernetics & AI
Faculty of Electrical Engineering & Information Technology
Technical University Kosice
Slovakia
E-mail: sincak@ccsun.tuke.sk

Abstract. Paper deals with supervised classification approach based on fuzzy sets. The fuzzy approach represents a minor adaptation of supervised fuzzy classification which can be found in [2] and used formerly for speech recognition problems. This method was tested for classification of remotely sensed multi-spectral video imagery. Results of this classification approach seem to be promising in comparison with classical techniques and also with some neuro-fuzzy approaches.

1 Introduction

Remote sensing technology is an important approach in various kinds of applications. Video R–G–B imagery is a cheap and efficient type of data and processing of this data is important for further processing and integration of achieved thematic images into map update procedures. Accuracy of processing is an important issue and study of various more accurate methods is very useful direction of research. Studied method based on fuzzy sets is an example of utilizing fuzzy sets in classification procedures.

2 Short description of Supervised Classification approach

Lets have a classification task to discriminate data into **m-classes**

$$(c_1, \ldots, c_m)$$

by the help of n-dimensional feature space. Description of these 2 steps is as follows:

- **Learning:** So, initially each pattern can be described as follows:

$$\mathbf{x} = \{x_1, \ldots, x_n\}, \tag{1}$$

---

* This paper was supported with USA-Slovak grant # 94077, awarded by Slovak–USA International Agency for Science and Technology in 1994. Cooperation is underway with Dr. Howard Veregin from University of Minnesota, USA – Co-PI of the project.

where x_i is a feature from the n–dimensional feature space. Generally we can consider a following training set structure

$$\underbrace{x^1, \ldots, x^{k_1}}_{c_1}, \ldots \underbrace{x^1, \ldots, x^{k_m}}_{c_m}.$$

That means that we assume, that training sets consists of m classes of interest and so e.g. class 1 has k_1 patterns, class 2 has k_2 and etc. Each pattern is described in the sense of equation 1. Now for each pattern in training set we can calculate the following value $\mu(l, j, i)$. This value represents a membership function of $l - th$ pattern **to class** j considering feature i . So definition intervals are for $l \in < 1, k_j >$, $j \in < 1, m >$ and for $i \in < 1, n >$.

$$\mu(l, i, j) = \left\{ 1 + |\frac{\hat{x}_i(j) - x_i^l(j)}{E}|^F \right\}^\alpha, \tag{2}$$

where $\hat{x}_i(j)$ is a mean value of the $i - th$ feature of overall training vectors in class j, $x_i^l(j)$ is particular value of feature i concerning $l - th$ pattern, which belongs to class j. Also coefficients E,F are important for membership function determination a α is usually set to -1. So in fact we are constructing a new feature space and pattern x from class j in equation 1 is now transformed into new feature space as follows:

$$x = \{\mu(l, 1, j), \mu(l, 2, j) \ldots, \mu(l, n, j)\}, \tag{3}$$

where l pattern x^l in class j. So each class j of m has its own k_j patterns x described now by its membership function vector according to equation 3.
Classification: this part of supervised classification is based on discrimination rule procedure. The discrimination rule is based on calculation of so-called **similarity vector** between unknown vector **y** and class j in the n-dimensional feature space (or better each pattern of training set of class j)

$$s(y, l_j) = \{s(y, l_j, 1), s(y, l_j, 2), \ldots, s(y, l_j, n)\}, \tag{4}$$

where e.g. $s(y, l_j, 1)$ is a value which represents similarity between unknown vector y and $l-th$ pattern in $j-th$ class, considering 1-st feature. More detail information will be provide in full paper. According to above consideration we end–up with

$$k_1 \text{ values of } |s(y, l_1)| \quad l_1 \in < 1, k_1 >$$
$$k_2 \text{ values of } |s(y, l_2)| \quad l_2 \in < 1, k_2 >$$
$$\ldots$$
$$\ldots$$
$$\ldots$$
$$k_m \text{ values of } |s(y, l_m)| \quad l_3 \in < 1, k_m >$$

So if we have

$$K = \sum_{j=1}^{m} k_j$$

then we have K various modules of vectors for single unknown vector y. Further decision is more-less clear – so we look for

$$max\{|s(y,l)|\} \qquad (5)$$

If we find a **max** we can decide that if

$$max\{|s(y,l)|\} \geq T, \qquad (6)$$

where T is selected threshold, then unknown vector y belongs to class to which belongs vector l. If equation 6 is not true than unknown vector does not belong to any of classes of interest $\{1, \ldots, m\}$.

3   Experiments on R–G–B imagery

Our experiments were done on remotely sensed R–G–B video data over North-Eastern Ohio, USA. So number of features was $n = 3$ and number of classes of interest was $m = 8$. R–G–B imagery was 410 per 1100 8–bits pixels large. There was over 8 000 pixels determined in representative set and from this number aprx. 4400 was used as training set and rest of it for testing purposes. Classes of interest are on table 1.

class	defined as
A	grass–stubble
B	grass
C	asphalt
D	concrete
E	shingle roofs
F	water
G	shadow
H	deciduous trees

Table 1. List of classes classified on studied area of North-Eastern, Ohio – USA

513

4 Experimental Results

Studied and adapted methods was tested on training and testing sites of the image. These sites were determined by expert geographer. Evaluation was done with contingency table (CT) approach and accuracy assessment is in tables # 2 and # 3. Contingency table and weighted PCC are described in [5]. In this project further 2 methods were also studied. Fuzzy ARTMAP has been tested on these images. Results of these processing can be found in CT–table # 4 on training sites and in table # 5 on test sites. Modular Neural network in sense of modularity described in [1] has been tested, too. Results can be found in tables # 6 and # 7 respectively. Comparison analysis shows that results of supervised approach based on fuzzy sets are much better than classical classification approach [4] and regarding a fuzzy ARTMAP results are very similar, but supervised classification based on fuzzy sets is much faster. Modular neural network gives very bad results. In further research has been investigated an approach to integrate all 3 results into final conflation classification thematic image. More about this technique can be found in [3].

5 Conclusion

It is showed that supervised approach based on fuzzy sets is a promising approach for image processing application. Further study of this method will be lead in direction of membership function determination for each training class respectively. It is believed that results can be improved with more sophisticated membership function design. Further research is still underway.

References

1. Simon Haykin. *Neural Network - A Comprehensive Foundation*. ISBN 0 02 352761 7. Macmillian College Publishing Company, New York, USA, 1994. 680pp.
2. Yoh-Han Pao. *Adaptive Pattern Recognition and Neural Network*. Adison-Wesley, New York, USA, 1989. ISBN 0 201 12584 6.
3. P. Sinčák, H. Veregin, and N. Kopčo. Conflation techniques to improve image classification accuracy. *under review in Remote Sensing of Enviroment*, 1996.
4. H. Veregin, P. Sinčák, K. Gregory, and L. Davis. Integration of highresolution viodeo imagery and urban storm water runoff modelling. In *Proceeding of 15-th Biennial Workshop on Videography and Color Photography in Resource Assesment*, pages 182–191, Terre Haute, Indiana, USA, 1995. American Society for Photogrametry and Remote Sensing. ISBN 1-57083-024-X.
5. Howard Veregin. A taxonomy of error on spatial databases. Technical Report 89-12, National Center for Geographic Information and Analysis, University of California, Santa Barbara, USA, 1989.

Appendix

In appendix are presented contingency tables for particular classification results and a thematic image achieved by supervised classification based on fuzzy sets approach.

	Classified Classes A–H Training sites A'–H'								$\sum$
	A'	B'	C'	D'	E'	F'	G'	H'	
A	209	0	0	0	0	1	3	0	213
B	1	105	0	0	0	0	0	0	106
C	0	0	1224	0	0	0	0	0	1224
D	0	1	0	405	1	0	0	0	407
E	0	0	0	0	873	0	0	42	915
F	0	0	0	0	6	676	1	1	684
G	3	0	0	0	0	0	559	0	562
H	0	0	0	0	17	0	2	750	769
$\sum$	213	106	1224	405	897	677	565	793	4880
Weighted PCC = 0.985									

Table 2. CT for supervised classification based on fuzzy sets (training sites)

	Classified Image A–H Test sites A'–H'								$\sum$
	A'	B'	C'	D'	E'	F'	G'	H'	
A	14	1	0	0	2	3	4	2	26
B	0	13	0	4	0	0	0	0	17
C	0	7	24	1	0	0	0	0	32
D	2	4	1	19	1	1	0	0	28
E	0	0	0	1	21	0	0	3	25
F	3	0	0	0	0	17	3	2	25
G	6	0	0	0	0	4	18	1	29
H	0	0	0	0	1	0	0	17	18
$\sum$	25	25	25	25	25	25	25	25	200
Weighted PCC = 0.713									

Table 3. CT for supervised classification based on fuzzy sets(test sites)

	Classified Classes A–H Training Sites A'–H'								$\sum$
	A'	B'	C'	D'	E'	F'	G'	H'	
A	213	0	0	0	0	0	0	0	213
B	0	106	0	0	0	0	0	0	106
C	0	0	1224	0	0	0	0	0	1224
D	0	0	0	405	0	0	0	0	407
E	0	0	0	0	890	0	0	25	915
F	0	0	0	0	0	677	1	0	678
G	0	0	0	0	0	0	564	0	564
H	0	0	0	0	7	0	0	768	775
$\sum$	213	106	1224	405	897	677	565	793	4880
PCC = 0.99324, Kappa = 0.99190									

Table 4. CT for Fuzzy ARTMAP Neural Network Classification on Training Sites

	Classified Image A-H Test Sites A'-H'								Σ
	A'	B'	C'	D'	E'	F'	G'	H'	
A	16	1	0	0	2	0	1	0	20
B	0	14	0	3	0	0	0	0	17
C	0	6	24	0	0	0	0	0	30
D	1	2	1	21	1	1	0	0	27
E	1	2	0	1	20	1	0	11	36
F	2	0	0	0	1	18	3	1	25
G	5	0	0	0	0	5	21	0	31
H	0	0	0	0	1	0	0	13	14
Σ	25	25	25	25	25	25	25	25	200
PCC = 0.73500, Kappa= 0.69714									

Table 5. CT for Fuzzy ARTMAP Neural Network Classification on Test Sites

	Classified Classes A-H Training Sites A'-H'								Σ
	A'	B'	C'	D'	E'	F'	G'	H'	
A	53	7	3	11	73	71	185	17	420
B	51	35	6	25	111	158	103	68	557
C	1	1	1200	1	0	0	0	2	1205
D	12	31	1	282	35	13	35	2	411
E	43	14	1	36	459	239	72	163	1027
F	10	2	0	2	111	125	9	24	283
G	6	14	10	39	12	19	36	6	142
H	37	2	3	9	96	52	125	511	835
Σ	213	106	1224	405	897	677	565	793	4880
PCC = 0.55348, Kappa = 0.47279									

Table 6. CT for Modular Neural Network Classification on Training Sites

	Classified Image A-H Test Sites A'-H'								Σ
	A'	B'	C'	D'	E'	F'	G'	H'	
A	5	3	0	0	2	1	10	2	23
B	8	9	3	1	3	9	6	1	40
C	0	3	18	0	0	0	0	0	21
D	0	5	3	22	1	1	1	0	33
E	3	2	0	2	10	3	2	12	34
F	1	0	0	0	3	2	0	3	9
G	4	3	1	0	4	0	1	1	14
H	4	0	0	0	2	9	5	6	26
Σ	25	25	25	25	25	25	25	25	200
PCC = 0.36500, Kappa=0.27429									

Table 7. CT for Modular Neural Network Classification on Test Sites

516

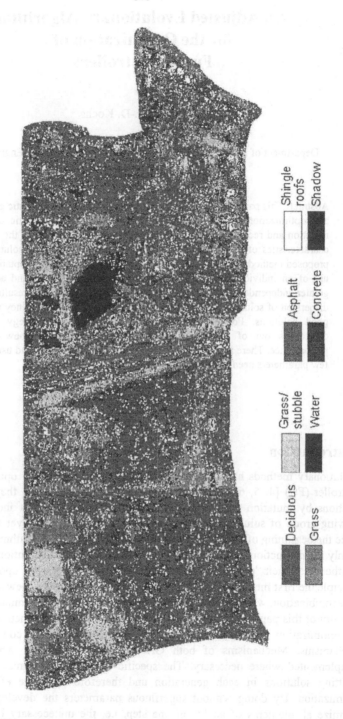

Shingle roofs

Shadow

Asphalt

Concrete

Grass/ stubble

Water

Deciduous

Grass

Fig. 1. Thematic image classification result achieved by supervised classification based on fuzzy sets

An Adjusted Evolutionary Algorithm
for the Optimization of
Fuzzy Controllers

S. Wagner, H.-D. Kochs

Gerhard-Mercator-University - GH Duisburg
Department of Information Processing / Faculty of Mechanical Engineering

Abstract. This paper describes an evolutionary method for automatic generation and optimization of fuzzy controllers (FC). The typical genetic operations mutation and recombination are designed with special respect to the structural characteristics of an FC and permit an effective generation of solutions. The proposed method goes without global parameters for step size adaptation by the use of an individual adaptive mutation mechanism selection and achieves a greater independence of the individuals of a population. The resulting broad dispersion of solutions in the search space leads to a high efficiency in finding good solutions. For the shown control examples the strategy generates controllers out of randomly occupied individuals which show excellent performance. There is high reliability in finding good solutions and usually very few parameters need to be tuned manually.

1 Introduction

Evolutionary methods have proven to be efficient techniques for optimizing a fuzzy controller (FC) [4, 5, 6, 7]. Just like evolution of living beings they build up new solutions by mutation and recombination. Selection of the best individuals of the growing group of solutions causes a continuous improvement over the generations. Since the beginning of the research concerning evolutionary algorithms there has been mainly the distinction between genetic algorithms (GA) and evolution strategies (ES) together with their 'programming variants' GP and EP. Basing upon the evolution principle the first method uses binary coding and tries to evolve new solutions mainly by recombination, whereas ES use real-valued vectors and prefer mutation.

The aim of this paper is to show the design of an evolutionary method especially for the optimization of fuzzy controllers by adequate design of coding and genetic mechanisms. Mechanisms of both GA and ES are utilized where useful and complemented where necessary. The specific operations minimize the number of unfitting solutions in each generation und therefore raise the effectivity of the optimization. By doing without superfluous parameters the developed system can optimize all elements of an FC in one step, i.e. the unnecessary but mostly done separation between optimization of rulebase first and linguistic variables afterwards could be avoided.

The proposed approach varies the possibility of chosing one of its mutation mechanisms dependant on its success. So the individuals of a population become more autonomous, i.e. they favour different search paths according to the actual adjustment of the mutation mechanism probabilities and increase the dispersion of solutions in the search space this way. This results in a method of high global safety in finding good solutions because broad search in search space is supported. The method gives up global step size control and forces convergence not by smaller step sizes but through increasement of the mutation rate of the necessary optimization parameters. So all individuals are independent from one another concerning their convergence. A step size suitable for the best individual will not be forced on another, which would be the case in most standard applications of adaptive mutation rates.

Simulations show that the developed optimization strategy leads to good results in spite of accidental initialization of the rule base and simultaneous adaptation of membership functions and rule base. Examples concerning the generation of fuzzy controllers for different dynamic systems are shown.

2 Development of an adjusted optimization strategy

Following the basics of our method, i.e. parameter reduction, genetic mechanisms and adaptive mutation mechanism selection, are described. Similarities and differences to GA and ES are shown.

2.1 Basic aspects

Both results of biological evolution and its variety reflect the effectivity of the evolutionary optimization principle. Therefore ES and GA try to transfer this principle to other optimization problems, i.e. they attempt to model biological mechanisms by suitable algorithms. Genetic algorithms further try to reproduce the coding of the DNA. When one takes into consideration, that evolution brings out optimized results, but its mechanisms undergo no evolution themselves, then one may say, that maybe other mechanisms could be more suitable for an optimization.

According to this our strategy utilizes just some of the principles of biological evolution by imitating their effects and brings in some new problem specific mechanisms with no biological equivalent. The well known mechanisms fixed this way are:

- Mutation: It generates new variants by modifying existing solutions. Mutation mechanisms should be on the one hand flexible to guarantee a wide scatter of solutions and on the other hand they should lead frequently to good solutions by implementing some process knowledge to be efficient.

- Recombination: It creates new variants by mixing existing ones. Recombinations should be useful and problem specific, i.e. they should take into consideration the structure of the solutions to reduce the number of unnecessary trials.

- Selection: It choses the solutions coming into the next generation, where the best individual of each generation is taken over in our approach to ensure convergence. The selection mechanism should further cause a good mixture of the populations to avoid premature convergence.

Additional degrees of freedom and supplementary mechanisms for coding lead to a reduction of parameters used for coding of the linguistic terms, more effective mutation mechanisms and adaptive mutation mechanism choice, which are special for our method.

2.2 Coding and parameter reduction

Optimization with the help of evolutionary algorithms has to start with a coding of the problem to be optimized. In our case the parts of a fuzzy controller relevant to optimization have to be encoded. We have chosen to optimize size and position of the membership functions as well as rule base entries, whereas number of membership functions and fuzzy inference mechanisms are fixed. GA carry out a binary coding, which is especially useful for dicrete optimization. Our approach uses real valued coding of the optimization variables, just like in ES. Further the number of free parameters is reduced to a maximum extent, because the more values are to be optimized the longer the optimization runs take and the higher the risk of getting stuck in local minima is.

Usually rule base and fuzzy sets are optimized separately to reduce the number of parameters to code but this approach ignores the dependencies between those main elements of a fuzzy controller. We have carried out tests which show that a controller, which rule base has been optimized for initially homogeneously distributed fuzzy sets, can only be slightly improved by modifications of these fuzzy sets. This led to our decision to optimize rule base and linguistic terms simultaneously.

The rule base gives no opportunity to reduce the number of parameters needed for coding, because each of its elements represents one rule and therefore one optimization parameter. That is why reduction of parameters is only possible for modifications of the linguistic terms. In one extreme case the coding of the fuzzy terms can be done by coding each of their support values, which results in the largest possible number of parameters. In the case of trapezoid variables for example there would be four parameters needed for each membership function. This approach can lead to fuzzy sets, which can take on every possible form during optimization, so that linguistic interpretation usually becomes impossible. The other extreme case is the multiplication of homogeneously distributed membership functions with variable scaling factors. In this case the number of parameters underlying coding reduces to the number of input and output values of the controller, but the distribution of the membership functions remains homogeneous for all scaling factors, only the controller gain is modified. For many applications however it is useful to enlarge the controllers sensitivity to special areas of the input space by condensing respectively loosen up the membership functions. So the finally chosen approach tries to reduce the number of fuzzy set parameters to code but gives nevertheless the opportunity to change position, support and tolerance of the membership functions.

The assumptions made are:

- There are n_i membership functions for the representation of each linguistic variable i,

- membership functions are characterized by symmetric trapezoids with modal value m_{ij}, tolerance b_{ij} and ascent e_{ij}, with j as the number of the respective fuzzy set,
- the universe of discourse of each FC-input/output i lies between $y_{i,\,min}$ and $y_{i,\,max}$,
- the modal value m_{i1} of the first membership function each is set to $y_{i,\,min}$, that of the last one is set to $y_{i,\,max}$.

The modal values and characteristic forms of the trapezoids are now defined as functions of the number j of the respective membership function:

$$
\left. \begin{aligned}
m_{ij} &= F_{1,\,i}(j) \\
b_{ij} &= F_{2,\,i}(j) \\
e_{ij} &= F_{3,\,i}(j)
\end{aligned} \right\} \text{ for } j = 1, \dots, n_i
$$

The utilization of these form functions $F_{1,\,i}$ to $F_{3,\,i}$ determines position and shape of each membership function. The shape of the form functions themselves is characterized by the six coding parameters $p_{1,\,i}$ to $p_{6,\,i}$.

$p_{1,\,i}$ determines the centre of $F_{1,\,i}$ and thus the modal value of the fuzzy set in the middle and $p_{2,\,i}$ defines the gradient of $F_{1,\,i}$ in its centre, which can be seen in figure 1 for the determination of 5 trapezoidal membership functions. Smaller values of $p_{2,\,i}$ draw the membership functions closer to the middle, larger values push them away to the outside.

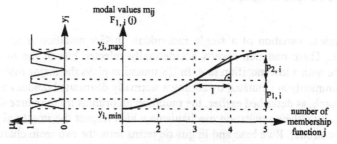

Fig. 1: Coding of the modal values

Tolerance b_{ij} and ascent e_{ij} are described by the piecewise linear functions $F_{2,\,i}$ and $F_{3,\,i}$. The coding parameters $p_{3,\,i}$ respectively $p_{5,\,i}$ immediately set the characteristics of the membership function in the middle, $p_{4,\,i}$ and $p_{6,\,i}$ set those of the fuzzy sets at the edges. With the help of the form functions $F_{2,\,i}$ and $F_{3,\,i}$ the b_{ij} and e_{ij} of the other membership functions can be calculated. For different relations between $p_{3,\,i}$ and $p_{4,\,i}$, respectively $p_{5,\,i}$ and $p_{6,\,i}$ the supports of the membership functions can increase, decrease or remain constant from the centre to the outside. The shown configuration expands the fuzzy sets the more they are apart from the centre (fig. 2).

A coding of linguistic terms in this way surely means a considerable restriction of degrees of freedom but by using this approach the fuzzy sets belonging to each linguistic variable are described by only six parameters, no matter how many membership functions are used. Furthermore there are still enough possibilities to

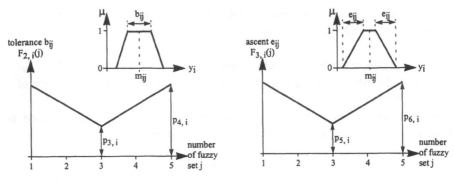

Fig. 2: Coding of tolerances and ascents

influence the structure of the controller and the generation of linguistically nonsensical controllers is avoided without further measures.

Singletons are used as consequence membership functions, so that $p_{1, i}$ and $p_{2, i}$ are enough to describe them. So the linguistic terms of an exemplary MISO-controller with two inputs and one output are fully described by only 14 parameters.

2.3 Adjusted genetic mechanisms

This section describes the realization of the mutation and recombination operations and their peculiarities. Both mechanisms are designed with respect to the special requirements on optimizing an FC.

Mutation

Mutation means variation of a single individual. In GA mutation mostly plays an inferior role. There mutation inverts accidentially selected bits of the binary coded chromosome with a low mutation rate. In ES mutation plays the major role and object variables commonly are mutated by adding a normally distributed random number.

In our approach, as described earlier, the mutation algorithms try to reduce the number of bad trials by utilizing mutation mechanisms which respect the structural properties of a fuzzy controller. Rule base and linguistic terms form the two main characteristical elements of a fuzzy controller, so that two classes of mutation mechanisms are built up in our approach:

Concerning the mechanisms for the *mutation of linguistic terms*, mutation always varies exactly one value out of the amount of all possible parameters of the linguistic terms. This means, that according to the actual mutation rate one linguistic parameter is chosen by chance and mutated by adding a gaussian distributed random number. The restriction to one parameter allows conclusions concerning the efficiency of the mutation of this parameter, which furthermore enables possibilities for an intelligent mutation mechanism selection (see 2.4).

The *mutation of the rule base* belongs to a specially designed catalog of possible mechanisms which respect the neighbourhood relationships between the single rule base entries. Figure 3 shows the developed mutation mechanisms for mutating the rule base of a two-input-one-output FC.

The areas which are underlying variation are emphasized; the transformation of an

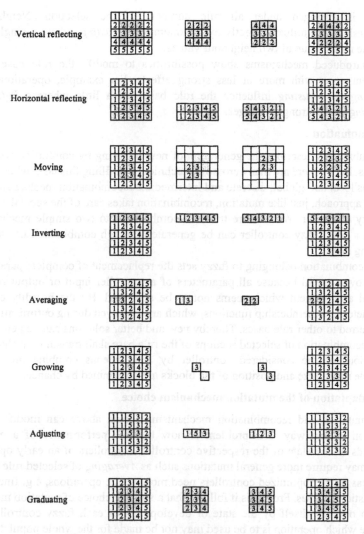

Fig. 3: Mutation mechanisms of the rule base

original individual into its mutant happens from left to right. First the relevant sections of the rule base are determined. For *Growing* and *Adjusting* this means the determination of a single rule and one of its neighbours, in all other cases these sections are rule blocks of any size, where *Averaging* requires a section of at least two rules. *Horizontal* and *Vertical Reflecting* change the selected rules by corresponding exchange along the axis of the respective section. *Moving* copies all rule elements of a section in the same arrangement into another position of the rule base. *Growing* fulfills a similar task with the restriction of copying only one element. The mechanisms *Inverting*, *Graduating* and *Averaging* replace the number of each consequent rule by the respective 'opposite', slightly raised respectively lowered value or by the value, which is averaged under all rule numbers of the selection. Neighbourhood relationships are utilized directly by the operation *Adjusting*, where a single element gets the mean value of two neighbour values.

The introduced mechanisms show possibilities to modify the rule base in many different ways with more or less strong effects. For example, operations such as *Growing* or *Adjusting* influence the rule base just a little whereas *Reflecting* or *Inverting* cause stronger changes.

Recombination

Recombination describes the generation of new offspring by combining two or more parents. In GA there are many crossover techniques fulfilling this task, whereas ES use various local and global, discrete and intermediate recombination mechanisms.

In our approach, just like mutation, recombination takes care of the special structure of a fuzzy controller. We have restricted recombination to two simple mechanisms by which a new fuzzy controller can be generated through combination of two already existing ones.

For recombination belonging to fuzzy sets the replacement of complete parameter sets has proven useful because all parameters of a controller input or output represent a special arrangement which seems not to be separated. By the heredity of coupled parameter sets membership functions, which are improved during optimization, can be transferred to other rule bases. Thereby new and better solutions can be generated.

The recombination of selected sections of the rule base shall transmit the characteristic behaviour of the considered controller by simultaneous combination with other suitable rules. Size and position of the blocks are determined by chance.

2.4 Adaptation of the mutation mechanism choice

The mutation and recombination mechanisms shown above can modify an initial solution in any way. The problem is now that the performance of a mechanism depends on the state of the respective controller: controllers of an early optimization state may require more general mutations such as *Averaging* of selected rule base areas whereas further optimized controllers need more gentle operations, e.g. fine tuning of linguistic variables. From this it follows that a suitable choice of mutation mechanisms has to orientate itself by the state of development of each fuzzy controller. So the choice which operation is to be used may not be made for the whole population.

This condition is realized by variable frequencies for the mutation mechanisms, which are different for each individual. The probability for the choice of each mechanism is varied depending on its success rate, i. e. mutation mechanisms which lead to an increase in fitness will be chosen more frequently in the future. To adapt the probability values to the state of development each value is additionally decreased by a certain amount after each mutation. On the whole this procedure causes an adaptation of the mutation mechanism choice, which tries to adjust itself to the fuzzy controller to be tuned.

This means that an individual is no longer described by its state variables only but also by its choice probabilities, similar to the coding of mutation variances as strategy parameters in evolution strategies. The mutation mechanism choice adaptation and the giving up of a global convergence parameter has the following advantageous effects:

- Individuals differ not only by their actual structure but also by their tendency to change. After an accidental initialization of all controller components the more successful mechanisms determine the further look of a controller because they will be used more often.

- The higher independence of the individuals leads to a broader variance inside the search space because the search direction is not controlled globally. While some individuals tend towards a local minimum others can reach better minima. A global step size would have restricted the agility of the latter individuals.

- In general, mutations of the rule base are more successful in early states while after some generations the optimization of the fuzzy sets becomes more important. The adaptation strategy makes the simultaneous optimization of both components possible because its adjusts itself to the state of development.

In contrast to the adaptation of the choice of mutation mechanisms the recombination operations are determined according to a fixed frequency distribution. With the above a mutation happens as follows: If an individual gets the request to mutate, first an adaptive choice of one out of ten possible mutation mechanisms (8 mutation mechanisms for the rule base, mutation of a premise fuzzy set or mutation of a consequence singleton) takes place. If mutation of the rule base is chosen, range and position of the sections to mutate are selected randomly. If fuzzy sets are to be changed, an adaptive choice of the corresponding variable and their respective parameter happens. Finally the variation of this parameter is the mutation of the chosen linguistic variable.

2.5 Synthesis of the single aspects

The main elements of the approach - coding of the linguistic terms, mutation and recombination mechanisms and their adaptive choice - have been introduced. In the following the realized algorithm as one possible solution out of many combinations of these mechanisms is described. Figure 4 shows the complete procedure in a schematically representation to show the course of mutation, recombination and selection.

The three used populations A, R and B are shown as circles, the corresponding arrows illustrate transitions of individuals from one population to another. The main population A is randomly initialized with a individuals at the beginning of the optimization process. A newly generated individual is immediately tested and evaluated after each mutation or recombination. This direct test run enables the application of the above described adaptive mutation mechanism choice. This additional testing slows down the computation of one generation cycle but rises the convergence velocity by useful setting of the adaptation values. So it is problem dependant if the additional test runs are worth the extra expenditure: If the time consumption in computing the test runs is sufficiently low, this pays in faster optimization convergence, so that less generations have to be run through.

One population cycle starts with the selection of the b best individuals of the main population. Each of these individuals runs through a sequence of n^* mutations, in which the mutant of every step only replaces the original individual if it has been improved. The sense of these sequential mutations is to guarantee a high variance of

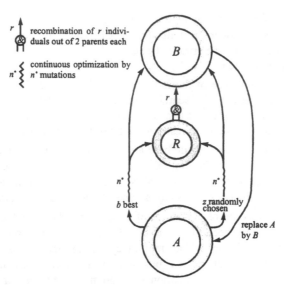

Fig. 4: Algorithm scheme

solutions inside the search space. The sequential mutations give the individuals enough time to travel inside the search space corresponding to their tendencies. They shall develop properties which distinguish them from other individuals. Frequent recombination or selection would be working against this effect. This simulates in some manner the effect of natures niche populations, because it gives individuals with smaller fitness the chance to improve without influence by other individuals, so that they may improve instead of dying prematurely.

After this the improved individuals are added to both recombination population R and target population B. Because of the survival of the whole population B, the b best individuals of A survive - perhaps in improved, mutated condition. In a similar way z randomly selected individuals of A run through a sequence of mutations and the results are also added to R and B. Finally r new solutions are generated by recombination of two each parents of R. These individuals form the new population A of the next generation together with the individuals of B. R and B are now deleted and the cycle starts all over again.

The parameter n^* underlies no restrictions, whereas b, r and z must amount to a. The values of r and b have to be much smaller than z to avoid a flood of B with only the best individuals and this way maybe an early fixing of the optimization direction. The sum of b and z has to be only slightly smaller than a to reject just a few individuals out of A. In this way the selection pressure becomes not too large, which reduces the convergence speed but enables the algorithm to develop better solutions.

Because of the transferring of the b best individuals from A to B - possibly after an improvement - the quality of the best individual will increase from generation to generation without step backs. The problem is to decide how independent the evolution processes shall be. More independent processes produce individuals, which spread more widely but may get stuck in local minima because of their missing information exchange. Dependencies between the solutions enable an abort of wrong ways and the start of others, but they must not be too large to not restrict the search space too strong. The introduced method tries to find a good compromise but further research concerning the influence of n^*, a, b, z and r on the optimization course is necessary. The new approach looks at first sight more computing intensive than conventional methods but it works with smaller populations and converges very fast, what is more than a compensation.

3 Example: Generation of fuzzy controllers

The performance of the proposed algorithm is tested on the following P_{T2}-systems

$$G_1(s) = \frac{19.6}{s^2 + 0.42s + 0.49} \quad , \quad G_2(s) = \frac{40.0}{s^2 + 1.0} \quad .$$

System G_1 is an example for a damped, oszillating system and G_2 represents a system at its stability margin. The settings for the optimization runs are shown in table 1.

Optimization	Population size	20
	Parameters	b=2, z=16, r=2, n*=30
	Generations	100
Fuzzy controller	Universes of discourse	Input 1 (control value): [-80, 80]
		Input 2 (change in control value): [-5, 5]
		Output (control signal): [-5, 5]
	Number of fuzzy sets	5 fuzzy sets for each linguistic variable
Adapation mechanisms	Initial mutation probability indices for the adaptive mutation mechanism choice	Mutation of the rule base: 50
		Mutation of a fuzzy set: 300
		Mutation of singletons: 150
		Mutation of membership function shapes: e=1000, b=20 and m=1000
Test conditions	Performance criterion	Sum of control error
	Simulation clock rates	System: 0.001, Controller: 0.5

Tab. 1: Settings for the generation of a fuzzy controller

The fitness function is designed to follow a given step with small reach time and little overshoot. The FC used is a standard Mamdani Min-Max controller. The final controller behaviour is shown in figure 5. Step response, membership functions and rule base for the generated controllers are also shown.

4 Summary and conclusion

An evolutionary method for the automatical generation of fuzzy controllers has been presented. This approach is characterized by its mechanisms and their arrangements,

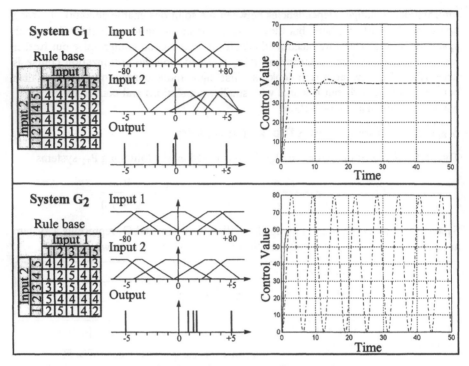

Fig. 5: Results of the fuzzy controller generation

making it different from genetic algorithms or evolution strategies. The adjusted evolutionary method highly takes the pecularities of fuzzy controllers into consideration and allows the simultaneous optimization of both fuzzy sets and rule base. The results show the effectiveness and high reliability of the approach. The described examples concerning the generation of fuzzy controller for simple dynamic systems have proved this. Other examples concerning nonlinear MIMO-systems will follow in the future, but the proposed method seems to work quite well also for more complex problems.

An important criterion for the evaluation of the algorithm is its reliability in finding good solutions. The algorithm has found indeed useful solutions in all test runs. This is remarkable, especially because of the random initialization of the individuals and the simultaneous optimization of linguistic terms and rule base. 100 generations with 30 optimization steps each seem to be more time consuming at first glance but the small population size of 20 individuals has to be considered, which lies below the usual size of 100 up to 150 individuals. The test runs show furthermore, that there is fast convergence and good results are got after about 20 generations. Together with the high reliability and less necessary test runs the approach seems to be justified.

Finally the break with conventional types of evolutionary techniques has not led to poor results. On the contrary, the development of an algorithm especially designed for fuzzy controllers seems to be more effective and reliable.

References

1. Bäck, T. and Hoffmeister, F.: Extended selection mechanisms in genetic algorithms. Proc. of the 4th Int. Conf. on Genetic Algorithms, 92-99, 1991.
2. Darwin, C.: Die Entstehung der Arten, Josef Singer Verlag, Charlottenburg und Leipzig.
3. Holland, J.: Artificial genetic adaption in computer control systems, PhD thesis, University of Michigan, 1971.
4. Karr, C.: Fuzzy control of pH using genetic algorithms, IEEE Trans. on Fuzzy Systems 1, 46-53, 1993.
5. Karr, C.: Genetic algorithms for fuzzy controllers, AI-Expert , 27-33, 1995.
6. Kropp, K.: Optimization of fuzzy logic controller inference rules using a genetic algorithm, Proc. EUFIT '93, 1090-1096, 1993.
7. Lee, A. und Takagi, H.: Integrating design stages of fuzzy systems using genetic algorithms, Proc. 2nd IEEE Int. Conf. on Fuzzy Systems, 612-617, 1993.
8. Maeda, M. und Murakami, S.: A self-tuning fuzzy controller, Fuzzy Sets and Systems 51, 29-40, 1992.
9. Zadeh, L. A.: Fuzzy-Sets, Information and Control 8, 1965.

Power Factor Correction in Power Systems Having Load Asymmetry by Using Fuzzy Logic Controller

Muğdeşem Tanriöven   Celal Kocatepe   Adem Ünal

Yıldız Technical University,Electric-Electronic Faculty
Electrical Eng. Dept. 80750 Yıldız-İstanbul/TURKEY

Abstract. As known, the studies on power factor correction have been carried out for many years. But in these studies, generally, phase powers received from supply are assumed to be balanced. In the paper, power systems having unbalanced load are considered.

The fuzzy logic controller based on the fuzzy set theory provides a useful tool for converting the linguistic control from the expert knowledge into automatic control rules. By using fuzzy automatic rules from the heuristic or mathematical strategies, complex processes can be controlled effectively in many situations.[1] But the most important and difficult point is how to obtain the proper control rules for a given system.

In this paper, power factor correction is corrected to demand value by using fuzzy logic controller for the electric power system containing current unbalance i.e. load asymmetry. As well as this, generalized fuzzy logic controller was resembled by writing its software in Q - basic programming language. Consequently, the software of the system's simulation results were given for power factor correction in the power systems having load asymmetry. In addition to that, the developed method is compared according to conventional compensation systems.

1 Introduction

Line losses of a power system are computed after load flow analysis is carried out. Line losses of the system are the function of busbar-powers, busbar voltages and tap ratios of transformers. In other words, the values of active and reactive powers received from the network determine line losses. It is tried to minimize line losses of power systems using various methods. The main purpose of optimal power flow discussed in 1968 and the concerning algorithms is to minimize line losses, i.e. to cause loss minimization.

In power systems it is tried to minimize both constant and variable costs. The minimization of line losses, variable costs, must be created with reference to optimal operation of power system. For minimization of line losses, power factors of busbars in power systems must be great, or between 0.9 and 1.0 . In the last twenty years,

There has been an increase in the current unbalances of power systems due to new technologies such as arc furnaces etc. If the three-phase load received from the supply is unbalanced because of like above conditions, in this case a new approach must be used to determine compensation power for loss-minimization or optimal operation.

2 The Theoretical Analysis Of Reactive Power Compensation

2.1 Simple Reactive Power Compensation

Let's examine the power factor correction on the vector diagram. If the capacitor is connected to circuit as a parallel, the state of figure 1a,b,c,d would occurred [2].

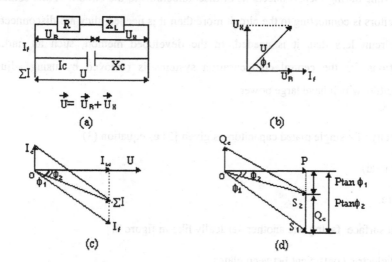

Fig. 1 Power factor correction at the single line.

Where,

$$\vartheta_c = P (\tan \phi_1 - \tan \phi_2) \tag{1}$$

ϑ_c : capacitor powers for desired power factor of the systems (VAr)
P   : Systems active power (kW)
ϕ_1 :Phase angle before correction
ϕ_2 : Phase angle after correction

On the other hand

$$\vartheta_c = u^2 \, w \, c \, 10^{-6} \qquad\qquad (2)$$

Where ;

c   : Capacity (μF)
ϑ_c :Capacitor power (VAr)
u   : Circuit voltage (V)
w   : Angular speed (rad /sec)

As known, in the central conventional compensation system, there are certain capacitor levels which are connected or disconnected to circuit according to change in load by system's contactors and relays. Capacitors can't be suitable for optimum level at all time during their connections or disconnections to the circuit. While sometimes, capacitors is connecting to the circuit more than it is needed, they are disconnecting to the circuit less than it is needed. In the developed method, such an undemand situations for the central compensation systems is removed by using adjustable capacitors which have large power.

Capacity of a single plated capacitors is given [3] as equation (3).

$$c = \varepsilon \, (S/d) \qquad\qquad (3)$$

Where,

S is a surface facing one another vertically like in figure 2.

ε : Dielectric coefficient between plates.

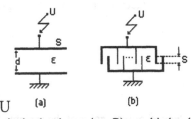

Fig.2. a) a single plated capacitor. B) a multi plated capacitor.

For needing of compensation power relative to plate surface (S) is given by equation (4). Compensation power ;

$$\vartheta_c = \Delta\vartheta$$
$$\vartheta_c = u^2 \, w \, \Delta c$$
$$\vartheta_c = u^2 \, w \, (c_1 - c_2)$$
$$\vartheta_c = u^2 \, w \, (\varepsilon/d) \, (S_1 - S_2) \tag{4}$$

As shown in the figure 2, for the power factor, $\cos\phi_2 = 0,95$, lets write $\cos\phi$ as fallow :

$$0,95 = \cos\arctan((\vartheta - \vartheta_c)/P) \text{ than,}$$
$$\vartheta_c = \vartheta - 0,328P \tag{5}$$

is obtained

3 Proposed Method For Power Factor Correction

The block diagram of the power factor correction system is given in the figure 3.

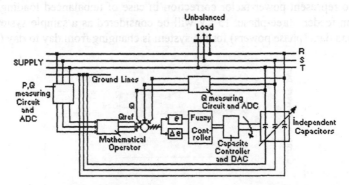

Fig. 3. Block diagram of the power factor correction system

In the block diagram, independently operating capacitors per phase, which arc ablc to adjustable by changing their surface is facing one another like in figure 2. The data active and reactive power measured from the system is used as an input value of mathematical operator for finding reference value of fuzzy controller.

3.1 Fuzzy Membership Functions and Rules

Figure 3, shows the input and output membership functions. And some fuzzy rules control rules as fallow [4] :

If error (e) is P and change in error (ce) is P than ΔQ is PL
If (e) is P and (ce) is Z than ΔQ is P
If (e) is P and (ce) is N than ΔQ is Z
If (e) is Z and (ce) is P than ΔQ is NS
If (e) is Z and (ce) is Z than ΔQ is Z
If (e) is Z and (ce) is N than ΔQ is PS
If (e) is N and (ce) is P than ΔQ is S

If (e) is N and (ce) is Z than ΔQ is NS
If (e) is N and (ce) is N than ΔQ is NL

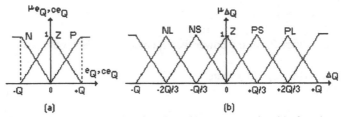

(a)

(b)

Fig. 3. (a) input membership functions (b) output membership functions.

4 Comparison of Developed Method with Conventional System and Conclusions

In order to represent power-factor correction in case of unbalanced loading, a radial distribution feeder, three-phase, 15 kV will be considered as a sample system. In the system, load data (phase powers) for the system is changing from day to day (fig. 4).

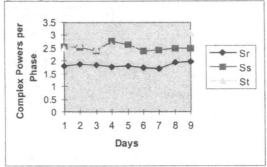

Fig.4 Phase power of sample system for 9 days.

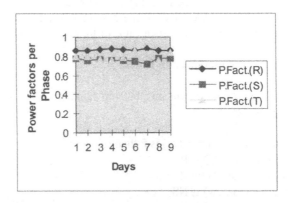

Fig.5 Systems power-factor before correction

In our system, the developed method was compared with the fallow conventional system :

- System is assumed to be balanced in application. For this reason, firstly power-factor is corrected by choosing one phase power as a reference

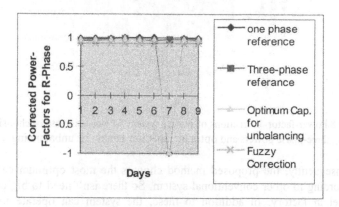

Fig.6 Power-factor corrections of sample system's R phase in case of choosing one phase (S), three-phase powers and optimal capacitor banks for unbalancing as a reference

- Secondly power-factor is corrected by choosing three-phase power as a reference.

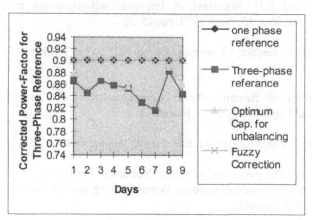

Fig.7 Power-factor corrections of sample system's S phase in case of choosing one phase (S), three-phase powers and optimal capacitor banks for unbalancing as a reference.

- Thirdly an optimal capacitor banks is determined for power-factor correction by using symmetrical components in case of having load asymmetry. [5]

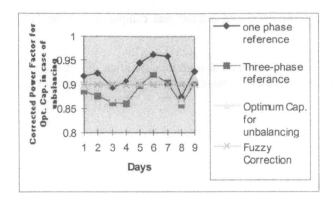

Fig.8 Power-factor corrections of sample system's T phase in case of choosing one phase (S), three-phase powers and optimal capacitor banks for unbalancing as a reference

- Consequently, the proposed method chooses the most optimum capacitor level according to other conventional system. So there isn't need to big compensation panel in factory. In addition to these, the system can operate in the central automation system. The adaptation of the system to load changing is perfect.

References

1. Procky and E.H. Mamdani, A linguistic self-organising process controller, Automatica, Vol. 15, No.1 1979, pp. 15-30

2. Türkmen, C. Geçtan, " Command Circuits (2)", Yeni Yol Press, İzmir, September 1992.

3. The Notes of Reactive Power Compensation, TMMOB, The Chamber of Electrical Engineering, 1983, İstanbul.

4. Timothy, " Fuzzy Logic with Engineering Applications", U.S.A., 1995.

5. Ay Selim, The periodical of Yıldız Technical University, "A new approach for power factor correction in power systems having load asymmetry", vol., page (59 - 66), 1993 Istanbul.

Fuzzy Partitions in Learning from Examples

Wieslaw Traczyk

Warsaw University of Technology, Warsaw, Poland

Abstract. Learning from examples is a popular methodology giving the set of rules (or decision trees) able to properly classify objects from predefined set. One of the main problems with this methodology is discretization – the process of converting continuous values of used attributes into more practical discrete values. Fuzzy partitions, introduced in this paper, can be viewed as a convenient way for expressing uncertainty in both: membership to discrete value and classification of cases, absent in the initial training set.

1 Introduction

A learning system is a computer program that makes decisions based on the accumulated experience contained in successfully solved cases. If the goal is to classify cases described by a sequence of attribute-value pairs, experience is usually presented as a set of examples (*training set*) of the following form:

$$\langle case, attr1, val1, attr2, val2, ..., attrn, valn, class \rangle$$

Training set is then transformed to a general description of each class, usually expressed as a set of decision rules or as a decision tree. If description of classes is good enough – it is able to classify properly each new case (from predefined set) with known attribute values.

Many different methods have been developed to address this issue. One more approach presented here is justified by its generality: extended notion of partition and fuzzy partition proved to be a good tool for design of systems with ideal data (quantitative and qualitative), as well as with uncertain, incomplete and imprecise information. The method deals with crisp, fuzzy and rough sets in a unified manner, what simplifies an implementation. This paper presents only these aspects of the methodology that concerns fuzzy partitions, or help to understand applications of fuzzy partitions.

2 Basic notions

2.1 Foundations of Learning from Examples

Let us consider the set X^* of *cases* (objects, situations), described by *attributes* from the set $A = \{a_1, a_2, ... a_N\}$. Attribute a_i can have *discrete values* from $U_i = \{u_1, u_2, ... u_K\}_i$ or *continuous values* – usually real numbers ($U_i \subset \mathbb{R}$ and

$u_{ij} \in U_i$). Strings of attribute values for each case are then taken from the set $\mathbf{U} = U_1 \times U_2 \times \ldots \times U_N$ and will be presented as $\mathbf{u} \in \mathbf{U}$.

Since number of possible values significantly influence complexity and duration of classification and reasoning – *primary* (original) values u are assembled into groups, usually described by linguistic name v, and each attribute has a set of *secondary values* $V_i = \{v_1, v_2, \ldots v_L\}_i$. Bounds of sets v are established according to norms or conventions existing in the field of application (and are given), or are selected in such a manner that the usage of a group name instead of the original value has as small as possible impact on the result of classification.

Let $\mathbf{V} = V_1 \times V_2 \times \ldots \times V_N$ and $\mathbf{v} \in \mathbf{V}$. When each component of $\mathbf{u}$ is covered by appropriate component of $\mathbf{v}$ then $\mathbf{u} \subseteq \mathbf{v}$. If all cases are to be classified into Q classes from the set $C = \{c_1, c_2, \ldots c_Q\}$, relations between important elements of ideal learning system may be defined by the following mappings:

Val: $X^* \rightarrow \mathbf{U}$ – primary valuation,

Dscr: $\mathbf{U} \rightarrow \mathbf{V}$ – discretization or coding,

Rec: $\mathbf{V} \rightarrow C$ – class recognition,

Cl: $X^* \rightarrow C$ – classification.

In reality – class recognition and classification are often ambiguous, and appropriate mappings have the form: Rec: $\mathbf{V} \rightarrow \mathcal{P}(C)$ and Cl: $X^* \rightarrow \mathcal{P}(C)$, where $\mathcal{P}(C)$ – a power set.

The process of learning is based on the training set – taken from experience information about a part (X) of all cases (X^*). For $X \subset X^*$ and all elements from $X = \{x_1, x_2, \ldots x_M\}$ we know what are affiliations to classes c and primary attribute values u:

$$\mathbf{u}_x = \langle a_1(x),\ a_2(x), \ldots, a_M(x) \rangle$$

The task of learning is to assign mappings Dscr and Rec and to present principles of classification in the form of rules or decision trees. Then, for a new case x^* with known primary valuation, one can find a proper class $Cl(x^*)$ from the equation

$$Cl(x^*) = \mathrm{Rec}(\mathrm{Dscr}(\mathrm{Val}(x^*))) \ . \tag{1}$$

The main problem with mappings Dscr and Rec is that the knowledge concerning only a part of all possible cases should be extended on entire space of cases.

2.2 Partitions

Approach suggested here attempts to utilize partitions, the effective calculus proposed for the first time (in this application) in [2]. We will present conventional definitions and some useful extensions, leading to fuzzy partitions.

Partition $\pi(E)$ of a finite set $E = \{e_1, e_2, \ldots\}$ is a set of mutually disjoint subsets (*blocks*) B:

$$\pi(E) = \{B_1, B_2, \ldots, B_R\}$$

such that $\forall ij \ B_i \cap B_j = \emptyset$ and $\bigcup_{i=1}^{R} B_i = E$

The *product* of partitions is defined as

$$\pi(E).\pi'(E) = \{B_{ij} | B_{ij} = B_i \cap B_j,\ B_i \in \pi, B_j \in \pi'\} \ . \tag{2}$$

The relation $\pi(E) \leq \pi'(E)$ is described by the formula

$$\pi(E) \leq \pi'(E) \Leftrightarrow \forall B_i \in \pi \exists B_j \in \pi' \; B_i \subseteq B_j \; . \tag{3}$$

If two sets are related by a function $f : E \to F$, and there exists some equivalence relation on the set F (for instance – equality, membership to the same block of partition), then set F (with $|F| < |E|$) can *impose* partition $\pi_F(E)$ on E, where

$$\pi_F(E) = \{B \mid (e_i, e_j \in B) \; \Leftrightarrow \; f(e_i) \equiv f(e_j)\} \; . \tag{4}$$

Since each block of this partition is determined by some class of the set F, blocks can be labeled by the symbols of appropriate class (B_1^α, B_2^β etc.).

In the case of ordered set (for instance $E_0 = \langle e_1, e_2, \ldots, e_7 \rangle$) partitions preserve ordering, e.g.

$\pi(E_0) = \langle e_1, e_2, e_3 \parallel e_4, e_5, e_6 \parallel e_7 \rangle = \langle B_1, B_2, B_3 \rangle$

If the set E contains real numbers, partially ordered from the smallest to the largest, it is assigned by special symbols, e.g.

$\pi(E_0) = \langle\langle e_1, e_2, e_3 \parallel e_4, e_5, e_6 \parallel e_7 \rangle\rangle$

Sometimes blocks of such partition represent bounded sets of continuous values – *intervals*. Assigning by $[m, M]$ interval with bounds m and M, one can describe the previous example by

$\tilde{\pi}(E_0) = \langle\langle [e_1, e_3], [e_4, e_6], [e_7] \rangle\rangle$

This kind of transition from discrete blocks to crisp intervals leaves some values from continuous universum outside the set of intervals. When there is no precise information about the gaps but given subset of real numbers (e.g. $E_0 = [e_0, e_8]$) should be covered – one can use *fuzzy intervals (fuzzy sets)* instead of crisp intervals.

We will assume that a membership function for an element between two crisp intervals is a linear function of the distance to each interval, what gives the reasons for trapezoidal form of this membership function. Using a 4-parameter description of a fuzzy set ($[a,b,c,d]$) we can present the fuzzy partitions of the set E_0 as

$\tilde{\pi}(E_0) = \langle\langle [e_0, e_0, e_3, e_4], [e_3, e_4, e_6, e_7], [e_6, e_7, e_8, e_8] \rangle\rangle$

More generally – *fuzzy partition* of a continuous set $E = [e_0, e_Z]$ is a set of fuzzy subsets (fuzzy intervals) $\tilde{B}$:

$\tilde{\pi}(E) = \{\tilde{B}_1, \tilde{B}_2, \ldots, \tilde{B}_R\}$

such that $\forall e \in E \; \sum_{i=1}^{R} \mu_{B_i}(e) = 1$.

The product of fuzzy partitions is defined as

$$\tilde{\pi}(E).\tilde{\pi}'(E) = \{\tilde{B}_{ij} \mid \mu_{\tilde{B}_{ij}}(e) = \min(\mu_{B_i}(e), \mu_{B_j}(e)), \; B_i \in \pi, B_j \in \pi'\} \; . \tag{5}$$

The relation $\leq$ between fuzzy partitions is described by

$$\tilde{\pi}(E) \leq \tilde{\pi}'(E) \; \Leftrightarrow \; \forall B_i \in \pi \; \exists B_j \in \pi' \; \mu_{B_i}(e) \leq \mu_{B_j}(e) \; . \tag{6}$$

3 Learning from Examples

3.1 Examples with Consistent Data Set

The case with discrete values. To illustrate the possibilities of the suggested approach we will first consider simpler case with discrete attribute values.

The training set determines classes for all cases from X, thus equal values given by $\mathrm{Cl}(X)$ impose partition on the set of cases X:

$$\pi_{\mathrm{Cl}}(X) = \{B_1^{c_1}, B_2^{c_2}, \dots, B_G^{c_G}\}$$

with blocks labeled by names of adequate classes.

For simplicity – we will use the symbol a as a name of an attribute, as well as a function $a(x)$, returning the value u of an attribute. Let

$$a_i(X) = \langle a_i(x_1), a_i(x_2), \dots, a_i(x_M)\rangle.$$

Equal values from this set can impose partition on the set of cases:

$$\pi_{a_i}(X) = \{B_{1i}^{u_{1i}}, B_{2i}^{u_{2i}}, \dots\}$$

Sometimes such an attribute exists that its blocks are relevant to classes, and can be used for class identification. More formally:

$$\text{if } \exists i \ \pi_{a_i}(X) \le \pi_{Cl}(X) \quad \text{then} \quad \forall m \ B_m^{c_m} = B_{ni}^{u_{ni}} \cup B_{oi}^{u_{oi}} \cup \dots \ . \qquad (7)$$

This means that, when the condition is satisfied, each class c_m can be recognized by the following rule:

$$\text{IF } a_i(x^*) = u_{ni} \text{ OR } a_i(x^*) = u_{oi} \text{ OR } \dots \text{ THEN } \mathrm{Cl}(x^*) = c_m.$$

or, using secondary values $v_{ni} = u_{ni} \cup u_{oi} \cup \dots$, by simpler rule:

$$\text{IF } a_i(x^*) = v_{ni} \text{ THEN } \mathrm{Cl}(x^*) = c_m.$$

Set of rules like this (for all classes c_i) enables complete classification of the set X^*.

Usually the attribute fulfilling conditions from (7) does not exist, and its role should be taken by the (as small as possible) set of attributes. Let $P = \{p, q, \dots\}$ and $\mathbf{U}_P = U_p \times U_q \times \dots$ with $\mathbf{u}_j \in \mathbf{U}_P$ and $\mathbf{u}_j = \langle u_{pj}, u_{qj}, \dots\rangle$. Just as in the previous case – one can formulate the rule:

$$\text{if } \prod_{i \in P} \pi_{a_i}(X) \le \pi_{Cl}(X) \quad \text{and} \quad \prod_{i \in P} \pi_{a_i}(X) = \{B_1^{\mathbf{u}_1}, B_2^{\mathbf{u}_2}, \dots\}$$

$$\text{then} \quad B_m^{c_m} = B_n^{\mathbf{u}_n} \cup B_o^{\mathbf{u}_o} \cup \dots$$

When these conditions are satisfied – classification can be done by the set of production rules:

$$\text{IF } a_p(x^*) = u_{pn} \text{ AND } a_q(x^*) = u_{qn} \text{ AND } \dots \text{ THEN } \mathrm{Cl}(x^*) = c_m$$
$$\text{IF } a_p(x^*) = u_{po} \text{ AND } a_q(x^*) = u_{qo} \text{ AND } \dots \text{ THEN } \mathrm{Cl}(x^*) = c_m$$

$$\dots$$

Two values of a_i with the same function in these rules can be joined into a secondary value v.

The case with Continuous Values. When $a_i \in \mathbb{R}$ – intervals of this value can help in class recognition. Let

$$a_i(X) = \langle\langle a_{ri}(x_r), a_{si}(x_s), \ldots \rangle\rangle$$

Partitions and blocks of attribute values will be, for distinction, assigned by τ and D. If values of classes are able to impose partition on values of a_i, blocks of this partition point out values of classes:

$$\tau_{\text{Cl}}(a_i(X)) = \langle\langle D_{1i}^{c_{1i}}, D_{2i}^{c_{2i}}, \ldots \rangle\rangle$$

Blocks with different real values are not helpful in the process of class recognition and therefore should be transformed to intervals. If number of samples in the training set is properly chosen, we can assume that all values between the extreme values in the block have equal impact on classification, and blocks describe intervals. Intervals constructed in this way leaves gaps with unknown membership. Since there are no reasons for the location of a crisp boundary line splitting the gap into two regions allocated to neighbouring intervals, introduction of fuzzy intervals seems to be fully justified.

Fuzzyfication of the crisp partition τ_{Cl} gives the set of fuzzy blocks that can be used as secondary values of considered attribute:

$$\tilde{\tau}_{\text{Cl}}(a_i(X)) = \langle\langle \tilde{D}_{1i}^{c_{1i}}, \tilde{D}_{2i}^{c_{2i}}, \ldots \rangle\rangle = \langle\langle \tilde{v}_{1i}, \tilde{v}_{2i}, \ldots \rangle\rangle$$

The gap between two crisp intervals is covered by two fuzzy intervals, so value $a_i(x^*)$ of a new case has two membership functions (μ_α describes membership to $\tilde{v}_\alpha$). In this case blocks of $\tilde{\tau}_{\text{Cl}}(a_i(X))$ are directly connected with names of classes, and therefore membership to block (secondary value) may be understood as certainty measure of appropriate class. If $\tilde{v}_{zi}$ and $\tilde{v}_{z'i}$ are neighbouring values, classification can be done with the help of the following rule:

IF $a_i(x^*) \in [\tilde{v}_{zi}/\mu_{zi}, \tilde{v}_{z'i}/1 - \mu_{zi}]$ THEN $\text{Cl}(x) = [c_{zi}/\mu_{zi}, c_{z'i}/1 - \mu_{zi}]$

or, introducing new symbols, by the rule:

IF $a_i(x^*) \in \tilde{w}_{zi}$ THEN $\text{Cl}(x) = \eta(\tilde{w}_{zi})$

If partition of values of one attribute has too many blocks or does not separate classes properly – we should consider values of the set of attributes. Since multiplication of two partitions with m and n blocks generate mn blocks, it is reasonable to use several attributes with small number of blocks. When two or more adjoining blocks D from $\tau_{Cl}(a_i(X))$ are substituted by one block $\mathbf{D}$, a new partition τ^i arises, such that

$$\tau^i \geq \tau_{Cl}(a_i(X)) \text{ and } \tau^i = \langle\langle \mathbf{D}_{1i}^{\alpha_i}, \mathbf{D}_{2i}^{\beta_i}, \ldots \rangle\rangle$$

A set of partitions τ_i, with $i \in P = \{p, q, \ldots\}$ and $\mathbf{v}_m = \langle v_p, v_q, \ldots \rangle$, can separate classes if imposed partitions on X are in proper relation with π_{Cl}:

$$\text{if} \quad \prod_{i \in P} \pi_{\tau^i}(X) \leq \tau_{Cl}(a_i(X)) \quad \text{and} \quad \prod_{i \in P} \pi_{\tau^i}(X) = \{B_1^{\mathbf{v}_1}, B_2^{\mathbf{v}_2}, \ldots\} \tag{8}$$

then there are some blocks $B_n^{\mathbf{v}_n}, B_o^{\mathbf{v}_o}, \ldots \}$ such that

$$B_m^{c_n} = \{B_n^{\mathbf{v}_n}, B_o^{\mathbf{v}_o}, \ldots\} . \tag{9}$$

Now, when we know which partitions are useful, they can be fuzzyfied into the form:

$$\tau_{\text{Cl}}(a_i(X)) = \langle\langle \tilde{\mathbf{D}}_{1i}^{\alpha_i}, \tilde{\mathbf{D}}_{2i}^{\beta_i}, \ldots \rangle\rangle = \langle\langle \tilde{v}_{1i}, \tilde{v}_{2i}, \ldots \rangle\rangle$$

Appropriate rules look as follows:

IF $a_p(x^*) = \tilde{w}_{pn}$ AND $a_q(x^*) = \tilde{w}_{qn}$ AND ... THEN $\text{Cl}(x^*) = \eta(\tilde{w}_{pn}, \tilde{w}_{qn}, \ldots)$

IF $a_p(x^*) = \tilde{w}_{po}$ AND $a_q(x^*) = \tilde{w}_{qo}$ AND ... THEN $\text{Cl}(x^*) = \eta(\tilde{w}_{po}, \tilde{w}_{qo}, \ldots)$

$$\ldots$$

Formal description of the function η is more complicated (because of many indices) than practical calculations, therefore we will explain the role of η on a simple example.

Example 1. Let $P = \{p, q\}$, $V_p = \{A, B\}$, $V_q = \{Y, Z\}$
and $\text{Rec}(A, Y) = c_1$, $\text{Rec}(A, Z) = c_2$, $\text{Rec}(B, Y) = c_2$, $\text{Rec}(B, Z) = c_1$.
If $a_p(x^*) = [A/0.7, \ B/0.3]$ and $a_q(x^*) = [Y/0.9, \ Z/0.1]$ then

$$\text{Cl}(x^*) = [c_1/0.7 \times 0.9, \ c_2/0.7 \times 0.1, \ c_2/0.3 \times 0.9, \ c_1/0.3 \times 0.1]$$
$$= [c_1/0.63, \ c_2/0.7, \ c_2/0.27, \ c_1/0.3]$$

and finally: $\text{Cl}(x^*) = [c_1/0.66, \ c_2/0.34]$.

3.2 Examples with Inconsistent Data Set

When secondary values of all attributes are given and it is impossible to find values for $\mathbf{v}$ such that $\pi_{\mathbf{v}}(X) \leq \pi_{\text{Cl}}(X)$ – values are inconsistent and classification is ambiguous. If we only want to know which cases are classified correctly and what are bounds of the indiscernible examples – partitions can help to determine *rough sets* with

$X_{\{c_i\}} = \{B^{\mathbf{v}} \mid B^{\mathbf{v}} \subseteq B^{c_i}\}$ – lower approximation,

$X^{\{c_i\}} = \{B^{\mathbf{v}} \mid B^{\mathbf{v}} \cap B^{c_i} \neq \emptyset\}$ – upper approximation,

$X_{\Theta} = \bigcap_{c_i \in \Theta} X^{c_i}$ – cases with common classes.

Generic rules can be defined as follows:

$$\text{IF } x^* \in B^{\mathbf{v}} \text{ AND } B^{\mathbf{v}} \subseteq X_{c_i} \text{ THEN } \text{Cl}(x^*) = c_i$$

$$\text{IF } x^* \in B^{\mathbf{v}} \text{ AND } B^{\mathbf{v}} \subseteq X_{\Theta} \text{ THEN } \text{Cl}(x^*) = \Theta$$

If all attributes are discrete – such a rough approximation is the only information we can get, but if some attributes are continuous – we can try to limit indiscernibility redefining secondary values of these attributes. This can be done by totally new process of discretization – in accordance with the methodology described above, or by partial correction of some secondary values.

When $x_A, x_B \in X_{c_a, c_b}$ then for all i: $a_i(x_A \in v_i$, $a_i(x_B \in v_i$ and $\text{Cl}(x_A \neq \text{Cl}(x_B$. Value v_i should be substituted by two fuzzy values $\tilde{v}_A$, $\tilde{v}_B$, with regard to other cases from the set X_{c_a} and X_{c_b}.

542

4 Conclusions

Partitions can be viewed as a convenient tool for learning from examples when data set is consistent or inconsistent, and attribute values are discrete or continuous. Fuzzy partitions, introduced here, are specially useful in the process of discretization. The method of discretization is *global* because it considers all attributes when choosing the best way for classification.

References

1. Chmielewski M.R., Grzymala-Busse J.W.: Global discretization of continuous attributes as preprocessing for machine learning. International Journal of Approximate Reasoning **15** (1996) 320–331
2. Grzymala-Busse J.W.: On the reduction of instance space in learning from examples. Methodologies of Intelligent Systems, 5. Elsevier Co.(1990) 388–395
3. Klir G.J., Folger T.A.: Fuzzy sets, uncertainty, and information. Prentice Hall. (1988)
4. Traczyk W.: Partitions in knowledge discovery and learning. Foundations of Computing and Decision Sciences **19** (1994) 145–156

Economic Forecasting Using Genetic Algorithms

Adriana Agapie, Alexandru Agapie*

Institute of Economic Forecasting, Bucharest, Romania
* National Institute of Microtechnology, Bucharest, Romania
agapie@oblio.imt.pub.ro

Abstract. When one deals with short-length, non-stationary time series with seasonal components, the statistical procedures of forecasting or even the neural networks prove to be unsatisfactory. We propose in this paper a Genetic Algorithm method for finding the optimal parameters involved in the classical Holt-Winters model of time series forecasting.

1. Introduction

In the last years a great number of intelligent techniques - such as different types of Neural Networks (NN), Fuzzy Logic (FL) based algorithms, or the State Space Reconstruction method - promised insights that the traditional approaches to the very old problem of forecasting cannot provide. Although, the most difficult situation appears when the length of the time series is too short: in this case one can apply neither the traditional statistical procedures for parameter estimation, nor the NN algorithms (even the high competitive models from [4]).

Based on the mechanics of natural genetics and natural selection, GA are probabilistic algorithms which combine survival of the fittest among string structures (called *chromosomes*) with a structured yet randomized information exchange to form a robust search algorithm, [2].

2. The Holt-Winters Seasonal Model. Experimental Results

The exponential smoothing forecasting procedure may be extended in order to incorporate the local linear trend and the seasonal component. Non-stationary time series with seasonal components are usually fitted with local polynomials. The main idea is that the forecasts have to be adaptive, thus the low-order polynomials used for extrapolation must have different coefficients at each observation, [3].

When the data exhibit the seasonal behavior the direct extension of the Holt's method - due to Winters -may be applied. Thus, the forecasting function becomes:

$F(t+k)=[a_0(t)+ka_1(t)]C(t+k-s)$

where $C(t)$ is a multiplication seasonal effect and s is the number of observation points in the whole year. The corresponding updating formulae are as follows:

$a_0(t)=\alpha_1 y(t)/C(t-s)+(1-\alpha_1)[a_0(t-1)+a_1(t-1)]$

$a_1(t)=\alpha_2[a_0(t)-a_0(t-1)]+(1-\alpha_2)a_1(t-1)$

$C(t)=\alpha_3 y(t)/a_0(t)+(1-\alpha_3)C(t-s)$

As in the two-parameters model, the values of α_1, α_2, α_3, and the initial values of a_0, a_1 and C are required at start (as in [1]). In the examples below we extend the use of GA to the optimal choice of s (and the initial values C(1),.., C(s) respectively).

We analyze the efficiency of the GA method proposed on two examples of short-length, non-stationary time series with seasonal components. Each time series was shared in two parts: a training set and a test set (as in the NN methods). The chromosome consists of more building blocks, corresponding to the values of s, α_1, α_2, α_3, $a_0(s)$, $a_1(s)$, C(1), ..., C(s) - for the Holt-Winters model with variable s.

In fig.1 is presented the evolution of a GDP index.

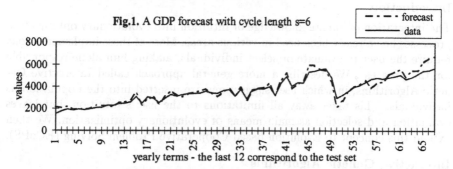

Fig.1. A GDP forecast with cycle length s=6

- - - - forecast
——— data

yearly terms - the last 12 correspond to the test set

In fig.2 is depicted the agricultural component of the same GDP. The two cycle lengths obtained are very close - this has an economic reason: in the GDP considered, the agricultural component is prevalent.

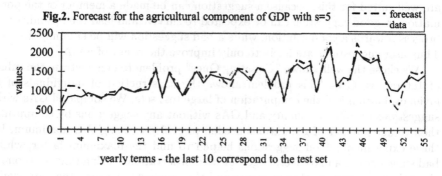

Fig.2. Forecast for the agricultural component of GDP with s=5

- - - - forecast
——— data

yearly terms - the last 10 correspond to the test set

Short References

1. Adriana Agapie, A. Agapie: Parameter Estimation in Time Series Forecasting Using Genetic Algorithms, EUFIT'96, Aachen, Germany, 1996, 2155-2158
2. D.E.Goldberg: Genetic Algorithms in Search, Optimization and Machine Learning, Addison Wesley Publishing Comp., 1989
3. M.Kendall: Time Series, Edward Arnold Publishing Comp., 1993, 123-143
4. A.S.Weigend, N.A.Gershenfeld (Eds.): Time Series Prediction: Forecasting the Future and Understanding the Past, Addison Wesley Publishing Comp., 1994

Interactive Genetic Algorithms and Evolution Based Cooperative Problem-Solving

J. Albert, J. Schoof

Lehrstuhl für Informatik II, Universität Würzburg, Am Hubland, D-97074 Würzburg
schoof@informatik.uni-wuerzburg.de

Introduction

The integration of human knowledge or intuition into evolutionary optimization processes has already been used in various areas. Most of these implementations require the user to evaluate or select individuals, making him alone responsible for the progress. We describe a more general approach called Interactive Genetic Algorithms, in which user suggestions are inserted into the population as individuals. This takes away all limitations to the user's intuition, but leaves evaluation and selection as main means of evolutionary optimization. We then extend this concept to Evolution based Cooperative Problem Solving (ECoPS).

Interactive Genetic Algorithms

The idea of using human knowledge or intuition as part of an optimization process is described by e.g. Krolak, Felts and Marble (in Communications of the ACM, Vol. 14, No. 5). GAs and other population based optimization procedures are well suited for this, because a suggestion can be made a member of the population. A good suggestion can be expected to survive and lead the optimization process towards a good solution, while a bad suggestion will be eliminated soon. Thus user suggestions are likely to only improve the result of a GA.

We chose the well-known "Counting Ones" problem to demonstrate the value of Interactive GAs. This problem allows the construction of individuals of a given evaluation and the computation of large test sets. We compared GAs with suggestions of different quality and GAs without any suggestions by computing the average number of generations needed to evolve to the global maximum. In these experiments good suggestions helped to find the maximum faster, while bad suggestions had no effect. Together with observations during runs it becomes clear that the number of suggestions and the moments at which they are made effective are important for the progress of the GA. The main problem in our opinion is to keep the number of insertions to a GA at a reasonable number.

Evolution based Cooperative Problem-Solving

Evolution based cooperative problem-solving is quite similar to Interactive GAs. The main difference is that suggestions can also be generated by other programs in addition to the human participation. An optimization tool for simulations of material design processes has been equipped with a program for statistical design of experiments, which provides the GA with hints. We also use a database to collect all evaluations done so far for further use. This combination's solutions are better and significantly more reliable than the the ones of each of its parts.

A Fuzzy Model Based Control of Well Mixed Reactor with Cooling Jacket

Alpbaz, M., Zeybek, Z., Özbek, L., Aliev, F.

Faculty of Science, Ankara University 06100 Tandogan, Ankara /Turkey

Abstract. Fuzzy logic control uses linguistic rather than crisp numerical rules to control the industrial processes. Fuzzy logic control may be attractive when the process is either difficult to control or to model by conventional methods. Fuzzy control of processes is an alternative when systems cannot be well controlled by classical or modern control techniques which are based on crisp mathematical models and also reply heavily on measurements to indicate variation in process conditions. For many processes, models and measurements are very difficult to obtain correctly and they show nonlinear behavior. Chemical reactors, pH neutralization process and waste water treatment are the examples for these processes. In such a systems, control rules can be developed depending on whether process is to be controllable. If the operators knowledge and experience can be explained in words, then linguistic rules can be written easily. In such cases, fuzzy model refers to the description of the operators input/output control actions using fuzzy implications.

In the present work inlet temperature of a well mixed reactor with cooling jacked was controlled by using fuzzy model based control algorithm at desired value. Inlet and outlet temperature of the cooling water of the jacket were measured with two thermocouples which are connected to the on-line computer. Cooling water was pumped to the system at a certain temperature. Heat input from immersed heater was given to the system with definite values of heat input and cooling flow rate system shows steady-state behavior. With some physical properties these system are observed nonlinearities. Heat input was adjusted by Triack Module and it was chosen as an manipulated variable.

The reference sets and scaling factors are used for mixing tank temperature and heat input deviation variables from steady state values. Necessary data was generated by applying the pseudo random uniform effects in every 15 minutes. The deviations from steady state of heat input are distributed between ±9000 cal/sec. Six thousand data points were recorded for run time. The identification algorithm was applied and the model relation matrix was evaluated.

The developed model is used to predict the output for possible control actions. Nine allowable control changes in heat input are taken as the values of 0.0, ±500, ±1000, ±4000, ±9000 cal/sec. These were added in turn to the current values of heat input and fed to the model together with the current values of tank temperature. The model calculates the expected value of tank temperature at the reset control interval. The decision maker then selects the most favorable action to take the one which results in the smallest error. The selected control action is then applied to the process and the this produce is repeated every control interval.

Fuzzy control of the reactor temperature was realized experimentally. Also theoretically work has been done using simulation program. It was observed that theoretical result gave very good agreement with experimental datas.

Key words: Fuzzy Logic, Well Mixed Reactor, Model Based Control.

Optimizing Video Signal Processing Algorithms by Evolution Strategies

H. Blume, O. Franzen, M. Schmidt

Universität Dortmund
Lehrstuhl für Nachrichtentechnik, AG Schaltungen der Informationsverarbeitung
44221 Dortmund
bl@nt.e-technik.uni-dortmund.de

Abstract

Today many kinds of postprocessing are used in digital TV receivers or multimedia terminals for video signals to enhance the picture quality. To achieve this the properties of human visual perception have to be regarded. Because of the nonlinear nature of human visual perception (e.g. perception of edges and objects) many algorithms have been developed and optimized by heuristic methods or by application of rough image models.

This is a severe problem as there are sometimes contradictory demands (e.g. detail resolution and alias suppression) and there are many optimization problems which cannot be solved analytically. Furthermore the simulations which have to be carried out in the field of video processing have to take into account a great variety of test sequences and therefore possess a heavy simulation load.

In this paper we present the results of evolution strategies (ES) applied to develop and optimize some modules of a video signal processing feature box. The modules we have analyzed are as follows:

A **proscan conversion** module is required to convert incoming interlaced TV signals into a progressive format which is obligatory for computer monitors or LCD and DMD devices (e.g. projectors) as they cannot display interlaced signals [1].

Further **linear and nonlinear filter techniques** are required for spatial conversion techniques like zooming or a picture in picture reproduction.

For high quality **temporal scan conversion techniques** (e.g. 50 Hz interlace to 100 Hz interlace scan conversion reducing annoying artifacts as large area or detail flicker) motion vector based video processing is state of the art [1]. The motion information is generated by **motion estimation** algorithms.

Optimizing a nonlinear median based proscan conversion

Applying nonlinear filters like rank order filters for picture processing the design process is much more complicated than for linear filters. Whereas linear filters can be described analytically in the frequency domain this is not possible for nonlinear filters. But nonlinear filters offer a lot of advantages like edge preservation, spike removal or robustness which make them very attractive for image processing. There are generally two methods for designing nonlinear filters. These are

- design by root signals, which are invariant concerning filtering with a given nonlinear filter,
- design by investigation of the output distribution function of a given nonlinear filter.

But each method has its own limitations well known by the experienced user so that still many algorithms are developed by heuristic methods, sequence simulations and subjective evaluations [1].

One of the most interesting applications for nonlinear filters (especially weighted median filters) in the video processing domain is a proscan conversion that means the interpolation of missing lines for interlaced video signals. Many proscan conversion algorithms make use of median filters [1]. We have optimized such a nonlinear median based proscan conversion by evolution strategies.

The evolution strategy implemented for this problem represents a median filter arrangement by a vector of possible median window assemblies for proscan conversion. For each possible median window position a weighting factor has to be chosen which is an integer value in the range between 0 and 10. In summary with a total window size (sum of all possible pixel positions within three subwindows) of 50 this yields a search space of about 11^{50} different configurations. Regarding the time and memory consuming simulation for each individual (filter configuration) it becomes clear that only a tiny fraction of these configurations can be regarded in practice so that a sophisticated search algorithm is required.

Fig. 1 depicts the results of an optimization run for a nonlinear proscan conversion. The objective function which is used here for evaluating each individual is the PSNR (Peak Signal to Noise Ratio) which is computed by a comparison between a filtered test sequence and an ideal reference sequence.

The evolution strategy which is used for the optimization process of Fig. 1 is a (μ,λ)-ES [3] with μ=10 parents and λ=70 offsprings which are generated each time by mutation and discrete/intermediary recombination of two parent individuals. From the resulting offspring population the best 10 individuals are selected as new parent individuals by their corresponding PSNR values. After that selection the evolutionary loop of mutation and recombination, evaluation and selection starts again (for an exhaustive survey of evolution strategies see also [3]).

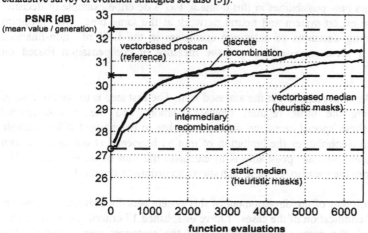

Fig. 1: *Simulation results for an optimization process with a (10, 70)-ES for a proscan conversion*

In Fig. 1 also the simulation results of a static (non vectorbased) median proscan algorithm [1], the results of a vectorbased median algorithm with a median mask which has been found heuristically and the results of a vectorbased reference method [2] are depicted. It can be seen that after 6500 function evaluations the new median configuration yields significantly better results than the static and the vectorbased approach with heuristic masks and approaches more and more the reference method which requires much more implementation effort than the method which has been optimized here.

Besides this proscan conversion we have investigated further video processing specific problems. Evolution strategies yielded also very good results in optimizing these problems.

References

[1] *Blume, H.; Schröder, H.*: "Image Format Conversion - Algorithms, Architectures, Applications" Proc. of the IEEE ProRISC Workshop, pp. 19-37, Mierlo, Netherlands 27./28. 11. 1996

[2] Ivanov, K.V. : "Ein neues Verfahren zur Konversion von Fernsehbildsignalen für die progressive Wiedergabe", (in German) FKT-Magazin, 48. Jhg.; Nr. 10, pp. 1-7, 1994

[3] *Schwefel, H.P.; Bäck, T.*: "Evolution Strategies I/II" chap. 6/7 in "Genetic Algorithms in Engineering and Computer Science" ed. by J. Périaux, G. Winter; Wiley & sons, New York, 1995

An Example for Decision Support in Insurance

István Borgulya
Janus Pannonius University, Faculty of Business and Economics
H-7621 Pécs, Rákoczi u 80.

There are several judicial sentences concerning non-financial compensation. Most of the cases are about road accidents and the clients frequently seek the help of legislation because they find the sum of compensation too low. We have got lots of such cases available to be able to predict future sentences. For more than twenty years statistics and graphs have been used for assessment. The ever more popular method of artificial intelligence offers new possibilities in this domain, too. I describe example of decision support that use expert system and neural network at the same time. The program aims at supporting decisions of insurance specialists in the field of non-financial compensation: it gives an assessment of the measure of compensation (based on previous cases).

This method takes two steps to get the assessed sum. The first step is to see the case as 'typical', meaning the most regular, typical situations and the compensational assessment. At these 'special cases' the sum assessed has to be changed if for example the injured is an alcoholic, or the limitation of motion happened at old age, the sum granted is reduced. Another possibility is to calculate the sum of an allowance. The system determines the measure of these alterations according to some rules.

The most difficult part of the development of this system was the elaboration, uniform description and codification of the cases. I have established 17 criteria gathering all the factors by which the cases were described and the sentences were justified. The description of a case is done with these 17 criteria in the program and each criterion has to be given a score. If there is a factor in the case that is not listed in the corresponding criterion the relative score has to be assessed.

The program assessing compensation is made up of three parts. The parts and their tasks are the following:
1. a neural network, that assess the award (using the Neuroshell 2 program) as a typical case,
2. an expert system, that modifies the sum assessed at special case (using the Level 5 Object shell),
3. the Excel (main) program, that links neural network to expert system.

The main program guarantees the connection with the user and uses a client/server architecture for the communication between programs.

The catalogue of standard cases helps to determine the cases, as it lists the possible damages grouped by the criteria. If a situation is not listed in the catalogue the user has to decide about the value of the relative score. Thus the program supports the final decision and requires constant consideration.

Fuzzy Identification of Distributed Components

E. Damiani ** M.G. Fugini*
** Università di Milano, Polo di Crema
* Politecnico di Milano
Dipartimento di Elettronica e Informazione

Abstract. Several software component identification problems require evaluation of the fitness of a *candidate* on the basis of the information attached to it by a *classification model*. Fuzzy query algebras can effectively deal with these problems, choosing query execution mechanisms on the basis of the semantics selected by the user.

1   Introduction

Some of the current issues related to the evolution of software technology, namely product-intensive, reuse-aware development and distributed execution environments, exhibit common features such as software components classification for reuse or re- engineering [1], or identification of services over a network. Modern *development environments* and *distributed execution environments* can take advantage of fuzzy component classification and query systems. For example, in the framework of the increasing influence of WWW and Java/CORBA technology, an effective identification of server objects would allow for a user-centric computing model where the desktop client is a generic navigation and command center, activating the user actions via a dynamic remote invocation mechanism.

2   A Fuzzy Component Trading System

In both scenarios sketched above, a query to a *Component Trading System - CTS* (*Trader* in the CORBA standard) is a list of the *offer properties*, functional and non-functional, that the desired component should possess. This property list must be compared with similar lists, called *software descriptors*, previously attached to available objects by some consistent classification mechanism ([1]). Usually, not all properties stand on an equal basis; this difference may be expressed through *weights* associated to them by the classification mechanism. Thus, a CTS can be outlined as a single relational table, whose schema is simply   offer property, ObjectID, weight. Using a standard extension of the relational model [2], we interpret the weights as values of the usual membership function μ, obtaining a fuzzy relation. Considering a query as another fuzzy table whose schema is property, the problem of computing a ranked list of ObjectID of candidates having a certain degree of fitness w.r.t. the desired properties can be cleanly stated as computing the fuzzy relational division between those tables. Although relational division is not a primitive operator [2], this way of stating the problem helps relating the algebraic setting to application semantics. In fact, a wide variety of such semantics can be expressed by our simple model: for instance, weights may represent the relative *importance* of the listed properties, but could also be used as *fulfilment* degrees of performance/cost ratios that must be guaranteed by the retrieved component. Applications may affect fuzzy division in many other ways; for instance the well-known *black box* component reuse technique imposes all requested properties to be present in the retrieved components while other standard techniques, called *white box*, allow for partial fulfillment of requirements.

A set of pre-defined application-oriented semantics may be used as a parameter in the query execution of our fuzzy CTS. We recall that given two relations S and R, the usual relational division $\div$ is defined as follows:

$$x \in R \div S \iff S \subseteq \Gamma^{-1}(x), \ where \ \Gamma^{-1}(x) = \{a|(x,a) \in R\} \qquad (1)$$

If S and R are fuzzy relations, we have the same expression, but different kinds of fuzzy inclusions may be used in the right hand side of formula (1). This results in different semantics. For instance, a fuzzy-implication based inclusion, $A \subseteq B \iff \forall x(x \in A) \rightarrow (x \in B)) \iff \forall x(\mu_A[x] \rightarrow \mu_B[x])$ resulting in $min_x(\mu_A[x] \rightarrow \mu_B[x])$ can be associated to black-box technique. It is easy to verify that if this inclusion is used in (1), the lack of even one of the requested properties will result in the exclusion of the corresponding component from the result of the fuzzy division. Different kinds of white-box semantics can in turn be modelled by adopting a compensatory approach to inclusion, defined as

$$A \subseteq B = \frac{\sum_x T(\mu_A[x], \mu_B[x])}{\sum_x T(\mu_A[x])} \qquad (2)$$

where T is any T-norm. In our CTS, the user will choose the desired application-oriented semantics at the interface level; the choice will be translated in a fuzzy division definition (e.g., by choosing the T-norm to be used in (2)) and will result in a custom query execution mechanism. A complete fuzzy *Component Query Language* for our CTS based on this approach is currently being experimented, and a prototype environment is already in place.

References

1. Bosc, P., Pivert, O., SQLf: A Relational database Language for Fuzzy Querying, IEEE Trans. On Fuzzy Systems, vol.3, n.1 Feb. 1995.
2. Damiani, E., Fugini, M.G., Bellettini, C., A Hierarchy-aware Approach to Faceted Classification of O-O Libraries, Technical Report, Politecnico di Milano, 001-97

Neural Control of a Nonlinear Elastic
Two-Mass System

Michael Englert, Rainer Trapp, Richard de Klerk

Forschungszentrum Karlsruhe, Hauptabteilung Ingenieurtechnik
Postfach 3640, D-76021 Karlsruhe, FRG

In designing and developing telerobotic systems for the *Minimal Invasive Surgery* (MIS) various control methods have been designed. These control concepts take the high influence of nonlinear friction on the closed loop controlled system into consideration.

Static and dynamic deviations as well as stick-slip effects, caused by nonlinear friction, restrict the performance of conventional motion control systems. Therefore in addition to observer and fuzzy based control concepts a neural network based motion control system was developed to compensate especially the influence of nonlinear friction on the closed loop system.

The neural control structure which is presented here can be used to improve performance of electro-mechanical systems with non-linear friction and backlash. Simulations and tests on a real-time test bench show that the neural controller is able to compensate the influence of friction and to remove stick-slip effects. A neural network is used in parallel to a conventional feedback controller. A *Multilayer Perceptron* with one hidden layer, four inputs and one output neuron has been used. The network is trained online with the backpropagation learning algorithm, which enables it to adapt immediately to disturbances or changes in system parameters.

The developed *Direct Learning Controller* is able to improve performance of the nonlinear system. Static and dynamic errors are much reduced and stick-slip effects can be removed.

This neural control concept can be applied to systems which suffer from nonlinear friction and require slow movements or precise positioning.

All simulations were carried out with the *MATRIX$_X$/Xmath* tool. For the integration of Neural Networks into the simulation of any linear and nonlinear dynamic system the software module *NIMAX* (Neural Interface for *MATRIX$_X$*) was developed, which allows the use of nearly all *SNNS* (Stuttgart Neural Network Simulator) functions and algorithms in an easy way in the *MATRIX$_X$* simulation environment. So, the simulation of neural as well as hybrid control systems is possible by using a standard simulation environment.

A Learning Fuzzy System for Predicting Overshootings in Process Control

Bernd Freisleben[1] and Stephan Strelen[2]

[1]Department of Electrical Engineering & Computer Science, University of Siegen
Hölderlinstr. 3, D–57068 Siegen, Germany
E-Mail: freisleb@informatik.uni-siegen.de

[2]OKA Spezialmaschinenfabrik GmbH & Co KG,
Grenzallee 5, D–64297 Darmstadt, Germany
E-Mail: strelen@t-online.de

1 Problem Description

The task of a typical controller is to generate a control signal $u(t)$ that forces the system output $y(t)$ to follow the command input $w(t)$ as good as possible [1]. If the command input $w(t)$ abruptly changes its value (unit-step input), it takes some time until the system output $y(t)$ reaches the desired value of $w(t)$. During this time, it can happen that the value of $y(t)$ is higher than $w(t)$. This is called an *overshooting* [2]. An overshooting might be tolerable or not. If it is too high, it may cause damage to the system or its environment. In order to have enough time to stop the process, an early detection of the risk of an intolerable overshooting is desired. In this paper, a fuzzy system is presented which estimates the probability of an intolerable overshooting appearing before the process output has settled and learns from its own experience.

2 The Learning Fuzzy System

In the particular application scenario considered, there is a control system and a controlled process; both of them change over time. After each changing period, a unit step input is given to the controller and the system is observed over the next 400 sampling points. The system is supposed to estimate at each sampling point whether there will be an overshooting higher than an allowed limit or not. At each of these sampling points, particular values (e.g. the current value and the derivative of the current system output $y(t)$) are measured and all fuzzy inference rules are determined. An example of such an inference rule is

IF (current_overshoot = big *AND* derivative_y = very big) *THEN* P(over) = high

where $P(over)$ is the probability of an overshooting that is larger than an allowed limit and appears within the next 400 sampling points after a unit step input. The inference computation over all the rules is performed after each sampling point, and the exact value of the error probability is calculated by a gravity method [4]. If this probability is in a particular range, the system recommends to stop the process.

As already mentioned, the fuzzy system is designed to learn from its experience. The aim of this learning is to correct the linguistic terms of the conclusions. Before the learning phase is started, the conclusions of each fuzzy rule must be verified. The verification is based on calculating the relative probability of whether the *IF*-part of a fuzzy rule indicates the occurrence of an intolerable overshooting or not. The calculation is realized by counting how often a conclusion of a rule was activated and the overshooting appeared before the system output settled, and how often this conclusion was activated and no overshooting occured.

After some time, we are able to estimate how probable it is that the conclusion of the rule is activated and that the overshooting will appear soon, and how probable it is that the conclusion of the rule is activated and the error will not appear soon. If the first probability is called $prob\_1$ and the second $prob\_2$, then we can compute $r = prob\_1/prob\_2$ which gives us an idea of whether the rule indicates the appearance of the overshooting or whether it indicates that it will not occur. Based on r, we are then able to adapt the linguistic term of each rule.

3 Results

At the start of each experiment, only neutral rules in which the centre of gravity is at zero were generated. Consequently, all predictions indicated initially that an overshooting error would not happen, and the system was unable to recommend a stop of the controlled process. It took only a short time until the system predicted all overshootings that appeared correctly, but at this early stage most of these correct predictions came very late and the system often recommended a stop although an overshooting did not occur. While the system evolved, the number of wrong predictions decreased and the correct predictions were made earlier. After several hundred experiments, all overshooting errors were predicted correctly, and these correct predictions were made, on the average, after 20% of the time period when an overshooting occured (e.g. after 8 sample points an overshooting was predicted that actually happened after 40 sample points). Furthermore, in only about 25% of all predicted overshootings the system predicted an overshooting which later did not occur.

References

1. Aström, K. J. and B. Wittenmark. *Computer-Controlled Systems.* Prentice-Hall, 1990.
2. Kuo, B.C. *Automatic Control Systems.* Prentice-Hall, 1995.
3. Isermann, R. *Digitale Regelsysteme I.* Springer-Verlag, 1988.
4. Zimmermann, H.J. *Fuzzy Set Theory and its Applications.* Kluwer Academic, 1991.

ANN - based Power Unit Protective System

A. Halinka, M. Szewczyk, B. Witek
Silesian Technical University of Gliwice, Poland

Numerous changes introduced to the input and measuring elements of modern digital protective arrangements considerably improved their reliability and access to the extended information about processes in the protected object, they did not cause however any essential improvement of the protection decision-making. To extend the possibilities of the decision-making units of the protective systems the ANN structures are proposed for the proper fault detection, on the base of the measuring signals delivered from the protected object and its environment.

In the generator-transformer units the criterial quantities (voltage and current) and their components measurement is performed for the protected object classification to one of the event classes:
- protected unit normal condition,
- external fault,
- internal fault.

It has to be pointed out that the proposed DS-ANN structure is assigned only for electro-magnetic (high-current) faults (i.e. three-phase, phase-to-phase and phase-to-phase-to-earth faults) diagnosis. As a next step other elements of the complete, ANN-based power unit protective system are foreseen, dedicated also for other types of faults detection e.g.: phase-to-earth fault, out-of-step, loss of excitation etc.

For the above described function realisation, the protective relay may be assigned as a pattern recognition unit, where the *pattern* is a set of measuring signals, delivered to the net in form of the limited data window. The advantage of the ANNs as a decision-making units in the protective relays is first of all the ability of the network adaptation for the proper classification of earlier unknown patterns (i.e. knowledge generalization ability).

The hierarchical structure of the decision-making system is presented in Fig.1. The primary circuit consists the generator (G), unit -transformer (T), transmission line (L) and the equivalent EPS. The secondary signals of the current and voltage transformers are filtered in anti-aliasing analogue filter, converted to digital signals and pre-processed in DSP units and finally delivered to the sub-nets inputs. The assumed sampling frequency is 5kHz. Each subnet uses other combination of input signals for the decision-making, to achieve fast, reliable and selective fault detection. As a result the main net gives an information about the fault detection inside or outside the protected zone.

The network is learned to analyse dynamically the state of the protected object, by introducing to its input the discrete, pre-processed signals. Thus, before the network simulation two items must have been realised:
- creation of data-base for the ANN learning,
- ANN structure choice and optimization.

Collected in the data-base discrete current and voltage values consider various states of the protected unit:

- normal states,
- fault states (fault or post-fault),
- transient states (state before fault and initial fault condition).

Each group consists current and voltage waveforms considering various unit loads, different magnitudes and phase angles of the generator and equivalent system voltages.

The ANN structure is assumed as a multi-layer perceptron (MLP) with three layers for all subnets and one layer for the main net. Subnets input layers are additionally assumed as input signals normalizing systems. Initial synaptic weights values have been taken randomly. Hyperbolic tangent sigmoid is assumed as a transfer function and back propagation (BP) method with Levenberg-Marquardt was used in the learning process.

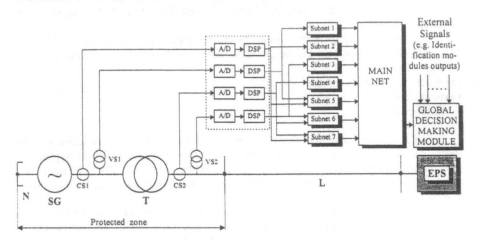

SG - synchronous generator, T unit transformer,
EPS - electric power system,
CS1, CS2 - current signals, VS1, VS2 - voltage signals
A/D - analog/digital converter, DSP - digital signal processing

Fig.1: Generator-transformer unit protection architecture

In consequence of learning, testing and optimization process we can say, that all subnets correctly qualifying the input signals into two event classes:
  1) fault in the protected zone,
  2) normal condition or fault out of the protected zone.
The advantage of the hierarchical structure of the decision-making system, is that bad decisions made by the individual subnets have no influence on the output of the main net, developing a global decision about the state of the protected unit. Moreover, the decision is made fast an surely for all samples defined in the data-base for the ANN testing process. This model can be easily extended by adding new subnets which can recognize other states of the protected unit than the high current faults e.g.: phase-to-earth fault, out-of-step, loss of excitation etc.

A Scheduling Problem with Multicriteria Evaluation

Ileana Hamburg
IAT, Wissenschaftszentrum NRW
Munscheidstr. 14, Gelsenkirchen, Germany

Lungu Marin
Universitatea Craiova
Str. A. I. Cuza 13, Craiova, Romania

In real time processes we often have situations in which two or more tasks are "ready" to run, that means to occupy the central processor unit (CPU) of a computer or other machine. A planner - decision maker - is concerned with making the decision which task to run first, using a suitable algorithm [4]. It has to take into consideration not only one but more criteria (as forms of attributes or objectives), some of them being qualitative and contradictory.

A scheduling algorithm based on a multicriteria evaluation was developed within a cooperation project between the University of Craiova, Romania and the IAT Gelsenkirchen, Germany. One version of our algorithm [2] uses stochastic correlation coefficients which are extensions of those introduced by Spearman [3] and the second version uses fuzzy correlation coefficients. The algorithm gives a suitable weight for the waiting time in the queue in order to rule out a process being postponed repeatedly due to its lower priorities relative to other criteria. In this abstract we present only the fuzzy version of the algorithm.

We consider a system of n tasks $S_1,... S_n$ and k criteria $C_1,...C_k$. In the case of a unique valuation, the order of tasks taking into consideration only the criterion C_i, is described by a permutation $s_i; N_n \to N_n$ of the set $N_n = \{1,...n\}$ where $s_i(j)$ is the position of the task S_j.

Because of the subjective priorities given to tasks by the users [1], often this order is not determined "unique": it could be characterized using a fuzzy permutation $p_i: N_n \times N_n \to I = [0,1]$, where $p_i(j, l)$ shows the degree in which the task S_j is right to occupy the position l in the queue, taking into consideration only the criterion C_i

p_i has to satisfy the condition $\sum_{j=1}^{n} p_i(j, l) = \sum_{l=1}^{n} p_i(j, l) = 1$ for each i in N_n.

In the case of a unique valuation, the permutations p_i are crisp.

The steps of the scheduling algorithm are the following:

1. The elements of the 3D matrix $p = (p_i(j, l))$ are calculated.

2. Fuzzy correlation coefficients CR(i, h) are calculated for every pair of criteria C_i and C_h:

$$CR(i, h) = 1 - 6/n(n-1) \sum_{l=1}^{n} \sum_{j=1}^{n} l^2 [p_i(j, l) - p_h(j, l)]^2.$$

3. Fuzzy components a_i (which represents the importance of the criterion C_i relative to the other k-1 criteria) are calculated. $a_i = \sum_{h=1}^{k} CR^2(i, h)$.

4. The degree CSI(j, l) in which the task S_j is right to occupy the position l, taking into consideration all the criteria, is calculated using the fuzzy weights w(i) which the user gives to the criterion C_i: $CSI(j, l) = \sum_{i=1}^{k} a_i p_i(j, l) w(i)$.

6. SG = S(CSI) selects for each l = 1,...,n the task j with the greatest value of CSI(j, l).

A software implementation of a planner (Fig. 2. [3]) using the scheduling algorithm based on a multicriteria evaluation is important for computer systems which are linked to dymanical processes where the order of tasks has to be flexible taking into consideration external parameters and asynchronous events. When an internal or external event (of type hardware or software) appears, an interruption takes place and a routine CM based on the scheduling algorithm modifies the running order of the "ready" tasks. Between two events, the CPU is allocated using the round robin scheduling algorithm: each process is assigned a time interval (quantum), which it is allowed to run. If the process is still running at the end of the quantum, the CPU is preempted and given to another process; if the process has blocked or finished before the quantum has elapsed, the CPU switching is done.

When the quantum runs out on a process, this process is allocated a new position at the end of the list (Fig. 1. - [5]).

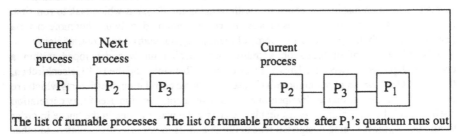

Fig. 1. Round robin scheduling

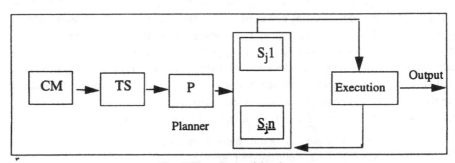

Fig. 2. The scheme of the planner

References

1. I. Hamburg: The Potential of Fuzzy Logic Methods to Support Multicriteria Group Decision-Making in the Engineering Design Process. In: B. Bouchon-Meunier, R. Yager (eds.): Information Processing and Management of Uncertainty in Knowledge-Based Systems. Paris 1994, pp. 633-638.
2. M. Lungu: Sistem in timp real pentru implementarea unei methode der calcul ierarhizat. In: SINTES 6: Simpozionul National de Teoria Sistemelor. Craiova 1996, pp. 152-160.
3. G. Saporta: Probabilité. Analyse des données en statistique. Paris: Editions techniques, 1990, pp. 141-143.
4. A. Tanenbaum: Modern Operating Systems. New Jersey: Prentice Hall, 1992, pp. 64-65.

Hardware Implementation of a Mixed Analog-Digital Neural Network

R. Izak, K. Trott, Th. Zahn and U. Markl

Technical University, Ilmenau 98684, Germany

Abstract. This paper describes a hardware implementation of a firing neural network based on the models of Gerstner. It mainly consists of analog building blocks (neurons, synapses), but because of their digital interface and controlling it is a mixed-mode structure. The complete physical implementation of all components allows massive parallel and real time computation. The input data processing is located off-chip to enlarge the number of implementable neural networks.

1 Overview

The internal information processing of synapses and neurons is locally distributed. This simplifies the interconnection problem of the building blocks which is resolved by choosing a classical array architecture. This enables the realization of a wide variety of network topologies, ranging from fully connected toward multi-layer and locally connected architectures.

A neuron receives spatially and temporal added current pulses from its synapses placed in the column above. The neural activity A is a result of the comparison between the threshold and the capacitor voltages, charged by incoming current pulses. This binary activity is distributed to all synapses in a row. Aditionally, every neuron generates the pre- and postsynaptic history potentials H and H_d, respectively. Both are used in the synapses to implant a modified Hebbian learning algorithm to change the synaptic weights.

Each firing of the neuron is followed by a refractory period modeled by the refractionary circuit and lifting the firing threshold during a certain amount of time.

Every synapse consists of a learning circuitry (multiplier and charge pump), a weight storage capacitor with its refresh unit and a voltage-to-current converter. A single ended voltage-to-current converter generates the current output and the activity multiplier samples it thus generating the output pulses of the synapse.

The system has been currently simulated in its components at the algorithmic and electric domains using synthetic data. The synapse circuit has been initally implanted in a $2.4\mu m$ CMOS technology for verification and measurement. Implementations of a complete synapse and neuron in a $0.5\mu m$ CMOS are under way. Our aim is to develop a tool kit for design and implementation of free configurable neural networks including automatic place and route.

This work is part of the "Graduiertenkolleg GRK 164/1-96" granted by the "Deutsche Forschungsgemeinschaft".

A Fuzzy Segmentation Method for Digital Substract Angiograms Analysis

MC. Jaulent(1), V. Bombardier(2), A. Bubel(2) and J. Brémont(2).

(1) Service d'Informatique Médicale, Hôpital Broussais 96 rue Didot 75014 Paris, FRANCE, Tel : 1 43 95 91 66; Fax : 1 43 95 92 09; E-mail : jaulent@hbroussais.fr

(2) Centre de Recherche en Automatique de Nancy, CNRS URA 821, Equipe d'Electricité et d'Automatique, Université de Nancy I-Faculté des Sciences 54506 Vandoeuvre Cedex FRANCE; Tel : 83 91 20 05 Fax : 83 91 24 15   E-mail : bombardier@cran.u-nancy.fr

The automatic evaluation of the severity of lesions in digital subtraction angiography depends on the quality of the segmentation process that provides the outlines of the arteries [1]. The automatisation is an important issue to reduce the inter and intra observer variability in the assessment of lesions. In classical approaches, a human intervention is generally needed to provide some Region of Interest (ROI) or at least some specific starting or key points in order to determine the median axis of the artery and to delineate the boundaries by using adaptive tracking techniques [2]. This abstract outlines a knowledge based approach for the automatic and reproducible identification of arteries boundaries in 2D renal angiograms. The procedure takes into account the different types of knowledge involved in the segmentation of renal angiograms. The knowledge representation is based on the fuzzy set theory [3] and the process is based on the cooperation of two fuzzy edge operators [4][5]. The segmentation procedure that provides the outlines of the renal arteries follows three steps:

The first step determines the outlines of the aorta (step 1). The object aorta is characterized by an important size, vertical outlines and an homogeneous texture. The first operator is based on a linguistic description of an edge and the definition of homogeneous and transition regions as fuzzy sets [4]. This edge detection operator takes into account the shape, thickness and the orientation of the aorta and allows to adapt the notion of homogeneous or transition region in relation to the image quality, that is, to modify their membership functions to obtain a better detection.

The second step of the method concerns the detection of the regions of interest that encompass the renal arteries. The knowledge related to the image expresses the fact that the two renal arteries are located almost at the same level of the aorta. The search for main discontinuities facing each other along the vertical outlines of the aorta leads to the location of the two renal arteries bifurcations (depicted by rectangles). The search procedure provides automatically the two regions of interest (step 2).

For the detection of renal arteries outlines the use of morphological criteria is not sufficient. Besides this, digital angiograms are always blurred and present low transitions of gray level. The choice of the second operator is an edge detection operator [5] with a fuzzy clustering based on the fuzzy-C-Means algorithm [6]. The second operator is applied on the detected ROI to get the arteries outlines (step 3). The results of the fuzzy edge operator can be further run by a simple tracking algorithm to finally get the precise outlines of the arteries.

The resulting procedure is robust in real situations since it is adaptable to various kind of perturbations in the source image. The method provides the outlines without any human intervention during the process. However, the current knowledge must be validated and the method must be evaluated on a large number of images.

References

[1]Wunderlich W., Linderer T., Backs B., FischerF., Schroder R. "Optimum edge detection in   quantitative coronary arteriography." In: Proc of CAR'93. H.U. Lemke, K.Inamura, C.C.   Jaffe and R Felix(eds). Springer-Verlag. 1993; pp303-308.

[2]Dumay A, Gerbrands J and Reiber H. Automated extraction, labelling and analysis of the coronary vasculature from arteriograms. International Journal of Cardiac Imaging, 1994; 10:205-215.

[3]Zadeh L.A.: "Fuzzy sets. Information and Control",1965; vol 8, pp338-353.

[4]Lemarquis B., Bubel A., Bombardier V., LevratE., Bremont J. "Integration de connaissances linguistiques pourun operateur de detection de contours par règles floues ",quatrième journee nationale sur les applications des ensembles flous, Lille, Dec 94.

[5]Levrat E., Bombardier V., Lamotte M.,Bremont J. "Multi-level image segmentation using fuzzy clusteringand local membership variation detection ", FUZZ-IEEE 92, IEEE international Conference on fuzzy system, San Diego,USA, 8-12 march 1992, Proceeding pp. 221-228.Bezdek J.C "Pattern recognition with fuzzy objective algorithms." Plenum Press New York, 1981.

[6]Bezdek J.C "Pattern recognition with fuzzy objective algorithms." Plenum Press New York, 1981.

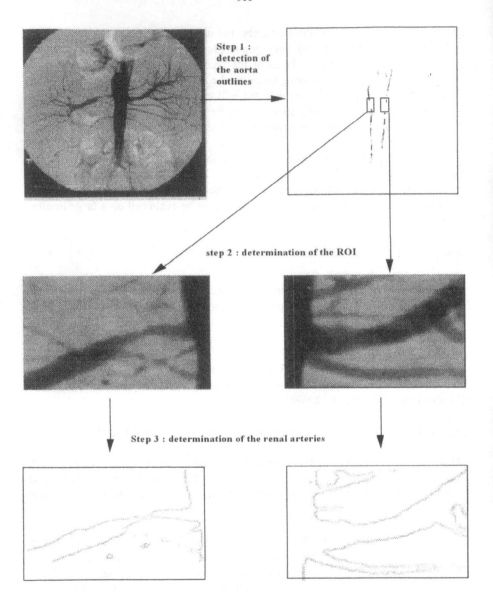

Step 1 : detection of the aorta outlines

step 2 : determination of the ROI

Step 3 : determination of the renal arteries

Application of Genetic Algorithms to Computer Assignment Problem in Distributed Hard Real-Time Systems

Jerzy Martyna

Jagiellonian University, Dept. of Computer Science,
ul. Nawojki 11, 30-072 Kraków, Poland

In this paper we consider some genetic algorithms for required optimization in the design and building of distributed hard real-time systems. Given that some of the component computers connected by typical or specialized local area network working in time constrained environment, it is desirable that there is a suitable method to obtain all main parameters of these systems.

First, we cover issues involved in distributed hard real-time systems [1]. We consider their parameters such as investment cost of computers, reliability and availability, mean response time of distributed system, etc., and also the crucial parameter for the design of these systems - the probability of dynamic failures of system in time-constrained environment [2] caused by the hardware and software failures, system reconfiguration, interference on the communication network as well as other causes such as missing the hard deadline. We argue that some genetic algorithms are effective for the solution of computer assignment problem in distributed hard real-time systems.

Next, we discuss the principal ways of organizing computation at the programming level with application of some genetic algorithms to computer assignment problem in distributed hard real-time systems. We present some details of these computations: fitness function, chromosome encoding and genetic operators that are used. Finally, we give the experimental results of proposed genetic algorithms.

In summary, the results indicate an advantage of using the genetic algorithm to computer assignment problem in the reasonable amount of time. Although, the results obtained are only approximated solutions, they are sufficiently precise for the assumed computation.

References

1. Burns, A.: Distributed Hard Real-Time Systems: What Restrictions are Necessary? in: H.S.M. Zedan (Ed.): Real-Time Systems, Theory and Applications, Elsevier Science Publ. B.V. (North-Holland), 1990, pp. 297 - 303
2. Krishna, C.M., Shin K.G.: Performance Measures for Multiprocessor Controllers, in: A.K. Agrawala, S.K. Tripathi (Eds.): Performance'83, North-Holland Publ. Comp., 1983, pp. 229 - 250

Local Search for Computing Normal Circumstances Models (Abstract)*

Bertrand Mazure, Lakhdar Saïs and Éric Grégoire

CRIL, Université d'Artois
rue de l'Université SP 16
F-62307 Lens Cedex, France
{mazure,sais,gregoire}@cril.univ-artois.fr

Abstract. In this paper, a new method is introduced to check several forms of logical consistency of nonmonotonic knowledge-bases (KBs). The knowledge representation language under consideration is full propositional logic, using "Abnormal" propositions to be minimized. Basically, the method is based on the use of local search techniques for SAT. Since these techniques are by nature logically incomplete, it is often believed that they can only show that a formula is consistent. Surprisingly enough, we find that they can allow inconsistency to be proved as well. To that end, some additional heuristic information about the work performed by local search algorithms is shown of prime practical importance. Adapting this heuristic and using some specific minimization policies, we propose some possible strategies to exhibit a "normal-circumstances" model or simply a model of the KB, or to show their non-existence.

Keywords: local search, tabu search, nonmonotonic reasoning, common-sense reasoning, SAT, consistency checking

Extended Abstract

Assume that a knowledge engineer has to assemble several logic-based proposi-tional knowledge modules, each of them describing one subpart or one specific view of a complex device. In order to make fault-diagnosis possible in the future, each knowledge module describes both normal functioning conditions and faulty ones. To this end, the ontology is enriched with McCarthy's additional "Abnor-mal" propositions (noted Ab_i) allowing default rules to be expressed together with faulty operating conditions. For instance, the rule asserting that, under normal circumstances, when the switch is on then the lights should be on is rep-resented by the formula $switch\_on \land \neg Ab_1 \Rightarrow lights\_on$, and in clausal form by $\neg switch\_on \lor Ab_1 \lor lights\_on$. In this very standard framework, Ab_i propositions are expected to be false under normal operating circumstances of the device. The knowledge based system (KB) is expected to be used in a nonmonotonic

---

* This work has been supported by the Ganymède II project.

way, in the sense that conclusions can be inferred when they are satisfied in some preferred models of the KB where Ab_i are **false**. Also, model-based diagnosis can be performed in order to localize faulty components in the presence of additional factual data.

The specific issues that we want to address in this framework is the following one. How can the knowledge engineer check that the global KB is consistent? Also, how can he exhibit (or show the non-existence of) one model of the KB translating normal functioning conditions of the device? We would like these questions to be answered for very large KBs and practical, tractable, methods be proposed.

Actually, consistency checking is a ubiquitous problem in artificial intelligence. First, deduction can be performed by refutation, using inconsistency checking methods. Also, many patterns of nonmonotonic reasoning rely on consistency testing in an implicit manner. Moreover, ensuring the logical consistency of logical KBs is essential. Indeed, inconsistency is a serious problem from a logical point of view since it is global under complete (standard) logical rules of deduction. Even a simple pair of logically conflicting pieces of information gives rise to global inconsistency: every formula (and its contrary) can be deduced from it. This problem is even more serious in the context of combining or interacting several knowledge-based components. Indeed, individually consistent components can exhibit global inconsistency, due to distributed conflicting data.

Unfortunately, even in the propositional framework, consistency checking is intractable in the general case, unless $P = NP$. Indeed, SAT (i.e., the problem of checking the consistency of a set of propositional clauses) is NP-complete. Recently, there has been some practical progress in addressing hard and large SAT instances. Most notably, simple new methods that are based on local search algorithms have proved very efficient in showing that large and hard SAT instances are consistent. However, these methods are logically incomplete in that they cannot prove that a set of clauses is inconsistent since they do not consider the whole set of possible interpretations.

However, we have discovered the following phenomenon recently. When the work performed by local search algorithms is traced when they fail to prove consistency, a very powerful heuristic can be extracted allowing us to locate probable inconsistent kernels extremely often. Accordingly, a new family of powerful logically complete and incomplete methods for SAT has been proposed.

In the full paper, we extend this previous work in order to address nonmonotonic propositional KBs, using the above "Abnormal" propositions that are expected to be false under normal circumstances. Using this preference for normal conditions, we guide the local search towards a possible "normal circumstances" model. When such a model is not found, several issues can be addressed. First, using the above heuristic and assuming that normal circumstances are satisfied, we propose a technique that allows us to prove (very often) the absence of such a model and to exhibit an inconsistent kernel. Then, dropping the special status of "Abnormal" propositions to several possible extent, we introduce various strategies for showing the consistency or inconsistency of the KB.

A Graphical Tool for Implementing Neural Networks for Digital Signal Processing on Parallel Computers

C. J. O'Driscoll*, J. G. Keating**

* Department of Electronics and Communications, Dublin Institute of Technology, Kevin Street, Dublin 8, Ireland.
** Department of Computer Science, St. Patrick's College, Maynooth, Co. Kildare, Ireland.

Abstract. An emerging and rapidly expanding Parallel Distributed Processing technology for Signal Processing is the Neural Network. Artificial Neural Networks (ANNs) have been effectively used in the solution of signal processing problems, such as optimisation, identification and prediction. A graphical entry tool, SoftDSP, designed for the simulation of DSP functions, has been used as a front end for a graphical compiler to develop ANNs for parallel systems. The compiled code is implemented in parallel C which can be mapped onto an array of transputers.

The graphical entry tool is used to create a schematic description of an arbitrary sized neural network architecture. Predefined components and the ability to define new library components, permit the description of complex networks. Initially the schematic is validated according to a set of basic graphical rules and outputs a netlist, containing interconnection information and component data.

A set of rules, a syntax, specific to neural network structures is then used to parse the netlist and generate a valid parse tree. Additional parameters required for the learning algorithm are evaluated prior to the final generation of the C source code.

Separate sections of the schematic can be compiled into different tasks. Additional routing functions are added to the tasks to handle data over one input and one output channel. This approach permits the mapping of the ANN onto multiple processors.

The Back Propagation (BP) learning algorithm is implemented as a separate C source module. Data generated by the compiler is used to define structures necessary for BP calculations. The algorithm is decimated into different sections, output section calculations, hidden layer calculations and updating of weights. This permits the mapping of the algorithm over a number of different processors.

This graphical compiler allows the design and implementation of ANNs which can then be mapped onto parallel processors. The compiler has been tested, verified and performance results have been generated for standard ANN problems.

Fuzzy Guidance Controller for Paraglider

Valentine Penev

Institute of Control and System Research - BAS,
Acad G. Bonchev str, Bl.2, Sofia, Bulgaria
E-mail: apsau@bgcict.acad.bg

Abstract. The general features of a fuzzy controlled paraglider are studied. The simulation results show that paraglider model and fuzzy controlled paraglider are appropriate for a variety of control and guidance applications.

1. Paraglider model

The model of paraglider gives an account of its non rigid structure. This model [1,2] presents an intermediate class paraglider realistically. The paraglider has a somewhat different azimuth control, which is bank and attack angles directed.

$$\dot{v}_t = \frac{-D}{m_t} - g \sin \gamma_t - w_x \cos \gamma_t - w_h \sin \gamma_t; \quad \dot{\gamma}_t = \frac{L \cos \phi}{m_t v_t} - \frac{g \cos \gamma_t}{v_t} - \frac{(w_x \sin \gamma_t - w_h \cos \gamma_t)}{v_t};$$

$$\dot{\psi}_t = \frac{L}{m_t v_t \cos \gamma_t} \cdot \sin \phi; \quad \dot{x}_t = v_t \cos \gamma_t \cos \psi_t; \quad \dot{y}_t = v_t \cos \gamma_t \sin \psi_t; \quad \dot{h}_t = v_t \sin \gamma_t;$$

The paraglider primary flight-control surfaces consist of horizontal stabilators capable of symmetric or differential movement. The model includes two identical actuators for stabilator surfaces.

2. Fuzzy control

Fuzzy controller rules are synthesized on the base of available knowledge about the system. They have three input variables LOS azimuth towards the goal φ^l, LOS pitch - γ^l and the difference between the altitude of goal and the altitude of the paraglider h^d and two outputs attack angle and bank angle respectively for pitch and yaw control loops.

Table 1. Rules for bank angle - yaw loop.

φ^l/h^d	ZO	PS	PM	PB	PVB
PB	pb	pb	nb	pb	pb
PM	pb	pm	nb	ps	pm
PS	pm	ps	ps	pm	ps
NS	ns	nm	ns	nm	ns
NM	nm	nb	nm	nb	nm
NB	nb	pb	nb	nb	nb

Table 2. Rules for atack angle - pitch loop.

γ^l/h^d	ZO	PS	PM	PB	PVB
PB	ns	ns	nb	ns	ns
PM	ns	nb	nb	nm	nb
PS	nb	nm	nm	nb	nm
NS	ns	nm	ns	nm	ns
NM	nm	nb	nm	nb	nm
NB	nb	ns	nb	nb	nb

Paraglider pitch control loop is similar to the yaw control loop and has the following differences. Symmetric stabilator command is changed with differential stabilator command.

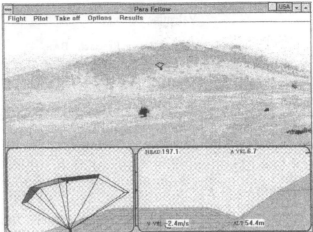

Fig. 1. The simulator view.

3. Conclusion

The most important feature of the fuzzy engines is the different dynamic response in the respect of altitude between paraglider and landing point. The connection between altitude and attack and bank angles has very simple linguistic expression:
IF altitude is big THEN bank angle is big and attack angle is big
IF altitude is small THEN bank angle is small and attack angle is small.
This feature make the paraglider more stable and decreases possibility of colaps. The results from this investigation may be used to design a real paraglider controller, which can by employed for surface surveillance by the Army or other user. velocity angle

References

1. Bodner V.A., Control systems for aircrafts, Mashinostroene, Moscow, 1973.

2. Penev V.N., Georgiev G.L., Petrov P.S., Karastoianov D.N., Fuzzy guidance controller for paraglider, HEMUS'96

An Efficient Algorithm Approximately Solving the General Routing Problem in Graphs

Peter Richter

Institute for Informatics, Potsdam University
PO box 601553, 14415 Potsdam, Germany

Abstract. As to the *General Connection Problem* in *Graphs* we present an efficient constructive algorithm that tackles in a simultaneous manner both the placement $\vec{\pi}$ of the entities M with respect to constraints $\Pi \subseteq M \times V(\mathbf{G})$ and the routing $\vec{\varphi}_{\vec{\pi}}$ with respect to connection rule $\Omega \subseteq M^2 \times \tilde{A} \times \tilde{N}$ resulting to an approximate layout ($\vec{\pi}_{appr}$:M$\rightarrow$V($\mathbf{G}$)), $\vec{\varphi}_{appr} \subseteq \Omega \times E(\mathbf{G})$) with a time effort $O(k^2 p \cdot m\ (k^2 + \log n))$-such the solution grants an upper cost bound $\tilde{C}((\vec{\pi}_{appr}, \vec{\varphi}_{appr})) / \tilde{C}((\vec{\pi}, \vec{\varphi}_{\vec{\pi}})) \leq \frac{\kappa_F}{\kappa_H} \cdot \frac{k-1}{\omega^\bullet} (2+\omega^\bullet - k)$. The layout ($\vec{\pi}_{appr}, \vec{\varphi}_{appr}$) is nearly optimal for sparse graphs!

κ_F, κ_H = inherent graph constants,

$\omega^\bullet = |\Omega^\bullet|$, k= |M|.

1 Introduction

The *General Connection Problem* in *Graphs* with respect to pure routing cost (*GCPG*$_{\text{ROUTING}}$).reads as follows:

Instance: Set of entities M, layout graph $\mathbf{G}$, ℓ: E($\mathbf{G}$) $\rightarrow R_+$, connection rule $\Omega \subseteq M^2 \times \tilde{A} \times \tilde{N}$ (connection type $\tilde{A}$, line numbers $\tilde{N} \subseteq N$), placement rule $\Pi \subseteq M \times V(\mathbf{G})$, specific connection cost μ: $M^2 \times \tilde{A} \rightarrow R_+$.

Task: To find a layout ($\vec{\pi}$, $\vec{\varphi}_{\vec{\pi}}$) such that

$$\tilde{C}((\vec{\pi}, \vec{\varphi}_{\vec{\pi}})) = \min_{\text{function}\, \pi \in \Pi} \left\{ \min_{\varphi_\pi \subseteq \Omega \times E(\mathbf{G})} \{\tilde{C}_1(\varphi_\pi)\} \right\} \text{ where for any}$$

$\varphi \subseteq \Omega \times E(\mathbf{G})$ it hols: $\tilde{C}_1(\varphi) := \sum_{((a,b),x,y)\in\Omega} \left\{ \sum_{k \in \varphi((a,b),x,y)} \ell(k)\, \mu(a,b,x) \right\}$.

Unfortunately, *GCPG*$_{\text{Routing}}$ is *NP*-hard. Thus, we look for an ε -approximate algorithm *R2* that solves the problem polynomially bounded in time granting a sufficient upper cost bound.

2 Advantage of Algorithm *R2*

We consider the simultaneous optimization of the ENTITIES' PLACEMENT with respect to MINIMUM LINE COST. We abandon the more conventional strategy where the determination of a pre-optimized placement via the *QAP* is followed by the determination of an approximate routing supposing that the entities' placement has been quite well

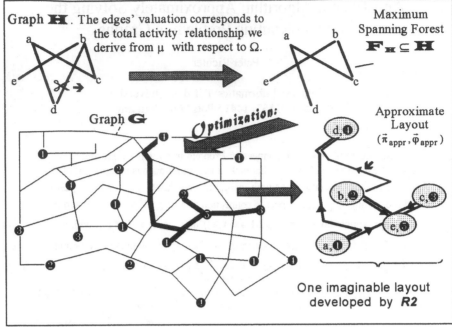

Fig. 1.   Rough description of the strategy of R2

done. The point is how the optimization strategy is to design that the μ -weighted
Ω-graph $\mathbf{H}$ will be embedded into layout graph $\mathbf{G}$ such that the routing cost are
minimum only with respect to the <u>maximum spanning forest</u> $\mathbf{F_H} \subseteq \mathbf{H}$ (<u>not</u> to
$\mathbf{H}$ because of the *NP*-hardness). We are able to state an upper cost bound (given
above) that proves near optimality with respect to $\mathbf{H}$ if it is sparse, i.e. $\omega^{\bullet}$ is near
k. Note, most of industrial connection rules Ω are sparse. Algorithm *R2* proceeds in
a constructive and deterministic manner. In its first optimization step, $\mathbf{F_H}$ is
analyzed in a bottom-up like proceeding where we consider all neighbors $\subseteq$
$V(\mathbf{F_H})$ for the current generation leaves $\subseteq V(\mathbf{F_H})$ of the current forest $\mathbf{F_H}$ in
order to determine the best connections via shortest paths to their leaves in $\mathbf{G}$.
Consider a certain tree $\mathbf{T} \subseteq \mathbf{F_H}$. Once, the current tree $\mathbf{T}$ consists of only one
node, *R2* starts a top-down like procedure that really finds the best placements for
the shortest μ-weighted $\mathbf{T}$-connections in $\mathbf{G}$. Thereby, <u>all</u> Ω -connections come
into consideration finally to yield an approximate layout ($\vec{\pi}_{appr}$, $\vec{\varphi}_{appr}$).

3   Conclusions

Time behavior and upper cost bound of algorithm *R2* justify a priority treatment
and introduction for the optimization of industrial connection structures. This yields
also for alternative approaches via simulated annealing, neuronal networks , and
evolution strategies where good starting structures essentially influence the solutions'
quality.

Design of a Method for Machine Scheduling for Core Blowers in Foundries

Martin Rüttgers

FIR - Research Institute for Operations Management
at Aachen University of Technology (RWTH)
Pontdriesch 14/16, 52062 Aachen, Germany

EXTENDET ABSTRACT: The increasing complexity of the production planing process lead to an increasing number of restrictions and nonlinear objective functions that has to be considered for optimizing scheduling problems. Many heuristic methods exist to solve this kind of problems, but for to reasons the application of this methods to practical problems ist questionable. First the underlying models are based on many restrictions not representing real situations, second the methods are unflexible so that they can not be used in case of variing restrictions.

For these reasons a new method has been developed to solve the problem of machine scheduling. The concept of this new method is very close to them of Evolutionary Algorithms: Scheduling plans were generated in form of indiviuals, which can evolve from one generation to the next by means of specified changing operators. By this process a global search over the entire objective function is initialized which allows the determination of an optimized solution. The main difference of this algorithm to existing Evolutionary Algorithms is the mutation-operator for changing the solutions from one iteration to the next. The construction of this operator is based on many imperical tests with different mutation operators which show clearly that the confergence process is as faster as more intelligent the „move“ of an individuum is. The basic idea in this case is that the mutation should take the slope of the objective function into account. That means if the objective function is flat than the individuals diverge, only when the function becomes steeper the search is concentrated on this point. This quality of the mutation operator is realized by taking the Hamilton-Distance of the different individuals into consideration. For this reason the method is called „Differential Evolution“.

The method of Differential Evolution has been implemented for the scheduling problem of core blowers. In the production process of foundries machine scheduling procedures for the core blowers are of major interest because of their strong connection to automatic molding plants. The cores are necessary for molding and for this reason the scheduling has to be very precise. The lack of cores causes very high costs on the automatic molding plants. The process of core production consists of four major parts. For each part special processors and tools are necessary which have limited capacities. For this reason the production is a four-stage production, and to optimize the entire process all stages must be taken into acccount. On each stage there are $m>2$ heterogenous processors, that can produce only one job at each time. The technological demands of the jobs can be so different that not all the jobs may be processed on every machine. For machine scheduling a lot of other restrictions like limited manpower, limited place for material storage and preparation times have to be considered. The number of jobs that have to be scheduled varies from 10 to several hundred, production times of the jobs between hours and several days. The objectice function of the problem represents the sum of the time differences between completion times and time demands of all scheduled jobs.

The method is tested with different scheduling problems of different complexity. The comparison of the results with algorithms using only stochastic mutation-operators show that the method of Differential Evolution produces solutions with higher quality and that convergence time is much shorter. Also the comparison with an adaptive simulated annealing as well as the annealed Nelder&Mead approach, both of which have a reputation of being very powerful, show that the method converges faster and with more certainty.

Paragen II : Evolving Parallel Transformation Rules

Conor Ryan and Paul Walsh *

Dept. of Computer Science and Information Systems
University of Limerick, Ireland.

Abstract

Traditionally, parallel programs were the reserve of large corporations or generously sponsored research institutions, with more modestly sized organisations not able to afford the speed up offered by parallel architectures. However, with the development with such software as Parallel Virtual Machine (PVM) which permits a group of (possibly heterogenous) machines, to work as though they were nodes in a single parallel machine, parallel processing is now affordable by practically every institution.

Most of the companies that could most benefit from systems like those described above, such as banks, insurance companies, those involved in simulation or engineering already have serial code written to carry out their tasks. To make the transition from serial to parallel, these companies are faced with a decision to either write the code from scratch - clearly an impractical solution, or to somehow convert their code from serial to parallel. Such a conversion is possible, but, as no automatic technique currently exists, is fraught with difficulties, requires much expertise in the area, and tends to be something of a hit and miss affair, with the quality of the final parallel program depending very much on the ability of the programmer.

There are few automatic techniques for parallelizing serial, and existing methods generally consist of a set of interactive tools to help programmers, or to identify common pieces of code that can easily be parallelized, e.g. matrix multiplication etc. The difficulty in autoparallelization, and parallelization in general, is twofold. Firstly, the identification of which areas of the code can have one or more standard transformations applied to them, and secondly discovering the optimal order in which to apply these transformations.

We describe a system, *Paragen II*, which automatically generates a list of transformations which, when applied to a serial program, produce a parallel version. Paragen II hybridizes Genetic Programming with existing parallelisation techniques, to evolve a list of parallel transformations. This list identifies both the transformations to use, and the order in which to apply them. Moreover, because the list employs standard parallel transformations, it can subsequently be used to prove that the parallel version of the program is functionally identical to the version.

---

* email Conor.Ryan@ul.ie

Alarm Optimising by Fuzzy Logic in Preterm Infants Monitoring

Dipl.-Ing. D. Schenk [a], Dr. M. Wolf [a], Prof. Dr. H.-U. Bucher [a],
Dr. Y. Lehareinger [b], Prof. Dr. P. Niederer [b]

[a] Clinic of Neonatology, University Hospital Zurich
[b] Institute of Biomedical Engineering and Medical Informatics, ETH Zurich
dan@fhk.smtp.usz.ch

Introduction

The control of oxygen status in preterm infants is a crucial task in intensive care. All newborns less than 34 weeks of gestational age should be monitored for apnoea and bradycardia. Accordingly the infants are monitored routinely by devices, such as electrocardiograph, pulse oximeter and transcutaneous partial oxygen pressure. Each monitor generates an alarm independently from the others whenever the measured parameter is below or above a constant threshold. However, practice shows that a considerable number of false alarms is triggered by movement artefacts. On an average, the proportion of false to significant alarms is as high as eight to one. Eliminating these false alarms would be a great help for the nursing staff.

Improving Alarm Signalisation Using a Fuzzy System

For this purpose a system based on a PC-Interfacecard with a transputer T425 was developed. The T425 collects and pre-processes on line the vital parameters from the different bedside monitors. A Windows application program manages in real time to display the data on the screen, to save it in a file and an implemented fuzzy system analyses the physiological status of the infant. Using only his/her medical knowledge to describe the physiologic status, an user with no technical background is assisted through the development of the fuzzy monitoring system by a sequence of dialogboxes. One of these dialogboxes allows the customisation of the monitor surface, which plots the data in different predefined types of display modules: F/t-diagrams, bar graphs, analog and digital meters or data lists. Other dialogboxes lead through the definition of the fuzzy variables and the set up of the rule base. Eventually, a first reliability test of the whole fuzzy system can be performed.

A first fuzzy system was created using the following signals: heart rate (HR), arterial oxygen saturation (SaO2), transcutaneous partial oxygen pressure (tcPO2). In addition the difference between heart and pulse rate and a regression of tcPO2 over the past 10 seconds were computed and used as supplementary input variables. Each variable was spread over two or three fuzzy sets. The system with its rule base consisting of 10 rules, was able to successfully discriminate movement, apnoeas and bradycardias.

First Results

In a first test with the data of almost 23 hours of 6 new-borns the system was able to suppress nearly all 175 false alarms due to movement artefacts. However all of the 32 critical situations (among 20 were serious) involving different types of apnoeas and bradycardias, isolated or in combination were detected reliably.

Supervised Learning of Fuzzy Tree Hypotheses

Christoph Schommer   Matthias Stemmler
J.W. Goethe University, Dept. of Computer Science
Robert-Mayer-Str. 11-15, 60325 Frankfurt am Main, Germany
Email: {schommer,stemmler}@dbis.informatik.uni-frankfurt.de
Phone: ++49.69.798.28823, Fax: ++49.69.747021

1 Introduction

A possible application scenario for knowledge discovery involves the ability to perform *classification* on datasets ([2]). Prototypical scenarios for classification are, e.g., target mailing, store location and credit approval: in the credit approval scenario, classification functions might be necessary to show customers' profile and are in addition strongly valid. However, approximately exact information processing does not always satisfy requirements for database marketing applications ([1]), and often it could be sufficient to have vague information that describe the companies' intentions; for the credit approval scenario, a bank could also be interested in approving credits to candidates that though not satisfy all criteria which are necessary belonging to class "credit-worthy" (e.g. because they do not have a desired income), but are otherwise "good" borrower in respect to some k percentage. Therefore, vagueness is a natural part of knowledge discovery because discovered information is naturally not absolutely true. Fuzziness describes the theory of classes with unsharp boundaries ([3]), and this is exactly the case if we look at the example given above.

2 Architecture of FTLearn

FTLearn is a supervised learning system that learns to infer fuzzy tree hypotheses (= classification functions) from a given training tree. Training tree and tree hypotheses consist of nodes which represent predicates of the underlying dataset. Each node has a *set of objects* which gives some kind of "strength" by the number of objects within its set. FTLearn uses a top-down search algorithm that generates and restructures tree hypotheses using *founding* and *developing* methods, and accepts or refuses tree hypotheses depending on their interest and quality. The interest of a tree hypothesis in respect to the training tree is expressed by a support value, that denotes the number of predicates (nodes) that both have in common; if a minimum support value sup_{min} is reached, the tree hypothesis is accepted, otherwise not. The quality is expressed by a similarity function, which refers to "good", if tree hypotheses correspond to the training tree by a sufficient percentage. It refers to "bad", if similarity decreases below a second similarity value called sim_{min}. "Possibly useful" denotes the grey-zone between these two values with the consequence that hypotheses need further to be processed. The architecture of FTLearn is shown in Figure 1.

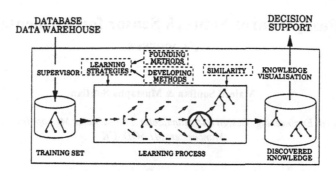

Fig. 1. Architecture of FTLearn

FTLearn consists of several steps, starting with obtaining data by a given database or a data warehouse. Learning denotes the process of finding appropriate hypotheses which result in similar trees of k percentage in respect to the training tree. The representation of knowledge aims at storing discovered information with scope on visualization for decision support needs. FTLearn starts by a nondeterministic selection of predicates which results in tree hypotheses containing only one node. This node is accepted if it is member of the training tree and then expanded to more complex tree structures. There must be both "enough" nodes (which refer to predicates of the tree hypothesis) on equivalent tree positions within tree hypotheses and training tree and approximately the same number of objects that are associated with a corresponding node. In addition, it is checked whether the quality has increased in respect to the ancestor hypothesis: if yes, the new tree hypothesis shows a more similar structure than the previous hypothesis and is therefore added to the knowledge base with a preferred mark. An important point is, that subsequent tree hypotheses converge either in acceptance or refusal: all tree hypotheses become either more similar or distinct. Similarity is therefore a strong pruning method, since if a tree hypothesis is detected as "bad", it needs not further be processed (as well as its successors). The algorithm stops if no new tree hypotheses are found to be of interest and quality.

Currently we test the algorithm on a bank customer dataset in order to discover fuzzy tree hypotheses that represent customers' credit behavior.

References

[1] J. Gessaroli, "Data Mining: A powerful technology for database marketing", Telemarketing, Vol. 13, No. 11, May 1995, pp. 64-68.

[2] G. Piatetsky-Shapiro, W. Frawley (Eds.): *Knowledge Discovery in Databases,* AAAI Press, The MIT Press, California, 1991.

[3] L. A. Zadeh, "Fuzzy Logic, Neural Networks, and Soft Computing", Communications of the ACM, March 1994, Vol. 37, No. 3, pp. 77-84.

A Fuzzy Neural Network Sensor for Fermentation Processes

Majeed Soufian & Mustapha Soufian

Mechanical Engineering, Design and Manufacture, Manchester Metropolitan University, Manchester, M1 5GD, UK

Abstract. *In this paper computational intelligence has been considered as a tool (software sensor) in state-estimation and prediction of biomass concentration in a fermentation process. An optimised fuzzy system based on genetic algorithm, an artificial neural network and an integration of fuzzy logic and neural network paradigms have been introduced as computational engines. The constructed computational engine 'infer' the production of biomass from variables easily measured on-line. The construction and the application of each paradigm are presented and sum of squared errors and graphical fit are used to compare the performance of each structure.*

1. Introduction

Both Artificial Neural Networks (ANN) and Fuzzy Logic Systems (FLS) have some drawbacks when used on their own. The ANN can produce mapping rules from empirical training set through learning but the mapping rules in the network are not visible and are difficult to understand. On the other hand, since the FLS does not have learning capability, it is difficult to tune the rules and optimise its performance.

2. Integration of Fuzzy Logic and Neural Networks

In order to solve these difficulties, an integrated paradigm (*Fuzzy Neural Networks*) of these two evolving disciplines is designed for non-linear identification of a target system. In the integrated paradigm, as illustrated in figure 1, the learning capability of neural networks is added to the principle-base of FLS to learn mapping rules $(R^{(1)},...,R^{(M)})$ of *principle* base from input and output of fuzzified training set. In this configuration, input/output fuzzy sets ($'F^1,F^2,...F^I'$ and $'G^1,G^2,...G^I'$) are estimated from expert knowledge. After training the ANN, parameters of fuzzy rules are stored in the weights of the neural networks and will play the role of *principle* base. However, it may be observed that the operations involved in a fuzzy inference engine show a similar mathematical behaviour in a neuron, i.e., summation of products as an *s-t norm*. Therefore it will be a correct conclusion to state that by injecting the input fuzzy sets, $'F^1,F^2,...F^I'$ to the trained ANN, the output of ANN will be the inferred output fuzzy sets $'G^1,G^2,...G^I'$. So the ANN, not only plays the role of *principle* base and stores the rules as its weights, but also, it can be employed as the inference engine.

578

3. Conclusion

In this paper, various modelling techniques, are applied to a fed- batch fermentation process. The identifications based on ANN and genetically optimised FLS techniques are shown to give a good performance, although genetically optimised FLS performed less accurately than the ANN model. The best result is obtained by integration of these two paradigms (fuzzy neural network) to act as a 'software sensor' for the biomass concentration.

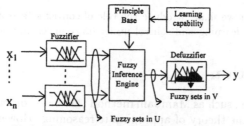

Fig.1. A Fuzzy logic controller with learning capability.

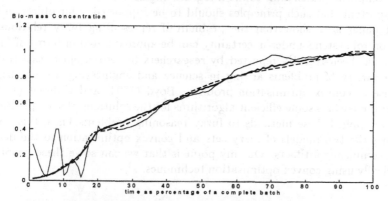

Fig.2. Comparison of predicted biomass concentration from the GAs-based optimised-FLS (thin solid line) with actual data (thick solid line) and the ANN (---) output.

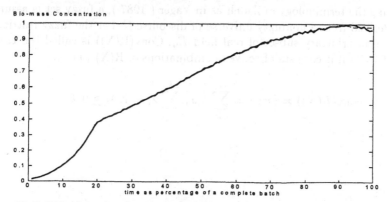

Fig.3. Comparison of predicted biomass concentration from the fuzzy neural networks (---) with actual biomass concentration (thick line).

Convex Fuzzy Restriction

Appolo Tankeh and E. H. Mamdani

Dept. Electrical Eng, Imperial College , London, SW7 2BT

Abstract

In this paper we shall introduce the role of convex sets as a means for fuzzy inference, learning and control under fuzzy uncertainties.

1   Introduction

Most fuzzy inference , such as Mamdani method , are based on *modus ponens* and other principles in the theory of approximate reasoning. However, in the spirit of soft computing which aims to the integrate linguistic and numerical methods, it is important that such principles should to be supported. Our object in the present paper is to show that the problem of representing fuzzy relations in fuzzy control systems under uncertainty can be approximated in terms of fuzzy sets. It has been recently reported by researchers in convex optimisation that a large variety of problems arising in science and engineering are not usally recognise as convex optimisation problems Boyd (1994) and Gahinet (1995) and yet there exists some efficient algorithms for the solutions of such problems. We have applied these methods to fuzzy reasoning problems.Thus, this paper combines the two models of fuzzy sets and convex optimisation to the design of soft computing artifacts. Our my point is that we can solve these problems numerically using convex optimisation techniques.

2   Convex fuzzy restriction

Following the terminology of Zadeh as in Yager (1987), a fuzzy set is assumed to be derived from a nonfuzzy universe of discourse U. Let the fuzzy restriction $R(X)$ be an arbitrary subset of real field E_n. Conv(R(X)) is called the *convex hull* of R(X) if it consists of convex combinations of R(X) , i.e . .,

$$\text{conv}(R(X)) = \{x : x = \sum_i^k \lambda_i x_i, \sum_i^k \lambda_i = 1, \lambda_i \geq 0, k \geq 1\} \tag{1}$$

Indeed , Conv(R(X)) is the smallest convex set containing the fuzzy restriction R(X) Witzgall (1970). In fact ,it is the intersection of all convex set containing R(X) . If R(X) is convex then $R(X) = conv(R(X))$. Thus, we can construct a fuzzy set as a convex set with a membership function. By recoursing to a convex set to characterise the domain of our fuzzy sets, we can then be able to derive various fuzzy relations using convex sets. The difficulty that arises from this is the computation of the membership function. Given any collection of memberships functions on a convex fuzzy restriction, we can compute the hypographs of these collections Witzgall (1970). The infimum (or inf) of these hypographs gives the familiar result of the membership function of intersection, where the inf is the **min** operator in Mamdani inference as in Mamdani (1979). Similarly, the membership function of union of convex fuzzy sets can be derived. This principle has been extended in this work to the pinciple of composition rule of inference and other aspects of learning in neuro-fuzzy systems. We have also shown that if the state transition matrix A in a dynamical system depends on a parameter which takes its values in a convex fuzzy restriction then the question of quadratic stability and robust control can be solved using convex optimisation. The results have been impressive.

3   Conclusion

We have characterised fuzzy sets and relations in terms of convex sets. The process of inference and learning can be approximated as convex program. In this way, principles and methods of convexity and optimisation can be brought to bear on the task of characterising solutions to such problems. The result is direct relationship between fuzzy sets and corresponding convex set that circumscribes the space that contains them. Then, learning can then be approximated in this space in the process of iterative optimisation.

References

[1] Assilian S. and Mamdani E. H., An expirement in linguistic synthesis with a fuzzy logic controller, *Int. J. Man Machine-Studies 7(1)* (1979) 1-13.

[2] Witzgall Christoph and Stoer Josef ; Covexity and Optimization in finite dimensions I *Springer-Verlag* (1970.)

[3] R. R. Yager, S. Ovchinniko, R.M Tong and Nguyen H. T., Fuzzy sets and its applicaitions: *Selected papers by Zadeh , John Wiley and Sons* (1987).

[4] Gahinet P., Nemirovski A., Laub A. J., Chilali M., LMI Control Toolbox, *The Math Works Inc.* (1995).

Genetic Programming in Optimization of Algorithms

Piotr Wąsiewicz, Jan Mulawka
Institute of Electronic Fundamentals, Warsaw University of Technology
Nowowiejska 15/19, 00-665 Warsaw, Poland
E-mail: pwas@ipe.pw.edu.pl.; jml@ipe.pw.edu.pl.
phone: (04822)-660-5319, fax: (04822)-252300

Abstract. The problem of improving the efficiency of Genetic Algorithms to search global optimum is considered. An approach based on applying Genetic Programming methodology to find the best structure of Genetic Algorithms for global optimization is described. It allows to obtain better results in comparison with standard Genetic Algorithms.

1. Initial consideration and statement of the problem

Genetic Algorithms (GA), and Genetic Programming (GP) have been invented in order to optimize complex problems. These efficient heuristics are based on evolutionary methods derived from Nature.

GAs [2] are techniques that can be successfully applied to NP-hard optimization problems. Their search imitate Darwinian strife for survival. They start with randomly generated initial population. Each individual of this set represents a possible solution to the problem. Then, while termination condition is not true, the following cycle is performed: number of generation := number of generation + 1; select Population(t) from Population(t-1); recombine Population(t), using crossover and mutation operators; evaluate Population(t).

GP [1] has been recently developed as a methodology to solve problems by genetically breeding populations of computer programs. For a particular problem sets of functions and terminals are to be created. An initial population of LISP-like expressions is a collection of random compositions of functions and terminals. Each such expression can be viewed as a rooted tree. Each expression called also a program is evaluated against the problem. Genetic operators of selection and crossover are applied to create new populations of programs. Evolutionary process is continued until a maximum number of generations is reached.

In the following we assume that GAs solve the global optimization problem [3] for functions with one global and many local optima and cube constrains (l - a left constrain, r - a right constrain) which can be written as

$$\min(f(X)), \quad X = (x_1, x_2, \dots, x_N), \quad l_1 \le x_1 \le r_1, \dots, l_N \le x_N \le r_N$$

In the above notations each variable is restricted to a given interval. For such defined a search space feasible solutions are being found.

GAs are good at optimizing functions with one global and many local optima. Their standard structures are well known. But with Genetic Programming we find new better structures, which will be applied to find a global optimum of different test functions. We compare the efficiency of developed algorithms obtained during optimization of these functions.

In this paper, to solve the global optimization problem, we introduce a technique based on GA which is controlled by GP. It is difficult to find best receipt for constructing GAs, which can find a global optimum of different functions. In the approach to be presented a way of optimizing structures of GAs to search quickly for global optima is outlined.

2. Genetic Programming Approach

Applying GA to the global optimization we start with defining a single chromosome, representing a potential solution to the problem, in the given search space. A part of a chromosome has N genes, where N is a length of a machine word. It represents a coordinate of a point. A single gene can take values 1 or 0. There is a point v within an interval [l, r] and we transform this interval into a machine word interval [0, 2^N] and we receive a new binary representation c of the point v. After this operation we change c using Gray code and this is our new representation of a coordinate and a part of the chromosome. In this way we transform coordinates of feasible points in the search space into binary strings. Here a value of the given function in a given point is a fitness value of that point (individual, chromosome).

In the process of evolution we apply standard genetic operators such as mutation, crossover. The GA finds the individual with the best feasible solution (the highest value of a function) located in the defined space.

During execution of GA it is important which genetic operations (e.g. crossover, mutation e.t.c.) are performed. With GP we can create trees made of genetic operations. Thus, *these tree-like structures describe a GA way of execution during one generation (epoch). After each GA epoch the best solution is chosen.* Genetic operators of GP e.g. selection and crossover create new populations of algorithms. The evolutionary process of GP is continued until a maximum number of generations is reached. After this the best solution is displayed. The aim of GP is to process sets of genetic operators and special functions in order to find the most appropriate GA. It can explain, which order of operators is the best for a problem of global optimization. To be brief, the GP finds the best structure of the GA. Here fitness of a tree is the best feasible solution obtained by GA represented by this tree.

We develop a set F of independent operations on population, parameters and one terminal, which is just a probability of the particular operation. The terminal set is $T = \{x\}$. The function set is following

$F = \{SEL, SUS, SEI, SET, SPT, BST, MUT, MT1, MT2, CRS, DCR, PCR, SCR, UCR, INV, +, -, *, /, =, <\}.$

The functions have the following meaning: *(SEL x):* proportional selection, the method of „the roulette wheel", *(SUS x)* : stochastic selection with sample probing[2], *(SEI x)* : selection: in the next population only exists copies of the best individual from the previous population, *(SET x)* : tournament selection: from two at random chosen chromosomes the best wins, *(SPT x)* : tournament selection with probability: from two at random chosen chromosomes the best wins with given probability, *(BST x)* : *(1 - x)* best strings of population is remembered and will be copied after all operations-arguments of BST execution, *(MUT x)* : mutation with probability $x \cdot 100$, *(MT1 x)* : mutation with probability $x/10$, *(MT2 x)* : mutation with probability x, *(CRS x)* : crossover with probability x, *(DCR x)* : double crossover with probability x, *(PCR x)* : fivefold crossover with probability x, *(SCR x)* : „sweep" crossover with probability x [2], *(UCR x)* : „uniform" crossover with probability x, *(INV x)* : inversion of bits between two points with probability x, *(+ x)* : addition to probability a predefined constant $st1 : x + st1$, *(- x)* : subtraction from probability a predefined constant $st1 : x - st1$, *(* x)* : multiplication of probability by a predefined constant $st : x * st$, *(/ x)* : division of probability by a predefined constant $st : x / st$, *(= (CRS x) (MUT x))* : predefined sequence of two GAs operators executed on the whole population, e.g. mutation is executed first, crossover - second, *(< (CRS x) (MUT x))* : predefined sequence of two GAs operators executed on the divided population, e.g. on the first half of the population mutation is executed, on the second half of the population crossover is executed.

As an example consider the tree *(= (CRS x) (MT1 x))* which can be written in a file form as:
= c CRS c x o o n MT1 c x o o o o. In *Fig. 1* a physical structure of this tree is provided. Here *"c"* means *"child"* (a first argument of a genetic operator of GP), *"n"* means *"next"* (a second argument of a genetic operator of GP) and *"o"* -

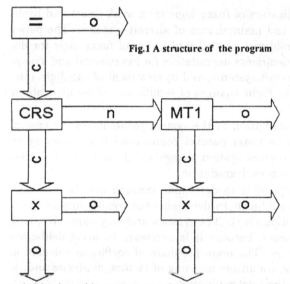

NULL (without an argument)[4]. As is seen operators of a typical tree are executed from right to left. Probability x comes from right operation to left operation.

Fig.1 A structure of the program

3. Results of experiments

With the above method of generating GA trees a lot of experiments with GP controlling these GA structures were carried out for global optimization problem. As follows from our considerations GP tries to choose the best sequence of genetic operators in tree forms. These chosen trees are like receipts and GAs do all what they order. Then it is more possible to find the best solution quicker. Thus, GA is not attached to only one scheme. There are a lot of other ones. They are often attached to a given type of functions.

Futher research in this field is conducted. With local optimizing procedures GA are quite powerful tool. Our experiments show that mentioned methodology is promising *in optimization of complex, multidimentional functions with many local and one global optimum* and in comparison with other methods good results are obtained.

4. Acknowlegdements

The simulated experiments have been conducted with use of a package: Genetic Programming in C++ v. 0.40 written by A.P. Fraser. This work was supported by the Polish State Committee for Scientific Research under Grant 8T11C 04611.

5. References

[1]. J.R. Koza: Genetic Programming, MIT Press 1992.
[2]. D.E. Goldberg: Genetic Algorithms in Search, Optimization and Machine Learning.
Addison-Wesley, Reading, MA, 1989.
[3]. A. Torn, A. Zilinskas: Global Optimization, Springer Verlag Berlin, 1989.
[4]. A. P. Fraser: An Introduction to Genetic Programming in C++ (Version 0.40), from ftp.io.com.

Application of Fuzzy Logic in Solving Problems from the Area of Reliability and Maintenance of Aircraft Systems

Milenko Živaljević and Boško Rašuo

Aeronautical Department, Faculty of Mechanical Engineering,
University of Belgrade, 27 Marta 80, 11 000 Belgrade, Yugoslavia

The paper deals with the application of fuzzy logic for a weak structured problems in the area of reliability and maintenance of aircraft systems. The paper presents a synthesis of two problems: a) the application of fuzzy logic for determination of performances coefficients degradation for kvazy-serial and kvazy-parallel connections for the aircraft systems, and b) treatment of the flight mission reliability, when during the flight changes of significance of components or flight safety system appear.

In order to analyze the contribution of the system reliability in the case of higher reliability (kvazy-serial or kvazy-parallel connection) it is necessary to determine the reliability of particular system elements and coefficient K_f, that represent the factor of performances degradation.

The elements reliability is possible to determine through defects statistical data collection, while it is very difficult to determine the combat airplane performances coefficients degradation (K_f). The reasons are: very expensive, long testing and high risk experiments, because it is necessary to make deliberate deficiency and defects simulation. The main problems of coefficient calculation are: complex factors dependency, multiple meaning of factors, unprecise and/or insufficiently clear criteria for their determination.

As a good example for performances coefficients degradation for kvazy-serial and kvazy-parallel connections, we consider the case of automatic breaking and stopping of combat airplane. As the first step, the breaking system reliability mathematical model was developed. The performances coefficients degradation for kvazy-serial and kvazy-parallel connections are stochastic values, that for airplane with specific take-off weight depend on pilot skills, wind intensity and environment temperature. Because of that, we applied fuzzy logic using expert knowledge of test center pilots.

For coefficients determination we used following fuzzy variables: O-pilot skills variable, V-wind intensity fuzzy variable, T-environment temperature fuzzy variable and E-breaking efficiency fuzzy variable.

Twenty-seven rules of approximate reasoning are established and the defuzzification is carried out using moment rule or center of gravity rule (Teodorović-Kikuchi).

In comparison of breaking efficiency values with breaking parachute obtained using fuzzy logic and values obtained through expert evaluations, we achieved good results compatibility.

Treatment of the flight mission reliability, when significant changes of components or system role on flight safety appear (task b) is carried out in a such way, that in function of meteorological conditions the system significance for flight safety is determined (as an input) on that basis, reliability correction of flight mission is carried out (as an output). The achieved results of the model using fuzzy logic are shown on the example of reliability of horizontal situation indicator of combat airplane.

References

[1] M. Živaljević, "The Contribution to the Research of the Fighter-Bomber Airplanes Systems Reliability Analysis and its Optimization", *These Doc. Ing.*, VTA, Belgrade, 1996. (in serbian)

[2] D. Teodorović and S. Kikuchi, "Introduction to Fuzzy Set Theory with Applications in Transportation Engineering", *Faculty of Transport and Traffic Engineering*, University of Belgrade, 1991.

[3] R. Belman and L. Zadeh, "Decision Making in a Fuzzy Environment", *Management Sci.*, 17, 1970.

[4] A. Kaufmann and M. Gupta, "Introduction to Fuzzy Arithmetic", *Van Nostrand Reinhold*, New York, 1985.

[5] T. J. Ross, "Fuzzy Logic with Engineering Applications", *McGraw-Hill*, New York, 1995.

[6] H-H. Bothe, "Fuzzy Logic", Einführung in Theorie und Anwendungen, *Springer-Verlag*, Berlin, 1995.

A Fuzzy Based Texture Retrieval System That Considers Psychological Aspects

Mario Köppen and Javier Ruiz-del-Solar
Department of Pattern Recognition
Fraunhofer-Institut IPK Berlin
Pascalstr. 8-9, D-10587 Berlin, Germany
email:{mario.koeppen l javier}@ipk.fhg.de

Abstract

Texture perception plays an important role in human vision. It is used to detect and distinguish objects, to infer surface orientation and perspective, and to determine shape in 3D scenes. An interesting psychological observation is the fact that humans' beings are not able to describe textures clearly and objectively, but only subjectively by using a fuzzy characterisation of them. On the other hand, with the new advances in communication and multimedia computing technologies, accessing mass amounts of digital visual information (image databases) is becoming a reality. In this context, textures, due to their aesthetical properties, play today an important role in the consumer-oriented design, marketing, selling and exchange of products and/or product information. For this reason, systems that allow the search and retrieval of textures in image databases, the so called *Texture Retrieval Systems*, are of increasing interest.

The propose is to describe a new Texture Retrieval System[1], which is based on the use of Fuzzy Logic, Neuro-Fuzzy Networks and Morphological Operators in the processes of *Qualitative to Quantitative Textural Properties Transformation*, *Color* and *Textural Features Extraction*, *Features Fusion* and *Feature Similarity Matching*. One important aspect of the proposed system is that it considers psychological aspects of description and perception of textures. A textural retrieval process can be divided into two main steps: a) off-line generation of the image annotations in the database, and b) on-line retrieval of textured images from the database.

The proposed retrieval system, whose block diagram is shown in Fig. 1, is made of the **Q²TPT** (Qualitative to Quantitative Textural Properties Transformation), the **FSM** (Feature Similarity Matching) and the **FTD** (Features/Texture Database) modules.

---

[1] Demo program under the following electronic address:
http://strauss.ipk.fhg.de/Textursynthese/texql.html

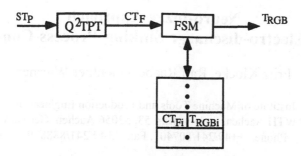

Fig 1. Block diagram of the proposed Texture Retrieval System.

The on-line phase of the texture retrieval process works as follows: A human user makes a query of a texture by using subjective textural properties (ST_P). The Q^2TPT module interprets the query and translates it into a quantitative texture description that we call Color-Textural Features (CT_F). The FSM module performs the search of the requested texture into the FTD by doing a similarity matching between the CT_F query and the CT_F database attribute. An important process in the training of the Q^2TPT module and in the construction of the FTD is the extraction of Color-Textural Features (CT_F) from textures (textured images). This process is performed by the FE (Features Extraction) module.

Neuro-fuzzy Approach
for Electro-discharge Sinking Process Control

Fritz Klocke, Ralf Raabe, Guenther Wiesner

Institute of Machine Tools and Production Engineering
RWTH Aachen, Steinbachstr. 53, 52056 Aachen, Germany
Phone: ++49/241/807401, Fax: ++49/241/8888293

Due to the increasing demands on modern manufacturing processes, traditional process and optimization systems are reaching their limits and reveal serious shortcomings. Intelligent technologies like fuzzy logic and neural networks can help to increase the economical viability of manufacturing. This article therefore presents an application in the field of electro-discharge machining (EDM).

Electro-discharge machining (EDM) is a reproductive metal forming process based on the thermal removal of material by means of temporally and spatially discrete discharges (sparks) between a tool electrode and a workpiece electrode. The main field of applications for the process is in one-off and small-batch production, e.g. in tool- and diemaking. The electro-discharge process is influenced by a number of technological parameters. "Classic" adaptive control produces unsatisfactory results in the case of electro-discharge machining, owing to the stochastic nature of the process and the unknown transmission behavior of the controlled system. Adaptive process optimization systems based on this traditional control help to improve the finished result but fail to exploit the full performance potential of such machines. During the process cycle, problems occur due to constantly varying machining parameters or unexpected process disturbances. Gap width controllers and adaptive controllers for the generator parameters usually work separately, with frequent negative influences between the two systems. Optimal process control, especially in difficult machining situations, is possible only to a limited extent, even with the aid of an experienced operator.

The Institute of Machine Tools and Production Engineering (WZL) has developed an optimization system based on fuzzy logic and neural networks, which formulates rules to integrate operator know-how in process control and to combine individual controllers in a multi-variable control system.

The core of the optimization system is a fuzzy expert system capable of acquiring machine operator experience both on- and off-line. The system uses rule bases to optimize controller parameters for gap width, generator and flushing in the ongoing process. All important machine parameters are measured and recorded and all process parameters are analyzed during the ED process, allowing the effects of parameters settings and fine adjustments on the process cycle to be assessed.

The combination of fuzzy logic and neural networks in a neuro-fuzzy system turns a purely knowledge-based optimization system into a learning capable system, in which individual rules are trained continuously, and in which new rule bases or attribution functions are learnt or automatically adapted to the process.

The advantages of fuzzy logic under difficult electro-discharge machining conditions have been demonstrated by a fuzzy gap width control and fuzzy optimization of motion flushing. For example: the use of fuzzy optimization reduces process time for a 40 mm sinking depth of a narrow wedge electrode by 15 % without arc danger. It has been shown that fuzzy logic can help the machine operator to get better machining results.

References

1. Klocke, F.,   Adaptive Control of EDM Sinking Process Using Fuzzy Logic,
   König, W.,   Eufit '96, Proceedings Fourth European Congress on Intelligent
   Raabe, R.:   Technics and Soft Computing Aachen, Vol. 2 (1996), S. 1085 -
                    1089.

2. Klocke, F.,   Steigerung der Wirtschaftlichkeit und Prozeßsicherheit von Fun-
   König, W.,   kenerosionsmaschinen mit Neuro-Fuzzy-Control,
   Raabe, R.:   Tagungsband: Fachtagung "Leistungselektronische Aktoren und
                    intelligente Bewegungssteuerungen", Otto-von-Guericke-Univer-
                    sität Magdeburg, 1996.

Simulation of Analogue Electronic Circuits Using Fuzzy Curves

Christoph Reich
reich@ai-lab.fh-furtwangen.de

University of Technology Furtwangen, Department of Computer Science,
78120 Furtwangen, Germany

The first step in the design process is drawing a topology of the circuit (circuit diagram) which is supposed to meet the specifications. This first design consists of a mix of connected discrete components and black-box-descriptions of well-known sub-circuits. The designer is modeling the behaviour of these sub-circuits as exact as it is needed. Of course this circuit topology drawn on paper is the first shot and is by no means correct. To become clear that the behaviour of the circuit is correct the given specified static electrical value informations and electrical signal sketches are added to the topology description of the circuit design. The signal sketches (graphs) are important add-ons to explain and to visualize the dynamic behaviour of the circuit. It gives the circuit designer a rough idea about the correct behaviour of the circuit. To determine the behaviour of the circuit the design engineer is propagating the specified signal sketches through all intermittent sub-circuits and components until the complete circuit is simulated. Thus after the first stage of the design process the design engineer produced a paper drawn circuit prototype (circuit paper prototype).

It is impossible to estimate with accuracy the parameters of the circuit since a model of the real circuit does not exist yet. Our approach is trying to lay out the fuzzy theory to model the uncertainty involved at this first design step.

Fuzzy numbers are used for uncertain static specifications and uncertainty concerning the model of the components and sub-circuits. We represent a fuzzy number as an ordered set of confidence intervals, each of them providing the related numerical value at a given presumption level $\alpha \in [0, 1]$ (α-cut set).

The input signals feeding the circuit are either chosen from a database (e.g. triangular signal) or drawn freehand by the designer. Sections of the signal which are of significance for the designer can be labeled with quantitative and/or qualitative values. Drawing signals is like writing letters. Each designer draws signals differently. Specified signals are more or less certain. These input signals drawn by the designer and the internal generated signals of the prototype are represented by fuzzy curves. Fuzzy curves represent signals piecewise by a set of discrete linear fuzzifying functions. With fuzzifying functions the point of view is not that of mapping a fuzzy number x with a regular function on Y but that of mapping x to Y through a fuzzy set of functions. Thereby is x a non-fuzzy variable. Mathematical approximation is avoided by this representation which is of advantage. First it is hard to tell what mathematical function should be used to approximate the signal since we do not know the designer's intention when he drew the signal. Second we would mislead an exactness of the signal which is of no means available at this stage of the circuit design.

Fuzzy variables are worth mentioning since they can interact in real world applications. It is possible to define strong- or non-interaction between two fuzzy variables. The multiple occurrence problem of fuzzy variables is taken into account.

To be able to propagate fuzzy curves through the circuit prototype modeled by a constraint network we have to extend standard mathematical concepts in order to deal with fuzzy quantities. Since we model the discrete fuzzy numbers and discrete fuzzy curves by α-cut fuzzy sets of which each α-level is one level of presumption, the fuzzy mathematics can be transfered to interval computation.

All together we have a tool which enables the designer to simulate a high-level description of a circuit using sketches of signals.

Correlated Activity Pruning (CAPing)

By C. M. Roadknight[1], D. Palmer-Brown[1] and G. E. Mills[2].
1. Novel Architectures Group, Department of Computing, The Nottingham Trent University, Burton Street, Nottingham NG1 4BU.
2. Department of Life Sciences, The Nottingham Trent University, Clifton Lane, Nottingham NG11 8NS.

Abstract.
The generalisation ability of an Artificial Neural Network (ANN) is dependent on its architecture. An ANN with the correct architecture will learn the task presented by the training set but also acquire rules that are general enough to correctly predict outputs for unseen test set examples. To obtain this optimum network architecture it is often necessary to apply a labourious 'trial and error' approach. One approach that helps to achieve optimum network architecture in a more intelligent way is pruning. Such methods benefit from the learning advantages of larger networks while reducing the amount of overtraining or memorisation within these networks. Sietsma and Dow (1988) describe an interactive pruning method that uses several heuristics to identify units that fail to contribute to the solution and therefore can be removed with no degradation in performance. This approach removes units with constant outputs over all the training patterns as these are not participating in the solution. Also, units with identical or opposite activations for all patterns can be combined. The approach to merging hidden units detailed in Sietsma and Dow's paper is useful, however, it only covers perfectly correlated, binary activations.

The method presented here generalises correlated activity pruning (CAPing) to all real valued positively and negatively, highly correlated activation sets from hidden neurons. There are several positive results to be gained by CAPing. Firstly, a speed up of the training and optimisation process can be achieved. Secondly, the weights of correlated units can be analysed and relationships between correlated features detected.

CAPing is initially applied to a theoretical curve fitting problem. Very little error is introduced until units with correlation coefficients of less than 0.8 are used. Analysis of the weights from the input units of the merged hidden units showed some sets of weights to be highly correlated but other sets to be non-correlated. This shows how CAPing removes two types of redundancy. Firstly when weight vectors represent similar features, and secondly differing features that occur for the same input patterns.

ANN's have been applied to the problem of predicting leaf damage to crop plants (Roadknight et al 1995, Balls et al 1996). ANN's trained to predict onset of leaf damage, with more hidden units than were required, were CAPed down to architectures with a near optimal number of hidden units.

Acknowledgements.
We would like to thank the UK Department of the Environment for financial support of the work in this project (Project number PECD 7/12/145)

References.
Balls GR, Palmer-Brown D, & Sanders GE. 1996. Investigating microclimate influences on ozone injury in clover (*Trifolium subterraneum*) using artificial neural networks. New Phytologist, 132, 271 -280

Roadknight CM, Palmer-Brown D & Sanders GE. 1995. Learning the equations of data. Proceedings of 3rd annual SNN symposium on neural networks (eds. Kappen B and Gielen S) Springer-Verlag. 253-257.

Sietsma J & Dow RJF. 1988. Neural net pruning - Why and how. Prc. IEEE Int. Conf. Neural Networks. Vol 1. p. 325-333.

Fuzzy Control Design of an Activated Sludge Plant

O. Georgieva, ICSR-BAS,
Acad. G. Bonchev Sr., Bl. 2, 1113 Sofia/Bulgaria

The biological stage of the waste water treatment requires maintenance of microorganism viability providing high intensity of the biochemical processes. That are tasks of analysis and synthesis of control system, which solution does not fit the commonly accepted mathematical frame of the traditional approach of the control synthesis because of the processes' complexity, uncertainty conditions and the presence of expert information.

The purpose of the paper is to develop control design method of the processes in biological waste water treatment using the available quantitative and qualitative information and under the uncertainty conditions.

In the first part of the paper a control algorithm of the considered processes is developed. The algorithm is based on the operating strategy, proposed by Joyce, Ortman and Zickefoose and relied upon the fast measurement and control of key plant variables. The realization of the operating strategy is based on the fuzzy sets theory which best fits for the process uncertainty and qualitative information. The values of the basic input variables:

- Foot/Biomass ratio (F/M);

- Respiration rate (RR);

- 5 minute settling volume (SV5),

which well define the current process state, are presented as fuzzy variables. Following fuzzy rule of inference, the values of the output process variables:

- Sludge conditioning time (SCT);

- Waste sludge flow (WSF);

- Returned sludge flow(RSF),

are maintained at their optimum range. The process data base contains 27 plant operating conditions (fuzzy rules), combined with specific operating instructions. The application of the control strategy is organized by use of the operating condition cubic matrix. Proper operation will keep the plant close to the center of the matrix, where F/M, RR and SV5 are all in the optimum range.

In order to account for the qualitative process information and to increase the system capability of adequate respond to the environment changes the developed algorithm is expanded in the second part of the paper. The number of the fuzzy sets describing the input and output variables is increased so that the expert information is fully used. The algorithm is supplied with procedure for on-line tuning of the membership functions which improves the system adaptability.

The results are connected with the development of new control design method for fuzzy on-line control of biological waste water treatment processes. The developed algorithms work as operator adviser. The basic application effects are summarized as following:

- Adaptable, prompt and efficient respond to the environment changes.

- Decreasing the operating consumption - economical regime of the oxygen supply and economical loading of the power capacities leading the input-output flows.

- Increasing the quality of the output treated water in according to the consisting of harm substances.

Key words: Waste Water treatment, Activated Sludge, Control Design, Fuzzy System, On-line Control

Use of Fuzzy Objects for Statistical Description of Discrete Systems with Complex Behavior

S. Barabash, P. O. Box 40, Dolgoprudnyy-1, Moscow region, 141701/Russia

This work illustrates the progress that can be made with the use of fuzzy notions in statistical study of systems whose dynamics cannot be adequately described with conventional methods. Different approaches are analyzed, and examples of properties which cannot be revealed by means of common statistics are given. The central object of the research is the Game of Life [1] which displays the main features typical of such systems and which actually was successfully studied in fuzzy way.

The Game of Life is a two-state 2D cellular automaton. Despite the simplicity of the rule of evolution[1], the Game of Life has an extremely varied dynamics and in classification made by S.Wolfram [3,4] it was characterized as a cellular automaton of class IV — an automaton with a complex dynamics, whose properties at any step depend on the initial configuration . The researchers studying the Game of Life usually seek colonies with specific or unusual behavior (for example see [1,2]), but statistical properties are also of extreme importance. First, Life is related to a lot of lately studied dynamic systems which may behave differently under similar initial conditions, but as shown below some features of the dynamics of Life make the common methods of physical statistics ineffective. Second, an automaton of class IV can work as a universal computer [3-5], and proper statistical description may be helpful in investigation of general properties in other computational systems (for instance neural networks).

Some of the statistical properties of the Game of Life and regularities in its behavior were already discussed [6-9, see also 18]. A circumstantial research of evolution of the density of colonies is given in [6]. This work describes the most probable value of the density at a time t as a function of the initial density, while the *original* configuration is generated randomly. However, even a random configuration in several turns acquires features typical of the colonies exactly in the Game of Life, and the density alone cannot describe this distinction. Consequently for statistical prediction of the evolution of populations from any certain moment of time one needs more detailed description of the state of the system. This conclusion reflects the fact that the dynamics of Life does not describe any transfer of matter (you may call such a system 'transferless') and therefore has no conservation laws unlike 'classical' dynamics of a system of N particles whose motion obey a set of 3N differential equations. The very nature of a 'transferless' system (not only of the Game of Life) makes mean field [10-12] and other statistical methods [13] ineffective.

It is possible to detail the information about the colony without introduction of fuzzy systems[2]. Different sets of variables have been considered as candidates for a proper set, basically they used structural characteristics ("density + fractal properties"[3] etc.) and diverse 'partial' densities for cells being in different conditions (depending on the vicinity of a cell). Employing some of these sets one can distinguish the stage of the colony's evolution and foresee changes of average density. However, statistical description implies prediction of the future values of *all the variables* that describe the system at the moment, and conventional approach accounts only for some of the set. Therefore, no satisfactory statistics can be introduced unless fuzzy objects are used.

The situation changes drastically when the problem is translated into fuzzy language. First of all, theoretical analysis of the rule of the game reveals some basic patterns which generate typical ways of behavior[4]; choosing these patterns as fuzzy references for density, one may introduce valid operators of interaction between subsystems and further find a kinetic equation of the system, which is the starting-point in any non-equilibrium statistical study. Unlike a conventional kinetic equation [13], this one contains only

---

[1] The Rule of Life: if among 9 neighbors of a certain cell (two-state cells are arranged on a 2D grid) there are exactly 3 'on' cells, the cell turns on, if there are 2 'on' neighbors it does not change its state, otherwise it turns off.

[2] actually statistics always uses fuzzy objects since it uses probability, but this fact is disregarded here

[3] the system may not have fractal properties (for example, the most colonies in the Game of Life do not exhibit self-similarity), but locally some 'size - density' dependence always exists and may be associated with fractal dimension at the point.

[4] e.g. horizontal and vertical lines tend to spread generating two similar objects, diagonal lines are apt to remain the same, and very dense structures generate blank spaces

these empirical operators and no 'hydrostatic' terms. The main drawback of such fuzzy discretization is that it uses information about microscopic state of a system, while statistics implies averaging at some scale. One may also use the dependence of density of cell 'clusters' on a size of a cluster (that can be considered as 'fuzzy fractality', see also footnote 3). This method exploits no microscopic patterns and is more adequate for our purpose though its predictions contain less information (at least in case of Life). However, both techniques utilize only arrangement of the elements and no 'dynamic' variables. On the contrary, statistical physics of conventional systems along with microscopic dynamics describes the state with the use of both positions and speeds of the particles[5].

Usual dynamic characteristics — birth and death rates — may be helpful in characterizing the stage of evolution but give no information about interaction between subsystems. Fuzzy representation allows us to define speed. Since fuzzy speed has no exact meaning, one may correlate any acts of birth or death in many ways so that the necessary operator of interaction is obtained. The specific choice can be maid to meet the requirement that the 'motion' must be observable at a 'physically infinitesimal' scale[6], not only at a microscopic one. In author's opinion, the most promising fact is that the particular procedure may be determined by theoretical study of the dynamics of a system. Mean field method, which is basically applied to discrete systems with complex or chaotic behavior, gives no chance to introduce any speed in a 'transferless' system theoretically. As yet, no profound analysis of the rules of Life from this point of view was made, and the above statement remains a mere hypothesis. In this research the fuzzy speed was introduced quite arbitrarily and its effectiveness was rather limited in comparison with fuzzy discretization mentioned above. However this approach derives a kinetic equation completely satisfying the principles of statistical study. Apparently it may be developed with the use of fuzzy differential calculus [14-16].

Introduction of fuzzy speed definitely shifts the focus of our attention from the arrangement of the cells to a new object associated with some average value. This object by itself has a rule of evolution (a kinetic equation) and may be considered as a probability cellular automaton (a 'child') generated by the initial ('parent') system. The child automaton can serve to verify the statistical model: the model would be correct if properties of the child statistically tally with those of the parent. The studying the child's behavior, provided the above condition is true, may help to discover some properties of the 'parent' automaton which are usually disguised by variety of microscopic patterns. In particular, it may elucidate different aspects of self-organization in the system (see also [17-19]).

Here is a concrete example of how the developed approaches may be applied for the investigation of evolution of particular configurations. Irreversible changes of the state of a system are always accompanied by the increase in entropy[7]. Fuzzy statistical description makes it possible to define entropy and analyze the development of the system. This analysis confirms the previously found regularities [6] and spotlights specific changes otherwise noticeable only by visual observations. The common practice of identifying probability with average density of cellular automata [18] is artificial and reveals no correlation between entropy and evolution of the system.

Finally it is essential to clarify the following: one could doubt acceptability of fuzzy language in statistical study basing on the fact that fuzzy description uses logical operators and any logic seems inappropriate for a quantitative statistics. Similar reasoning would be well-grounded for the systems with 'continuous' dynamics, but in the case of discrete systems it is unsound. Discontinuous rules of evolution must be substituted by continuous statistical variables and that requires some intermediary; fuzzy logic appears to be the best solution. As shown above, it can be used not only for an empirical research, but also as a tool in theoretical analysis. Spin glasses, chemical reaction models and many other discrete problems widely studied in recent times present a wide range of possible applications of the method.

---

[5] obviously there is no first derivative of a 'coordinate' in a discrete system like the Game of Life

[6] physically infinitesimal scale is defined as a scale large enough to have many particles and be described by statistical values, but infinitesimal in comparison with the whole system

[7] meaning statistical entropy $S = - <\ln w>$ (where w is probability) [13], not fuzzy entropy [20]

597

References

[1]Garner M. Wheels, Life and Other Mathematical Amusements. - San Francisco: Freeman, 1982.
[2]Berlekamp E.R. et al. Winning Ways for your mathematical plays. V.2. - N.Y.: Academic Press, 1982.
[3]Wolfram S. // Rev. Mod. Phys. 1983. V.55. P.601.
[4]Wolfram S. // Physica D. 1986. V.10. P.1-35.
[5] Wolfram S. // Nature. 1984. V.311. P.419.
[6]Franco Bagnoli et al. // Physica A. 1991. V.171. P.249.
[7] Langton C.G.//Physica D. 1986. V.22. P.120.
[8]Gosper R.Wm. //Physica D. 1984. V.10. P.75.
[9]Wentian Li et al. //Physica D. 1990. V.45. P.78.
[10]L.S.Schulman et al. //J.Stat.Phys. 1978. V.19. P.293.
[11]H.Gutowitz//Complex Systems. 1978. V.1. P.57.
[12] William K.Wooters et al. //Physica D. 1990. V.45. P.95.
[13]Klimontovich Yu.L. Statistical Physics - N.Y. :Harwood Academic Publ., 1986.
[14]Dubois D.J..Prade H.M. // Fuzzy Sets and Syst. 1982. V.8. P.1.
[15]Dubois D.J..Prade H.M. // Fuzzy Sets and Syst. 1982. V.8. P.105.
[16]Dubois D.J..Prade H.M. // Fuzzy Sets and Syst. 1982. V.8. P.225.
[17]P.Bak, C.Tang, K.Wiesenfeld //Phys.Rev.Lett. 1987. V.59. P. 381.
[18]Wentian Li et al. //Physica D .1990.V.45. P.78.
[19]Stefano Zapperi et al. //Phys.Rev.Lett. 1995.V.75.P.4071
[20]Loo S.G.//Cybernetica. 1977. V.20.P.201.

On Applying Genetic Algorithms to an Intelligent Sliding-Mode Control System Training

M. Cistelecan

Technical University of Cluj-Napoca, Dept. of Automation, 15 C. Daicoviciu St., RO-3400 Cluj-Napoca/Romania

1. INTRODUCTION

Sliding-mode (SM) control is a well-known control technique, able to provide robust control systems against the disturbances and parameter variations. Unfortunately, the SM control has two disadvantages - the chattering and the discontinuous nature of the control - which have been considered as serious obstacles for its application in practice. In order to reduce the chattering a neural sliding-mode (N-SM) controller was considered in (Cistelecan, 1995) and an I-SM controller was considered in (Cistelecan and Trifa, 1996). The N-SM and also the I-SM controller have to cope with the problem of a long learning time for the FNN configuration. The learning stage involves multiple epochs, each epoch being defined as a complete simulation of the control system using different initial condition - related to the state space - or using different reference values - related to the error state space. Each epoch is labelled by a different start vector (initial conditions or reference). This procedure for finding a FNN configuration which can cope acceptable - a robust FNN - with all possible start vectors is not the best one because the convergence of the learning cannot be guaranteed. For this reason a new methodology is proposed based on the genetic algorithms (GA) optimisation concept. The problem of finding a robust FNN is critical when the start vector can be changed inside an imposed range of values, at any time.

2. THE STRUCTURE OF THE I-SM CONTROL SYSTEM

The I-SM controller integrates the SM control concept with the neural and fuzzy control (Cistelecan and Trifa, 1996), based on the linear feedback with switched gains control law (Hung et. al., 1993). It is known that the neural and also the fuzzy control systems approaches are due to their capability to deal with nonlinearities and to treat the empirical information. However, the lack of a systematic design procedure and the impossibility to always guarantee the stability of the fuzzy and neural controllers restricts the use of these control concepts. In this context the advantages of the N-SM and I-SM hybrid control systems is that the SM theory provides a systematic design procedure for the involved neural and fuzzy-neural (FNN) modules. The SM theory guides the choice of the appropriate architecture for the neural and FNN modules, ensures the possibility to impose the control system performances, and also provides a rough partitioning of the input and output spaces of the controller (control spaces). The SM control is designed based on an approximative model of the plant and thus it seems to be convenient to express the SM partitioning as a fuzzy partitioning. With or without fuzzyfication, from the point of view of neural control, the control spaces partitioning based on SM provides a measure for rejecting the unnecessary information in the design of the neural controller and thus improves the learning of the neural modules. Moreover, for these hybrid control systems (N-SM, F-SM and I-SM) the stability is guaranteed by the SM control theory.

The input of any hybrid SM controller must contain explicitly the sliding variable (SV). The choice of the SV as input for the controller is the essence of any hybrid SM control system. Aiming to handle more information about the system dynamics, the previous SV value is used as another controller input. The state vector also can be used as another input vector for I-SM controller. Another useful input information is the distance, d, from the operating point to the state space origin. This information may handle the quality of the steady state.

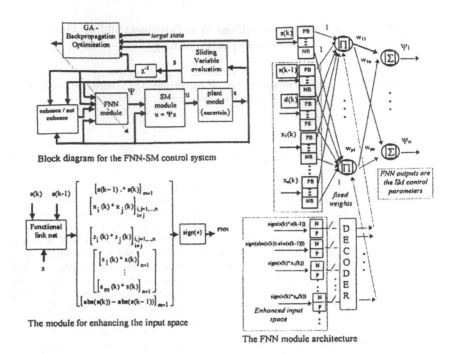

Block diagram for the FNN-SM control system

The module for enhancing the input space

The FNN module architecture

The I-SM controller involves a SM module and a FNN module (Cistelecan and Trifa, 1996a). The FNN structure is similar to that presented in (Fukuda and Shibata, 1992). The membership functions are on the Gaussian type and have two kinds of parameters: the center value and the width. The output layer contains linear sigma neurons. The FNN configuration is defined by the center values and the widths of the Gaussian functions and also by the weights of the links between the hidden and the output layers. The concept of enhanced input space is considered for the I-SM controller, (Cistelecan and Trifa, 1996a,b).

Prior to the on-line connection, the I-SM controller has to learn a robust configuration. Based on the SM methodology an initial fuzzy partitioning of the control space can be done and thus the FNN can be initialised. Then, the FNN is trained based on numerical simulations. In (Cistelecan, 1995; Cistelecan and Trifa, 1996a) an optimisation module governess the tuning of the FNN configuration. The aim of the optimisation is to reduce the chattering as much as possible, on the condition of the satisfaction of the control constraints provided by the SM theory. The backpropagation algorithm was used for updating the parameters. The SV was considered the error to be backpropagated through the plant model (possible uncertain) and also through the SM module to the FNN. The FNN configuration is updated only when some conditions are met, not in each iteration. This fact partially resolve the problem of an arbitrary choice of the learning parameter. Thus, the I-SM controller is adapted continuously, on-line, by numerical simulations to learn how to treat the uncertainties and nonlinearities of the plant, and how to be able to reject the disturbances of the control system. After the convergence of a pseudo-fixed FNN configuration is obtained the I-SM controller can be integrated in the real control system. It is important, that this pseudo-fixed FNN does not loose its adaptive character when operating on-line in the real control system. Small adjustments are to be done in order to preserve the robustness of the control system. For this case the learning parameter is very important, too.

When the FNN is trained with only one start vector one can obtain an optimal FNN configuration. However, an optimal FNN configuration may be not as effective under other start vector. Thus, in the cases when the controller have to cope with many arbitrary admissible start vectors a robust controller have to be estimated. Even when the range of the admissible start vectors is small there is no reason to believe that after many epochs labelled by different start vectors the controller convergence can be obtained. It is known that oscillations may occur even for very simple cases. The controller convergence also depends very much on the chosen learning parameter.

For this reason the aim of this paper is to develop a training algorithm for the FNN, using GA, in order to take into account for a variety of start vectors. Moreover, the appropriate dynamics of the learning parameters during an epoch is estimated.

When a change of the reference occurs, a translation of the error state space must be performed in order to update the two used state error vectors. For this reason, finding a robust FNN configuration is critical.

3. APPLICATION OF THE GA TO THE FNN TRAINING

The robust FNN configuration is searched by collecting in the first sample (start sample) GA solutions (individuals) for a range of different start vectors. The robust FNN configuration is defined as that which, although not optimum for any one epoch, performs well across a range of start vectors. A very close idea was met in [Markham, 1996].

For the first sample many populations (groups of individuals) are defined. All these groups have identical individuals and each group is labelled by its start vector. For the first sample all the genetic operations are done only into these groups. The aim of the first sample of the training optimisation is to find a number nn of proper individuals for each group of individuals. A proper individual steers the system from the initial state to the appropriate partition of the state space (or error state space). Thus, for the first sample the search is finished only when the nn proper individuals are obtained for each group. After the search is finished, np individuals are selected among the proper individuals nn of each group and then these are putted together into a colective mating pool. The next samples use only a group of individuals.

In all samples the individuals which are to be combined come from many different sources. First, the individuals are obtained from the previous sample by selection based on the fitness function Then, many other individuals are obtained by updating the FNN configuration obtained from the previous sample based on backpropagation of the SV and using different learning parameters.

The theoretical assertions are sustained by numerical simulation results.

REFERENCES

Cistelecan, R.M. (1995). A Neural - Sliding Mode Controller. Preprints Intelligent Manufacturing Systems. *Proceedings of IMS'95*, 3 rd IFAC/IFIP/IFORS Workshop, Bucharest, Vol. 2, pp. 483-488.
Cistelecan R.M., Trifa, V. (1996a). On Integrating Sliding-Mode, Neural and Fuzzy Control Concepts. *Proceedings of EUFIT'96*, 1010-1014.
Cistelecan R.M., Trifa, V. (1996b). A Neural Sliding -Mode Controller with Enhanced Input Space for An Inverted Pendulum via A Gear Train. Proceedings of PEMC'96, Vol. 3, pp. 177-181.
Davis, L. *Handbook of Genetic Algorithms*. International Thomson Computer Press, 1996.
Fukuda, T. and Shibata, T. (1992). Theory and Applications of Neural Networks for Industrial Control Systems. *IEEE Trans. on Industrial Electronics*, Vol. 39, No. 6, pp. 472-487.
Hung, J., et. al., (1993). Variable Structure Control A Survey. *IEEE Trans. on Industrial Electronics*, no. 1 Vol. 40, pp. 2-22.
Markham, K. (1996). Tuning Fuzzy Logic Systems for Crane Control. Proceedings EUFIT'96, 1070-1074.

Neural Networks and Edge Detection

P. Heirman, R. Serneels.

Limburgs Universitair Centrum, 3590 Diepenbeek, Belgium.

1. Introduction.

Using simple training sets we will study how two types of neural networks develop to identify future edges. Afterwards we discuss a multilayer edge detector with just two hidden neurons and an edge detector based on the fuzzy min-max description as described by Simpson[1]. In all cases edge detection is performed by scanning an image with a 5x5 window. To understand how a neural net edge detector works we consider in section 2 very simple 5x5 training examples with just two horizontal homogeneous regions. The central pixel of the window is part of the edge if its sides touch both homogeneous regions.

2.1. A multilayer network with two hidden neurons.

We start with a multilayer network with two hidden neurons. To represent the weights we choose a 5x5 matrix structure for the weights to a hidden neuron so that the geometric relation to the 5x5 input neurons is clear. The training stops when the training error was of the order 10^{-4} and typically 4000 training examples were offered in 100 epochs. After training the weight matrix shows the characteristics of the Prewitt operator: all elements on the central and outer rows are zero, in row 2 all elements are close to a value -8 and in row 4 all elements are nearly equal to the value 9. The bias values have different sign: $b1 = -1.2$ and $b2 = 0.4$ and the weights to the output neuron from the hidden neurons are $w1 = 11.8$, and $w2 = -11.8$, with bias 2.5. With these values the system recognises an edge.

2.2 A perceptron type network with non monotonic transfer function.

An alternative architecture is to use a perceptron network structure without hidden units. Such a system may work provided we consider a special non monotonic transfer function with the shape of a well. Similar work with a non monotonic transfer function was already reported in connection with the exclusive OR problem [2]. The well shaped transfer function $f(x,a,b)$ has two parameters a and b which describe the range of the wedge. When $x \gg a$, b or the reverse $x \ll a$, b $f(x,a,b)$ is approximately one and for values within the interval (a,b) a small output is generated, the shape resembles a well.

We included the coefficients a and b in the training and for horizontal edges we found b=–a=2 and again a weight matrix resembling the Prewitt operator. This time the understanding of the weight matrix is more clear and corresponds exactly with the common practice of convolving the input window with the Prewitt filter.

3.1 The well trained multilayer neural network.

If neural nets have to compete with classical detectors the training set should be improved. Although several attempts with synthetic images have been reported [3,4] this is still a major issue. To check out what is needed in the ultimate training set we considered training windows consisting of two homogeneous regions as discussed above but with a separating line in different orientations. Training sets with Gaussian noise were also studied. No attempt to include special X or Y type corners was made. On testing we introduced a pre-processing stage where the testing window was rotated over a set of angles and an edge character was decided if for some orientation a sufficiently high output was found. Situations with 1, 4 and 16 orientation variants were considered. This pre-processing may be considered similar to adding more training images. To check the efficiency we applied the neural net detector on the classic Lena image and compared with the Canny operator. The results show that the neural nets are very adequate even with this non optimal training set.

3.2 The fuzzy min-max neural network edge detector.

An alternative more sophisticated use of neural networks was described by P. K.Simpson
and is known as the fuzzy min-max neural network [1]. He described a supervised neural network classifier with a three layer neural network where the number of outputs corresponds to the number of classes and the second layer has as many units as the number of hyperboxes used to describe all the training examples. When after training a new input window is presented the membership functions of all the hyperboxes are evaluated, the hyperbox with maximal membership is selected and a class is attributed corresponding to the class related to the hyperbox. In our application we used two output units corresponding to edge and non edge and used the 5x5 neighbourhood window of each pixel of the image as inputs. We trained the system with 10000 edge/non edge examples and found 2000 hyperboxes. We used a maximum hyperbox

size of 0.05 and a parameter $g=10$ which describes how fast a membership function decreases. Application of the fuzzy min-max edge detector to the Lena picture shows good results.

4. Statistical performance.

We checked the performance of the network edge detectors on two new artificial images for which the edges are known: one image with four circles of different shade of grey and different size and another one with horizontal stripes with varying intensity. The circle was chosen to allow a check for a running edge line in all possible directions. To simulate real images we considered also an image version with Gaussian noise with a standard deviation of 30 units. To discuss the efficiency of an edge detector we calculated ROC curves where the probability of false and correct edge detection are displayed in terms of the clipping parameter which finally has to be used to identify an edge or not. In general neural edge detectors perform better then the classical ones which were studied.

Acknowledgements.
P. Heirman acknowledges an FWO grant which made this work possible.

References.

1. P. K. Simpson. IEEE Trans. Neur. Netw. 3, 776, 1992.
2. G.J. Bex and R. Serneels. Fuzzy Logic and Intell. Techn., Mol, Belgium, ed. E. Kerre et al, 68, 1994.
3. A. J. Pinho and L. B. Almeida. Proc. IEEE ICIP'94, Texas, 1994.
4. T. Law, H. Itoh and H. Seki. Patt. Anal. Mach. Intell. 18, 481,1996.

Design Elements of EHW Using GA with Local Improvement of Chromosomes

Mircea Gh. Negoita; Adrian Horia Dediu; Dan Mihaila

Department of Developments in Infotmation Technologies, National Research and Development Institute of Microtechnology
PO Box 38 -160, 72225 Bucharest, Romania
E - mail: negom@oblio.imt.pub.ro; dediuh@oblio.imt.pub.ro;danm@oblio.imt.pub.ro

This paper treats a GA with local improvement of chromosomes used to optimize the evolved configuration of an EHW (*evolvable hardware*) architecture. EHW applications are specific to autonomous agents for characterizing the behavior output pattern as a mapping problem between sensor outputs and actuator commands.

The first part introduces briefly the problem of real - time adaptivity in hardware, the second part presents the architecture of a typical PLD circuit allowing the technological support of the *hardware reconfigurable block* (RLD) of the EHW architectures.The general flow of the GA with local improvement of chromosomes emphasizing the Nagoya mutation is tackled in the third part. Chapter 4 describes our variant of GA using the Nagoya mutation (NGA) compared with Goldberg GA (SGA) to establish (evolve) the connection matrix of a PLD macro - cell for the evolution of an XOR circuit with four inputs.The chromosome is represented by the concatenation of the connection matrix having 8x16x2 binary values,and the evaluation function is: $f_e = \sum_{input=0}^{max\,input} err(input)$,where $err(input_{16}) := \overset{8}{\underset{i=1}{OR}}$ [(input$_{16}$ or not matr(I).v) and (not input$_{16}$ or not matr(I).vnot) = \$ffff] $\neq f_{requested}$; input$_{16}$ represents the paralel combination of the 16 inputs.

The above mentioned coding uses four states for a connection point in the matrix: A; notA; 0 - for simultaneously connected inputs A and notA;1 - for disconnected inputs. It is a convenient but redundant coding leading to a 2^{256} search space. For further research we will use a tri - state coding leading to a $3^{128}x2^8$ search space, that being an economy of 10^{11} points. The table of the experimental results is:

Errs	Rngs. SGA	Rngs. NGA	Rngs. SGA	Rngs. NGA	Rngs. SGA	Rngs. NGA	Rngs. SGA	Rngs. NGA
6			4				2	1
5		1	4	4	8	2	5	1
4	3	6	1	5	1	4	3	4
3	6	3		1	1	4		3
2	1							1
Eval	$16x10^3$	$64x10^4$	$16x10^3$	$64x10^4$	$4x10^4$	$16x10^5$	$4x10^4$	$16x10^5$
pc	0.8	0.8	0.4	0.4	0.8	0.8	0.4	0.4
pm	0.05	0.05	10^{-3}	10^{-3}	10^{-3}	10^{-3}	10^{-3}	10^{-3}

Fig.1 The best error during the sets of 10 runings in 4 variants of GA parameter settings

ANN and Fuzzy Logic Applied to Control of Disturbances in Mechatronics System

Tadeusz S. Matuszek and Szymon Grymek

Faculty of Mech. Engineering, Technical University of Gdansk,
PL- 80-952 Gdansk, Poland

Abstract. In this work a purpose of the ANN and Fuzzy Logic for integrated logic control of disturbances in mechatronics system has been considered. The main task of this study was to investigate whether electronic logic designed into various aspects of mechatronics system control, can provide and improve flexibility as well as control capability, to fully automate and optimise process control with programming options. A neural controller, based on feedforward neural net with back propagation algorithm, was designed and implemented. The controller was compared to a classic PID controller for disturbances in motion control loop. There results were very promising.

1. Introduction

The design of control of disturbances in mechatronics system is usually described from modelling, identification to adaptive control for nonlinear systems, and is after followed by solution of supervision tasks with fault diagnosis. In many of mechatronic systems the combination of motion together with transfer forces and torques can be observed. Nowadays, a development of strongly integrated engineering systems consisting of mechanics, electronics and information technology, should be analysed and considered as a simultaneous everyday engineering experience set up for designing and control of mechatronics procedure. Early studies carried out in this field revealed that the most powerful data analysis available for use with electronic devices are artificial neural network (ANN) which can be taught to recognise the sensors responses used to differentiate various types of disturbances in motion control of mechatronics system.

2. Methods

A feedforward artificial neutral network was used as a neural controller in the mechatronic system applied to the motion control of a robot. In experiment the system was represented by a simple discrete computer model. The designed neutral network had one output with linear neuron, four inputs in the input layer and one hidden layer with six non-linear neurons. The neural controller was trained by an error backpropagation learning algorithm using Levenberg-Marquardt optimisation method and momentum, and over training.

3. Tests and Results

The results were based on the tests provided for an estimation of the differences calculated as an error between theoretical optimal conditions for motion control of the chosen part of robot and disturbances existing in the transfer of the real signal values between inner and outer loop, for neural control and PID control. As it is known the motion in general is classified according to its stiffness with two extreme states: infinite and zero stiffness in motion control. Controller based on disturbances control in both cases tested is not an exception.

4. ConcludingRemarks

The achieved results of applying the neural network disturbances control in motion in the case of robot taken as a part of mechatronic system, shows advantages of the neural network approach as compared to standard methods.

In view of the above mentioned method and experiments it can be concluded that ANN and Fuzzy Logic application to the control of disturbances in mechatronics system is very promising due to the following reasons: they are easy to develop, better for complicated or non-linear systems, more reliable, and can provide better performance.

However, it is understandable, that in order to realize various aspects of disturbances in motion control, more variable stiffness in the components of mechatronics system should be taken into account.

References

1. Ohnishi, N. Matsui, Y. Hori, Estimation, Identification, and Sensories Control in Motion Control System, Proceedings of IEEE vol.82, No.8, pp.1253-1265(1994)

2. Warwick, G.W. Irwin, K.J. Hunt, Neural networks for control and systems, Peter Peregrinus Ltd., London, United Kingdom (1992)

3. Ohnishi, N. Matsui, Y. Hori, Estimation, Identification, and Sensories Control in Motion Control System, Proceedings of IEEE vol.82, No.8, pp.1253-1265(1994)

4. Nagasawa, E. Yokoyama, Precision Motor Control System for VTR Using Disturbance and Velocity Observer, Trans. on IEEJ, vol. 114-D, No. 1 pp. 25-32 (1993)

5. Sz. Grymek, T. Matuszek: A case study for the ANN process control, Copernicus Programme, 1st Main Meeting, Porto, Portugal (1995)

Author Index

Springer
and the
environment

 Springer

Lecture Notes in Computer Science

For information about Vols. 1–1143

please contact your bookseller or Springer-Verlag